ANIMAL SCIENCE, ISSUES AND RESEARCH

BIODIVERSITY OF THE BULGARIAN-ROMANIAN SECTION OF THE LOWER DANUBE

Animal Science, Issues and Research

Additional books and e-books in this series can be found on Nova's website under the Series tab.

ANIMAL SCIENCE, ISSUES AND RESEARCH

BIODIVERSITY OF THE BULGARIAN-ROMANIAN SECTION OF THE LOWER DANUBE

PETER STANISLAVOV SHURULINKOV
ZDRAVKO HUBENOV
STOYAN BESHKOV
AND
GEORGI POPGEORGIEV
EDITORS

nova
science publishers
New York

Library of Congress Cataloging-in-Publication Data

ISBN: 978-1-53615-663-8
Library of Congress Control Number:2019943051

Published by Nova Science Publishers, Inc. † New York

Contents

PREFACE

The Danube River is the second longest river in Europe; it flows through ten countries and is 2,850 km in length. For 470 km, the river acts as a border between Bulgaria and Romania and provides a refuge for very rich, yet insufficiently studied, biodiversity.

The book is comprised of 12 separate scientific articles (chapters) that each contain specialized information about key organism groups forming the typical Lower Danube ecosystem. Vertebrate fauna of the study region includes 392 species out of which 83 species are fishes, 19 are amphibians, 15 are reptiles, 201 are birds, and 74 are mammals. Birds are presented in three separate articles with a focus on heron colonies, water birds in the marshlands, and diurnal raptors. Among the invertebrates, several key groups which are comparatively well-studied, are presented, including mollusks; dragonflies and damselflies (in overall 52 species); dipterans of suborder Nematocera (153 species); mayflies (52 species); stoneflies (3 species); and caddisflies (49 species), butterflies (153 species). An article about habitats and typical plants of the Danube plain is also included.

The articles all contain information on the status, distribution and ecology of the study's organism groups, and the threats they face. Overall numbers and population trends of some of the presented taxa are also included; 281 maps, 29 graphs, 21 tables and 49 photographs help illustrate various aspects of the collected data. The authors of the articles are well-known and experienced experts on specific organism groups, working mainly out of two institutions of the Bulgarian Academy of Sciences: the National History Museum (Sofia) and the Institute of Biodiversity and Ecosystem Research (Sofia).

Chapter 1 - Important Plant Areas in Bulgaria are determined using a complex of various criteria of flora and habitats. The aim of this study is to make a critical evaluation of 12 chosen IPAs along the Bulgarian stretch of the Danube River. As a result, the criteria for their determination have been complemented. The expansion of some of the IPAs and the designation of one new site – Kalimok-Brashlen – is also proposed. The Bulgarian IPAs network is generally representative of the flora and habitat diversity found along the

Danube River; however, their argumentation could be improved after more intensive studies concerning the diverse Danube's flora and vegetation.

Chapter 2 - The current review presents the composition, distribution, zoogeographical characteristics, some ecological traits, and conservation status of the mollusc fauna (Gastropoda and Bivalvia) in the Bulgarian sector of the Danube River and the adjacent marshes and backwaters. A total of 67 mollusc species (47 gastropods and 20 bivalves), about 13% of the Bulgarian mollusc fauna, have been reported until now in the available literature. The reported species belonged to 14 families (ten of gastropods and four of bivalves). The highest species richness was recorded in the family Planorbidae, followed by Sphaeridae and Lymnaeidae. The mollusc species were classified in 19 zoogeographical categories. The highest number of the molluscs (79.10%) was of Palaearctic distribution. Five species (*Theodoxus transversalis*, *Anisus vorticulus*, *Anodonta cygnea*, *Pseudanodonta complanata*, and *Unio crassus*) have been considered of higher conservation concern according to the IUCN Red List, European and national legislation. The globally known major threats to freshwater biodiversity, including loss and degradation of habitats, overexploitation, pollution, introduction of invasive alien species, and climate change, have affected also the mollusc fauna in the Bulgarian sector of the Danube River and adjacent wetlands. Therefore, in order to protect the native mollusc fauna appropriate conservation and management measures directed to the restoration and protection of the mussel populations and to the restoration of the damaged aquatic habitats, including their catchments, are recommended.

Chapter 3 - This paper collates original, unpublished, and literature data on mayflies (Ephemeroptera), stoneflies (Plecoptera), and caddisflies (Trichoptera) from the Bulgarian section of the Danube River and nearby territories. It focuses on their aquatic immature and aerial adult stages at more than 100 localities. A total of 104 taxa from the three insect orders are known from the study area: 52 mayflies, three stoneflies, and 49 caddisflies. They belong to 27 families and 47/48 genera that encompass13 families and 20 genera of Ephemeroptera, two families and three genera of Plecoptera, and 12 families and 24/25 genera of Trichoptera. Six mayfly species are reported only for the Danube River area and almost all of them are listed in the Red Data Book of Bulgaria as regionally extinct and critically endangered. In addition to these six, another two mayfly species are listed as endangered, three as vulnerable, and seven as rare or very rare. Only three stonefly species are known from the study area and they are typical for the Bulgarian stretch of the Danube River; however, during the last century, their populations have dramatically decreased and these species are currently considered extinct. Eight of the caddis taxa are confirmed only for the Danube River area, while 15 of the registered species are considered rare or very rare. Newly recorded for the region are four species of mayflies and two/three genera and two species of caddis. Further, the authors report new localities and data for 19 species of Ephemeroptera and six taxa of Trichoptera.

Chapter 4 - This study intends to collect and present all available information on seasonal and habitat distribution of Odonata (dragonflies and damselflies) in the Bulgarian stretch of the Danube River between chainages 845 (at the Timok River estuary) and 375 (town of Silistra) and adjoining wetlands,(up to 20 km from the river bank) included in 109 MGRS UTM 10x10 km quadrants. To date, 52 species of Odonata have been recorded in the study area, 51 of which were identified in previous publications, including by this study's author, who recorded 46 of the 51 species. From the total number of species, four have not been confirmed as present around the Danube River, one has been removed because of change in taxonomy, and one new species has been recorded in the study area.

Chapter 5 - The Nematocera suborder of the Danube River, the Danubian islands and the nearby parts of the Bulgarian Danube River bank are scrutinized. A total of 153 species from 14 families have been established. Three families are better studied – Culicidae (20 species – 13.1%), Ceratopogonidae (36 species – 23.5%) and Chironomidae (63 species – 41.2%). The established species belong to 52 areographical categories. The fauna can be divided into 2 main groups: 1) species with Mediterranean type of distribution (12 species – 7.8%) – more thermophilic and distributed mainly in the southern parts of the Palaearctic. Four species of southern type, distributed in the Palaearctic and beyond it, can be formally related to this group as well; 2) species with Palaearctic and Eurosiberian type of distribution (141 species – 92.2%) – more cold-resistant and widely distributed in the Palaearctic. Sixty-four species of northern type, distributed in the Palaearctic and beyond it, can be formally related to this group as well. Of the established taxa, 14 are pests on the forestry or agriculture, and 13 species have a human or veterinary medical significance.

Chapter 6 - This review is a much-needed update of the known distribution of butterflies (Lepidoptera: Papilionoidea) in the least-known part of Bulgaria: the Danubian Plain and much of the Fore-Balkan hilly country north of the Stara Planina mountain chain (in total, over a third of the country's territory). The author combines previously published records with results from their own extensive field research in the area, their revision of public collections, and other sources. A brief history of lepidopterological research in the area is provided. A revised systematic checklist and distribution maps of all species are provided, with notes on the more notable species. In total, 153 species are recorded (70% of Bulgarian butterflies) of which two (*Euphydryas maturna* and *Lopinga achine*) are not found elsewhere in Bulgaria. Ten of the recorded species are legally protected by EU and/or national legislation. The results underscore the importance of the area for nature conservation and invite further intensive butterfly research to fill in the remaining, unfortunately still extensive, blank spots.

Chapter 7 - The ichthyofauna of Bulgarian stretch of Danube River and lower courses of its tributaries is reviewed based on the published and original data. A total of 83 fish species belonging to 21 families have been reported. Of these, 69 species are native for Danube River basin and 12 are alien and introduced. The status of two species is unclear.

Eight species are considered endemic in the Danube River basin. Full faunistic list with notes on distribution, origin and conservation status of the fish species is given.

Chapter 8 - Here the authors present the first attempt to combine the collected data, map localities, and analyze the distribution of the herpetofauna in the area stretching ca. 10 km south of the Bulgarian Danube River. The authors identified 687 records from publications that they attributed to 62 cells of the 10 × 10 km MGRS grid. The authors included previously unpublished data collected through various visits throughout the past ca. 13 years, comprising 1918 records with real coordinates, which fall in 1269 cells of the 1 × 1 MGRS grid. The authors have confirmed 34 native species – 15 amphibians (4 salamanders and 11 frogs) and 19 reptiles (1 turtle, 2 tortoises, 8 lizards and 8 snakes), and one invasive turtle. Although finding other species is unlikely, additional field studies could confirm the presence of *Pseudopus apodus* (for now, this species cannot be accepted as part of the regional fauna). Further detailed studies are needed, especially in areas that were likely not sampled, and in the detection of cryptic species such as *Pelophylax lessonae*. The authors also caution that future changes to the river's hydrodynamics must really consider the highly likely multitude of negative effects to the biodiversity and specifically to the herpetofauna.

Chapter 9 - Distribution and numbers of colonially nesting herons, cormorants, Eurasian Spoonbills and Glossy Ibises were studied along the whole Bulgarian –Romanian section of the Danube River in the period 2010-2014. A total of 15 colonies, 12 of them mixed, were located and their species composition and trends in numbers were followed during the study. The total numbers by species differed considerably between years as follows: Great Cormorant (*Phalacrocorax carbo*) - 1573 -2299 pairs, Pigmy Cormorant (*Microcarbo pygmaeus*) – 283 - 877 pairs, Grey Heron (*Ardea cinerea*) – 237-399 pairs, Purple Heron (*Ardea purpurea*) – 4 - 18 pairs, Great White Heron (*Ardea alba*) – 6 - 12 pairs, Squacco Heron (*Ardeolla ralloides*) – 60- 181 pairs, Night Heron (*Nycticorax nycticorax*) – 234 - 484 pairs, Eurasian Spoonbill (*Platalea leucorodia*) – 89 -147 pairs, Glossy Ibis (*Plegadis falcinellus*) – 10 - 186 pairs. The most numerous colonies were at Ibisha Island, small islet off Golyama Burzina Island (on Romanian territory) and in Srebarna Lake. Trends of decrease in numbers for the period 2006 -2014 were detected for Pigmy Cormorants and Grey Herons. Increasing numbers were recorded for Night Herons, Eurasian Spoonbills and Glossy Ibises. The remaining species showed more or less stable population numbers.

Chapter 10 - Distribution and numbers of colonially nesting herons, cormorants, Eurasian Spoonbills and Glossy Ibises were studied along the whole Bulgarian –Romanian section of the Danube River in the period 2010-2014. A total of 15 colonies, 12 of them mixed, were located and their species composition and trends in numbers were followed during the study. The total numbers by species differed considerably between years as follows: Great Cormorant (*Phalacrocorax carbo*) - 1573 -2299 pairs, Pigmy Cormorant (*Microcarbo pygmaeus*) – 283 - 877 pairs, Grey Heron (*Ardea cinerea*) – 237-399 pairs,

Purple Heron (*Ardea purpurea*) – 4 - 18 pairs, Great White Heron (*Ardea alba*) – 6 - 12 pairs, Squacco Heron (*Ardeolla ralloides*) – 60- 181 pairs, Night Heron (*Nycticorax nycticorax*) – 234 - 484 pairs, Eurasian Spoonbill (*Platalea leucorodia*) – 89 -147 pairs, Glossy Ibis (*Plegadis falcinellus*) – 10 - 186 pairs. The most numerous colonies were at Ibisha Island, small islet off Golyama Burzina Island (on Romanian territory) and in Srebarna Lake. Trends of decrease in numbers for the period 2006 -2014 were detected for Pigmy Cormorants and Grey Herons. Increasing numbers were recorded for Night Herons, Eurasian Spoonbills and Glossy Ibises. The remaining species showed more or less stable population numbers.

Chapter 11 - A total of 28 diurnal birds of prey species and Black Storks were recorded during intensive research along the Bulgarian section of the Danube River between 2008 and 2018. Out of these, 20 species are breeding. The Danube River and its neighbouring hills, forests, steppes, and the lowest sections of its tributaries hold significant populations of species of high conservation importance; including the Lesser Spotted Eagle (*Clanga pomarina*),40-50 breeding pairs (br.p.); Booted Eagle (*Hieraaetus pennatus*),10-15 br.p.; Short –toed Eagle (*Circaetus gallicus*),50-70 br.p.; European Honey Buzzard (*Pernis apivorus*),17-25 br.p.; Levant Sparrowhawk (*Accipiter brevipes*), 35-60 br.p.; Long-legged Buzzard (*Buteo rufinus*),75-130 br.p.; White-tailed Sea Eagle (*Haliaeetus albicilla*),12 br.p.; Marsh Harrier (*Circus aeruginosus*), 25-40 br.p.; and Black Stork (*Ciconia nigra*),80-150 br.p.. During migration, the most numerous species are the Marsh Harrier, Black Stork, European Honey Buzzard, Common Buzzard (*Buteo buteo*), Eurasian Sparrowhawk (*Accipiter nisus*) and Lesser Spotted Eagle. Migration is stronger during autumn and in central and eastern parts of the study area.

Chapter 12 - A review of the biodiversity of mammals in the Danubian-Dobrudzha region (DDR) of the Lower Danubian Province (LDR) of Bulgaria was made. Seventy-four mammalian species occur in the area. Recent data on the distribution of 57 species were used to compile models of habitat suitability throughout the country. The modeling results for LDR are presented in detail. Analyses have been made to assess the territorial differentiation of habitat suitability and species richness within particular ecological groups (orders or group of orders). The suitability of habitats and species richness are unevenly distributed in the area, and spots with high suitability and species richness largely overlap in the ecological groups. They are mainly associated with forests, wetlands, and karst areas. The spots with high species diversity are connected with ecological corridors and “stepping-stones.” They follow the river valleys and are formed mainly by the few preserved riparian forests. The role of the Natura 2000 network in the area has been assessed. It covers most of the identified patches of high habitat suitability and species richness as well as the eco-corridors. Habitat suitability for most mammalian species is higher in the protected areas.

ACKNOWLEDGMENTS

We are grateful to our colleagues and friends from WWF- Danube-Carpathian Program Ivan Hristov, Veselina Kavrakova, Maria Kachamakova, Rayna Popova, Stela Apostolova for their assistance and financial support for collecting materials and preparing that book. We also thank to our colleagues Ivailo Zafirov, Assen Ignatov, Boris Belchev, Matej Polačik, Lyubomir Andreev and Svetoslav Spasov who supplied some key photographs for that book.

INTRODUCTION

***Peter Stanislavov Shurulinkov*[1] *and Rossen Tzonev*[2]**
[1]Natural Museum of Natural History, Bulgarian Academy of Sciences
Sofia, Bulgaria
[2]Department of Ecology and Environmental Protection, Faculty of Biology,
Sofia University "St. Kliment Ohridski", Sofia, Bulgaria

In its lower stretches, the Danube River passes through large lowlands and remains an almost intact natural ecosystem. The flora and fauna of this specific area, not far from the Danube Delta ecosystem, and its environmental importance are the topic of this book.

There are no dams or other significant hydrotechnical infrastructure on the mentioned stretch of the River and downstream to the Danube Delta. Therefore, there is nothing to cause a barrier effect to migratory fish species or other animals. Unfortunately, fish migration is already a rare occurrence in many of Europe's largest rivers due to human alteration and physical barriers. Furthermore the special regime on the Romanian-Bulgarian border during the communist-era regime and partly afterwards has contributed for the preservation of the wildlife from the human impacts.

The River is very rich in small to large-sized islands and archipelagos that are mainly covered by diverse alluvial forests. These island forests are flooded by river water for four to five months in the spring and summer. Many small swamps and channels are formed inside the islands, creating excellent conditions for spawning of many fish species. These remote places are also of great importance as feeding grounds for many bird and mammal species living along the river.

The depth of the river is highly variable with a complicated relief offering many bottom thresholds and pits. When the water level is low, many small sandy or gravel islands and sandpits appear above the water. The shores of the river are often covered by riparian forests which flood during the spring and summer, when the Danube reaches its highest water levels. These forests are dominated mainly by willows (*Salix* spp., mostly *Salix alba* and *S. purpurea*) and poplars (*Populus nigra, P.alba*). Rarer are elm (*Ulmus laevis*), ash (*Fraxinus oxycarpa*), and White oak (*Quercus robur*). Unfortunately, forestry practices along the river and also on the islands have led to significant forest degradation and alteration of the primary forests into forest plantations mainly dominated by hybrid poplars. Nevertheless, approximately 40-50% of these forests still represent primary habitats and continue to hold great importance for the preservation of rich biodiversity. The largest areas of typical alluvial forests in the Bulgarian-Romanian section of the River can be found on the Persin (Belene) Island archipelago (2050 ha), the Kalimok Island complex (1000 ha, which includes forests on both river banks), Nikopol's islands (481 ha), Malak and Golyam Kosuy islands (338 ha), Vardim Island (270 ha), Aleko Island (255 ha), Byalata Prust and Komluka islands (255 ha), and several other islands.

Like the forests, the wetland ecosystems in the Danube's lowlands have also been negatively impacted in recent years. Drainage, water and air pollution, poaching (fish, birds, mammals), illegal logging, rural and infrastructural development, and hydrotechnical construction are among the biggest threats, both past and current, to the wetlands. Between 1920 and 1970, a great majority of the Danube's marshes were drained for agricultural development. Marshes covering thousands of hectares, including Svishtov's Balta, the Karaboaz Marshes, Kozloduy Marsh, and Aydemir Marsh, disappeared from maps, together with their impressive and rich wildlife. In addition, dikes constructed along the river and around Persin Island in the middle of the 20th century led to the degradation of wetland areas, as they were gradually cut off from the river. The river dikes also instigated a process of rapid secondary degradation and eutrophication among many of the partially remaining wetlands,

The main study area is the river stretch between the Timok River mouth and the town of Silistra, with a focus on the Bulgarian river bank and islands. Also included is the Danube Plain at a distance of up to 10-15 km south of the river on Bulgarian territory. Many wetlands are still intact in the river lowlands and, in part, on the islands. Most of them are natural lakes, marshes or swamps, including Srebarna Lake, Persin Marshes (Peschina, Murtvo and Dulyova Bara), Garvan Marsh, Ostrov Swamp, Dobri Dol Marsh, and Kaykusha Marsh. Others wetland of natural origin were later turned into fishponds such as Kalimok, Orsoya, and Mechka Fishponds. These are no longer functioning; however Bulgarian authorities are currently restoring the natural habitats in some of the Danube's fishponds, including Kalimok.

The largest part of Danube Plain included in the study area is covered mainly by agricultural lands, steppe vegetation, and oak and silver lime forests. Many of the Danube's right tributaries, including the rivers Lom, Ogosta, Iskar, Vit, Osam, and Yantra, also cross this plain. These river valleys are key for biodiversity and maintaining green corridors, particularly the Roussenski Lom River system which forms a number of spectacular karst canyons. Other important areas include the lower stretches of the Yantra and Osam Rivers.

The biodiversity of the Bulgarian and Romanian Lower Danube stretch has been actively studied, to an extent, since the beginning of 20th century. Previously, there were only occasional reports from travelers familiar with some of the animals and plants along the study area. During the last 100 years, the knowledge on many taxonomic groups of organisms, specifically along the Lower Danube, has increased. This includes information on plants, birds, amphibians, reptiles, fishes and other hydrobionts, many groups of invertebrates, among others.

In the present book we gathered part of that knowledge including some better studied taxa. It is far from an exhaustive study but it includes groups for which we have more actual records. All vertebrate classes were included in the book. Vertebrate fauna of the study region includes 392 species comprised of 83 species of fishes, 19 of amphibians, 15 of reptiles, 201 of birds, and 74 of mammals. We also included vulnerable groups directly threatened by local development.

Unfortunately, numerous and important groups of organisms, mainly invertebrates, were not included because of lack of sufficient studies. This book presents important findings on records and localities of rare and threatened taxa, with the intention of publishing this data to be a useful tool for the conservation of threatened habitats, flora and fauna along the Lower Danube.

In recent decades, the importance of the River's ecosystems has been evaluated in a positive light. Several projects for the restoration of marshes have been implemented; for example, Srebarna Lake, Persin Island Marshes, Kaykusha Marsh and Kalimok Fishponds. Similar projects are necessary to ensure the future of other Danube marshes such as the Ostrov Marsh, Garvan Marsh, Dobri Dol - Orsoya Floodplain and Karaboaz Floodplain.

The Danube's biodiversity continues to be in danger from a wide variety of threats including intensive logging (including on the islands), overfishing, dragging of the river bottom, and artificial changes to water levels caused by the Iron Gates Dams. The Dams were constructed nearly 50 years ago on the territories of Serbia and Romania and has had a significant impact on all of the Danube's ecosystems. One very serious ongoing threat to these ecosystems is the plan for future hydrotechnical improvements to ensure more intensive shipping along particular sections of the River. The realization of such a project will drastically change the Lower Danube ecosystem, as the proposed engineering activities involve intensive dragging and construction of chevrons, groins and bank protection structures. These heavily damage important riverine plant communities and protected species, including those in the Habitat and Bird Directives and protected territories designated according to the Bulgarian Protected Areas Act and NATURA 2000 sites. We believe that the information presented in this book will contribute to the assessment of the real impacts of planned developments on the Danube's ecosystems.

Lastly, through this book we hope to introduce readers to some of the most interesting creatures living along the Danube River. The unique "Great River" ecosystem, a product of millions years of evolution, should be better known and appreciated for its uniqueness and diversity. We believe that a better understanding of the wildlife of the Danube River will ultimately lead to a recognition of its importance and efforts for the protection of the Danube's ecosystems.

In: Biodiversity of the Bulgarian-Romanian Section
Editor: P. S. Shurulinkov, Z. Hubenov et al.
ISBN: 978-1-53615-663-8

Chapter 1

Important Plant Areas along the Bulgarian Stretch of the Danube River: Contemporary Knowledge and Gaps in Their Determination

***Rossen Tzonev*[1,*], *Chavdar Gussev*[2], *Valeri Georgiev*[3] and *Sonya Tsoneva*[2]**

[1]Department of Ecology and Environmental Protection, Faculty of Biology, Sofia University "St. Kliment Ohridski", Sofia, Bulgaria
[2]Division of Applied Botany, Department of Plant and Fungal Diversity and Resources, Institute of Biodiversity and Ecosystem Research, Bulgarian Academy of Sciences, Sofia, Bulgaria
[3]Division of Flora and Vegetation, Department of Plant and Fungal Diversity and Resources, Institute of Biodiversity and Ecosystem Research, Bulgarian Academy of Sciences, Sofia, Bulgaria

Abstract

Important Plant Areas in Bulgaria are determined using a complex of various criteria of flora and habitats. The aim of this study is to make a critical evaluation of 12 chosen IPAs along the Bulgarian stretch of the Danube River. As a result, the criteria for their determination have been complemented. The expansion of some of the IPAs and the designation of one new site – Kalimok-Brashlen – is also proposed. The Bulgarian IPAs network is generally representative of the flora and habitat diversity found along the Danube River; however, their argumentation could be improved after more intensive studies concerning the diverse Danube's flora and vegetation.

* Corresponding Author's Email: rossentzonev@abv.bg.

Keywords: riverine vegetation, flora, habitats, conservation significance

INTRODUCTION

Important Plant Areas (IPAs) are natural or semi-natural sites exhibiting exceptional botanical richness and/or supporting an outstanding assemblage of rare, threatened and/or endemic plant species and/or vegetation of high botanical value. IPA is a programme set up in the United Kingdom (UK), by the organisation Plantlife, to provide a framework for identifying and maintaining the richest sites for plant life, possibly within existing protected areas, though the protection of the IPA itself is not legally enforced (www.plantlife.org.uk). IPAs are designated using a system of criteria (Anderson 2002, Darbyshire et al. 2017). In 2002, the countries that ratified the Convention on Biological Diversity also accepted the Global Strategy for Plant Conservation. This strategy contributes to sustainable development goals and halt the processes of biodiversity loss. It was comprised of 12 goals to be achieved by 2010; Goal 5 calls for the protection of 50% of the most important areas for plant biodiversity (Gussev & Petrova 2015).

The Bulgarian IPA network was established from 2006 to 2009; 125 sites were determined to be IPAs, totalling an area of approximately 1403770 ha. Large part of the IPA network (70%) overlaps with NATURA 2000 sites in the country. This overlap is understandable considering the similar criteria used for the selection of the NATURA and IPA sites, including species richness, habitat diversity and conservation importance (Peev et al. 2009, Peev & Petrova 2012). Of the 125 designated IPAs, only 12 are situated along the Danube River and the Danube lowlands, covering a total area of 12148.68 ha. These 12 sites account for 0.86% of the entire IPA coverage in Bulgaria. According to Bulgarian legislation (Biodiversity Act), the IPAs are part of the national ecological network. Their legal definition: "Important site for plants is an area of international significance for the conservation of plant species, described by a standard methodology of Plantlife International."

The aim of this study is to make a critical analysis of the IPA network along the Bulgarian stretch of the Danube River and summarize current knowledge on these sites, as well as any gaps in knowledge that may exist in their selection. There are numerous papers on the fauna of this part of Bulgaria (Paspaleva-Antonova 1961, Russev & Naidenow (Eds) 1978, Michev et al. 1981, Michev (ed.) 1995), but the floristic diversity is still poorly known. Most of the studies are outdated or cover only small areas of the river's stretch. The most important of them are focused on the wetlands with notes on their vegetation (Bonchev 1929, Michev & Stoyneva 2007). Especially concentrated on the flora and vegetation of the Danube's wetlands are the works of Petkoff (1911) and Kochev & Yordanov (1976, 1981).

There are also studies on the islands and their alluvial forests (Petkov 1940, Stoyanov 1927, 1948, Tchernyavski & Pavlov 1955, Tzanov 1992). Only Srebarna Lake, which is a UNESCO World Heritage Site and reserve, and several neighboring protected areas were objects of more intensive floristic and vegetation studies (Kochev 1987, Baeva 1985, 1986, 1987, 1988a, b, 1995, Petrova & Velev 1998, Valchev et al. 2012). There are also studies on macrophytes along the Danube stretch (Yurukova 2002, Valchev et al. 2006, Valchev et al. 2018), as well as floristic studies on parts of the Danube Plain close to the river (Urumov 1917, 1926, 1935, Stoyanov 1947, Yordanov 1946-1947, Yordanov & Kochev 1973, Tzonev & Sumberova 2004, Tzonev 2005, Vladimirov & Tsoneva 2006, Nedelcheva & Tzonev 2006), vegetation types (Kochev et al. 1987, Tzonev 2002), and protected areas (Kochev & Tsolova 1984, Dister et al. 2009, Tzonev 2009) along the Danube. New information for many macrophytes, including those found along the Danube River, is included in the Atlas of Aquatic and Wetland Plants in Bulgaria (Tsoneva et al. 2012).

Material and Methods

Study Area

The Danube River and its islands, lowlands and surrounding hills of the river's terraces with steppe vegetation) were included in the study. The overall study area was comprised of a 5 km-wide strip in the southern part of the River containing habitats and their flora formed directly or indirectly by the river's activities.

Criteria for IPAs

The criteria for the identification of IPA sites were summarized by Anderson (2002) and applied to Bulgaria by Peev et al. (2009); the selected sites were published by Peev et al. (Eds) (2012). According to the cited publications, the following general criteria were applied for selection of IPAs: Criterion A – The site holds significant populations of one or more species that are of global or European conservation concern; Criterion B – The site has an exceptionally rich flora in European context in relation to its biogeographic zone; Criterion C – The site is an outstanding example of a habitat type of global or European plant conservation and botanical importance. The Bulgarian IPAs along the Danube River were determined using criteria A and C (Figure 1).

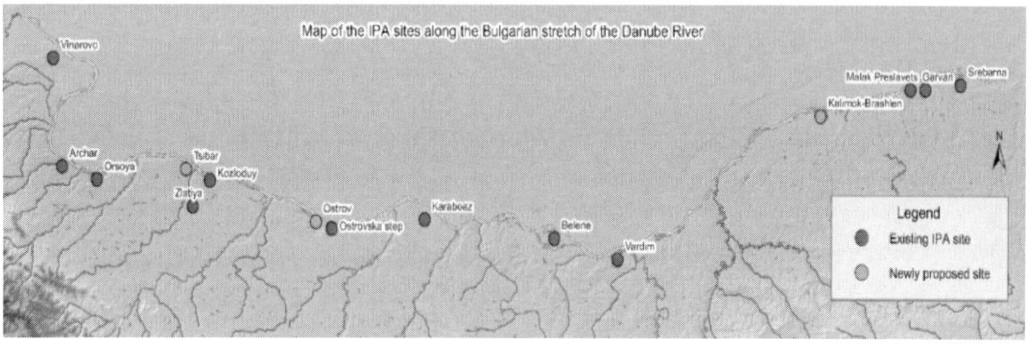

Figure 1. Map of IPA sites along the Bulgarian stretch of the Danube River.

Criterion A: Threatened Species

It has four categories: A(i) Globally threatened plant species; A(ii) Plant species threatened on European level; A(iii) Threatened endemics, not covered by the above two categories; and A(iv) Threatened near endemic (limited range) species, not covered by the first two categories.

Criterion C: Threatened Habitats

The IPAs identify the sites with threatened habitat types. For the IPA program in Europe, threatened habitats are listed in Annex I of the Habitats Directive and in Resolution 4 of the meeting of 16th Standing Committee of Bern Convention. This criterion is further divided into two categories: C (i) Priority Threatened Habitats is based on the priority habitats of Annex I of the Habitats Directive and any corresponding to the Bern Convention habitats and C (ii) Threatened Habitats is based on the threatened habitats contained in Annex I of the Habitats Directive and the Bern Convention not covered by C (i). C (ii) and C (i) habitats are equally important; the creation of two categories provides a way to track the development of the sites, identified in programs as Natura 2000. For the application of Criterion C, national teams compiled a list of threatened habitats based on the combination of habitats in the appendices of the Habitats Directive and the Bern Convention.

In the present study the application of IPA criteria is used to propose some improvement of IPAs network along the Danube's stretch. Besides the plant species that meet Criteria A (ii, iii, iv), those included in categories "critically endangered" (CR), "endangered" (EN), "vulnerable" (VU) and "near threatened" (NT) according to the Red List of Bulgarian plants (Petrova & Vladimirov 2009), as well as some not evaluated but very rare or little known species are given in the site's descriptions.

Species and habitats, for which the sites are designated among the top 5 in Bulgaria, were noted with an exclamation mark in brackets (!).

RESULTS

The following vascular plant species found in the IPAs along the Danube River meet the following Criteria A categories: A (ii): *Aldrovanda vesiculosa*, *Armoracia macrocarpa* (Figure 2), *Eleocharis carniolica*, *Marsilea quadrifolia* (Figure 3), *Salvinia natans*, *Trapa natans* (Figure 4); A (iii): *Limonium bulgaricum* (Figure 5); or A (iv): *Centaurea rumelica* (Figure 6).

There are other species which have limited range only along the Danube River and are closely tied to the river's habitats. These species could also be suitable for inclusion in the checklist of Criterion A (iv) and include the following: *Tragopogon floccosus*, *Rindera umbellata*, *Stachys arenariiformis*, and *Achillea ochroleuca*.

Figure 2. *Armoracia macrocarpa* – endemic only to the Danube stretch. Photo by Sonya Tsoneva.

The Danube River stretch is a key area (50% or more from their whole area in Bulgaria) for the preservation of following habitats presented below. They are characterized more completely in the Red Data Book of Bulgaria, vol. 2 (see Biserkov et al. (eds.) 2015).

Figure 3. *Marsilea quadrifolia* – critically endangered species in Bulgaria. Photo by Rossen Tzonev.

Figure 4. Community of *Trapa natans* in IPA Belene. Photo by Rossen Tzonev.

Figure 5. *Limonium bulgaricum* – regional endemic for Middle Danube Plain in Bulgaria. Photo by Rossen Tzonev.

Figure 6. *Centaurea rumelica* – regional endemic for the northern part of Danube Plain in Bulgaria. Photo by Rossen Tzonev.

2340 Pannonic Inland Sand Dunes

This habitat (Figure 7) is only present in the Danube lowlands, with the largest distribution in the Bregovo-Novo Selo, Orsoya, Tsibar, Ostrov and Karaboaz lowlands. Many of these dunes have been ploughed and are now covered by ruderalized communities. Typical species include Kochia laniflora, Plantago scabra, Centaurea arenaria, Alkanna tinctoria and Seseli tortuosum. It is one of the highest priority target habitats, now limited only to one narrow strip along the river.

Figure 7. Inland dunes (2340) in IPA Orsoya. On the forefront the Danube's endemic *Tragopogon floccosus*. Photo by Sonya Tsoneva.

3130 Oligotrophic to Mesotrophic Standing Waters With Vegetation of the *Littorelletea uniflorae* and/or Isoeto-Nanojuncetea

This habitat mainly occurs along the Danube River because it is characterized by dried up alluvial deposits in the river stream, which are sandy or sandy-loam, and their vegetation is dominated by short hygrophytes. The most typical species are Dichostlylis michelianus, Gnaphalium uliginosum, Heliotropium supinum, Eleocharis acicularis, Amaranthus lividus, Cyperus fuscus, Lindernia dubia, Persicaria lapathifolia, Plantago altissima, Astragalus contortuplicatus, Crypsis alopecuroides and Cyperus glomeratus. This habitat also occurs along other large rivers in Bulgaria like the Maritsa River, as well as along several large dams; however, these areas are significantly smaller than the one present along the Danube River.

Figure 8. Eutrophic lake (3150) with diverse plant communities (*Phragmites australis, Nymphaea alba, Stratiotes aloides*) in IPA Srebarna. Photo by Rossen Tzonev.

3150 Natural Eutrophic Lakes with Magnopotamion or Hydrocharition – Type Vegetation

In the beginning of the 20th century, the Danube's lowlands had not yet been subjected to draining (see Bonchev 1929) and, as a result, were the most important region in Bulgaria for wetlands (marshes and lakes) with their very rich fauna and flora. Bulgarian wetlands are now 1/20 the size they were in the early 20th century (Kochev & Yordanov 1981, Michev (ed.) 1995). The largest wetlands were situated in the Svishtov-Belene lowland, Karaboaz, Kalimok-Brashlen, Srebarna, and Aydemir. Most of them were completely or partially destroyed due to drainage and other negative activities. However, some of the most important wetlands or parts of them still remain in the Danube River area, including the Srebarna Reserve (World Heritages site) (Figure 8), Persina Island, Ostrov, Garvan-Popina, Malak Preslavets (Figure 9), and Kalimok-Brashlen. These wetlands with their aquatic flora and vegetation are of significant natural value to the Danube's ecosystems.

3270 Rivers with Muddy Banks with *Chenopodion rubri p.p.* and *Bidention p.p.* Vegetation

This habitat is often in a complex with 3130. However, these communities are located on polluted alluvial deposits where the number of nitrophilous species increases. Such species are *Bidens cernua, Bidens frondosa, Bidens tripartita, Chenopodium ambrosioides, Chenopodium glaucum, Chenopodium rubrum, Myosoton aquaticum, Persicaria hydropiper, Persicaria lapathifolia, Potentilla supina, Rumex conglomeratus, Rumex maritimus, Veronica anagallis-aquatica, and Xanthium italicum*. The habitat is also common along some large rivers in the interior of the country like the Maritsa, Iskar, and Ogosta Rivers. As a result, the Danube stretch is not of exclusive importance for the conservation of the habitat.

Figure 9. Eutrophic lake (3150) with community of *Nymphaea alba* in IPA Malak Preslavets. Photo by Rossen Tzonev.

6260 *Pannonic Sand Steppes

It is one of the rarest habitats in the country; its range is restricted to the sandy-loess hills between the villages of Archar and Dobri Dol (Vidin District). These communities are variable but mostly dominated by Festuca vaginata, Apera spica-venti, Psilurus aristatus and Achillea ochroleuca.

Figure 10. Loess walls with steppic vegetation in IPA Kozloduy. Photo by Rossen Tzonev.

6250 *Pannonic Loess Steppic Grasslands

This habitat type occurs on the elevations along the Danube River (Figure 10). It is also found 20-30 km south of the river but is most representative on the highest loess hills close to the Danube, including the loess "hunches" at Ostrov village and Oryahovo town, which reach heights of 150-200 m above the river's level. Closed and densely tufted herbaceous communities prevail on the richest and least-eroded soils. Their species composition includes many rare species such as Astragalus asper, Astragalus austriacus, Chamaecytisus kovacevii, Stachys arenariiformis, Centaurea rumelica and Rindera umbellata. It is one of the target habitats on the hills surrounding the Danube's lowlands.

6440 Alluvial Meadows of River Valleys of the *Cnidion dubii*

The continental flooding meadows were typical of and restricted to the lowlands of the Danube River, depending on continental climate conditions and flooding periods. The redirection of the river bed and draining of the swamps have caused the almost complete disappearance of these meadows. Very small fragments remain in places with high humidity that are sporadically flooded by the neighbouring drainage canals or high underground waters. These fragments are strongly ruderalized and have changed species composition. Typical species include *Agrostis stolonifera*, *Allium angulosum*, *Alopecurus pratensis*, *Althaea officinalis*, *Clematis integrifolia*, *Elymus repens*, *Euphorbia lucida*, *Festuca arundinacea*, *Lycopus exaltatus*, *Lythrum virgatum*, *Poa palustris*, *Ranunculus acris*, *Teucrium scordium*, *Thalictrum flavum*, *T. lucidum*, *Gratiola officinalis*, *Scutellaria hastifolia* and *Symphytum officinale*. Some of the Danube's lowlands are the only places in Bulgaria where small fragments of these meadows have survived.

Figure 11. Alluvial willow-poplar forests in Nikopol's Archipelago – IPA Belene. Photo by Rossen Tzonev.

91E0 *Alluvial Forests with *Alnus glutinosa* and *Fraxinus excelsior* (*Alno-Padion*, *Alnion incanae*, *Salicion albae*)

This habitat consists of riverine forests, most often willow-poplar, pure willow or poplar forests (Figure 11), in the lowlands and plains, and in the continental climatic conditions in northern Bulgaria. The riverine forests are widespread along many large and small rivers in Bulgaria but mostly as narrow strips with ruderalized species composition that are not in good ecological status. Their composition is most diverse along the Danube and its islands. Despite their changed floristic structure – mostly because of the penetration of many neophytes like *Amorpha fruticosa*, *Acer negundo*, *Sycios angulatus* and *Echinocystis lobata* – the alluvial forests only on the Danube's islands still cover significant areas and they are included as target habitat for the Danube stretch.

There are also other habitat types widespread along the Danube River, but the study area is not of key importance to their conservation on a national level. Such habitats are: 1530 *Pannonic salt steppes and salt marshes, 6240 *Sub-pannonic steppic grasslands; 6430 Hydrophilous tall herb fringe communities of plains and of the montane to alpine levels, 91F0 Riparian mixed forests of *Quercus robur*, *Ulmus laevis* and *Ulmus minor*, *Fraxinus excelsior* or *Fraxinus angustifolia*, along the great rivers (*Ulmenion minoris*); and 91Z0 Moesian Silver lime woods.

Here are presented the descriptions of all IPAs along the Danube River, as characterized by Apostolova & Meshinev (2012) and Meshinev & Apostolova (2012), and new proposals for improvement of their argumentation.

BGIPA002 Archar; 43° 48‘N, 22° 56‘E – 30 ha

Description: The site is not far from Archar village and is comprised of typical sandy-loess hills with psamophytic and semi-ruderal vegetation. The dominant species are *Achillea ochroleuca*, *Festuca vaginata*, *Apera spica-venti* and *Psilurus aristatus*. According to Meshinev & Apostolova (2012), the following species and habitats meet IPAs criteria: A (ii): *Armoracia macrocarpa*. C (i): 6260 *Pannonic sand steppes (!).

Proposals: The species *Armoracia macrocarpa* (Figure 7) inhabits alluvial forests, which also meet Criterion C (i) 91E0 *Alluvial forests with *Alnus glutinosa* and *Fraxinus excelsior* (*Alno-Padion*, *Alnion incanae*, *Salicion albae*) (!). As a result, the site should be expanded to include the Archar Lowland and Dovleka (Kukuruza) Island. In this case, the species A (ii) *Salvinia natans*, which occurs in ditches, could be added to the site characteristic. Other habitats with conservation significance in the site include the following: C (ii) 6440 Alluvial meadows of river valleys of the *Cnidion dubii* (one of the most important sites on national level) (!) and C (ii) 3130 Oligotrophic to mesotrophic standing waters with vegetation of the *Littorelletea uniflorae* and/or *Isoeto-Nanojuncetea* (!). Other species with conservation significance in the site are *Achillea ochroleuca* (CR), *Festuca vaginata* (VU), *Centaurea arenaria* (VU), *Dichostylis michelianus* (VU) and *Leucojum aestivum* (VU).

BGIPA006 Belene; 43° 40‘ N, 25° 11‘ E – 6897 ha

Description: The site includes the largest Bulgarian island in the Danube River, Persin (Belene),which is 45 sq. km, and other islands (Golyama Barzina, Lakat, Magariatsa, Belitsa, Gradina, Milka, Kitka, etc.) belonging to two archipelagos – Belene and Nikopol. There are three big island marshes, Peschina, Martvoto and Dulyova Bara (Staro Blato), and large areas covered by alluvial forests; the biggest one is in the eastern part of the

Persin Island. The site also includes the former Belene-Svishtov floodplain, a mostly agricultural landscape that has some surviving wetlands like Kaykusha marsh, slightly-salinized meadows and sandy-loess dunes. The water regime of the island marshes was restored; however, in the lowlands, the flooding is caused due to the rise in underground waters simultaneous with high water levels in the Danube River. The area is well-known for its rich wildlife, especially birds, with more than 60 species of conservation importance at the European level. According to Apostolova & Meshinev (2012), the species and habitats behind this site's designation as an IPA are as follows: Species A (ii) *Eleocharis carniolica* (!), *Salvinia natans* (!), *Trapa natans* (!) (Figure 8), *Marsilea quadrifolia* (Figure 9) (!), *Armoracia macrocarpa* (!); and A (iii) *Limonium bulgaricum* (!) (Figure 10); and Habitats C (i) 6240 *Sub-pannonic steppic grasslands (!), 91E0 *Alluvial forests with *Alnus glutinosa* and *Fraxinus excelsior* (*Alno-Padion*, *Alnion incanae*, *Salicion albae*) (!); and C (ii) 3130 Oligotrophic to mesotrophic standing waters with vegetation of the *Littorelletea uniflorae* and/or *Isoeto-Nanojuncetea* (!), 3150 Natural eutrophic lakes with *Magnopotamion* or *Hydrocharition – type vegetation* (!), 3270 Rivers with muddy banks with *Chenopodion rubri* p.p. and *Bidention* p.p. vegetation (!), 6430 Hydrophilous tall herb fringe communities of plains and of the montane to alpine levels (!), 91F0 Riparian mixed forests of *Quercus robur*, *Ulmus laevis* and *Ulmus minor*, *Fraxinus excelsior* or *Fraxinus angustifolia*, along the great rivers (*Ulmenion minoris*) (!).

Proposals: The following habitats are also very important: C (i) 2340 *Pannonic inland sand dunes (!), widespread on the Persin Island (see Tzonev 2009) and the Belene Lowland; C (i) 1530 *Pannonic *salt* steppes and *salt* marshes (!); in Belene Lowland; and C (i) 6250 *Pannonic loess steppe grasslands, on the hills surrounding the Belene floodplain.

Other species with conservation significance include the following: *Allium angulosum* (CR), *Utricularia australis* (CR); *Corispermum marschalii* (EN), *Nuphar lutea* (EN), *Nymphaea alba* (EN), *Nymphoides peltata* (EN), *Senecio paludosus* (EN), *Astragalus contortuplicatus* (VU), *Centaurea arenaria* (VU), *Dichostylis michelianus* (VU), *Leucojum aestivum* (VU), *Corispermum nitidum* (VU), *Euphorbia lucida* (VU), *Potamogeton trichoides* (VU), *Stachys arenariiformis* (VU), *Utricularia vulgaris* (VU), *Lemna gibba* (NT).

BGIPA013 Vardim; 43° 37′ N, 25° 26′ E – 1168 ha

Description: Vardim is one of the biggest islands in the Danube River. It was completely covered by flooded forests in the past; currently, many of the forests have been turned into poplar plantations. The surviving forests have a disturbed structure due to human activities.

According to Meshinev & Apostolova (2012), the site only meets the IPA criteria related to habitats. The habitat C (ii) 91F0 Riparian mixed forests of *Quercus robur*, *Ulmus* *laevis* and *Ulmus minor*, *Fraxinus excelsior* or *Fraxinus angustifolia*, along the great rivers (*Ulmenion minoris*) (!), is comparatively well preserved. It is represented by unique for the Bulgarian stretch of the Danube River old forests of *Quercus robur*, called by the local people "Vardim Oak", and is preserved in the "Stariya Dab" Protected Area. Another important habitat is C (ii) 91E0 *Alluvial forests with *Alnus glutinosa* and *Fraxinus excelsior* (*Alno-Padion, Alnion incanae, Salicion albae*) (!).

Proposals: The following species could be added to the site's characteristic: A (ii) *Salvinia natans*, occurring in the flooded depressions, and A (ii) *Trapa natans*, recorded from "Runtava Bara" marsh on the island. Habitats to be added are C (ii) 6430 Hydrophilous tall herb fringe communities of plains and of the montane to alpine levels. Other species with conservation significance found in this site are *Corispermum marschalii* (EN), *Leucojum aestivum* (VU), *Euphorbia lucida* (EN) and *Dichostylis michelianus* (VU). This site could provide a suitable habitat for *Armoracia macrocarpa* but the species has yet to be found in the area.

BGIPA016 Vinarovo; 44° 05‘ N, 22° 50‘ E – 24.7 ha

Description: According to Apostolova & Meshinev (2012), this site is situated in Northwest Bulgaria, not far from Vidin town. In this part of the Danube Plain, the loess sediments are thin, changed by sands that form inland dunes of various sizes. In its current boundaries, the site is relatively small. This is due to ongoing intensive agriculture, especially viticulture favored by the above-mentioned conditions. *Plantago scabra, Trifolium arvense, Centaurea arenaria, Cynodon dactylon, Silene alba* and *Xeranthemum annuum* are relatively abundant. Vinarovo was selected as an IPA due to the presence of the priority habitat C (i) 2340 *Pannonic inland dunes (!), and is one of the three best places for it in this country.

Proposals: None.

BGIPA025 Garvan; 44° 06‘ N, 26° 54‘ E – 210.69 ha

Description: The site is a marsh between the villages of Garvan and Popina (Silistra District). In the past, it had a connection with the Danube, but after the construction of a dike, this connection disappeared. The marsh is now water-fed by a small river flowing through the village and by underground waters. The water depth does not reach more than 0.5–0.7 m and is part of wetland dry up during the summer. It is mostly covered by communities of *Phragmites australis* and *Typha* spp. Small opened surfaces (mostly in the southernmost part) are covered by floating communities of *Nymphoides peltata, Hydrocharis morsus-ranae,* and *Salvinia natans*, among others. According to Apostolova

& Meshinev (2012), the diversity of birds is suitable for educational tourism. Changes of the water regime leading to eutrophication are the most important threats to the site. The IPA criteria for designation of this site are the species A (ii) *Salvinia natans* (!) and the habitats C (ii)3150 Natural eutrophic lakes with *Magnopotamion* or *Hydrocharition* – type vegetation (!) and 6430 Hydrophilous tall herb fringe communities of plains and of the montane to alpine levels.

Proposals: None. According to Apostolova & Meshinev (2012), *Nuphar lutea* also occurs in the marsh, but this was not confirmed by our studies. A relatively stable population of *Nymphaea alba* (EN) was found. The species *Nymphoides peltata* (EN) could also form communities in the marsh.

BGIPA044 Zlatiya; 43° 44' N, 23° 29' E – 40 ha

Description: This site includes the hills on the right shore of the Tsibritsa River between Zlatiya and Gorni Tsibar villages. The site was selected due to the habitat C (ii) 6250 *Pannonic loess steppic grasslands (!). However, according to Meshinev & Apostolova (2012), most of the area was terraced and used as arable land in the past. This has resulted in formation of areas covered by secondary grasslands in intermediate succession stage between the initial weedy and ruderal vegetation and more stable phases with domination of tussock grasses.

Proposals: The site is too small and the single target object (steppe vegetation) is not in good status, nor is representative. This site must be significantly expanded to include the lowland between the villages of Stanevo, Gorni and Dolni Tsibar, and Tsibar Island in the Danube River. This site would then include several important habitats, as C (ii) 3130 Oligotrophic to mesotrophic standing waters with vegetation of the *Littorelletea uniflorae* and/or *Isoeto-Nanojuncetea*, 3150 Natural eutrophic lakes with *Magnopotamion* or *Hydrocharition* – type vegetation, 3270 Rivers with muddy banks with *Chenopodion rubri* p.p.and *Bidention* p.p. vegetation (!), 6430 Hydrophilous tall herb fringe communities of plains and of the montane to alpine levels; and C (i) 91E0 *Alluvial forests with *Alnus glutinosa* and *Fraxinus excelsior* (*Alno-Padion*, *Alnion incanae*, *Salicion albae*), 2340 *Pannonic inland sand dunes (!), 1530 *Pannonic salt steppes and salt marshes, and 6250 *Pannonic loess steppe grasslands. Tsibar Lowland is one of the best five sites in Bulgaria for the habitat 2340; it also has species to meet criteria A (ii) *Salvinia natans* and A (iv) *Centaurea rumelica*. Other species with conservation significance are as follows: *Alkanna tinctoria* (EN), *Astragalus contortuplicatus* (VU), *Astragalus ponticus* (VU), *Centaurea arenaria* (VU), *Dichostylis michelianus* (VU), *Leucojum aestivum* (VU), *Euphorbia lucida* (VU) and *Lemna gibba* (NT). The proposed areas would increase the site by approximately 3000 ha. We also propose to change the name of this site to Tsibar, as its Natura 2000 site name is BG0000199 Tsibar.

BGIPA051 Karaboaz; 43° 42' N, 24° 38' E – 689.8 ha

Description: The Karaboaz Lowland is the biggest floodplain of the Danube River, with an area of approximately180 sq. km. It is situated in Pleven District, between the Vit and Iskar River mouths. Following drainage, the lowland has been mostly occupied by arable fields, draining ditches, sands with ruderal communities, and halophile grasslands. However, during high water levels of the Danube, and due to underground waters and abandoned channels, large parts of the floodplain close to the Danube River could be flooded. According to Meshinev & Apostolova (2012), the site was selected according to habitat criteria C (i) 1530 *Pannonic salt steppes and salt marshes (!) and 2340 *Pannonic inland dunes (!).

Proposals: The site could be better motivated with species A (ii) *Salvinia natans*, as well as other important habitats such as C (i) 91E0 *Alluvial forests with *Alnus glutinosa* and *Fraxinus excelsior* (*Alno-Padion*, *Alnion incanae*, *Salicion albae*); C(ii) 3130 Oligotrophic to mesotrophic standing waters with vegetation of the *Littorelletea uniflorae* and/or *Isoeto-Nanojuncetea* (!); C (ii) 3150 Natural eutrophic lakes with *Magnopotamion* or *Hydrocharition* – type vegetation; C (ii) 3270 Rivers with muddy banks with *Chenopodion rubri* p.p. and *Bidention* p.p. vegetation (!), 6430 Hydrophilous tall herb fringe communities of plains and of the montane to alpine levels; C (ii) 91F0 Riparian mixed forests of *Quercus robur*, *Ulmus laevis* and *Ulmus minor*, *Fraxinus excelsior* or *Fraxinus angustifolia*, along the great rivers (*Ulmenion minoris*) (!); and C (ii) 92D0 *Southern* riparian galleries and thickets (*Nerio-Tamaricetea* and *Securinegion tinctoriae*) (!). For the habitat 92D0, the site is its only location in northern Bulgaria. The habitat 91F0 is represented by very old oak forests in the "Genchov Orman" Protected Area; some of the oldest trees are over 300 years. The species with conservation significance are *Astragalus contortuplicatus* (VU), *Centaurea arenaria* (VU), *Dichostylis michelianus* (VU), *Leucojum aestivum* (VU), *Euphorbia lucida* (VU) and *Lemna gibba* (NT). Kochev & Yordanov (1981) have also recorded the following species in the site ("Trapkata" swamp near Gigen village): *Marsilea quadrifolia* (CR), *Nymphoides peltata* (VU) and *Nuphar lutea* (CR) (south of Zagrazhden village). There is, however, a strong possibility that these three species have become extinct in the site.

BGIPA057 Kozloduy; 43° 47' N, 23° 36' E – 12 ha

Description: The site is situated between the town of Kozloduy and Gorni Tsibar village (Vratsa District) and is comprised of high bare loess walls reaching towards the river terrace of the Danube. The highest parts are covered by typical steppe vegetation dominated by tufted grasses like *Chrysopogon gryllus*, *Stipa capillata* and *Festuca valesiaca*. According to Meshinev & Apostolova (2012), the site was selected using both criteria A (iv) *Centaurea rumelica* (!) and C (i) 6250 *Pannonic loess steppic grasslands (!). For this habitat, the site is one of the best and most representative places in the country.

Proposals: The following species could be added in Criterion A (iv) as regional endemics for the Danube's water basins: *Stachys arenariiformis* (VU) (Rossen Tzonev, personal observation) and *Rindera umbellata* (NE) (which was pointed for the place by Urumov 1935).

BGIPA067 Malak Preslavets; 44° 05' N, 26° 50' E – 92.7 ha

Description: The marsh is situated in the surroundings of Malak Preslavets village (Silistra District). It is a eutrophic wetland with a stable regime sustained by underground karst waters and rainfalls. A dike was constructed in the northern part of the marsh; the depth of the water basin is about 3 m. The shore is covered by communities of *Phragmites australis* and *Typha latifolia*. It is one of the largest localities in Bulgaria with monodominant communities of *Nymphaea alba*. The slopes above the marsh are occupied by forests mostly dominated by *Tilia tomentosa*. According to Meshinev & Apostolova (2012), the site was selected using both criteria: species A (ii) *Salvinia natans* (!); and habitats: C (ii) 3150 Natural eutrophic lakes with *Magnopotamion* or *Hydrocharition* – type vegetation (!); and C (ii) 91Z0 Moesian Silver lime woods. The site is one of the best five in the country for typical habitat 3150 Natural eutrophic lakes with *Magnopotamion* or *Hydrocharition* – type vegetation (!).

Proposals: The species A (ii) *Marsilea quadrifolia* could be mentioned in the species criteria because it was reported by Kochev & Yordanov (1981) for the site; however, it has not been confirmed at the site for many years. Another species with conservation significance is *Nymphaea alba* (EN), which population in the site is probably the most stable in Bulgaria. Some species pointed by Kochev & Yordanov (1981) like *Nuphar lutea* (EN), *Nymphoides peltata* (VU) and *Wolffia arrhiza* (VU) could also be added, but they have not been observed for the last two decades.

BGIPA079 Orsoya; 43° 46' N, 23° 04' E – 2460.6 ha

Description: The site includes a lowland of the Danube River, between the villages of Orsoya (Montana District) and Dobri Dol (Vidin District). The area is occupied by alluvial and sandy-loess dunes, and former wetlands, some of which were transformed into fishponds and later abandoned. During the spring high waters of the Danube River, a large part of the Orsoya Lowland remains flooded until midsummer; however, the dunes are dry and covered by typical psamophytic communities. The site is the most representative of the habitat 2340 *Pannonic inland dunes (an area of more than 600 ha) in Bulgaria. The dune communities are dominated by *Plantago scabra*, *Cynodon dactylon*, *Dichanthium ischaemum*, *Kochia laniflora*, *Festuca vaginata*, *Artemisia campestris* and *Polygonum pulchellum*. Typical species are *Seseli tortuosum*, *Alkanna tinctoria*, *Gypsophila paniculata* and *Linaria genistifolia*. According to Meshinev & Apostolova (2012), the site was selected using both criteria: species: A (ii) *Salvinia natans*; and habitats – C (i) 2340 *Pannonic inland dunes (!), 91E0 *Alluvial forests with *Alnus glutinosa* and *Fraxinus excelsior* (*Alno-Padion, Alnion incanae, Salicion albae*) (!); C(ii) 3130 Oligotrophic to mesotrophic standing waters of the *Littorelletea uniflorae* and/or *Isoeto-Nanojuncetea* (!); and C(ii) 3150 Natural eutrophic lakes with *Magnopotamion* or *Hydrocharition* – type vegetation (!).

Proposals: There are other habitats with a significant importance on the national level, including C (ii) 3270 Rivers with muddy banks with *Chenopodion rubri* p.p. and *Bidention* p.p. vegetation (!); C (ii) 6430 Hydrophilous tall herb fringe communities of plains and of the montane to alpine levels (!); and C (ii) 91F0 Riparian mixed forests of *Quercus robur*, *Ulmus laevis* and *Ulmus minor*, *Fraxinus excelsior* or *Fraxinus angustifolia*, along the great rivers (*Ulmenion minoris*). The species C (ii) *Marsilea quadrifolia* was found in the past in this site (Velev 1984). The species A (ii) *Armoracia macrocarpa* also occurs in the region of Archar, which is very close to this site. There are also other species with conservation significance such as *Tragopogon floccosus* (CR), which only exists in a single locality in the country in Orosya region (Vladimirov & Tsoneva 2006), *Alkanna tinctoria* (EN), *Festuca vaginata* (EN), *Nuphar lutea* (EN), *Nymphaea alba* (EN), *Nymphoides peltata* (EN), *Stachys leucoglossa* (EN), *Senecio paludosus* (EN), *Thelypteris palustris* (VU), *Astragalus contortuplicatus* (VU), *Centaurea arenaria* (CE), *Corispermum nitidum* (VU), *Dichostylis michelianus* (VU), *Euphorbia lucida* (VU), *Leucojum aestivum* (VU), and *Utricularia vulgaris* (VU).

BGIPA082 Ostrovska Step; 43° 40‘ N, 24° 09‘ E – 12 ha

Description: The IPA is situated east of the village of Ostrov (Vratsa District) and includes hills of loess deposits elevated to 200 m above the Danube River. The region is one of the most representative of the typical loess steppes. The grasslands are dominated by *Chrysopogon gryllus*, *Dichanthium ischaemum*, *Stipa capillata* and *Festuca valesiaca*. There are many steppe floral elements like *Astragalus austriacus* and *Astragalus asper*. According to Apostolova & Meshinev (2012), the selection of this site was predetermined by the existence of priority habitat C (i) 6250 *Pannonic loess steppic grasslands (!).

Proposals: The site is too small to be representative for the botanically important complexes in this region. Our proposal is for a significant expansion of the site to include valuable steppe complexes between the villages of Ostrov, Gorni and Dolni Vadin, as well as the loess hills between the villages of Galovo and Selanovtsi. These steppe communities could be determined among the first three places in Bulgaria for habitat 6250. The site could also be expanded to include the Ostrov Lowland. It is surrounded by the villages of Ostrov, Galovo and Leskovets, and includes the islands of Esperanto (Leskovets) and Ezika. The Ostrov Lowland is one of the few areas where the former swamps –"Staroto Blato" – the biggest one, and some smaller – "Kochumina" and "Gola Bara" – have not been completely drained. They are important for the preservation of typical aquatic fauna, as well as the following rare plants: *Nymphaea alba*, *Salvinia natans, Trapa natans* and *Stratiotes aloides*. Ezika Island has now merged with the shore and the channel has become a swamp; there is also a channel inside the island which becomes a swamp. The eastern, slightly elevated part of the lowland has one of the most representative alluvial dune complexes (habitat 2340 *Pannonic inland dunes). It was partially destroyed by the planting of orchards and *Robinia pseudoacacia* forest; however, the open places are occupied by typical psamophytic communities with *Centaurea arenaria, Kochia laniflora*, *Alkanna tinctoria*, *Plantago scabra* and *Polygonum pulchellum*. Unfortunately, this habitat was not included in the standard form of existing NATURA 2000 site BG0000334 Ostrov. It also has not been found and mapped in the project of MOEW "Mapping and assessment of conservation status of habitat and species – Phase 1", carried out in the period between 2011 and 2013. We propose that the site be renamed to "Ostrov" and include an area of approximately 4000 ha. This IPA will be designated using the following criteria: A (ii) *Salvinia natans* (!), *Trapa natans*; C (i) 2340 *Pannonic inland dunes (!), 6250 *Pannonic loess steppic grasslands (!), 91E0 *Alluvial forests with *Alnus glutinosa* and *Fraxinus excelsior* (*Alno-Padion*, *Alnion incanae*, *Salicion albae*); C (ii) 3130 Oligotrophic to mesotrophic standing waters with vegetation of the *Littorelletea uniflorae* and/or *Isoeto-Nanojuncetea*; C (ii) 3150 Natural eutrophic lakes with *Magnopotamion* or *Hydrocharition – type vegetation* (!); C (ii) 3270 Rivers with muddy banks with *Chenopodion rubri* p.p. and *Bidention* p.p. vegetation; C (ii) 6430 Hydrophilous tall herb fringe communities of plains and of the montane to alpine levels; and C (ii) 91F0 Riparian mixed forests of *Quercus robur*, *Ulmus laevis* and *Ulmus minor*, *Fraxinus excelsior* or *Fraxinus angustifolia*, along the great rivers (*Ulmenion minoris*).

Other species of conservation importance in Ostrov site include: *Nuphar lutea* (EN), *Nymphaea alba* (EN), *Nymphoides peltata* (EN), *Senecio paludosus* (EN), *Astragalus contortuplicatus* (VU), *Alkanna tinctoria* (EN), *Centaurea arenaria* (VU), *Euphorbia lucida* (VU), *Dichostylis michelianus* (VU), *Leucojum aestivum* (VU) and *Stachys arenariiformis* (VU). There are also data for the distribution of the following rare steppe species in this site (villages of Selanovtsi, Ostrov and Dolni Vadin): *Astragalus dasyanthus* (CR) and *Rindera umbellata* (NE) (see Urumov 1917, 1935).

BGIPA103 Srebarna; 44° 06‘ N, 27° 04‘ E – 511.19 ha

Description: Srebarna is a lake situated in Silistra District. The lake is located in a large karst sink and measures approximately 2.5 km in length from north to south and 2 km in width from east to west. There are also two islands in the Danube River, opposite to the lake: Komluka and Byalata Prast. The lake has been protected since 1948; since 1999, it has also been a managed reserve. The most important reason for its designation was water bird diversity, and because the lake is the home of only remaining colony of the Dalmatian Pelican (*Pelecanus crispus*) in Bulgaria. In 1975, the area was declared a wetland of international importance after the Ramsar Convention. Since 1977, it has been a biosphere reserve, and, in 1985, it was included in the World Heritage List of UNESCO. Srebarna Lake is also an Important Bird Area (IBA) and NATURA 2000 site according to both directives. The waters are considerably eutrophic as its connection with the Danube was severed by the construction of a dike from 1949 to 1994. This caused accumulation of slit and organic matter; with the construction of a channel to the Danube in 1994, the ecological conditions of the lake were restored, the food basis became richer and bird diversity rapidly increased. Most of the lake is covered with reed, *Phragmites autralis* which also make floating mats named "kochki". In the open waters, there are communities dominated by *Hydrocharis morsus-ranae, Stratiotes aloides, Salvinia natans, Potamogeton* spp., *Nymphaea alba,* and *Ceratophylum* spp. In the surroundings of the lake, there are abandoned agricultural fields, meadows, pastures with steppe grasslands, and woody vegetation dominated by *Tilia tomentosa*. According to Apostolova & Meshinev (2012), the IPA was selected due to the fact that it is the only natural locality in the country for *Aldrovanda vesiculosa*. It is also one of the best places for habitat C (ii) 3150 Natural eutrophic lakes with *Magnopotamion* or *Hydrocharition* – type vegetation (!). The Criterion A species in the habitat include A (ii) *Aldrovanda vesiculosa* (!), *Salvinia natans* (!) and the Criterion C habitats include C (i) 6250 *Pannonic loess steppic grasslands; C (ii) 3150 Natural eutrophic lakes with *Magnopotamionor Hydrocharition* – type vegetation (!); and C (ii) 6430 Hydrophilous tall herb fringe communities of plains and of the montane to alpine levels (!).

Proposals: The expansion of the IPA site should coincide with the SCI NATURA 2000 site, enabling it to include more habitats as IPA criteria as: C (i) 91E0 *Alluvial forests with *Alnus glutinosa* and *Fraxinus excelsior* (*Alno-Padion, Alnion incanae, Salicion albae*); C (ii) 3130 Oligotrophic to mesotrophic standing waters with vegetation of the *Littorelletea uniflorae* and/or *Isoeto-Nanojuncetea*; C (ii) 3270 Rivers with muddy banks with *Chenopodion rubri* p.p. and *Bidention* p.p. vegetation (!); C (ii) 91F0 Riparian mixed forests of *Quercus robur, Ulmus laevis* and *Ulmus minor, Fraxinus excelsior* or *Fraxinus angustifolia*, along the great rivers (*Ulmenion minoris*); and C (ii) 91Z0 Moesian silver lime woods. Species of conservation significance in the lake include the following: *Cicuta virosa* (CR), *Stratiotes aloides* (CR), *Utricularia australis* (CR), *Nuphar lutea* (EN),

Nymphaea alba (EN), *Nymphoides peltata* (EN), *Trapa natans* (EN), *Sonchus palustris* (EN), *Thelypteris palustris* (VU), *Dichostylis michelianus* (VU), *Leucojum aestivum* (VU) and *Utricularia vulgaris* (VU). Some of these species, such as *Cicuta virosa, Nuphar lutea* and *Trapa natans,* have not been confirmed in the lake region for many years.

As a result of analysis, we propose the following site to be designated as a new IPA in Bulgaria:

BGIPA126 Kalimok-Brashlen; 44° 01‘ N, 26° 25‘ E – 7550 ha

Description: Kalimok-Brashlen is located in the Tutrakan Lowland or Pobrezhie, one of the largest former Danube floodplains in Bulgaria. The site is located between the villages of Staro Selo, Nova Cherna and Brashlen. It was previously occupied by two large wetlands – Kalimok (Belitsa) and Brashlensko Blato; the former was transformed into the biggest fishponds in Bulgaria, which are abandoned at present, and the latter was partially drained. Despite drainage, Brashlensko Blato continued to flood and continued to have large communities of *Phragmites australis*. From 2004 to 2009, the water regime of the Kalimok Marsh was restored and the fishpond basins became re-occupied by typical aquatic vegetation, including communities of *Rorippa amphibia*, *Nymphoides peltata*, *Salvinia natans*, *Phragmites australis*, *Typha angustifolia* and many others. It is one of the three largest areas for both aquatic flora and habitats, the other two being Srebarna Lake and Persinski (Belenski) Marshes. Due to the restoration of the water regime, quick recovery of the aquatic flora and vegetation was also observed. The territory of Brashlensko Marsh is covered mostly by reeds but communities of *Nymphaea alba* also exist. The site includes several Danube islands such as Kalimok, Radetski, Golyam Brashlyan and Mishka, which are representative places for the alluvial poplar-willow forests.

Criteria: Species: A (ii) *Salvinia natans* (!), A (ii) *Marsilea quadrifolia* (!), A (ii) *Trapa natans* (!), A (ii) *Armoracia macrocarpa* (!). Habitats: C (i) 91E0 *Alluvial forests with *Alnus glutinosa* and *Fraxinus excelsior* (*Alno-Padion*, *Alnion incanae*, *Salicion albae*) (!); C (ii) 1530 *Pannonic salt steppes and salt marshes; C (ii) 3130 Oligotrophic to mesotrophic standing waters with vegetation of the *Littorelletea uniflorae* and/or *Isoeto-Nanojuncetea;* C (i) 3150 Natural eutrophic lakes with *Magnopotamion* or *Hydrocharition – type vegetation*; C (i) 3270 Rivers with muddy banks with *Chenopodion rubri* p.p. and *Bidention* p.p. vegetation (!); C (i) 6430 Hydrophilous tall herb fringe communities of plains and of the montane to alpine levels; C (i) 91F0 Riparian mixed forests of *Quercus robur*, *Ulmus laevis* and *Ulmus minor*, *Fraxinus excelsior* or *Fraxinus angustifolia*, along the great rivers (*Ulmenion minoris*); C (i) 6440 Alluvial meadows of river valleys of the *Cnidion dubii*. Other species with conservation significance in the site include *Utricularia australis* (CR), *Nuphar lutea* (EN), *Nymphaea alba* (EN), *Nymphoides peltata* (EN),

Senecio paludosus (EN), *Dichostylis michelianus* (VU), *Leucojum aestivum* (VU), *Euphorbia lucida* (VU), *Utricularia vulgaris* (VU) and *Lemna gibba* (NT).

DISCUSSION

The IPAs along the Bulgarian stretch of the Danube River comparatively support well the scientific basis well for the preservation of the area's rich flora, vegetation and habitat diversity. However, there are some recommendable changes which could improve the coverage of all significant plant areas. A serious problem that was identified during the course of the study was the lack of boundaries for the existing IPAs; as a result, we could only guess if these sites included some of the important floristic and vegetation complexes. This was especially true for the following sites: Archar, Belene, Karaboaz, Orsoya and Ostrovska Step. There are also sites which need to be significantly expanded such as Archar and Srebarna. For the sites Zlatiya and Ostrovska Step, their proposed expansion is a transformation into two new and renamed sites: Tsibar and Ostrov. One completely new site is proposed to be designated with the name Kalimok-Brashlen.

It is also necessary to discuss the existing network of protected areas and NATURA 2000 sites, which more or less overlap with many of the proposed and existing IPAs. More important are the NATURA 2000 SCI which have habitats as target objects of their conservation. Among the identified species, only *Marsilea quadrifolia*, *Eleocharis carniolica* and *Aldrovanda vesiculosa* are target objects. The following IPAs completely or partially overlap with SCI: Archar (BG0000497 Archar); Belene (BG0000396 Persina); Vardim (BG000204 Ostrov Vardim); Garvan and Malak Preslavets (BG0000530 Pozharevo-Garvan); Zlatiya (BG0000199 Tsibar); Karaboaz (BG0000335 Karaboaz); Kozloduy (BG0000527 Kozloduy); Orsoya (BG0000182 Orsoya); Ostrovska step (BG0000528 Ostrovska step-Vadin); and Srebarna (BG0000241 Srebarna).

The national protected areas are not a well-developed network when it comes to the Danube River. There are three maintained reserves, Srebarna (892.05 ha), Ibisha (34 ha) and Persinski Blata (385.2 ha); Ibisha is on the island of Dolni Tsibar and Srebarna Lake is also a biosphere reserve, a UNESCO heritage site, and an object of the Ramsar Convention. Persinski Blata Managed Reserve is located on Belene Island. There are several other protected areas in the Belene Complex; two small islands of the Belene archipelago are strict reserves for the preservation of typical alluvial forests – Kitka (27.6 ha) and Milka (38.9 ha). The eastern part of Belene Island is included in the Persina Iztok Protected Area (719 ha); the central part of the island is included in the Persin Protected Area (551.3). The Nikopol and Belene archipelagos and the Svishtov-Belene Lowland are included in the Persina Nature Park (21762 ha) – the only Nature Park along the Danube River in Bulgaria. There are also small protected areas which partially overlap with the IPAs, including Ribarnitsi Orsoya (147.18 ha); Ostrov Tsibar (141.6 ha); Genchov Orman

(26 ha); Katinata (28 ha); Stariya Dab (98.3 ha); Blatoto kray selo Malak Preslavets (154.94 ha), Garvanski Blata (280 ha); and Kalimok-Brashlen (5952.34 ha). Some of these protected areas are designated to preserve rare plants, plant communities (*Tamarix ramosissima*, *Quercus robur*) or wetland ecosystems. The protected areas Gola Bara (2 ha), Kalugerski Grad-Topolite (0.2 ha) and Kochumina (2.5 ha) are situated in the Ostrov Lowland to protect localities of *Nymphaea alba* and *Stratiotes aloides;* however, it is not known (at least from 20 years ago) if these species continue to occur in these localities.

According to the Bulgarian Biodiversity Act, IPAs are part of the national ecological network in Bulgaria. Despite this designation, it is necessary to determine some legal conservation category for sites that do not also fall under the NATURA 2000 protection. Overall, there are no real legal mechanisms to protect IPAs outside of the national ecological network. In addition, the lack of clear IPA borders (indicated topographically or geodesically) may pose serious challenges to the adoption of the Biodiversity Act. One improved characteristic of the existing IPAs includes also their well determined borders and scopes. Other serious problem is that the Danube Plain and the Danube River have been out of the scope of the most intensive botanical studies in Bulgaria. Better knowledge about these IPAs could promote further study and renewed conservation efforts regarding the rich and diverse flora and vegetation of the Danube River stretch in Bulgaria.

Acknowledgments

We want to express our thanks to the WWF-Danube-Carpathian program (Bulgaria) for the opportunity to publish this study.

References

Anderson S. 2002. *Identifying Important Plant Areas: a site selection manual for Europe.* Plantlife International, Salisbury. www.plantlife.org.uk/publications/identifying_important_plant_areas_a_site_selection_manual_for_Europe. Accessed 19 Sep. 2016.

Angelov E. & Ganchev S. 1957. Soils and Vegetation in the Garvan-Popina-Srebarna Region. *Nauchni Trudove na Nauchnoizsledovatelski institut za pochveni izsledvaniya "N. Pushkarov"* [*Scienific researches of the Institute for Soil Research "N. Pushkarov"*] 3: 341-389 (in Bulgarian).

Apostolova I. & Meshinev T. 2012. Archar; Belene; Vinarovo; Garvan; Ostrov; Srebarna. In: Peev D., Petrova A., Apostolova I. & Assyov B. (eds). *Important Plant Areas in Bulgaria.* Pensoft Publishers, Sofia, Pp.: 59-60; 71-75; 102-104; 131-134; 307-309; 381-384.

Baeva G. 1985. Rare and protected plants in the Srebarna Reserve. *Izvestiya na Sayuza na Nauchnite Rabotnitzi v Bulgaria - klon Silistra* [*Announcements of the Union of Scientific Workers in Bulgaria - Silistra Branch*] 4: 138-143 (in Bulgarian).

Baeva G. 1986. Phytocoenological characteristics of higher water vegetation in the Srebarna Biosphere Reserve. *Izvestiya na Sayuza na Nauchnite Rabotnitzi v Bulgaria - klon Silistra [Announcements of the Union of Scientific Workers in Bulgaria - Silistra Branch*] 5: 143-148 (in Bulgarian).

Baeva G. 1987. Studies on higher aquatic plants and vegetation in the Srebarna Biosphere Reserve. In: Nedialkov S., H. Kochev T. Michev A. Damyanova & Velev V. (eds.). Mezhdunaroden symposium "Rolyata na vlazhnite zoni za opazvane na genetichnia fond" [International Symposium "The role of wetlands in the conservation of the genetic material"] Srebarna, 8-12 October 1984, Collected works, BAS, Sofia, pp. 109-118 (in Bulgarian).

Baeva G. 1988a. Ecological characteristics of the higher flora and vegetation in the Srebarna Biosphere Reserve. Razgrad District. PhD Thesis, Sofia, 228 pp. (in Bulgarian).

Baeva G. 1988b. New species for the flora of North-Eastern Bulgaria. In: Natsionalna nauchna sesia na mladite ucheni na tema "Ekologia i opazvane na prirodnata sreda" [National scientific session of young scientists on theme "Ecology and environmental conservation]*"*, 24 – 25 November, 1988, Plovdiv, 44–46 (in Bulgarian).

Baeva G. 1991. Comparative study on the vascular aquatic plants in the wetlands of the villages of Garvan, Maluk Preslavets and Nova Cherna. In: *Izvestija na suyuza na uchenite v Bulgaria. Klon Silistra* [*Announcements of the Union of Scientific Workers in Bulgaria - Silistra Branch*] 5: 140-152 (in Bulgarian).

Baeva G. 1995. Changes in the vegetation of Srebarna Biosphere Reserve for the period 1988–1993. In: Tsankov G. (ed.). *Sbornik ot Jubileina Nauchna Konferencia "100 godini ot rozhdenieto na Akademik "Boris Stefanov (1894–1994)"* [*Proceeding of the Jubilee Scientific Conference "100 years form the birth of Academician B. Stefanov (1894–1994)"*] PSSA, Sofia. Vol. 2. Pp. 63–65 (in Bulgarian).

Biserkov V., Gussev Ch., Popov V., Hibaum G., Roussakova V., Pandurski I., Uzunov J., Dimitrov M., Tzonev R. & Tzoneva S. (eds.). 2015. *Red Data Book of the Republic of Bulgaria*, vol. 3. *Natural habitats,* MOEW-BAS, Sofia, 422 p.

Bonchev G. 1929. *The swamps in Bulgaria.* Ministerstvo na zemedelieto i darzhavnite imoti (Ministry of agricultures and state properties], Sofia 5-6: 26-75 (in Bulgarian).

Darbyshire I., Anderson S., Asatryan A., Byfield A., Cheek M., Clubbe C., Ghrabi Z., Harris T., Heatubun Ch., Kalema J., Magassouba S., McCarthy B., Milliken W., de Montmollin B., Nic Lughadha E., Onana J.-M., Saı¨dou D., Saˆrbu A., Shrestha K. & Radford E. 2017. Important Plant Areas: revised selection criteria for a global approach to plant conservation. *Biodiversity Conservation* 26(8): 1767–1800.

Dister E., Shneider E., Gunther-Diringer D., Obrdlik P., Schneider Ec., Engelke T., Harm N., Dopke M., Belev T., Marin S., Dimitrova L., Berberova D., Tsonev R. & Stojanov S. 2001. Final report of equip on the consultant assignment: Planning of the management of protected areas in the framework of GEF project "Reducing of pollution of Danube River with biogenic elements and restoration of wetlands" (WWF; 14.11.2001) (in Bulgarian).

Gussev Ch. & Petrova A. 2015. Protected areas, European ecological network Natura 2000, Important Plant Areas and legal acts. In: Peev D., Petrova A., Anchev M., Temniskova D., Denchev C., Ganeva A., Gussev Ch. & Vladimirov V. (Eds). *Red Data Book of the Republic of Bulgaria.* Vol. 1, *Plants and fungi.* MOEW-BAS: Pp. 17-21.

Kochev H. & Tzolova I. 1984. Phytocoenological studies on the "Persinski Blata" Reserve. *Suvremenni teoritichni i prilozhni aspekti na rastistelnata ekologia* [*Contemporary theoretical and applied aspects of the plant ecology*] 1: 148-156 (in Bulgarian).

Kochev H. & Yordanov D. 1981. *Vegetation of Bulgarian water bodies. Ecology, protection and economic importance.* BAS. Sofia. (in Bulgarian).

Kochev H. 1987. Phytocoenological characteristics of the Srebarna Biosphere Reserve, Silistra District. In: Nedialkov S., Kochev H., Michev T., Damyanova A. & Velev V. (eds.). Mezhdunaroden symposium "Rolyata na vlazhnite zoni za opazvane na genetichnia fond" [International Symposium "The role of wetlands in the conservation of the genetic material"], Srebarna, 8-12 October 1984, Collected works, BAS, Sofia, pp. 100-107 (in Bulgarian).

Kochev H. & Yordanov D. 1976. Macrophytic vegetation in the water basins of the Danubian riverside in Bulgaria, its changes under human impact and its protection. *Botanicheskiy Zhurnal* (*Botanical Journal*) 61: 1294-1297 (in Bulgarian).

Kochev H., Husak S. & Otahelova H. 1987. Materials about phytocoenological characteristics of water and swamp vegetation in the Eastern part of the Danube, Bulgaria. In: Nedialkov S., Kochev H., Michev T., Damyanova A. & Velev V. (eds.). Mezhdunaroden symposium "Rolyata na vlazhnite zoni za opazvane na genetichnia fond" [International Symposium "The role of wetlands in the conservation of the genetic material"], Srebarna, 8-12 October 1984, Collected works, BAS, Sofia, pp. 100-107 (in Bulgarian).

Marinov M. & Fakirov V. 1977. Vegetation of the Danube Islands in a view to the classification of the floodplains according to the period of their flooding. – *Gorskostopanska nauka* [*Forest science*] 14(6): 11-18 (in Bulgarian).

Meshinev T. & Apostolova I. 2012. Vardim; Zlatiya; Karaboaz; Kozloduy; Maluk Preslavets; Orsoya. In: Peev D., Petrova A., Apostolova I. & Assyov B. (Eds). *Important Plant Areas in Bulgaria.* Pensoft Publishers, Sofia, Pp.: 93-95; 191-192; 221-223; 236-239; 265-267.

Michev, T. (ed.). 1995. *National Action Plan for the Conservation of the most Important Wetlands in Bulgaria.* Ministry of Environment, 55 p.

Michev T., Pomakov V., Nankinov D. & Ivanov B. 1981. Wetlands of International Importance in Bulgaria. Proceed. Reg. Symp. under project 8; MAB, UNESCO.

Michev T. & Stoyneva M. (Eds). 2007. *Inventory of Bulgarian Wetlands and their Biodiversity. Part 1: Non-Lotic Wetlands*. Publ. House Svetlostrouy, Sofia, 364 pp. + CD supplement.

Mihaylov Ts. & Popov V. 1978. Geomorphology of Danube shore. In Russev B. & Naidenow V. (eds). *Limnology of the Bulgarian stretch of Danube River*. S., BAS: Pp: 7-27 (in Bulgarian).

Nedelcheva A. & Tzonev R. 2006. *Achillea ochroleuca* (*Asteraceae*): a new species for the Bulgarian flora. *Phytol. Balcan.* 12 (3): 371–376.

Paspaleva-Antonova M. 1961. Studies on the avifauna of the Bulgarian Danube riverside. PhD thesis. Institute of Zoology, Bulgarian Academy of Sciences, Sofia (in Bulgarian).

Peev D. & Petrova A. 2012. General characteristic of the Important Plant Areas Network in Bulgaria. In: Peev, D., Petrova, A., Apostolova, I. & Assyov, B. (eds). 2012. *Important Plant Areas in Bulgaria.* Pensoft Publishers, Sofia., Pp.: 52-53.

Peev D., Petrova A., Apostolova I. & Assyov B. (eds.). 2012. *Important Plant Areas in Bulgaria,* Pensoft Publishers, Sofia, 469 p.

Peev D., Petrova A., Apostolova I. & Delcheva M. 2009. Section III: Important Plant Area country reports – national overview of IPAs in six countries. In: Radford E & Ode B. (eds.). *Conserving Important plant Areas: Investing in the green gold of South East Europe.* Plantlife International, Salisbury, Pp.: 27-36.

Petkoff S. 1911. Preliminary studies on the swamp flora of the Bulgarian Danubian riverside. *Godishnik na Sofiyskya Universitet, Fiziko-Matematicheski Fakultet* [*Annual of Sofia University, Faculty of Physic and Mathematic*] 6: 3-45 (in Bulgarian).

Petkov N. 1940. The Danube islands. *Lesovudska Misal* [*Forestry ideas*], 9 (6): 424-435 (in Bulgarian).

Petrova A. & Velev V. 1998. Vascular plants. In: Michev T., Georgiev B., Petrova A. & Stoyneva M. (Eds). *Biodiversity of the Srebarna Biosphere Reserve. Checklist and bibliography.* Sofia: Context and Pensoft, pp. 51-53.

Petrova A. & Vladimirov V. (eds). 2009. Red List of the Bulgarian vascular plants. *Phytologia Balcanica* 15(1): 63–94.

Russev B. & Naidenov V. (eds). 1978. *Limnology of the Bulgarian stretch of Danube River.* S, BAS, 308 p. (in Bulgarian).

Stoyanov N. 1927. The grassland coverage of Belene Island. *Gorsko stopanstvo* [*Forestry*] 3(2-3): 21-28: (3-4): 39-46: (5-6): 166-173 (in Bulgarian).

Stoyanov N. 1947. Floristic materials. *Godishnik na Sofiyskya Universitet, Fiziko-Matematicheski Fakultet* [*Annual of Sofia University, Faculty of Physic and Mathematic*] 43 (3): 55-61 (in Bulgarian).

Stoyanov N. 1948. *The Vegetation on Our Danube Islands and its Economic Utilisation.* BAS, Sofia, 188 p. (in Bulgarian).

Tchernyavski P. & Pavlov P. 1955. On the matter about management of Danube islands. *Gorsko stopanstvo* [*Forestry*] 7: 295-305 (in Bulgarian).

Tsanov Ts. 1992. *Flood-plain Forests along Bulgarian Part of Danube*. Bulg. Acad. Sci. Publishing House, Sofia (in Bulgarian).

Tsoneva S., Georgiev V., Valchev V. & Ganeva A. 2012. *Atlas of aquatic and wetland plants in Bulgaria*. IBER, BAS, Sofia, 200 p. (in Bulgarian).

Tzonev R. & Sumberova K. 2005. New data about the chorology of some little known adventive species on the banks of the Danube River in Bulgaria. *Phytol. Balcan.* 10 (2-3): 207-209.

Tzonev R. 2002. Flora and Vegetation in Middle Danube Plain between the valleys of Vit and Studena Rivers. Unpubl. PhD. thesis. 166 p. (in Bulgarian).

Tzonev R. 2005. New data and summarised information on the chorology of some rare, threatened and endemic plants in the Middle Danube Plain and Balkan Foothill region. *Godishnik na Sofiyskya Universitet, Biologicheski fakultet* [*Annual of Sofia University, Faculty of Biology*] 97(2): 62-72.

Tzonev R. 2009. Plant communities, habitats and ecological changes in the vegetation on the territory of three protected areas along the Danube River. In: Ivanova, D. (ed.), Plant, fungal and habitat diversity investigation and conservation. Proceedings of IV Balkan Botanical Congress, Sofia, 20–26 June 2006. Pp. 321–331. Inst. Bot., Sofia.

Urumov I. 1917. Thirteenth contribution to the Bulgarian flora. *Sborn. Bulg. Akad. Nauk.* [*Proceeding of Bulgarian Academy of Science*] 7: 1-225 (in Bulgarian).

Urumov I. 1926. Fifteenth contribution to the Bulgarian flora. *Sborn. Bulg. Akad. Nauk.* [*Proceeding of Bulgarian Academy of Science*] 22: 1-138 (in Bulgarian).

Urumov I. 1935. Flora of the Vratsa District. *Sborn. Bulg. Akad. Nauk.* [*Proceeding of Bulgarian Academy of Science*] 29: 1-105 (in Bulgarian).

Valchev V., Tzonev R., Georgiev V. & Tsoneva S. 2012. Aquatic macrophytes: species composition and syntaxonomy. In: Uzunov Y., Georgiev B., Varadinova E., Ivanova N., Pehlivanov L. & Vasilev V. (eds.). *Ecosystems of the Biosphere Reserve Srebarna Lake*. Prof. Marin Drinov Acad. Publ. House, Sofia, Pp.: 69-76.

Valchev V., Georgiev V., Ivanova D., Tsoneva S. & Janauer G. 2006. Conservationally important macrophytes in the Bulgarian stretch of the Danube River and the near water bodies. 36th International Conference of IAD/Austrian Committee of Danube Research Vienna.

Valchev V., Georgiev V. & Tsoneva S. 2018. Macrophytes of the Danube reach in Bulgaria. In: Janauer G., Gaberscik A., Kvet J., Germ M. & Exler N. (Eds.) 2018. *Macrophytes of the River Danube Basin*. ACADEMIA, Praha, Pp: 294-302.

Velev V. 1984. *Marsilea quadrifolia* L. In: Velchev V. (ed.). *Red Data Book of PR Bulgaria*. Vol. 1. Plants. Publishing House of the Bulgarian Academy of Sciences, Sofia; Pp: 29 (in Bulgarian).

Vladimirov V. & Tsoneva S. 2006. *Tragopogon floccosus* (*Asteraceae: Cichorioideae*), a recently discovered species in the Bulgarian flora. *Phytol. Balcan.* 12(1): 67-70.

Yurukova L. 2002. Biodiversity of freshwater vascular plants along the Bulgarian Bank of the Danube – survey and threats. In: Brezeanu G. & Ştiucă R. (eds.) *Proceedings of the 34th Conference of IAD* (*International Association for Danube Research*), Tulcea, Romania, 2002, 34: 251-258.

Yordanov D. 1946-1947. Materials and critical notes on the Bulgarian flora 1943-1946 *Godishnik na Sofiyskya Universitet, Fiziko-Matematicheski Fakultet* [*Annual of Sofia University, Faculty of Physic and Mathematic*] 43: 97-111 (in Bulgarian).

Yordanov D. & Kochev H. 1973. Materials and critical notes for the flora in Bulgaria. *Izvestiya na Instituta po Botanika* 24: 241-245 (in Bulgarian).

In: Biodiversity of the Bulgarian-Romanian Section
Editor: Shurulinkov, Hubenov et al.
ISBN: 978-1-53615-663-8

Chapter 2

SPECIES COMPOSITION AND DISTRIBUTION OF MOLLUSCA (GASTROPODA AND BIVALVIA) IN THE BULGARIAN SECTOR OF THE DANUBE RIVER AND THE ADJACENT WETLANDS

Teodora Trichkova[1,*], Milcho Todorov[1], Dilian Georgiev[2] and Zdravko Hubenov[3]

[1]Institute of Biodiversity and Ecosystem Research, Bulgarian Academy of Sciences Sofia, Bulgaria

[2]Department of Ecology and Environmental Conservation, Faculty of Biology, Plovdiv University 'Paisii Hilendarski', Plovdiv, Bulgaria

[3]National Museum of Natural History, Bulgarian Academy of Sciences Sofia, Bulgaria

ABSTRACT

The current review presents the composition, distribution, zoogeographical characteristics, some ecological traits, and conservation status of the mollusc fauna (Gastropoda and Bivalvia) in the Bulgarian sector of the Danube River and the adjacent marshes and backwaters. A total of 67 mollusc species (47 gastropods and 20 bivalves), about 13% of the Bulgarian mollusc fauna, have been reported until now in the available literature. The reported species belonged to 14 families (ten of gastropods and four of bivalves). The highest species richness was recorded in the family Planorbidae, followed by Sphaeridae and Lymnaeidae. The mollusc species were classified in 19 zoogeographical categories. The highest number of the molluscs (79.10%) was of Palaearctic distribution.

* Corresponding Author's E-mail: trichkova@gmail.com.

Five species (*Theodoxus transversalis*, *Anisus vorticulus*, *Anodonta cygnea*, *Pseudanodonta complanata*, and *Unio crassus*) have been considered of higher conservation concern according to the IUCN Red List, European and national legislation. The globally known major threats to freshwater biodiversity, including loss and degradation of habitats, overexploitation, pollution, introduction of invasive alien species, and climate change, have affected also the mollusc fauna in the Bulgarian sector of the Danube River and adjacent wetlands. Therefore, in order to protect the native mollusc fauna appropriate conservation and management measures directed to the restoration and protection of the mussel populations and to the restoration of the damaged aquatic habitats, including their catchments, are recommended.

Keywords: molluscs, species composition, distribution, zoogeography, conservation status, Danube River main channel, marshes and backwaters

INTRODUCTION

The studies on the mollusc fauna of the Bulgarian sector of the Danube River has started about 155 years ago with the works of Bourguignat (1863-1870, 1876), Kreglinger (1871), Westerlund (1885, 1886), and Kobelt (1897). The first Bulgarian authors published on molluscs are Hristovich (1892) and Iurinitsh (1906). Data on the molluscs in the Bulgarian sector of the Danube River have been published also by Wohlberedt (1911), Haas (1911), and Büttner (1928), and later by Petrbok (1941) and Drensky (1947) in their comprehensive works on the Bulgarian freshwater mollusc fauna. A major contribution to the studies of the benthic invertebrates, including gastropods and bivalves, in the Bulgarian sector of the Danube River and adjacent marshes and backwaters is made with the faunistic, ecological and biocenological works of Russev (1957, 1959, 1963, 1966, 1967, 1968a, b, 1974, 1978, 1979), Russev & Kaneva-Abadjieva (1973), and Russev & Uzunov (1990, 1991a, b). Extensive information about taxonomy, distribution and ecology of molluscs is presented in the monographs of Frank et al. (1990) and Angelov (2000a). Recent reviews and checklists of molluscs, which include also new data have been published by Hubenov (2005b, 2006b, 2007), Georgiev & Hubenov (2013), and Georgiev (2013, 2014, 2017). Long-term and extensive studies are dedicated to the benthic fauna, including molluscs, in Srebarna Lake, a Nature Biosphere Reserve (Angelov 1998, 2000b, Uzunov et al. 2001, Varadinova et al. 2011, 2012). Some studies deal with mollusc species of concervation concern (Pavlova et al. 2013), while others present a description or re-description of new species (Glöer & Georgiev 2012, 2014, Boeters et al. 2015). At the same time, three new alien bivalve species have been recorded in the Bulgarian sector of the Danube River and some studies focus on their introduction, quick establishment and range expansion (Hubenov 2001, 2005a, 2006a, Hubenov & Trichkova 2007, Hubenov et al. 2012, 2013).

The aim of the present work was to make a review of available literature and present the composition, distribution, zoogeographical characteristics, some ecological traits, and conservation status of the mollusc fauna (Mollusca: Gastropoda and Bivalvia) in the Bulgarian sector of the Danube River and the adjacent marshes and backwaters.

MATERIALS AND METHODS

The Bulgarian sector of the Danube River extends from the estuary of the Timok River (845 rkm) to the town of Silistra (374 rkm) and has a length of 471 km. There are about 84 islands with a total surface area of about 115 km^2 in the Bulgarian sector. In the past, there were over 20 permanent marshes along the river with a surface area of about 100 km^2. Currently, only a few of those marshes exist, such as Srebarna Lake and Garvan Marsh, as well as some smaller ones (Michailov & Popov 1978).

Table 1. Localities of the studied molluscs in the Bulgarian sector of the Danube River, the Danube islands, and the adjacent wetlands (marshes and lakes), with corresponding river kilometres and UTM Codes

№	Locality: Settlement/Site	Locality: Danube Island	Locality: Danube Wetlands	River km (rkm)	UTM Code
1	Vrav Village			836	FP39
2	Novo Selo Village			833	FP49
3	Florentin Village			827	FP48
4	Yasen Village			825	FP58
5	Gomotartsi Village			817	FP58
6	Koshava Village			811	FP68
7		Koshava Island		807	FP68
8	Vidin Town			791	FP57
9		Bogdan-Sechan Island		782.5	FP46
10		Islands Golyam Bliznak and Malak Bliznak		779-777	FP45
11	Simeonovo Village			776	FP45
12	Botevo Village			774	FP45
13	Ratsiaria Site			772	FP55
14	Archar Village			769.5	FP55
15	Orsoya			755	FP65
16		Skomen Island		759	FP65
17	Lom Town			744	FP75
18	Dolno Linevo Village			735	FP85
19	Stanevo Village			723	FP95
20		Tsibar Island		718	GP05
21	Dolni Tsibar Village			717	GP05
22	Kozloduy Town			701	GP15
23		Kozloduy Island		697	GP15-GP25
24			* Kozloduy Marsh		GP14
25	Oryahovo Town			678	GP34
26		Leskovets Island		672	KJ64
27	Ostrov Village			661	KJ64 KJ74
28			* Ostrov Marsh		KJ64

Table 1. (Continued)

№	Locality: Settlement/Site	Locality: Danube Island	Locality: Danube Wetlands	River km (rkm)	UTM Code
29	Gorni Vadin Village			655	KJ74
30	Dolni Vadin Village			651	KJ84
31	Baykal Village			641	KJ94
32		Boril Island		640	KJ94
33	Mouth of the Iskar River			636.7	KJ94
34	Zagrazhden Village			625	LJ04
35	Dabovan Village			620	LJ04
36	Gulyantsi Village			611	LJ13
37	Somovit Village			607	LJ23-LJ24
38	Cherkvitsa Village			601	LJ24
39	Nikopol Town			597	LJ24-LJ34
40		Lakat Island		588	LJ34
41		Palets Island		585	LJ34
42	Belene - Hisarlaka Site			577	LJ43
43	Belene Town			573	LJ43
44		Belene Island		568	LJ53
45			* Belene Marsh		LJ52
46		Malak Kondur Island		563	LJ53
47	Svishtov Town			554	LJ63
48			* Svishtov Marsh		LJ63
49	Vardim Village			546	LJ73
50		Vardim Island		543	LJ73
51	Mouth of the Yantra River			536.5	LJ83
52	Krivina Village			536	LJ83
53	Batin Village			526	LJ93
54		Batin Island		523.5	LJ93
55	Mechka Village			520	MJ04
56	Stalpishte Village			516	MJ04
57	Rousse Town, Prista Site			502	MJ15
58	Mouth of the Roussenski Lom River			498	MJ15
59	Rousse Town			495	MJ15
60	Marten Village			482	MJ26
61	Sandrovo Village			478	MJ26
62		Aleko Island		473.5	MJ26-MJ36
63	Ryahovo Village			466	MJ37
64		Mishka Island		459	MJ47
65	Brashlen Village	Golyam Brashlen Island		450.5	MJ47
66	Tutrakan Town			432	MJ67
67			* Tutrakan Marsh		MJ67
68		Kossuj Island		426	MJ77
69	Malak Preslavets Lake			414	MJ88
70		Slavyanin Island		407.5	MJ98
71	Vetren Village			395	NJ08
72		Vetren Island		393	NJ08
73			Srebarna Lake	393	NJ08
74		Komluka Island		393	NJ08
75	Silistra Town			380	NJ28

* Water bodies, which do not exist at present

The review covered all available scientific literature on the mollusc fauna in the Bulgarian sector of the Danube River, the adjacent backwaters and marshes (existing in the past and at present), and the marshes on the islands in the Bulgarian sector. Only data affiliated to a particular locality were included. Table 1 presents a list of main localities (settlements/sites, Danube River islands and marshes/lakes) cited in the reviewed literature.

The review included data on all taxonomic groups of Mollusca (Gastropoda and Bivalvia) published after 1900. The taxonomy follows the World Register of Marine Species (WoRMS 2019).

The following information is presented for each species: 1) scientific name; 2) distribution in the Danube River; 3) distribution in the Danube wetlands; 4) general distribution; 5) ecological traits; 6) conservation status; and 7) notes (on taxonomy, origin, invasiveness, and others).

The zoogeographic classification of species was made based on the works of Zhadin (1952), Starobogatov (1970), Frank et al. (1990), Stadnichenko (1984, 1990), Anistratenko & Stadnichenko (1994), Korniushin (1996), Boeters (1998), Angelov (2000a), Anistratenko & Anistratenko (2001), and Georgiev (2014, 2017). Some species were related to more than one category according to the literature sources. The secondary category was given in brackets. The zoogeographic analysis was made based only on the first/main category of the species.

The conservation status of the mollusc species was determined according to: the Bern Convention (1979); Directive 92/43/EEC (EU 1992); Biological Diversity Act of Bulgaria (BDA 2002); and the IUCN Red List of Threatened species (IUCN 2019) (DD – Data Deficient, LC – Least Concern, NT – Near Threatened, LR – Lower Risk, VU – Vulnerable, EN – Endangered).

RESULTS AND DISCUSSION

A total of 67 mollusc species (47 gastropods and 20 bivalves) have been reported until now for the Bulgarian sector of the Danube River and the adjacent wetlands. This is about 13% of the Bulgarian mollusc fauna, which amounts to about 520 species (Hubenov 2005b, 2007). The reported species belonged to 14 families (ten of gastropods and four of bivalves). The highest species richness was recorded in the family Planorbidae (14 species), followed by Sphaeridae (10 species) and Lymnaeidae (8 species).

The results of our review and the list of reported species, with data on their distribution, ecological tratits and concervation status is presented below.

Gastropoda

Neritidae

Theodoxus danubialis **(Pfeiffer, 1828)**

Distribution in the Danube River: along the entire Bulgarian sector, from 845 to 375 rkm (Angelov 2000a, Russev 1959, 1967, 1978, Georgiev 2014); from 834 to 414 rkm (Russev 1966); at Novo Selo, Vidin, Lom, Stanevo, Kozloduy, Oryahovo, Ostrov, Dolni Vadin, Baykal, Zagrazhden, Nikopol, Svishtov, Vardim, Rousse, Marten, Tutrakan, Malak Preslavets, and Silistra (Petrbok 1941, Drensky 1947, Russev 1963, 1974, Frank et al. 1990, Russev & Uzunov 1990, 1991a, b, Georgiev 2013); with an abundance of 40-50 ind./m^2 (Angelov 2000a).

General distribution: Southeast European-Pontian; Caspian relict.

Ecology: gravel and stones, rarely on mud and clay; slow-flowing water (Angelov 2000a, Russev 1963, 1978); up to 8.6 m depth (Angelov 2000a); up to 800 m distance from the shore (Russev 1978).

Conservation status: IUCN (Europe – LC).

Theodoxus fluviatilis **(Linnaeus, 1758)**

Distribution in the Danube River: along the entire Bulgarian sector (Angelov 2000a); at Ostrov (Frank et al. 1990) and Vardim (Georgiev 2013).

General distribution: European-Anatolian.

Ecology: stones; flowing water (Angelov 2000a).

Conservation status: IUCN (LC).

Theodoxus transversalis **(Pfeiffer, 1828)**

Distribution in the Danube River: along the entire Bulgarian sector, from 845 to 375 rkm (Russev 1959, 1966, 1967, 1978, 1979, Angelov 2000a); at Vidin, Lom, Stanevo, Kozloduy, Baykal, Somovit, Nikopol, Svishtov, near the mouth of the Yantra River, Rousse, Ryahovo, Tutrakan, Malak Preslavets, and Silistra (Petrbok 1941, Russev 1963, 1974, 1978, Frank et al. 1990, Russev & Uzunov 1991a, Angelov 2000a, Georgiev 2014); at the foreheads of Komluka Island and Vardim Island (Pavlova et al. 2013); one of the most abundant and frequent species in the Danube River in the past (Angelov 2000a); with an abundance of up to 96-120 ind./m^2 (Russev 1966, 1978, Angelov 2000a).

General distribution: Central and Southeast European.

Ecology: gravel and stones, also on sand, clay and mud; slow-flowing water (Russev 1959, 1978, Angelov 2000a); at depths from 0.2 to 10.7 m (Russev 1966); up to 660-900 m distance from the shore, being most abundant from 300 to 600 m (Russev 1966, 1978, Angelov 2000a).

Conservation status: Directive 92/43/EEC –Annex II and IV (EU 1992); IUCN (EN).

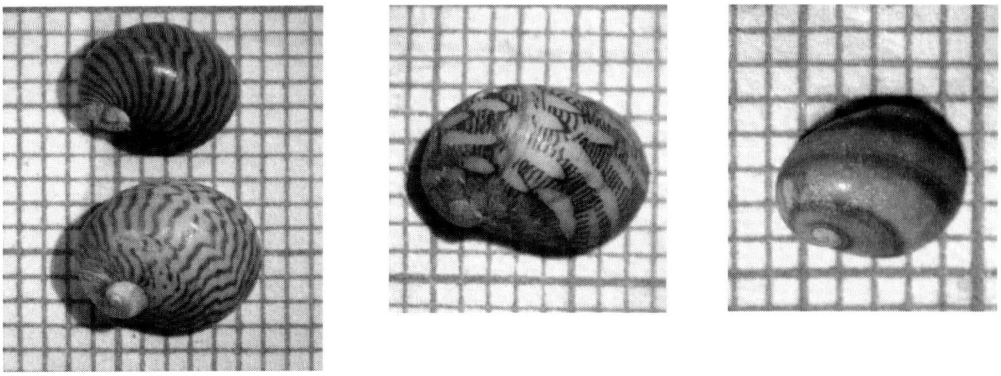

Figure 1. (a) *Theodoxus danubialis* ; (b) *Theodoxus fluviatilis*; and (c) *Theodoxus transversalis* in the Danube River at Vrav (2013) (Photos by Milcho Todorov).

Viviparidae

Viviparus acerosus **(Bourguignat, 1862)**

Distribution in the Danube River: along the entire Bulgarian sector, from 845 to 375 rkm (Russev 1979), at Vidin, Somovit, Svishtov, Vardim, Rousse, and Marten (Petrbok 1940, 1941, Georgiev 2013).

Distribution in the Danube wetlands: Srebarna Lake (Angelov 1998, 2000b), the adjacent marshes and backwaters of the Danube River near Vidin, Svishtov, and Rousse (Frank et al. 1990, Angelov 2000a), Belene Island, marshes on Persina Island (Russev & Kaneva-Abadjieva 1973), the big canal, former Svishtov Marsh (Georgiev 2013).

General distribution: Southeast European, introduced in Northern Europe.

Ecology: aquatic vegetation; shallow stagnant and slow-flowing water (Angelov 2000a, Georgiev 2014).

Conservation status: IUCN (LC).

Notes: Some authors report the occurrence of two separate species in the Danube River and the adjacent wetlands: *Viviparus acerosa* (Bourguignat, 1862), and *Viviparus danubialis* var. *penchinati* (Bourguignat, 1862) (Petrbok 1941); *Viviparus acerosus* (Bourguignat, 1862) and *Viviparus danubialis* (Bourguignat, 1862) (Frank et al. 1990, Angelov 2000a, Hubenov 2005b, 2006b, 2007); and *Viviparus acerosus* (Bourguignat, 1862), *Viviparus danubialis* var. *penchinati* (Bourguignat, 1862), and *Viviparus danubialis* var. *rumaenicus* (Bourguignat, 1862) (Russev 1979). According to Georgiev (2014), they belong to one and the same species, and *V. acerosa*, *V. danubialis*, *V. danubialis* var. *penchinati*, and *V. danubialis* var. *rumaenicus* are considered as synonyms of *V. acerosus* (Bourguignat, 1862).

Figure 2. *Viviparus acerosus* in the canal flowing into the Danube River at Orsoya (2016) (Photo by Milcho Todorov).

Viviparus contectus **(Millet, 1813)**

Distribution in the Danube River: near Ostrov and Dolni Vadin (Georgiev 2013).

Distribution in the Danube wetlands: Svishtov Marsh, Srebarna Lake (Drensky 1947, Angelov 1998, 2000a, b, Frank et al. 1990), and marshes on Belene Island and Batin Island (Russev & Kaneva-Abadjieva 1973).

General distribution: West Eurosiberian.

Ecology: mud and vegetation; stagnant and slow-flowing water (Angelov 2000a, Georgiev 2014).

Conservation status: IUCN (LC).

Figure 3. *Viviparus contectus*: (a) at Kalimok-Brashlen wetlands (2016); and (b) at Srebarna Lake (Photos by Teodora Trichkova).

Viviparus sphaeridius **(Bourguignat, 1880)**

Distribution in the Danube River: at Marten (Glöer & Georgiev 2014).

Distribution in the Danube wetlands: Brashlen Marsh (empty shells) (Glöer & Georgiev 2014).

General distribution: East European.

Ecology: sand with patches of aquatic vegetation (Glöer & Georgiev 2014).

Viviparus viviparus **(Linnaeus, 1758)**

Distribution in the Danube River: near Novo Selo, Vidin, Lom, Tutrakan and Silistra (Drensky 1947, Russev 1966, Russev & Uzunov 1991a).

Distribution in the Danube wetlands: at Srebarna Lake, marshes at Belene, Svishtov, and Tutrakan (Angelov 1998, 2000a, b, Frank et al. 1990), Srebarna Lake (Uzunov et al. 2001, Varadinova et al. 2011, 2012), marshes on the islands of Skomen, Tsibar, Kozloduy, Leskovets, Boril, Lakat, Palets, Belene, Vardim, and Vetren (Russev & Kaneva-Abadjieva 1973).

General distribution: European.

Ecology: gravel, mud, clay, and aquatic vegetation; stagnant and slow-flowing water; at depths from 0.5 to 7.9 m; up to 100 m distance from the shore (Russev 1966, Angelov 2000a)

Conservation status: IUCN (LC).

Figure 4. *Viviparus viviparus*: (a) shells in the Danube River at Koshava (2013); and (b) along the shore at Tutrakan (2014) (Photos by Milcho Todorov).

Melanopsidae

Esperiana esperi **(Férussac, 1823)**

Distribution in the Danube River: along the entire Bulgarian sector, from 845 to 375 rkm (Frank et al. 1990, Angelov 2000a, Russev 1959, 1979); from 834 to 432 rkm (Russev

1959, 1966); at Vidin, Lom, Rousse (Petrbok 1941, Drensky 1947); at Ostrov, Vardim, Sandrovo, and Brashlen (Georgiev 2013).

General distribution: Southeast European-Pontian; Caspian relict.

Ecology: gravel and stones; stagnant and flowing water (Angelov 2000a, Russev 1959, 1966, Georgiev 2013); a depth of 0.2 to 7 m; and up to 250 m from the shore (Russev 1966).

Notes: reported as *Melanopsis esperi* (Férussac, 1823) (Petrbok 1941, Drensky 1947) and *Fagotia esperi* (Férussac, 1823) (Frank et al. 1999, Angelov 2000a, Russev 1959, 1966, 1976, Hubenov 2005b, 2006b, 2007).

Microcolpia daudebartii **(Prevost, 1821)**

Distribution in the Danube River: along the entire Bulgarian sector, from 845 to 375 rkm (Frank et al. 1990, Angelov 2000a, Russev 1959, 1967, 1978, 1979); from 835 to 430 rkm (Russev 1966); at Rousse (Petrbok 1941), Lom, Svishtov (Drensky 1947), Stanevo (Russev 1963), Kozloduy, Baykal, and Lakat Island (Russev 1974), Tutrakan and Silistra (Russev & Uzunov 1991a), Kozloduy, Ostrov, Dolni Vadin, near the mouth of the Yantra River, at Marten and Brashlen (Georgiev 2013)

General distribution: Southeast European-Pontian; Caspian relict.

Ecology: gravel and stones, also on clay and mud; flowing water; up to 6.5 m depth; and up to 130 m distance from the shore (Russev 1959, 1966, 1967, 1974, 1978).

Conservation status: IUCN (LC).

Notes: reported as *Microcolpia acicularis* Fér. (Westerlund 1886), *Melanopsis acicularis* (Férussac) (Petrbok 1941), *Melanopsi (Microcolpia) acicularis* Fér. (Drensky 1947), *Fagotia acicularis* (Férussac, 1823) (Russev 1959, 1963, 1966, 1967, 1974, 1978, 1979, 1991a, Angelov 2000a, Hubenov, 2005b, 2006b, 2007), and *Esperiana (Microcolpia) daudebartii* (Prevost, 1821) (Georgiev 2013, 2014).

Holandriana holandrii **(C. Pfeiffer, 1828)**

Distribution in the Danube River: along the entire Bulgarian sector (Angelov 2000a), at Vidin, Lom, Kozloduy, Ostrov, Baykal, Rousse (Drensky 1947, Russev 1974, Frank et al. 1990), Kozloduy, Ostrov, Dolni Vadin, and Vardim (Georgiev 2013).

General distribution: Southeast European.

Ecology: stones; flowing water (Russev 1974, Angelov 2000a, Georgiev 2014).

Conservation status: IUCN (LC).

Notes: reported as *Melania crassa* Kutsch. var. *ovoidea* Bgt. (Westerlund 1886) and *Amphimelania holandri* (C. Pfeiffer, 1828) (Russev 1974, Frank et al. 1990, Angelov 2000a, Hubenov 2005b, 2006b, 2007).

Figure 5. *Esperiana esperi*, *Microcolpia daudebartii*, and *Holandriana holandrii* in a sample from the Danube River at Lom (2016) (Photo by Milcho Todorov).

Bithyniidae

Bithynia danubialis **(Glöer & Georgiev, 2012)**

Distribution in the Danube River: at Marten (Glöer & Georgiev 2012, 2014).

General distribution: Regional endemic.

Ecology: sand and mud; flowing water; in the littoral area (Georgiev 2014).

Bithynia leachii **(Sheppard, 1823)**

Distribution in the Danube River: at Somovit (Angelov 2000a).

General distribution: West Palearctic.

Ecology: aquatic vegetation, shallow slow-flowing and stagnant water (Angelov 2000a, Georgiev 2014).

Conservation status: IUCN (LC).

Bithynia tentaculata **(Linnaeus, 1758)**

Distribution in the Danube River: along the entire Bulgarian sector (Angelov 2000a), at Somovit (Drensky 1947), Vidin, Lom, Tsibar Island, Kozloduy, Svishtov, Vardim, Rousse, Silistra, (Frank et al. 1990), Novo Selo, Lom, Tutrakan (Russev & Uzunov 1991a), Novo Selo, Oryahovo (Russev 1966), near Belene, Vardim, and Marten (Georgiev 2013).

Distribution in the Danube wetlands: Srebarna Lake (Angelov 1998, 2000a, b, Frank et al. 1990, Uzunov et al. 2001, Varadinova et al. 2011, 2012, Vidinova et al. 2016), marshes on the islands of Lakat, Belene, Vardim and Kossuj (Russev & Kaneva-Abadjieva 1973), Svishtov Marsh (Frank et al. 1990), the big canal, former Svishtov Marsh (Georgiev 2013).

General distribution: West Palearctic.

Ecology: stones, sand, mud, and aquatic vegetation; stagnant and slow-flowing water; a depth of up to 5 m (Angelov 2000a, Georgiev 2014).

Conservation status: IUCN (LC).

Hydrobiidae

***Caspia milae* (Boeters, Glöer & Georgiev, 2015)**

Distribution in the Danube River: at Vardim Island (Boeters et al. 2015).

General distribution: Regional endemic.

***Turricaspia (Laevicaspia) lincta* (Milaschewitch, 1908)**

Distribution in the Danube River: at Oryahovo (Russev 1966).

General distribution: Pontian-Caspian; Caspian relict.

Conservation status: IUCN (LC).

Notes: reported as *Micromelania lincta* (Milaschewich, 1908); only shells; subfossil (Russev 1966, Hubenov 2005b, 2006b, 2007).

***Turricaspia variabilis* (Eichwald, 1838)**

Distribution in the Danube River: at Rousse; shells upstream of Lom and uptream of Silistra (Russev 1966).

General distribution: Pontian-Caspian; Caspian relict.

Conservation status: IUCN (LC).

Notes: reported as *Clessiniola variabilios* Eichwald, 1838 (Russev 1966, Hubenov 2005b, 2006b, 2007).

Lithoglyphidae

***Lithoglyphus naticoides* (C. Pfeiffer, 1828)**

Distribution in the Danube River: along the entire Bulgarian sector; from 845 to 375 rkm (Russev 1959, 1967, 1978, Angelov 2000a, Georgiev 2014); from 834 to 381 rkm (Russev 1966); at Vidin, Svishtov, Vardim, Rousse (Petrbok 1941), Lom and near the mouth of the Roussenski Lom River (Drensky 1947), at Stanevo, Kozloduy, Oryahovo, Baykal, Lakat Island, Tutrakan (Russev 1963, 1974), Novo Selo, Vidin, Archar, Lom, Kozloduy, Oryahovo, Somovit, Nikopol, Svishtov, Rousse, Tutrakan, and Silistra (Frank et al. 1990, Russev & Uzunov 1990, 1991a), 5 km northeast of Vidin, at Kozloduy, Brashlen, Sandrovo, Svishtov, Dolni Vadin, Belene, Ostrov, Vardim, Marten, and Zagrazhden (Georgiev 2013); with an abundance of up to 10 000 ind./m^2 (Russev 1978).

Distribution in the Danube wetlands: Marshes on the islands of Tsibar, Kozloduy, Palets, Batin, Aleko, Golyam Brashlen, Kossuj, and Vetren (Russev & Kaneva-Abadjieva 1973), at Srebarna Lake (Uzunov et al. 2001, Varadinova et al. 2011, 2012).

General distribution: Southeast European-Pontian; Caspian relict.

Ecology: stones, gravel, and mud; flowing water; up to about 12 m depth; and up to 875 m distance from the shore (Angelov 2000a, Russev 1959, 1963, 1966, 1978); prefers water rich in calcium, oxygen and a temperature of up to 22°C (Georgiev 2014).

Conservation status: IUCN (LC).

***Lithoglyphus pyramidatus* (Möllendorf, 1873)**

Distribution in the Danube River: at Lom (Drensky 1947, Angelov 2000a) and Brashlen (Georgiev 2014).

General distribution: Balkan endemic (? Balkan subendemic, East European-Anatolian).

Ecology: gravel and stones; flowing water (Angelov 2000a).

Valvatidae

***Borysthenia naticina* (Menke, 1845)**

Distribution in the Danube River: at Svishtov and Rousse (Frank et al. 1990, Angelov 2000a).

General distribution: Southeast European.

Ecology: gravel, stones, sand, and mud; flowing water (Angelov 2000a, Hubenov 2006a, 2007).

Conservation status: IUCN (LC).

Notes: reported as *Valvata naticina* Menke, 1845 (Frank et al. 1990, Angelov 2000a, Hubenov 2005b, 2006b, 2007).

***Valvata cristata* (O. F. Müller, 1774)**

Distribution in the Danube wetlands: Srebarna Lake (Angelov 1998, 2000a, b, Uzunov et al. 2001, Varadinova et al. 2011, 2012).

General distribution: West Eurosiberian.

Ecology: aquatic vegetation; slow-flowing and stagnant water (Angelov 2000a).

Conservation status: IUCN (Europe – LC).

***Valvata macrostoma* (Mörch, 1864)**

Distribution in the Danube River: at Kozloduy (Frank et al. 1990, Angelov 2000a).

General distribution: Disjunct Eurosiberian.

Ecology: aquatic vegetation; slow-flowing and stagnant water (Angelov 2000a).

Conservation status: IUCN (LC).

Notes: reported as *Valvata pulchella* Studer, 1820 (Frank et al. 1990, Angelov 2000a, Hubenov 2005b, 2006b, 2007).

Valvata piscinalis **(O. F. Müller, 1774)**

Distribution in the Danube River: along the entire Bulgarian sector; from 845 to 375 rkm (Russev 1959, Angelov 2000a), at Rousse (Petrbok 1941), Novo Selo and Svishtov (Russev 1966), Vidin, Lom, and Rousse (Frank et al. 1990), Tutrakan (Russev & Uzunov 1991a), Ostrov and Vardim (Georgiev 1913).

Distribution in the Danube wetlands: marshes on the islands of Kozloduy, Palets, Aleko, and Kossuj (Russev & Kaneva-Abadjieva 1973), Svishtov Marsh (Frank et al. 1990, Angelov 2000a), Srebarna Lake (Angelov 1998, 2000a, b, Uzunov et al. 2001, Varadinova et al. 2011, 2012).

General distribution: West and Central Palearctic (introduced in the Holarctic).

Ecology: gravel, sand, mud, and aquatic vegetation; slow-flowing and stagnant water; up to 9.2 m depth; and up to 857 m distance from the shore (Russev 1959, 1966, Angelov 2000a).

Conservation status: IUCN (LC).

Acroloxidae

Acroloxus lacustris **(Linnaeus, 1758)**

Distribution in the Danube wetlands: Srebarna Lake (Angelov 1998, 2000a, b, Uzunov et al. 2001, Varadinova et al. 2011, 2012).

General distribution: Holoeurosiberian.

Ecology: stagnat water with abundant aquatic vegetation (Angelov 2000a).

Conservation status: IUCN (LC).

Lymnaeidae

Ampullaceana balthica **(Linnaeus, 1758)**

Distribution in the Danube River: near Lom (Russev 1966) and at Vidin (Frank et al. 1990).

Distribution in the Danube wetlands: marshes along the Danube River (Angelov 2000a), marshes on the islands of Skomen, Tsibar, Kozloduy, Boril, Belene, Vardim, Batin, Aleko, Mishka, and Kossuj (Russev & Kaneva-Abadjieva 1973), and Srebarna Lake (Angelov 1998, 2000a, b, Uzunov et al. 2001, Varadinova et al. 2011, 2012).

General distribution: Holopalearctic.

Ecology: aquatic vegetation; stagnant and slow-flowing water (Angelov 2000a, Georgiev 2014).

Conservation status: IUCN (LC).

Notes: reported as *Radix ovata* (Draparnaud, 1801) (Frank et al. 1990, Russev 1966, Angelov 1998, 2000a, b, Russev & Kaneva-Abadjieva 1973, Uzunov et al. 2001, Hubenov 2005b, 2006b, 2007, Varadinova et al. 2011, 2012) and *Radix balthica* (Linnaeus, 1758) (Georgiev 2013, 2014).

***Galba truncatula* (O. F. Müller, 1774)**

Distribution in the Danube wetlands: Srebarna Lake (Angelov 1998, 2000b, Uzunov et al. 2001, Varadinova et al. 2011, 2012).

General distribution: Holarctic.

Ecology: clay and mud overgrown with aquatic vegetation; stagnant water (Georgiev 2014).

Conservation status: IUCN (LC).

***Lymnaea stagnalis* (Linnaeus, 1758)**

Distribution in the Danube River: along the entire Bulgarian sector; from 845 to 375 rkm (Russev 1959, Frank et al. 1990, Angelov 2000a), at Svishtov (Petrbok 1941), near Belene (Russev 1966), and near Vidin (Georgiev 2013).

Distribution in the Danube wetlands: Belene Marsh, Svishtov Marsh, Tutrakan Marsh, and Srebarna Lake (Drensky 1947), marshes on the islands of Bogdan-Sechan, Skomen, Tsibar, Kozloduy, Boril, Lakat, Belene, Vardim, Batin, Aleko, Mishka, Kossuj, Slavyanin, and Vetren (Russev & Kaneva-Abadjieva 1973), Srebarna Lake (Angelov 1998, 2000a, b, Uzunov et al. 2001, Varadinova et al. 2011, 2012), the big canal, former Svishtov Marsh (Georgiev 2013).

General distribution: Holarctic.

Ecology: mud and aquatic vegetation; stagnant and slow-flowing water (Angelov 2000a, Georgiev 2014); up to 3.7 m depth; and up to 857 m distance from the shore (Russev 1966).

Conservation status: IUCN (LC).

Figure 6. *Lymnaea stagnalis* at Kalimok-Brashlen wetlands (2014) (Photo by Teodora Trichkova).

Distribution in the Danube wetlands: marshes along the Danube River (Frank et al. 1990), marshes on the islands of Bogdan-Sechan, Tsibar, Palets, Batin, Aleko, Golyam Brashlen, Kossuj, and Vetren (Russev & Kaneva-Abadjieva 1973), and Srebarna Lake (Angelov 1998, 2000b, Uzunov et al. 2001, Varadinova et al. 2011, 2012).

General distribution: Subcosmopolitan (North American; introduced in other regions).

Ecology: stones, gravel, mud, and aquatic vegetation; stagnant and slow-flowing water (Russev 1974, Georgiev 2014, Hubenov 2006a, 2007).

Conservation status: IUCN (LC).

Notes: reported as *Physa acuta* Draparnaud (Russev 1966, 1974, Russev & Kaneva-Abadjieva 1973, Uzunov et al. 2001, Varadinova et al. 2011, 2012).

Figure 7. *Physella acuta* at Koshava Island (2016) (Photo by Milcho Todorov).

Planorbidae

Ancylus fluviatilis **(O. F. Müller, 1774)**

Distribution in the Danube River: at Novo Selo (Russev & Uzunov 1991a).

General distribution: West Palaearctic.

Ecology: stones; flowing water (Angelov 2000a, Georgiev 2014).

Conservation status: IUCN (LC).

Anisus leucostoma **(Millet, 1813)**

Distribution in the Danube River: at Svishtov (Petrbok 1941).

Distribution in the Danube wetlands: Srebarna lake (Angelov 1998, 2000a, b, Uzunov et al. 2001, Varadinova et al. 2011, 2012).

General distribution: West and Central Eurosiberian (? West and Central Palearctic).

Ecology: stagnant and slow-flowing water with abundant aquatic vegetation (Angelov 2000a).

Conservation status: IUCN (LC).

***Anisus septemgyratus* (Rossmässler, 1835)**

Distribution in the Danube wetlands: Srebarna Lake (Angelov 1998, 2000a, b).

General distribution: West Eurosiberian (? European).

Ecology: stagnant and slow-flowing water with aquatic vegetation (Angelov 2000a).

***Anisus spirorbis* (Linnaeus, 1758)**

Distribution in the Danube River: at Lom (Frank et al. 1990).

Distribution in the Danube wetlands: marshes on the islands of Lakat, Belene, and Vardim (Russev & Kaneva-Abadjieva 1973), Srebarna Lake (Angelov 1998, 2000a, b, Uzunov et al. 2001, Varadinova et al. 2011, 2012), at the mouth of the Yantra River (Georgiev 2013).

General distribution: West and Central Palearctic.

Ecology: stagnant and slow-flowing water with aquatic vegetation (Angelov 2000a).

Conservation status: IUCN (Europe – LC).

***Anisus vortex* (Linnaeus, 1758)**

Distribution in the Danube River: at Krivina Port (Russev 1966, Frank et al. 1990).

Distribution in the Danube wetlands: marshes along the Danube River (Angelov 2000a), marshes on the islands of Tsibar and Vetren (Russev & Kaneva-Abadjieva 1973), and Srebarna Lake (Angelov 1998, 2000a, b, Uzunov et al. 2001, Varadinova et al. 2011, 2012).

General distribution: West and Central Eurosiberian.

Ecology: stagnant and slow-flowing water with aquatic vegetation (Angelov 2000a, Georgiev 2014).

***Anisus vorticulus* (Troschel, 1834)**

Distribution in the Danube wetlands: Srebarna Lake (Angelov 1998, 2000a, b).

General distribution: West and Central Eurosiberian (? West Eurosiberian).

Ecology: stagnant and slow-flowing water with aquatic vegetation (Angelov 2000a, Georgiev 2014).

Conservation status: Directive 92/43/EEC –Annex II and IV (EU 1992); IUCN (Global – DD, Europe – NT).

Gyraulus albus **(O. F. Müller, 1774)**

Distribution in the Danube River: at Svishtov and Rousse (Petrbok 1941, Angelov 2000a).

Distribution in the Danube wetlands: Svishtov Marsh (Frank et al. 1990, Angelov 2000a) and Srebarna Lake (Angelov 1998, 2000a, b, Uzunov et al. 2001, Varadinova et al. 2011, 2012).

General distribution: Holarctic.

Ecology: stagnant and slow-flowing water with aquatic vegetation (Angelov 2000a).

Conservation status: IUCN (LC).

Gyraulus (Armiger) crista **(Linnaeus, 1758)**

Distribution in the Danube River: at Vidin (Frank et al. 1990).

Distribution in the Danube wetlands: Srebarna Lake (Angelov 1998, 2000a, b, Uzunov et al. 2001, Varadinova et al. 2011, 2012).

General distribution: Holarctic.

Ecology: stagnant and slow-flowing water with aquatic vegetation (Angelov 2000a).

Conservation status: IUCN (LC).

Gyraulus laevis **(Alder, 1838)**

Distribution in the Danube River: at Rousse (Frank et al. 1990, Angelov 2000a).

Distribution in the Danube wetlands: Srebarna Lake (Angelov 1998, 2000a, b, Uzunov et al. 2001, Varadinova et al. 2011, 2012, Vidinova et al. 2016).

General distribution: Holarctic (? European).

Ecology: stagnant and slow-flowing water with aquatic vegetation (Angelov 2000a).

Conservation status: IUCN (LC).

Hippeutis complanatus **(Linnaeus, 1758)**

Distribution in the Danube wetlands: Srebarna Lake (Angelov 1998, 2000a, b, Uzunov et al. 2001, Varadinova et al. 2011, 2012, Vidinova et al. 2016).

General distribution: West and Central Palearctic.

Ecology: mud and clay with aquatic vegetation; stagnant and slow-flowing water (Angelov 2000a, Georgiev 2014).

Conservation status: IUCN (LC).

Planorbarius corneus **(Linnaeus, 1758)**

Distribution in the Danube River: along the entire Bulgarian sector; from 845 to 375 rkm (Russev 1959, Frank et al. 1990, Angelov 2000a), at Nikopol, near Kozloduy, Tutrakan, and Silistra (Russev 1963, 1966), at Ostrov, the mouth of the Iskar River and Somovit, near Dolni Vadin, Belene, and Svishtov (Georgiev 2013).

Distribution in the Danube wetlands: marshes along the Danube River (Angelov 2000a), marshes of Kozloduy, Ostrov, Belene, Svishtov, Tutrakan, and Srebarna Lake (Drensky 1947), marshes on the islands of Kozloduy, Leskovets, Boril, Lakat, Belene, Vardim, Batin, Aleko, Mishka, Slavyanin, and Vetren (Russev & Kaneva-Abadjieva 1973), Srebarna Lake (Angelov 1998, 2000a, b, Uzunov et al. 2001, Varadinova et al. 2011, 2012, Vidinova et al. 2016), the big and small canals, former Svishtov Marsh (Georgiev 2013).

General distribution: West Eurosiberian.

Ecology: gravel, sand, clay, mud, and aquatic vegetation; stagnant and slow-flowing water; up to 9 m depth; and up to 45 m distance from the shore in the Danube River (Russev 1959, 1963, 1966, Angelov 2000a, Georgiev 2014).

Conservation status: IUCN (LC).

Notes: reported as *Planorbis (Coretus, Planorbarius) corneus* L. (Drensky 1947) and *Coretus corneus* (L.) (Russev 1959, 1963, 1966, Russev & Kaneva-Abadjieva 1973).

Figure 8. *Planorbarius corneus* at Kalimok-Brashlen wetlands (2014) (Photo by Teodora Trichkova).

Planorbis carinatus (O. F. Müller, 1774)

Distribution in the Danube River: along the entire Bulgarian sector (Angelov 2000a).

Distribution in the Danube wetlands: marshes along the Danube River (Frank et al. 1990, Angelov 2000a) and Srebarna Lake (Angelov 1998, 2000a, b, Uzunov et al. 2001, Varadinova et al. 2011, 2012).

General distribution: West Eurosiberian.

Ecology: mud and aquatic vegetation; stagnant and slow-flowing water (Angelov 2000a, Georgiev 2014).

Conservation status: IUCN (Global – DD, Europe – LC).

Planorbis planorbis **(Linnaeus, 1758)**

Distribution in the Danube River: along the entire Bulgarian sector (Frank et al. 1990, Angelov 2000a), at Vidin, Vardim, and Rousse (Petrbok 1941), near Vidin, at Ostrov, and near Belene (Georgiev 2013).

Distribution in the Danube wetlands: marshes along the Danube River (Angelov 2000a), Belene Marsh, Svishtov Marsh, and Srebarna Lake (Drensky 1947), Srebarna Lake (Angelov 2000a, b, Uzunov et al. 2001, Varadinova et al. 2011, 2012), the big and small canals, former Svishtov Marsh (Georgiev 2013).

General distribution: Holarctic.

Ecology: mud, aquatic vegetation; shallow stagnant and slow-flowing water (Angelov 2000a, Georgiev 2014).

Conservation status: IUCN (LC).

Notes: reported as *Tropidiscus marginalis* Draparnaud (Petrbok 1941).

Segmentina nitida **(O. F. Müller, 1774)**

Distribution in the Danube wetlands: Srebarna Lake (Angelov 1998, 2000a, b, Uzunov et al. 2001, Varadinova et al. 2011, 2012).

General distribution: West and Central Palearctic.

Ecology: stagnant water with aquatic vegetation (Angelov 2000a, Georgiev 2014).

Conservation status: IUCN (Europe – LC).

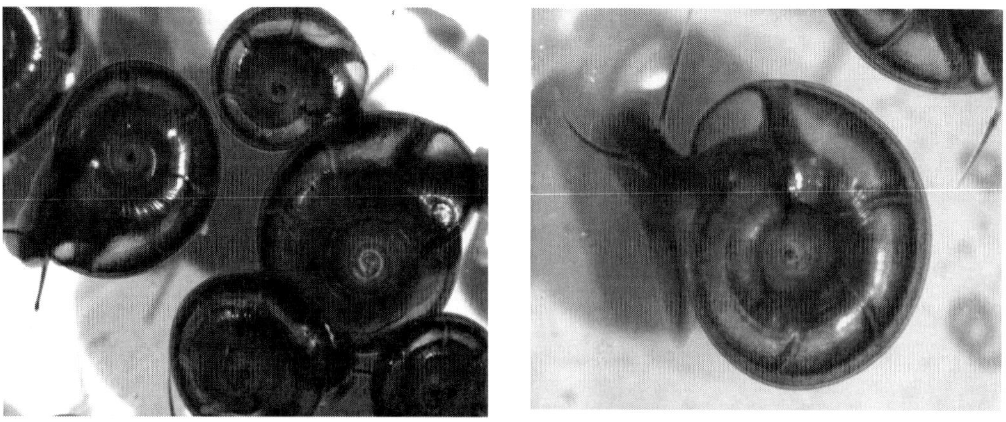

Figure 9. *Segmentina nitida* (Photos by Milcho Todorov).

Bivalvia

Unionidae

***Anodonta cygnea* (Linnaeus, 1758)**

Distribution in the Danube River: at Vidin (Wohlberedt 1911, Petrbok 1941, Frank et al. 1990).

Distribution in the Danube wetlands: in marshes and backwaters along the Danube River (Frank et al. 1990, Angelov 2000a), Belene Marsh, Svishtov Marsh (Drensky 1947), and Srebarna Lake (Angelov 1998, 2000b).

General distribution: European (? West Eurosiberian).

Ecology: mud; stagnant and slow-flowing water; close to the shore (Angelov 2000a, Hubenov 2006a, 2007).

Conservation status: IUCN (Global – LC, Europe – NT).

Notes: reported as *Anodonta cygnaea* var. *cygnea* Linné (Drensky 1947).

***Anodonta anatina* (Linnaeus, 1758)**

Distribution in the Danube River: along the entire Bulgarian sector; from 845 to 375 rkm (Frank et al. 1990, Angelov 2000a, Russev 1959), at Vidin (Petrbok 1941), near Lom (Russev 1966), at Lakat Island (Russev 1974), at Svishtov, and Rousse (as *Colletopterum letourneuxi* Bourguignat, 1881) (Frank et al. 1990, Angelov 2000a).

Distribution in the Danube wetlands: marshes along the Danube River (Frank et al. 1990, Angelov 2000a), marshes on the islands of Boril, Lakat, and Slavyanin (Russev & Kaneva Abadjieva 1973), and Srebarna Lake (Angelov 1998, 2000a, b)

General distribution: West Eurosiberian.

Ecology: gravel, sand and mud; stagnant and slow-flowing water; up to 2.9 m depth; and up to 100 m distance from the shore in the Danube River (Angelov 2000a, Russev 1966).

Conservation status: IUCN (LC).

Notes: reported as *Anodonta piscinalis* Nilsson (Petrbok 1941, Russev 1974). *Colletopterum letourneuxi* Bourguignat, 1881, has been published as a separate species by Frank et al. (1990), Russev (1966), Angelov (2000a), and Hubenov (2005b, 2006b, 2007). Bogatov et al. (2005) report this species as a synonym of *Colletopterum (Colletopterum) subcirculare* (Clessin, 1873), while Graf (2007) relates *C. (C.) subcirculare* (Clessin, 1873) to *A. anatina* (Linnaeus, 1758).

***Pseudanodonta complanata* (Rossmässler, 1835)**

Distribution in the Danube River: along the entire Bulgarian sector; from 845 to 375 rkm (Russev 1959, 1967, 1978, Angelov 2000a), at Svishtov and Rousse (Haas 1911, Wohlberedt 1911), Vardim (Petbok 1941), Lom (Drensky 1947), Stanevo, Nikopol

(Russev 1963), from 835 to 552 rkm, near Lom and at Oryahovo (Russev 1966), at Kozloduy (Russev 1974); with an abundance of up to 29 ind./m^2 (Russev 1966).

Distribution in the Danube wetlands: marshes along the Danube River (Angelov 2000a), marshes on the islands of Tsibar, Batin, Kossuj, and Vetren (Russev 1973), Svishtov Marsh (Frank et al. 1990), and Srebarna Lake (Frank et al. 1990, Angelov 1998, 2000a, b, Uzunov et al. 2001, Varadinova et al. 2011, 2012).

General distribution: European.

Ecology: gravel, sand, clay, mud; slow-flowing water; up to 9.2 m depth; and up to 124 m distance from the shore in the Danube River (Russev 1959, 1966).

Conservation status: IUCN (Global – VU; Europe – NT).

Notes: reported as *Pseudanodonta rossmässleri* Bourguignat (Haas 1911, Wohlberedt 1911, Petrbok 1941) and *Anodonta complanata* (Ziegler) (Russev 1959, 1963, 1973, 1974, 1978).

Figure 10. *Pseudanodonta complanata*: (a) in the Danube River at Batin (2011) (Photo by Teodora Trichkova); and (b) a sample taken at Dolni Tsibar dominated by *P. complanata* (2013) (Photo by Teodora Trichkova).

Sinanodonta woodiana (I. Lea, 1834)

Distribution in the Danube River: shells at Gorni Vadin, Baykal, Somovit, Belene – Hisarlaka, Vardim Island, and the mouth of the Roussenski Lom River (Hubenov 2006a), at Koshava Island, Vidin, Batin, Stalpishte, and Rousse (Hubenov et al. 2012).

General distribution: Introduced in the Transpalaearctic.

Ecology: mud; stagnant and slow-flowing water.

Conservation status: IUCN (Global – LC, Europe – NA).

Notes: reported as *Anodonta (Sinanodonta) woodiana* (Lea, 1834) (Hubenov 2006a, Hubenov et al. 2012). Considered as invasive alien species in Bulgaria and Europe; included in the 'black list' for the Danube River (Paunović & Csányi 2016).

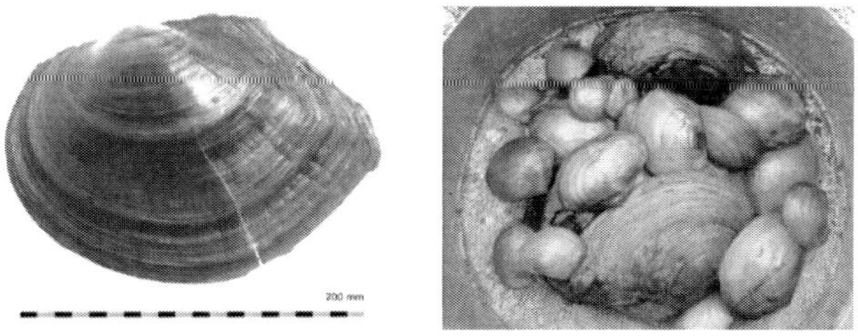

Figure 11. *Sinanodonta woodiana*: (a) the largest specimen reported in the Bulgarian sector of the Danube River at Vidin (2006) (Photo by Assen Ignatov) (Hubenov et al. 2012); and (b) a sample taken in the Danube River at Batin dominated by *S. woodiana* (2011) (Photo by Teodora Trichkova).

***Unio crassus* (Philipsson, 1788)**

Distribution in the Danube River: along the entire Bulgarian sector (Frank et al. 1990, Angelov 2000a), at Vidin, Svishtov, and Rousse (Haas 1911, Wohlberedt 1911, Petrbok 1941, Russev 1966), Silistra (Wohlberedt 1911), Lakat Island (Russev 1974), Tutrakan (Russev & Uzunov 1991a); with an average abundance of 80-90 ind./m^2 (Angelov 2000a).

Distribution in the Danube wetlands: marshes along the Danube River (Angelov 2000a) and Srebarna Lake (Angelov 1998, 2000a, b, Uzunov et al. 2001, Varadinova et al. 2011, 2012).

General distribution: European.

Ecology: gravel, sand, and mud; slow-flowing water; up to 1-1.5 m depth; and up to 10 m distance from the shore in the Danube River (Russev 1974, Angelov 2000a).

Conservation status: Bern Convention (1979); Directive 92/43/EEC – Annex II and IV (EU 1992); Biological Diversity Act of Bulgaria (BDA 2002); IUCN (Global – EN, Europe – VU).

Notes: reported as *Unio crassus cytherea* Küster, 1836 (Haas 1911, Wohlberedt 1911), *Unio batavus* Lam. (Wohlberedt 1911), *Unio batavus* Maton et Rackelt 1807 (Petrbok 1941), *Unio consentaneus* (Z.) Rossm. (Petrbok 1941), and *Unio crassus batavus* Maton u. Rackett, 1807 (Russev 1966).

Figure 12. *Unio crassus* from the Danube River at Dolni Tsibar (2013) (Photo by Teodora Trichkova).

Unio pictorum **(Linnaeus, 1758)**

Distribution in the Danube River: along the entire Bulgarian sector; from 845 to 375 rkm (Drensky 1947, Russev 1959, 1967, 1978, Frank et al. 1990, Angelov 2000a), at Vidin, Svishtov (Haas 1911, Wohlberedt 1911, Petrbok 1941), Kozloduy, Nikopol (Russev 1963), Lakat Island (Russev 1974), and Novo Selo (Russev & Uzunov 1991a); with an abundance of about 30 ind./m^2 (Russev 1978).

Distribution in the Danube wetlands: marshes along the Danube River (Angelov 2000a), marshes on Skomen Island and Kozloduy Island (Russev & Kaneva-Abadjieva 1973), and Srebarna Lake (Angelov 1998, 2000a, b, Uzunov et al. 2001, Varadinova et al. 2011, 2012).

General distribution: European.

Ecology: gravel, clay, mud, and sand; slow-flowing water; a depth from 0.2 to 11.2 m; and up to 900 m distance from the shore in the Danube River (Russev 1959, 1967, 1978, Angelov 2000a).

Conservation status: IUCN (LC).

Notes: reported as *Unio rostratus* Lam. (Wohlberedt 1911) and *Unio limosus* Nilsson (Petrbok 1941).

Figure 13. *Unio pictorum* from the Danube River at Dolni Tsibar (2013) (Photo by Teodora Trichkova).

Unio tumidus **(Philipsson, 1788)**

Distribution in the Danube River: along the entire Bulgarian sector; from 845 to 375 rkm (Drensky 1947, Russev 1959, 1967, 1978, Frank et al. 1990, Angelov 2000a), at Vidin and Rousse (Haas 1911, Wohlberedt 1911, Petrbok 1941), Nikopol (Russev 1959), from 834 to 420 rkm and at Vidin (Russev 1966), Lakat Island and Tutrakan (Russev 1974); an abundance of up to about 57 ind./m^2 (Russev 1978).

Distribution in the Danube wetlands: marshes along the Danube River (Angelov 2000a), Ostrov Marsh and Svishtov Marsh (Drensky 1947), marshes on the island of Skomen, Kozloduy, and Slavyanin (Russev & Kaneva-Abadjieva 1973), and Srebarna Lake (Drensky 1947, Angelov 1998, 2000a, b, Uzunov et al. 2001, Varadinova et al. 2011, 2012).

General distribution: European.

Ecology: clay, mud, and gravel; slow-flowing water; a depth from 0.2 to 9.2 m; and up to 138 m distance from the shore in the Danube River (Russev 1959, 1963, 1966, 1967, 1978, Angelov 2000a).

Conservation status: IUCN (LC).

Notes: reported as *Unio tumidus rostratus* Lamarck, 1819 (Russev 1966).

Dreissenidae

Dreissena polymorpha **(Pallas, 1771)**

Distribution in the Danube River: along the entire Bulgarian sector; from 845 to 375 rkm (Russev 1959, 1966, 1967, 1978, Frank et al. 1990, Angelov 2000a), at Lom and Rousse (Wohlberedt 1911, Petrbok 1941), at Lom, Nikopol, Svishtov, Rousse (Drensky 1947), Oryahovo, Nikopol, Tutrakan (Russev 1963), Kozloduy, Baykal, Lakat Island, Tutrakan, and along the sector from Malak Kondur Island to Belene Island (Russev 1974), at Rousse and Tutrakan (Lyutskanov 1981, 1983), Novo Selo, Lom, Oryahovo, Tutrakan, and Silistra (Russev & Uzunov 1991a); with an abundance of up to 867 ind./m^2 (Russev 1966).

Distribution in the Danube wetlands: marshes on the islands of Boril, Aleko, Mishka, and Kossuj (Russev & Kaneva-Abadjieva 1973), Srebarna Lake (Angelov 1998, 2000a, b, Uzunov et al. 2001, Varadinova et al. 2011, 2012).

General distribution: European (introduced in the Holarctic).

Ecology: stones, gravel, sand, and clay; stagnant and slow-flowing water; a depth from 0.2 to 15 m; and up to 900 m distance from the shore in the Danube River (Russev 1959, 1966, 1974, 1978, Angelov 2000a).

Conservation status: IUCN (Global – LC, Europe – NA).

Notes: reported as *Dreissena blancci* Westerlund (Wohlberedt 1911, Petrbok 1941) and *Dreissena polymorpha* var. *eximia* Bourguignat (Petrbok 1941). In Central and Western Europe and in North America considered as invasive alien species; listed as one of the 100 of the World's Worst Invasive Alien Species (Global Invasive Species Database, IUCN/SSC/ISSG, www.issg.org/worst100_species.html), and one of the 100 of the Worst Invasive Alien Species in Europe (DAISIE, www.europe-aliens.org/speciesTheWorst.do). Included in the 'black list' for the Danube River (Paunović & Csányi 2016).

***Dreissena bugensis* (Andrusov, 1897)**

Distribution in the Danube River: shells at Koshava, Archar, Rousse, and Sandrovo (Hubenov & Trichkova 2007), live specimens at Yasen, Batin, Rousse, Ryahovo, and Tutrakan (Trichkova et al., unpublished data).

General distribution: Southeast European (introduced in the Holarctic).

Ecology: stones, gravel, sand, and clay; stagnant and slow-flowing water (Trichkova et al., unpublished data).

Conservation status: IUCN (LC).

Notes: reported as *Dreissena (Pontodreissena) rostriformis bugensis* (Andrusov, 1897) (Hubenov & Trichkova 2007). Considered as invasive alien species in Bulgaria and Europe; included in the 'black list' for the Danube River (Paunović & Csányi 2016).

Figure 14. *Dreissena polymorpha* fouling on *Anodonta* sp. from the Danube River (2005) (Photo by Matej Polaиik).

Figure 15. *Dreissena* mixed colonies dominated by *D. bugensis* from the Danube River at Rousse (2011) (Photos by Lyubomir Andreev).

Cyrenidae

***Corbicula fluminea* (O. F. Müller, 1774)**

Distribution in the Danube River: shells at Vetren (Hubenov 2001), live speciments between Vrav and Novo Selo, at Novo Selo, Florentin, Yasen, Gomotartsi, Koshava, Koshava Island, Vidin upstream, Simeonovo upstream, islands Golyam Bliznak and Malak

Bliznak, Simeonovo, Botevo, Ratsiaria (at Archar), Orsoya, Lom upstream, Dolno Linevo, Dolni Tsibar, Kozloduy upstream, Oryahovo, Ostrov Village upstream, Baykal, Zagrazhden, Dabovan, Gulyantsi, Somovit downstream, Nikopol, Belene – Hisarlaka, Svishtov downstream, Vardim, Batin, Stalpishte, Rousse, Prista, Rousse, the Yacht Club, Rousse, near Riga hotel, Sandrovo, Ryahovo, Tutrakan, and Malak Preslavets (Hubenov et al. 2013); with an abundance of up to 16 560 ind./m^2 (Hubenov et al. 2013).

General distribution: Holarctic-Oriental (Southeast Palaearctic-Oriental, introduced in the Nearctic).

Ecology: coarse sand and sand with gravel; stagnant and slow-flowing water (Hubenov 2006a, 2007, Hubenov et al. 2013).

Conservation status: IUCN (Global – LC, Europe – NA).

Notes: in Europe and North America considered as invasive alien species; listed as one of the 100 of the Worst Invasive Alien Species in Europe (DAISIE, www.europe-aliens.org/speciesTheWorst.do). Included in the 'black list' for the Danube River (Paunović & Csányi 2016).

Figure 16. *Corbicula fluminea*: (a) the largest specimen reported from the Bulgarian sector of the Danube River at Batin in 2011 (a length of 40.79 mm) (Photo by Assen Ignatov) (Hubenov et al. 2013); (b) the Danube River shore at Svishtov covered with massive amounts of shell deposits (2012) (Photo by Teodora Trichkova); and (c) a sample taken from the Danube River at Dolni Tsibar (2012) (Photo by Teodora Trichkova).

Sphaeriidae

Euglesa casertana **(Poli, 1791)**

Distribution in the Danube River: at Rousse (Frank et al. 1990, Angelov 2000a), Archar and downstream of Kozloduy (Russev 1966).

Distribution in the Danube wetlands: Srebarna Lake (Angelov 1998, 2000a, b, Uzunov et al. 2001, Varadinova et al. 2011, 2012).

General distribution: Subcosmopolitan.

Ecology: sand, gravel; stagnant and flowing water; up to 8 m depth; and up to 40 m distance from the shore in the Danube River (Russev 1966, Angelov 2000a).

Conservation status: IUCN (LC).

Notes: reported as *Pisidium casertanum* (Poli, 1791) (Frank et al. 1990, Russev 1966, Angelov 1998, 2000a, b, Hubenov 2005b, 2006b, 2007, Uzunov et al. 2001, Varadinova et al. 2011, 2012, Georgiev 2017).

***Musculium lacustre* (O. F. Müller, 1774)**

Distribution in the Danube River: along the entire Bulgarian sector (Angelov 2000a).

Distribution in the Danube wetlands: marshes along the Danube River (Angelov 2000a), marshes on the islands of Tsibar, Lakat, Vardim, and Vetren (Russev & Kaneva-Abadjieva 1973), and Srebarna Lake (Angelov 1998, 2000b).

General distribution: Transpalaearctic (Holarctic).

Ecology: aquatic vegetation; shallow stagnant and slow-flowing water (Hubenov 2006a, 2007, Georgiev 2017).

Conservation status: IUCN (LC).

Notes: reported as *Sphaerium (Musculium) lacustre* (Müller) (Russev & Kaneva-Abadjieva 1973).

***Odhneripisidium moitessierianum* (Paladilhe, 1866)**

Distribution in the Danube River: at the mouth of the Archar River (Angelov 1976, 2001a).

General distribution: West Eurosiberian.

Ecology: fine sand and mud; stagnant and slow-flowing water (Georgiev 2017).

Conservation status: IUCN (Europe – LC).

Notes: reported as *Pisidium* (*Neopisidium*) *moitessierianum* (Paladilhe, 1866) (Angelov 1976, 2001a, Hubenov 2005b, 2006b, 2007, Georgiev 2017).

***Pisidium amnicum* (O. F. Müller, 1774)**

Distribution in the Danube River: along the entire Bulgarian sector; from 845 to 375 rkm (Frank et al. 1990, Russev 1959, Angelov 2000a, Georgiev 2017), at Archar, upstream of Lom, and at Stanevo (Russev 1966); with an abundance of up to 150 ind./m^2 (Angelov 2000a).

Distribution in the Danube wetlands: marshes along the Danube River (Frank et al. 1990, Angelov 2000a).

General distribution: Holopalearctic.

Ecology: gravel; stagnant and flowing water; up to 7 m depth and 45 m distance from the shore in the Danube River (Russev 1966, Angelov 2000a).

Conservation status: IUCN (Europe – LC).

Pisidium henslowanum **(Sheppard, 1825)**

Distribution in the Danube River: along the entire Bulgarian sector (Angelov 2000a), at Ostrov (1966), and at the mouth of the Archar River (Angelov 1976); with an abundance of 9 ind./m^2; (Russev 1966).

General distribution: Holopalearctic.

Ecology: mud; slow-flowing water; up to 6.4 m depth (Russev 1966, Angelov 2000a).

Conservation status: IUCN (LC).

Pisidium personatum **(Malm, 1855)**

Distribution in the Danube wetlands: Srebarna Lake (Angelov 1998, 2000a, b, Uzunov et al. 2001, Varadinova et al. 2011, 2012).

General distribution: West Palearctic (? European).

Ecology: stagnant and flowing water (Georgiev 2017).

Conservation status: IUCN (Europe – LC).

Pisidium **(*Pseudosphaerium*)** ***pseudosphaerium*** **(Schlesch, 1947)**

Distribution in the Danube wetlands: Srebarna Lake (Angelov 1998, 2000a, b, Uzunov et al. 2001, Varadinova et al. 2011, 2012).

General distribution: West Eurosiberian.

Ecology: stagnant water with aquatic vegetation (Angelov 2000a, Georgiev 2017).

Conservation status: IUCN (LC).

Pisidium supinum **(Schmidt, 1850)**

Distribution in the Danube River: upstream Lom (Russev 1966).

General distribution: West Eurosiberian.

Ecology: mud; a depth of 3.9 m; 132 m distance from the shore (Russev 1966).

Conservation status: IUCN (LC).

Notes: considered as a synonym of *Pisidium* (*Neopisidium*) *moitessierianum* (Paladilhe, 1866) by Georgiev (2017).

Sphaerium corneum **(Linnaeus, 1758)**

Distribution in the Danube River: along the entire Bulgarian sector (Frank et al. 1990, Angelov 2000a), at Lom and Silistra (Russev & Uzunov 1991a).

Distribution in the Danube wetlands: marshes along the Danube River (Frank et al. 1990, Angelov 2000a) and Srebarna Lake (Frank et al. 1990, Angelov 1998, 2000a, b, Uzunov et al. 2001, Varadinova et al. 2011, 2012).

General distribution: Holarctic (? Transpalaearctic).

Ecology: mud and aquatic vegetation; stagnant and slow-flowing water (Angelov 2000a, Hubenov 2006a, 2007, Georgiev 2017).

Conservation status: IUCN (LC).

Sphaerium rivicola **(Lamarck, 1818)**

Distribution in the Danube River: along the entire Bulgarian sector (Frank et al. 1990, Angelov 2000a), near Lom (Russev 1966).

General distribution: Central and East European (? European).

Ecology: sand, mud, and gravel; slow-flowing water; up to 4 m depth; and up to 106 m distance from the shore (Russev 1966, Angelov 2000a).

Conservation status: IUCN (Global – VU, Europe – LC).

Notes: reported as *Sphaerium (Sphaerium) scaldianum* (Normand) (Russev 1966).

Compared to the previous study of Angelov (2000a), the total number of the reported mollusc species for the Bulgarian sector of the Danube River and the adjacent wetlands has increased from 65 to 67 species. However, there are certain changes within the list. Nine new species were added. Three of them are not included by Angelov (2000a) in his catalogue although reported before that by Russev (1966) (*T. (L.) lincta* and *T. variabilis*) and by Russev & Uzunov (1991a) (*A. fluviatilis*). These species are found only at single sites in the Danube River, and one of them (*T. (L.) lincta*) occurs only as shells. One of the newly listed species (*V. sphaeridius*) is recorded at Marten Village and re-described by Glöer & Georgiev (2014). Two species (*B. danubialis* and *C. milae*) are newly described (Glöer & Georgiev 2012, 2014, Boeters et al. 2015) and also added to the list. Since 2000, three alien bivalve species (*C. fluminea*, *D. bugensis*, and *S. woodiana*) have been introduced to the Bulgarian sector of the Danube River and have rapidly increased their range within the river, the adjacent wetlands, and the inland water bodies of Bulgaria (Hubenov 2001, 2005a, 2006a, Hubenov & Trichkova 2007, Hubenov et al. 2012, 2013). The other changes in the presented species list have resulted from taxonomic revision of certain taxa.

Some of the gastropod species have been reported only from the Danube River, e.g., the representatives of the families Neritidae, Melanopsidae, Hydrobiidae, most of the Bithyniidae and some of the species belonging to the Lithoglyphidae. Other gastropod species are recorded only in stagnant water, of them *V. cristata*, *A. lacustris*, *G. truncatula*, *A. septemgyratus*, *A. vorticulus*, *H. complanatus*, and *S. nitida* have been reported only from Srebarna Lake (Angelov 1998, 2000a, b, Uzunov et al. 2001, Varadinova et al. 2011, 2012). The rest of the gastropod species are found in both the Danube River and the adjacent wetlands, some of them showing preferences to stagnant and slow-flowing water, e.g., representatives of the families Viviparidae, Lymnaeidae and Planorbidae. Among the mussels, the native representavies of the family Unionidae and *D. polymorpha* have been reported from both the Danube River and the adjacent marshes and backwaters. The alien species have been reported only from the Danube River, showing preferences to different substrates: mud (*S. woodiana*), stones, gravel, sand, and clay (*D. bugensis*), and coarse sand and sand with gravel (*C. fluminea*) (Hubenov 2006a, Hubenov & Trichkova 2007, Hubenov et al. 2013). The representatives of the family Sphaeriidae occur in stagnant and flowing water, some of them reported only from the Danube River (*O. moitessierianum*, *P. henslowanum*, and *P. supinum*) (Russev 1966, Angelov 2000a), while others recorded only in Srebarna Lake (*P. personatum*, *P.* (*P.*) *pseudosphaerium*) (Angelov 1998, 2000a, b, Uzunov et al. 2001, Varadinova et al. 2011, 2012).

The mollusc species reported from the Danube River and the adjacent wetlands were classified in 19 zoogeographical categories (Table 2).

Eleven of the reported mollusc species (16.42%) were considered as distributed within and outside the Palaearctic (Table 2). This group includes the subcosmopolitan *P. acuta* and *E. casertana*, as well as the olarctic-Oriental species *C. fluminea*, which has an original Southeast Palaearcticriental distribution. The rest of the Holarctic species belong to the families Lymnaeidae, Physidae, and Planorbidae. The species *D. polymorpha* (with an original European distribution) and *D. bugensis* (with an original Southeast European) have secondary Holarctic distribution.

The highest number of the mollusc species in the Bulgarian sector of the Danube River and the adjacent wetlands was classified of Palaearctic distribution (53 species or 79.10%) (Table 2). Of them, 17 species (25.37%) belong to more than one Palaearctic subgroup. They include mostly representatives of the families Lymnaeidae and Sphaeriidae. The rest of the species belong to only one Palaearctic subgroup: the Eurosiberian type (26 species, 38.80%) or the Mediterranean type (10 species, 14.92%). Most numerous in the Eurosiberian subgroup are the mollusc species of West Eurosiberian and European distribution (Table 2). All native representatives of the family Unionidae, many species of the Planorbidae and Sphaeridae belong to the Eurosiberian subgroup.

The Mediterranean species have Ponto-Mediterranean type of distribution and included species with Southeast European-Pontian (4), Southeast European (4), and Pontian-Caspian (2) range (Table 2). Six species of this group have been considered as Caspian relict species, affiliated to the Ponto-Caspian brackish area, and have Ponto-Caspian and Pontic ranges (Starobogatov 1970, Russev et al. 1998, Hubenov 2005b, 2006b). The Caspian relict fauna is represented by the following species: *T. danubialis*, *E. esperi*, *M. daudebartii*, and *L. naticoides* (of Southeast European-Pontian distribution), and *T. (L.) lincta* and *T. variabilis* (of Pontian-Caspian distribution) (Starobogatov 1970, Russev et al. 1998, Hubenov 2005b, 2006b).

The group of endemic species included species, whose distribution is restricted within the Balkan Peninsula (Hubenov 2005b, 2006b). Three endemic species have been reported in the Bulgarian sector of the Danube River: two regional endemics (*B. danubialis* and *C. milae*) described recently (Glöer & Georgiev 2012, Boeters et al. 2015), and one Balkan endemic (*L. pyramidatus*), for which East European-Anatolian distribution has also been supposed (Hubenov 2005b, 2006b). Some authors also consider as endemic the species *H. holandrii* known only from the Danube River and the Nishava River in Bulgaria (Hubenov 2005b, 2006b).

Most of the mollusc species (52 or 77.61%) in the Bulgarian sector of the Danube River and the adjacent wetlands have been listed under the category Least Concern in the IUCN Red List of *Threatened Species* (IUCN 2019). Seven of the species are not included in the list, while three species are not assessed at European level (IUCN 2019).

Five species have been considered of higher conservation concern according to the IUCN Red List, of them: three species (*A. vorticulus*, *A. cygnea*, *P. complanata*) are listed as Near Threaned; one species (*U. crassus*) as Vulnerable, and one species (*T. transversalis*) as Endangered (IUCN 2019). *Theodoxus transversalis*, *A. vorticulus*, and *U. crassus* are protected in Europe and included in the Annexes II and IV of the Directive 92/43/EEC (EU 1992). Furthemore, *U. crassus* is protected at national level under the Biological Diversity Act of Bulgaria (BDA 2002).

Table 2. Zoogeographic classification of Mollusca from the Bulgarian sector of the Danube River and the adjacent wetlands

Zoogeographic category	Number of species	Percentage, %
Species distributed within and outside the Palaearctic	11	16.42
Subcosmopolitan	2	2.98
Holarctic-Oriental	1	1.49
Holarctic	8	11.94
Species with Palaearctic distribution	53	79.10
Palaearctic type	17	25.37
Holopalaearctic	5	7.46
Transpalaearctic	3	4.48
West and Central Palearctic	4	5.97
West Palearctic	5	7.46
Eurosiberian type	26	38.80
Holoeurosiberian	1	1.49
West and Central Eurosiberian	3	4.48
West Eurosiberian	9	13.43
Disjunct Eurosiberian	1	1.49
European-Anatolian	1	1.49
European	8	11.94
Central and East European	1	1.49
Central and Southeast European	1	1.49
East Europaean	1	1.49
Mediterranean type	10	14.92
Southeast European-Pontian	4	5.97
Southeast European	4	5.97
Pontian-Caspian	2	2.98
Endemic species	3	4.48
Regional endemics	2	2.98
Balkan endemics	1	1.49
Total:	67	100

The main threats to the mollusc fauna in the Bulgarian sector of the Danube River and the adjacent wetlands are the same as the globally known major threats to freshwater biodiversity, including loss and degradation of habitats, overexploitation, pollution, introduction of invasive alien species, and climate change (Dudgeon et al. 2006, Geist 2011). Throughout the years, the water of the Danube River have been contaminated by different sources located upstream of the Bulgarian sector or along the Bulgarian and Romanian shores, such as urban and agricultural areas, different industries, navigation, etc. (Russev & Naidenow 1978). The introduction of nutrients and fine sediment, and the pollution with heavy metals and other toxic substances may severly affect the growth, reproduction and survival of molluscs (Lopes-Lima et al. 2017). The drainage and loss of large areas of former marshes along the Danube River, which provided suitable habitats for the molluscs, also had negative impact on their populations. The decrease in native fish, which carry mussels' glochidia, due to overexploitation and loss of suitable nursing and feeding habitats may also have a huge indirect negative impact on mussel populations of the family Unionidae (Lopes-Lima et al. 2017). The major threat in recent years has been the introduction, establishment and spread of invasive alien species in the Danube River (Paunović & Csányi 2016). In the last 15 years, owing to the increase in abundance and biomass of the newly introduced invasive alien mussels *C. fluminea*, *S. woodiana*, and *D. bugensis*, the benthic habitats in the Bulgarian sector of the Danube River, have been completely changed (Hubenov 2006a, Hubenov & Trichkova 2007, Hubenov et al. 2013). The invasive mussels may also adversely impact the native mollusc species directly through competition and fouling (*D. bugensis*). Therefore, in order to protect the native mollusc fauna in the Danube River and the adjacent wetlands appropriate conservation and management measures need to be applied and directed to the restoration and protection of the mussel populations, on the one hand, and to the restoration of the damaged aquatic habitats, including their catchments, on the other hand (Dudgeon et al. 2006, Geist 2011, 2014, 2015, Lopes-Lima et al. 2017). These may include the strict observation of the existing binding instruments, development and applying of effective water management and conservation action plans, as well as activities towards raising of public awareness about the conservation importance of native mussels and their current threats.

Acknowledgment

The study was made within the frame of the International Association for Danube Research (IAD), supported by the DIAS network.

References

Angelov A. 1976. Revision der Fam. Pisidiidae Gray, 1857 (Bivalvia – Mollusca) in Bulgarien. [Revision of the family Pisidiidae Gray, 1857 (Bivalvia – Mollusca) in Bulgaria.] *Annuaire de l'Université de Sofia, Faculté de Biologie (Zoologie)* [*Annual of Sofia University, Faculty of Biology (Zoology)*] 69: 109-119. (In Bulgarian, German summary).

Angelov A. 1998. Clas Gastropoda (Aquatic gastropods). Clas Bivalvia. In: Michev T., Georgiev B., Petrova A. & Stoyneva M. (Eds.) *Biodiversity of the Srebarna Biosphere Reserve. Checklist and bibliography.* Sofia: Context & Pensoft, 63-64.

Angelov A. 2000a. Mollusca (Gastropoda et Bivalvia) aquae dulcis. In: *Catalogus faunae bulgaricae* 4: Sofia-Leiden: Pensoft & Backhuys Publishers BV, 57 p.

Angelov A. 2000b. Freshwater molluscs (Mollusca: Gastropoda and Bivalvia) from the Srebarna Lake, Northeastern Bulgaria. *Historia naturalis bulgarica* 11: 133-138.

Anistratenko V. & Stadnichenko A. 1994. *Fauna Ukraine. 29 (1-2). Mollyuski (Littoriniformes et Rissoiformes).* [*Fauna Ukraine. 29 (1-2). Mollusca (Littoriniformes and Rissoiformes)*]. Kiev: Naukova Dumka, 174 p.

Anistratenko V. & Anistratenko O. 2001. *Fauna Ukraine. 29 (1-1). Class Polyplacophora or Chitons, class Gastropoda – Cyclobranchia, Scutibranchia and Pectinibranchia (part).* Kiev: Veles, 240 p.

BDA. 2002. Biological Diversity Act, *State Gazette*, 77: 99, August 2002.

Bern Convention. 1979. Council of Europe, Convention on the Conservation of European Wildlife and Natural Heritage. Bern, Switzerland. Available at: https://www.coe.int/en/web/bern-convention/home.

Boeters H. 1998. *Süßwasserfauna von Mitteleuropa. 5. Mollusca. Gastropoda. 1/2. Superfamilie Rissooidea.* [*Freshwater fauna of Central Europe. 5. Mollusca. Gastropoda.1/2. Superfamily Rissooidea*]. Stuttgart: G. Fischer, 77 p.

Boeters H. D., Glöer P., Georgiev D. & Dedov I. 2015. A new species of *Caspia* Clessin et W. Dybowski, 1887 (Gastropoda: Truncateloidea: Hydrobiidae) in the Danube of Bulgaria. *Folia Malacologica* 23 (3): 177-186.

Bogatov V. V., Starobogatov Y. I. & Prozorova L. A. 2005. Mollusks of the genus *Colletopterum* (Anodontinae, Bivalvia) from Russia and adjacent territories. *Zoologicheskiy Zhurnal* [*Zoological Journal*] 84 (9): 1050-1063. (In Russian, English summary).

Bourguignat J. 1863-1870. Mollusques nouveaux, litigieux ou peu connus. [New, problematic or not well known molluscs]. Paris, *Savy 1*: 1-324, Taf. 1-40 (1863-1868), Taf. 1-5 (1870).

Bourguignat J. 1876. Species novissimae molluscorum in Europaeuo systemati detectae, notis diagnosticis succinctis breviter descriptae. *Lutetiae* 1-80.

Büttner K. 1928. Beitrag zur Molluskenfauna Bulgariens. [Contribution to the mollusc fauna of Bulgaria.] *Jahresbericht des Vereins fur Naturkunde zu Zwickau, [Annual Report of the Natural History Society in Zwickau]* 1926/1928: 12-20.

Drensky P. 1947. Synopsis and distribution of freshwater Mollusca in Bulgaria. *Annuaire de l'Université de Sofia, Faculté de Physique et Mathématiques* [*Annual of the Sofia University, Faculty of Physics and Mathematics*] 43 (3): 33-54. (In Bulgarian, English summary).

Dudgeon D., Arthington A. H., Gessner M. O., Kawabata Z., Knowler D., L′ev^eque C., Naiman R. J., Prieur-Richard A. H., Soto D. & Stiassny M. L. J. 2006. Freshwater biodiversity: importance, threats, status, and conservation challenges. *Biological Reviews* 81: 163-182.

EU. 1992. Council Directive 92/43/EEC of 21 May 1992 on the conservation of natural habitats and of wild fauna and flora. *OJ L, 206, 22.7.1992, pp. 7-50.*

Frank C., Jungbluth J. & Richnovszky A. 1990. *Die Mollusken der Donau vom Schwarzwald bis zum Schwarzen Meer.* [*The molluscs of the Danube River from the Schwarzwald to the Black Sea*]. Budapest: Akaprint, 142 p.

Geist J. 2011. Integrative freshwater ecology and biodiversity conservation. *Ecological Indicators 11* (6): 1507-1516.

Geist J. 2014. Trends and directions in water quality and habitat management in the context of the European water framework directive. *Fisheries* 39: 219-220.

Geist J. 2015. Seven steps towards improving freshwater conservation. *Aquatic Conservation: Marine and Freshwater Ecosystems* 25: 447-453.

Georgiev D. 2013. Freshwater snails (Mollusca, Gastropoda aquatic) of Bulgaria and adjacent territories. DSc Thesis, Plovdiv University 'Paisii Hilendarski', Plovdiv, 329 p. (In Bulgarian).

Georgiev D. & Hubenov Z. 2013. Freshwater snails (Mollusca: Gastropoda) of Bulgaria: An updated annotated checklist. *Folia Malacologica* 21 (4): 237-263.

Georgiev D. 2014. *Freshwater snails of Bulgaria.* Plovdiv: Plovdiv University Press 'Paisii Hilendarski', 266 p. (In Bulgarian).

Georgiev D. 2017. The freshwater clams from the family Sphaeriidae (Mollusca: Bivalvia) in Bulgaria. Plovdiv: Plovdiv University Press 'Paisii Hilendarski', *ZooNotes* (Supplement 4), 63 p. (In Bulgarian, English summary)

Glöer P. & Georgiev D. 2012. *Bithynia danubialis*, a new species from the Bulgarian Danube (Gastropoda: Rissooidea: Bithyniidae). *Acta Zoologica Academiae Scientiarum Hungaricae* 58 (2): 193-197.

Glöer P. & Georgiev D. 2014. Redescription of *Viviparus sphaeridius* Bourguignat 1880 with an identification key of the European *Viviparus* species (Gastropoda: Viviparidae). *Ecologica Montenegrina* 1 (2): 96-102.

Graf D. L. 2007. Palearctic freshwater mussel (Mollusca: Bivalvia: Unionoida) diversity and the Comparatory Method as a species concept. BioOne Complete, *Proceedings of the Academy of Natural Sciences of Philadelphia* 156 (1): 71-89.

Haas F. 1911. Bulgarische Najaden. [Bulgarian Unionid bivalve molluscs.] *Abhandlungen der Naturforschenden Gesellschaft zu Görlitz* [*Reports of the Society of Naturalists in Görlitz*] 27: 235-238.

Hristovich G. 1892. Materials to study the Bulgarian fauna. *Collection of popular works* 8: 337-346. (In Bulgarian).

Hubenov Z. 2001. Corbiculidae – a new family for the Bulgarian recent malacofauna (Mollusca: Bivalvia). *Acta Zoologica Bulgarica* 53 (3): 61-66.

Hubenov Z. 2005a. *Dreissena* (Bivalvia: Dreissenidae) – systematics, autochthonous and anthropogenic areas. *Acta Zoologica Bulgarica* 57 (3): 259-268.

Hubenov Z. 2005b. Malacofaunistic diversity of Bulgaria. In: Petrova A. (Ed.) *Current state of Bulgarian biodiversity – problems and perspectives*. Bulgarian Bioplatform, Sofia, 199-246. (In Bulgarian, English summary).

Hubenov Z. 2006a. *Anodonta* (*Sinanodonta*) *woodiana* (Lea, 1834) (Mollusca: Bivalvia: Unionidae) – a new invasive species for the Bulgarian malacofauna. *Acta Zoologica Bulgarica* 58 (1): 35-40.

Hubenov Z. 2006b. Malacofaunistic diversity of the Bulgarian wetlands (Mollusca: Gastropoda et Bivalvia). *Lauterbornia* 56: 61-83.

Hubenov Z. 2007. Fauna and zoogeography of marine, freshwater, and terrestrial mollusks (Mollusca) in Bulgaria. In: Fet V. & A. Popov (Eds.). *Biogeography and ecology of Bulgaria*. Dodrecht: Springer, 141-198.

Hubenov Z. & Trichkova T. 2007. *Dreissena bugensis* (Mollusca: Bivalvia: Dreissenidae) – new invasive species to the Bulgarian malacofauna. *Acta Zoologica Bulgarica* 59 (2): 203-209.

Hubenov Z. K., Trichkova T. A., Kenderov L. A. & Kozuharov D. S. 2012. Recent distribution of invasive alien mussels *Anodonta woodiana* and *Corbicula fluminea* (Mollusca: Bivalvia: Unionidae & Corbiculidae) In Bulgaria. *Journal of International Scientific Publications: Ecology & Safety* 6 (2): 269-284.

Hubenov Z., Trichkova T., Kenderov L. & Kozuharov D. 2013. Distribution of *Corbicula fluminea* (Mollusca: Corbiculidae) over an eleven-year period of its invasion in Bulgaria. *Acta Zoologica Bulgarica* 65 (3): 315-326.

IUCN 2019. *The IUCN Red List of Threatened Species. Version 2018-2*. http://www.iucnredlist.org. Downloaded on January 2019.

Iurinitsh S. 1906. Notes sur les Bivalves d'eau douce de Bulgarie. [Notes on the Bulgarian freshwater Bivalvia] *Annuaire de l'Université de Sofia, Faculté de Physique et Mathématiques* [*Annual of Sofia University, Faculty of Physics and Mathematics*] 2: 61-67.

Kobelt W. 1897-1898. *Studien zur Zoogeographie 1. Die Fauna der palaearktischen Region.* [*Zoogeographical studies 1. The fauna of the Palearctic Region.*] Wiesbaden, 344 p.

Korniushin A. 1996. *Bivalve molluscs of the superfamily Pisidioidea in the Palaearctic region (Fauna, systematics, phylogeny).* Kiev: National Academy of Sciences of Ukraine, 175 p. (In Russian, English summary).

Kreglinger C. 1871. *Systematische Verzeichnis der in Deutschland lebenden Binnenmollusken.* [*Systematic list of the freshwater molluscs in Germany.*] Wiesbaden.

Lopes-Lima M., Sousa R., Geist J., Aldridge D., Araujo R., Bergengren J., Bespalaya Y., Bódis E., Burlakova L., Van Damme D., Douda K., Froufe E., Georgiev D., Gumpinger C., Karatayev A., Kebapçi Ü., Killeen I., Lajtner J., Larsen B., Lauceri R., Legakis A., Lois S., Lundberg S., Moorkens E., Motte G., Nagel K. O., Ondina P., Outeiro A., Paunovic M., Prié V., Von Proschwitz T., Riccardi N., Rudzīte M., Rudzītis M., Scheder C., Seddon M., Şereflişan H., Simić V., Sokolova S., Stoeckl K., Taskinen J., Teixeira A., Thielen F., Trichkova T., Varandas S., Vicentini H., Zajac K., Zajac T. & Zogaris S. 2017. Conservation status of freshwater mussels in Europe: State of the art and future challenges. *Biological Reviews 92* (1): 572-607.

Lyutskanov D. 1981. Morphological, taxonomic and ecological research of recent and fossil representatives of *Dreissena polymorpha* Pall. in Bulgaria. PhD Thesis, Chemical-Biological Faculty, Plovdiv University 'Paisii Hilendarski', Plovdiv, 253 p. (In Bulgarian).

Lyutskanov D. 1983. Notes on the ecology of *Dreissena polymorpha* Pall. in Bulgaria. *Travaux Scientifique de l'Université de Plovdiv* [*Scientific Works of Plovdiv University*] 21 (4) : 443-460. (In Bulgarian).

Michailov T. & Popov W. 1978. Geomorphologie des Donauufers. [Geomorphology of the Danube River shore.] In: Russev B. & Naidenow W. (Eds.). *Limnologie des Bulgarischen Donauabschnitts,* [*Limnology of the Bulgarian sector of the Danube River.*] Sofia: Bulgarian Academy of Sciences, pp. 7-28. (In Bulgarian, German summary).

Paunović M. & Csányi B. 2016. *Guidance document on Invasive Alien Species (IAS) within the Danube River Basin.* Working Document, International Commission for the Protection of the Danube River (ICPDR).

Pavlova M., Ihtimanska M., Dedov I., Biserkov V., Uzunov Y. & Pehlivanov L. 2013. New localities of *Theodoxus transversalis* (C. Pfeiffer, 1828) within European Natura 2000 Network on the Islands of the Lower Danube River. *Acta Zoologica Bulgarica* 65 (1): 121-123.

Petrbok J. 1940. Zweiter Beitrag zur Kenntnis der Mollusken-Fauna von Bulgarien. [Second contribution to the mollusc fauna of Bulgaria.] *Archiv für Molluskenkunde* [*Archive for Mollusc Science*] 72: 85-88.

Petrbok J. 1941. Posttertiaria nonmarina mollusca bulgarica. [Post-tertiary non-marine Bulgarian molluscs.] *Věstník Královské české Společnosti Náuk. Třida Mathematicko-Přírodovědecká* [*Bulletin of the Royal Czech Society of Sciences. Mathematics and Natural Science Class*] 1-39.

Russev B. 1957. On the bottom life of the Danube River near our banks. *Priroda [Nature]* 2: 44-49. (In Bulgarian).

Russev B. 1959. Beitrag zur Erforschung des Macrobenthos der Donau am bulgarischen Ufer. [Contribution to the study of the macrobenthos at the Bulgarian shore of the Danube River.] *Comptes rendus de l'Academie bulgare des Sciences* [*Proceedings of the Bulgarian Academy of Sciences*] 12 (4): 345-348.

Russev B. 1963. Anthropogene lithoreophile Biozönose in der Donau vor dem bulgarischen Ufer. [Anthropogenic lithoreophilic biocenosis at the Bulgarian shore of the Danube River.] *Comptes rendus de l'Academie bulgare des Sciences* [*Proceedings of the Bulgarian Academy of Sciences*] 16 (4): 545-548.

Russev B. 1966. Das Zoobenthos der Donau zwischen dem 845 und 357 Flusskilometer. I. Zusammensetzung, Verteilung und Ökologie. [The zoobenthos of the Danube River between the 845th and 357th river kilometers. I. Composition, distribution and ecology] *Bulletin de l'Institut de zoologie et Musée* [*Bulletin of the Institute of Zoology and Museum*] 20: 55-131. (In Bulgarian, Russian and German summaries).

Russev B. 1967. Das Zoobenthos der Donau zwischen dem 845 und 375 Flusskilometer. II. Biozönologie und Dinamik. [The zoobenthos of the Danube River between the 845th and 357th river kilometers. II. Biocenology and dinamics] *Bulletin de l'Institut de zoologie et Musée* [*Bulletin of the Institute of Zoology and Museum*] 23: 33-78. (In Bulgarian, Russian and German summaries).

Russev B. 1968a. Saprobiologische Characteristic des Donauzuflusses Jantra. [Saprobiological characteristics of the Yantra River, the Danube tributary] In: Limnologische Berichte der X. Jubiläumstagung Donauforschung [Limnological Reports of the X. Anniversary Conference on the Danube Research] 10-20 October 1966, Sofia: Bulgarische Akademie der Wissenschaften, Zoologisches Institut mit Museum, Societas Internationalis Limnologiae, Arbeitsgemeinschaft Donauforschung [Bulgarian Academy of Sciences, Institute of Zoology with Museum, International Society of Limnology, Working Group on the Danube Research,], pp. 461-465.

Russev B. 1968b. Saprobiologische Übersicht über die Donau und ihre Nebenflüsse zwischen dem 845 und 375 Flusskm. [Saprobiological overview of the Danube and ist tributaries between the 845th and 357th river kilometers] In: Limnologische Berichte der X. Jubiläumstagung Donauforschung [Limnological Reports of the X. Anniversary Conference on the Danube Research] 10-20 October 1966, Sofia: Bulgarische Akademie der Wissenschaften, Zoologisches Institut mit Museum, Societas Internationalis Limnologiae, Arbeitsgemeinschaft Donauforschung [Bulgarian

Academy of Sciences, Institute of Zoology with Museum, International Society of Limnology, Working Group on the Danube Research], pp. 481-489.

Russev B. & Kaneva-Abadjieva V. 1973. Sur la faune du benthos des marais de quelques îles Bulgares du Danube. [On the benthic fauna in the marshes of some Bulgarian islands in the Danube River] *Hidrobiologia [Hydrobiology] (Bucharest)* 14: 255-268.

Russev B. 1974. Das Zoobenthos der Donau zwischen dem 845ten und dem 375ten Stromkilometer. III Dichte and Biomasse. [The zoobenthos of the Danube River between the 845th and 357th river kilometers. III. Density and biomass] *Bulletin de l'Institut de Zoologie et Musée* [*Bulletin of the Institute of Zoology and Museum*] 40: 175-194. (In Bulgarian, Russian and German summaries).

Russev B. 1978. Besonderheiten und Bedeutung des Zoobenthos der Donau zwischen Strom-km 845 und 357. [Characteristics and significance of the zoobenthos in the Danube River between the 845th and 357th river kilometers] In: Russev B. K. & Naidenow W. T. (Eds.*), Limnologie des bulgarischen Dunauabschnitts,* [*Limnology of the Bulgarian Sector of the Danube River*] Sofia: Bulgarian Academy of Sciences, 145-200. (In Bulgarian, Russian and German summaries).

Russev B. K. & Naidenow W. T. (Eds.) 1978. *Limnologie des bulgarischen Dunauabschnitts.* [*Limnology of the Bulgarian Sector of the Danube River*] Sofia, Bulgarian Academy of Sciences, 308 p. (In Bulgarian).

Russev B. 1979. Gegenwärtige Kenntnisse über die Artenzusammensetzung des Zoobenthos der Donau. [Current knowledge about species composition of the zoobenthos in the Danube River] In: Jubuläeumstagung der Internationalen Arbeitsgemeinschaft Donauforschung. Limnologische Berichte, Sofia: Verlag der Bulgarische Akademie der Wissenschaften, [Anniversary Conference of the International Association for Danube Research. Limnological Reports, Sofia: Publishing House of the Bulgarian Academy of Sciences], pp. 306-339.

Russev B. K. & Uzunov J. I. 1990. Das Zoobenthos und der saprobiologische Zustand der Donau während der internationalen Expedition im März 1988. [The zoobenthos and the saprobiological state of the Danube River during the International expedition in March 1988] Ergebnisse der Donauexpedition 1988, [Results of the Danube River expedition in 1988,] Wien: Eigenverlag der IAD, [Vienna: IAD Publishing] Wienna], pp. 209-220.

Russev B. K. & Uzunov J. I. 1991a. Der Zustand der benthalen Zoozönosen im bulgarisch-rumänischen Donauabschnitt (Oktober 1986). [Condition of the benthic zoocenoses in the Bulgarian-Romanian Danube stretch (October 1986)] Hidrobiologia [Hydrobiology] (Sofia) 37: 24-33.

Russev B. K. & Uzunov J. I. 1991b. Zoobenthos und saprobiologischer Zustand der Donau während der im März 1988 durchgeführten internationalen Expedition. [The zoobenthos and the saprobiological state of the Danube River during the International expedition in March 1988] Hidrobiologia [Hydrobiology] (Sofia) 37: 34-40.

Russev B., Petrova A., Yaneva I. & Andreev S. 1998. Diversity of zooplankton and zoobenthos in the Danube River, its tributaries, and adjacent water bodies. In: Meine C. (Ed.), Washington; Sakalian M. (Ed.) (Bulgarian Version), *Bulgaria's Biological Diversity: Conservation Status and Needs Assessment.*, Vols. I-II. D. C.: Biodiversity Support Program. Sofia: Pensoft, 263-292.

Stadnichenko A. 1984. Fauna Ukrainy. 29 (9). Mollyuski (Unionidae, Cycladidae). [Fauna Ukraine., *29* (9). Mollusca (Unionidae, Cycladidae).] Kiev: Naukova Dumka, 382 p.

Stadnichenko A. 1990. *Fauna Ukrainy. 29 (4). Mollyuski (Lymnaeiformes). [Fauna Ukraine.Mollusca (Lymnaeiformes).*] Kiev: Naukova Dumka, 290 p.

Starobogatov J. 1970. *Fauna of mollusks and zoogeographical zones of the inland waters worldwide.* Leningrad: Nauka, 372 p. (In Russian).

Uzunov Y., Tsankova V., Todorov I. & Varadinova E. 2001. The macrozoobenthic fauna of the Biosphere Reserve Srebarna Lake in North-Eastern Bulgaria. *Lauterbornia* 40: 43-51.

Varadinova E., Borisova P., Pehlivanov L. & Uzunov Y. 2012. Macroinvertebrate communities of the Srebarna Lake Biosphere Reserve: species diversity, abundance and modelling of the ecological status. In: Uzunov Y. Georgiev B., Varadinova E., Ivanova N., Pehlivanov L. & Vasilev V. (Eds.). *Ecosystems of the Biosphere Reserve Srebarna Lake.* Sofia: Professor Marin Drinov Academic Publishing House, pp. 93-113.

Varadinova E. D., Pehlivanov L. Z., Stoichev S. A. & Uzunov Y. I. 2011. Recovering and succession of the species diversity of macrozoobenthos in Srebarna Biosphere Reserve (North-East Bulgaria). *Acta Zoologica Bulgarica* 63: 85-95.

Vidinova Y., Tjufekchieva V., Varadinova E., Stoichev S., Kenderov L., Dedov I. & Uzunov Y. 2016. Taxonomic list of benthic macroinvertebrate communities of inland standing water bodies in Bulgaria. *Acta Zoologica Bulgarica* 68 (2): 147-158.

Westerlund C. A. 1885. Fauna der in der paläarctischen Region (Europa, Kaukasien, Sibirien, Turan, Persien, Kurdistan, Armenien, Mesopotamien, Kleinasien, Syrien, Arabien, Egypten, Tripolis, Tunesien, Algerien und Marocco) lebenden Binnenconchylien. V. Fam. Succinidae, Auriculidae, Limnaeidae, Cyclostomidae & Hydrocenidae, [Fauna of the freshwater molluscs in the Palearctic Region (Europe, Caucasus, Siberia, Turan, Persia, Kurdistan, Armenia, Mesopotamia, Asia Minor, Syria, Arabia, Egypt, Tripoli, Tunisia, Algeria and Morroco). V. Families Succinidae, Auriculidae, Limnaeidae, Cyclostomidae & Hydrocenidae] pp. 1-135: 1-14.

Westerlund C. A. 1886. Fauna der in der paläarctischen Region (Europa, Kaukasien, Sibirien, Turan, Persien, Kurdistan, Armenien, Mesopotamien, Kleinasien, Syrien, Arabien, Egypten, Tripolis, Tunesien, Algerien und Marocco) lebenden Binnenconchylien. VI. Fam. Ampullaridae, Paludinidae, Hydrobiidae, Melanidae, Valvatidae & Neritidae, [Fauna of the freshwater molluscs in the Palearctic Region (Europe, Caucasus, Siberia, Turan, Persia, Kurdistan, Armenia, Mesopotamia, Asia

Minor, Syria, Arabia, Egypt, Tripoli, Tunisia, Algeria and Morroco). VI. Families Ampullaridae, Paludinidae, Hydrobiidae, Melanidae, Valvatidae & Neritidae] 156 p.

Wohlberedt O. 1911. Zur Molluskenfauna von Bulgarien. [On the mollusc fauna of Bulgaria] *Abhandlungen der Naturforschenden Gesellschaft zu Gorlitz* [*Reports of the Society of Naturalists in Görlitz*] 27: 167-234.

WoRMS. 2019. World Register of Marine Species. Available from http://www.marinespecies.org at VLIZ. Accessed 2019-01. doi: 10.14284/170.

Zhadin V. I. 1952. *Mollusks of fresh and brackish waters of the USSR*. Moscow & Leningrad: Akademii Nauk SSSR, [Academy of Sciences of USSR], 376 p. (In Russian).

In: Biodiversity of the Bulgarian-Romanian Section
Editor: Shurulinkov, Hubenov et al.
ISBN: 978-1-53615-663-8

Chapter 3

EPHEMEROPTERA, PLECOPTERA, AND TRICHOPTERA (INSECTA) FROM THE BULGARIAN STRETCH OF THE DANUBE RIVER AND ADJACENT TERRITORIES

Vesela Evtimova**, *Yanka Vidinova and Violeta Tyufekchieva
Institute of Biodiversity and Ecosystem Research, Bulgarian Academy of Sciences, Sofia, Bulgaria

ABSTRACT

This paper collates original, unpublished, and literature data on mayflies (Ephemeroptera), stoneflies (Plecoptera), and caddisflies (Trichoptera) from the Bulgarian section of the Danube River and nearby territories. It focuses on their aquatic immature and aerial adult stages at more than 100 localities. A total of 104 taxa from the three insect orders are known from the study area: 52 mayflies, three stoneflies, and 49 caddisflies. They belong to 27 families and 47/48 genera that encompass13 families and 20 genera of Ephemeroptera, two families and three genera of Plecoptera, and 12 families and 24/25 genera of Trichoptera. Six mayfly species are reported only for the Danube River area and almost all of them are listed in the Red Data Book of Bulgaria as regionally extinct and critically endangered. In addition to these six, another two mayfly species are listed as endangered, three as vulnerable, and seven as rare or very rare. Only three stonefly species are known from the study area and they are typical for the Bulgarian stretch of the Danube River; however, during the last century, their populations have dramatically decreased and these species are currently considered extinct. Eight of the caddis taxa are confirmed only for the Danube River area, while 15 of the registered species are considered rare or very rare. Newly recorded for the region are four species of mayflies and two/three genera and

* Corresponding Author's E-mail: vesela.evtimova@gmail.com.

two species of caddis. Further, we report new localities and data for 19 species of Ephemeroptera and six taxa of Trichoptera.

Keywords: lower Danube River region, EPT taxa inventory, distribution, Bulgaria

Introduction

Ephemeroptera (E), Plecoptera (P), and Trichoptera (T) are orders of semi-aquatic insects (Arthropoda, Insecta). Their representatives are known from the early age of zoological sciences and have been reported in "Systema Naturae" by the father of modern taxonomy, Linnaeus (1758). Their scientific names have a Greek origin and end in "-ptera", which comes from "*πτερον*" (*pteron*) or wing. "Ephemero" originates from the Greek *"φήμερα"* (*ephemera*) or "short life" and is related to their short imaginal stage. Burmeister (1839) proposed the name Plecoptera, from the Greek *"πλέκω" (pleko)* or "to fold, braid, weave", owing to the pleated hind wings of the adults, folded under the front wings. The name Trichoptera comes from the Greek *"τρίχα" (thricha)* or "hair" and literally means "hairy wings".

According to Edmunds & McCafferty (1988), mayflies (Ephemeroptera) are the most ancient of the extant insects. The first mayflies have been dated to the Late Carboniferous or the Early Permian/290 Mya (Barber-James et al. 2008). The oldest known fossil stoneflies (Plecoptera) also date back to the Early Permian of the Ural Mountains and are about 258 to 263 My old (Rohdendorf & Ranitsyn 1980). The "youngest" of the three orders, Trichoptera (caddisflies), dates back to the Middle-Late Triassic or 234 Mya (Malm et al. 2013), the period when the first dinosaurs started walking on Earth. However, recent records of Early Permian fossils from marine environments in present Brazil resemble caddis cases and imply they could be even more ancient than previously thought (Mouro et al. 2015). Some studies suggest that these insect orders flourished during the Jurassic, when they were numerous and quite diverse, but the recorded diversity and richness could be influenced by changes in habitat preferences, body parts durability, etc. (Sinitschenkova 1985, 1995, Rasnitsyn & Quicke 2002) and should be taken with caution.

These insects have not only survived for millions of years, but they could be found literally everywhere on the globe, on each continent except for Antarctica and some remote islands. There are about 21,400 described EPT taxa: > 3200 mayflies, > 3700 stoneflies, and almost 14,400 caddisflies (Zhang 2011). Mayflies, stoneflies, and caddisflies undergo metamorphosis: eggs and immature stages (larvae, nymphs, and/or pupae) are typically aquatic (rarely semi-aquatic or as an exception terrestrial), while all three have aerial (winged) adult stages. Their larval stages usually inhabit very diverse freshwater habitats (some taxa tolerate even lower salinity), with some of the species being particularly finicky about the environments they inhabit (Barbour et al. 1996, Wallace et al. 1996, Sandin &

Johnson 2000, Evtimova 2013, Evtimova & Donohue 2014). These are some of the reasons why EPT taxa are often used as bioindicators and various EPT metrics have been developed to evaluate ecosystem health and environmental conditions in freshwater systems (Lenat 1988, Eaton & Lenat 1991, Wallace et al. 1996). Thus, these insect orders are often studied together. Furthermore, aquatic (stages of) insects (including the EPT taxa) play an important role in the productivity of lowland large rivers in Europe. For instance, Russev (1968) pointed out the role of *Palingenia longicauda* and *Ephoron virgo* in the 1970s with regards to the importance of the macroinvertebrates in the total productivity of the Danube River ecosystem. As a result, researchers are often interested not only in the biodiversity of Ephemeroptera, Plecoptera, and Trichoptera but also in their habitat preferences and ecological spectra.

Bulgarian scientists began regular hydrobiological studies on the Danube River during the second half of the 20th century, when the International Association for Danube Research was founded (Vidinova et al. 2014). During the second half of the last century, Russev (1979, 1998) summarized data on macroinvertebrates from the entire Danube River. In his work from 1979, Russev listed 51 taxa of Ephemeroptera (24 in the Bulgarian section of the Danube River and two from non-Danubian localities on the territory of Bulgaria), 50 of Plecoptera (three in the Bulgarian section of the river), and 77 taxa of Trichoptera (18 for Bulgaria and one from non-Danubian locality on the territory of Bulgaria). Later, Russev (1998) listed 58 taxa of Ephemeroptera (three from the Bulgarian stretch), 29 taxa of Plecoptera (0 from the Bulgarian stretch), and 151 taxa of Trichoptera (four from the Bulgarian section) for the whole of the Danube River. In 2001, the International Commission for the Protection of the Danube River initiated the Joint Danube Survey (JDS) along the entire river length (www.icpdr.org/main/activities-projects/joint-danube-survey-1); it was followed by the JDS 2 in 2007 and JDS 3 in 2013. According to the final report of the JDS 3, 31 taxa of Ephemeroptera, one of Plecoptera, and 42 of Trichoptera were recorded in the Danube River area in 2013 (Graf et al. 2014). This report presents no detailed information on macroinvertebrate communities per country.

The aim of our paper is to summarize available data on EPT taxa from the Bulgarian section of the Danube River and its nearby territories, including both longitudinal and temporal scales. Thus, we present all of the available data on species, as well as some unpublished data, including new taxon records or additional information on the distribution of already listed taxa within the study area.

MATERIALS AND METHODS

The study area covers diverse habitats and includes the Bulgarian stretch of the Danube River and non-Danubian sites that lie within approximately 20 km. Apart from the Lower Danube River, nearby inundated zones, swamps, and lakes are also included, as well as the mouths of Bulgarian tributaries (large- and medium-sized lowland rivers) of the Danube River (Figure 1, Table 1).

This paper collates literature and original data on adults and larvae (abbreviated in the results as "La") of Ephemeroptera, Plecoptera, and Trichoptera from the Bulgarian section of the Lower Danube River and the adjacent areas. Original data are based on the aquatic immature stages. Samples were collected using a hand-held net with a mesh size of 500 μm. Each species record is linked to a corresponding literature source or, when there are original data, all additional information is provided. When there is no specific location, only the UTM code number is provided and these records are not explicitly referred in the location list (Table 1) but are marked on the map (Figure 1).

The nomenclature and systematics of Ephemeroptera follow Bauernfeind & Soldan (2012); subgenera and species are given in alphabetical order. The geographical distribution in Europe is according to Puthz (1978), Bauernfeind & Soldan (2012), and/or Belfiore & Thomas (2017, http://www.faunaeur.org/). Nomenclature and systematic for Plecoptera follow Muranyi (2008) and DeWalt et al. (2018). Taxonomy and classification hierarchy for Trichoptera follow Kumanski (1985, 1988), Holzenthal (2011), and Morse (2018).

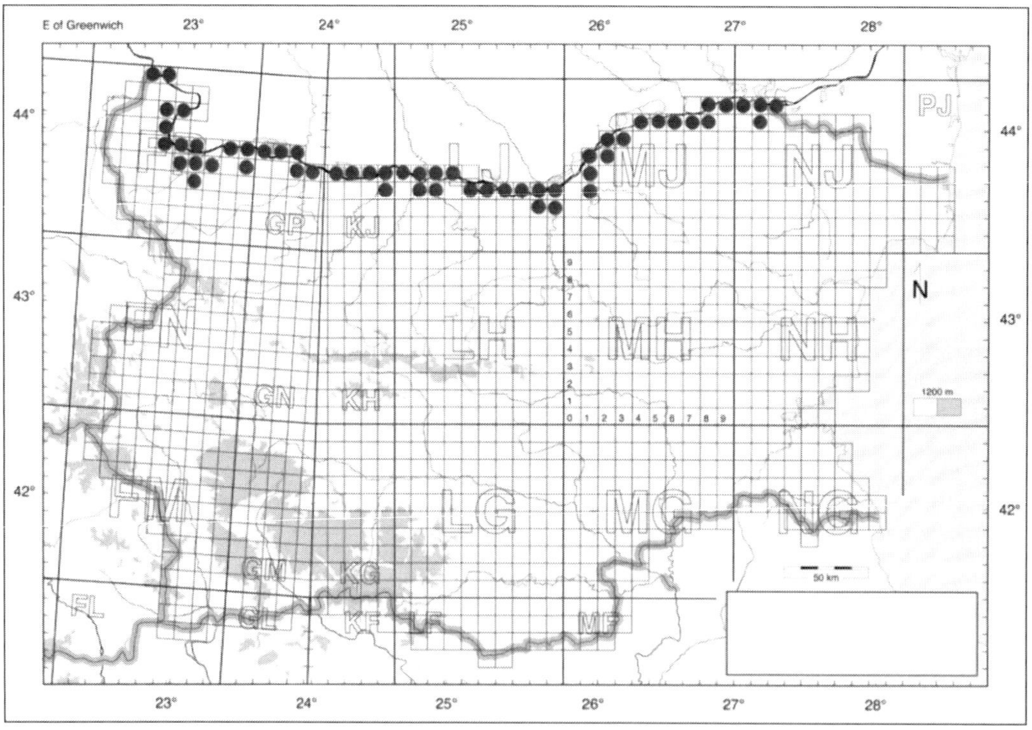

Figure 1. UTM grid (10x10, Michev 1999) distribution of EPT taxa within the Lower Danube River region on the territory of Bulgaria.

Table 1. List with localities/stations with taxa of Ephemeroptera, Plecoptera or Trichoptera from the Bulgarian section of the Danube River and the adjacent territories

№	Water body	Locality/Station	UTM Code	Altitude, m	N	E
1	Timok River	Bregovo Town	FP39	44	44.15902	22.62301
2	Danube River	775 river km, between Simeonovo and Botevo Villages	FP45	32	43.842587	22.858373
3	Archar River	Darzhanitsa Village	FP45	64	43.809226	22.836220
4	Danube River	783 r. km, Bogdan-Setchan Island	FP46	33	43.918298	22.851381
5	Topolovets River	downstream Gradets Village	FP47	50	44.027178	22.769836
6	Topolovets River	Vidin Town	FP46	47	43.98338	22.81728
7	Voynishka River	Tarnyane Village	FP46	40	43.92621	22.79877
8	Voynishka River	Dunavtsi Town	FP46	37	43.924355	22.813891
9	Vidbol River	Sinagovtsi Village	FP46	59	43.901764	22.761274
10	Vidbol River	Dunavtsi Town	FP46	42	43.90288	22.80512
11	Danube River	834 r. km, Novo selo Village	FP49	33	44.168955	22.779614
12	Skomlya River	Septemvriytsi Village	FP54	55	43.75835	22.94181
13	Archar River	Archar Village	FP55	37	43.81282	22.92031
14	Barzartsi Brook	near Archar Village	FP55	45	43.804101	22.923918
15	Danube River	770 r. km, near Archar Village	FP55	32	43.821304	22.908939
16	Danube River	799 r. km, Kutovo Village	FP57	32	44.013890	22.976284
17	Danube River	790 r. km, Vidin Town	FP57	32	43.971994	22.872771
18	Nechinska bara River	dam of Hristo Smirnenski Reservoir	FP63	139	43.617372	23.013208
19	Skomlya River	west of Dobri Dol Village, river mouth	FP64	38	43.78646	22.98025
20	Orsoya wetland	fishpond	FP65	31	43.791517	23.092539
21	Orsoya wetland	peaty marsh	FP65	30	43.770747	23.014481
22	Orsoya wetland	Chanel 2	FP65	30	43.794511	23.122981
23	Danube River	759 r. km, Skomen Island	FP65	31	43.790324	23.038355
24	Lom River	Traykovo Village	FP74	50	43.754148	23.195187
25	Lom River	Zamphir Village	FP85	38	43.786242	23.237619
26	Lom River	upstream Lom Town	FP85	33	43.80043	23.24489
27	Danube River	747 r. km, upstream Lom Town	FP85	28	43.822779	23.190810
28	Danube River	740 r. km, after the inflow of Lom River	FP85	28	43.840375	23.255260
29	Danube River	724 r. km, Stanevo Village	FP95	28	43.844963	23.441268
30	Danube River	715 r. km, Gorni Tsibar Village	GP05	27	43.809804	23.535769
31	Tsibritsa River	Vulchedrum Town	FP94	67	43.688217	23.449769
32	Tsibritsa River	Dolni Tsibar Village	GP05	36	43.80683	23.51137
33	Tsibritsa River	mouth	GP05	30	43.811989	23.517253
34	Danube River	704 r. km, Kozloduy Town, ponton	GP15	27	43.798551	23.667537

Table 1. (Continued)

№	Water body	Locality/Station	UTM Code	Altitude, m	N	E
35	Ogosta River	Glozhene Village	GP24	29	43.690304	23.824289
36	Skut River	downstream Miziya Town	GP24	33	43.70932	23.85071
37	Danube River	698 r. km, Kozloduy Island	GP25	31	43.795540	23.742602
38	Danube River	693 r. km, around Kozloduy Island	GP25	27	43.771764	23.788704
39	Danube River	690 r. km	GP25	27	23.788704	23.815754
40	Danube River	685 r. km, mouth of Ogosta River	GP34	26	43.748096	23.887676
41	Danube River	682 r. km, between the mouth of Ogosta River and Oryahovo	GP34	26	43.747011	23.901024
42	Danube River	678 r. km, Oryahovo Town	GP34	26	43.742721	23.962112
43	Ogosta River	mouth	GP34	29	43.745640	23.885287
44	Danube River	671 r. km, Leskovets Island (old name Orech)	KJ64	25	43.714705	24.057039
45	Danube River	665 r. km, against Popadiya Island	KJ64	26	43.690943	24.133829
46	Danube River	661 r. km, Ostrov Village, ponton	KJ74	25	43.684800	24.150716
47	Danube River	651 r. km, Dolni Vadin Village	KJ74/ KJ84	26	43.692790	24.280029
48	Iskar River	Krushovene Village	KJ93	30	43.643901	24.424132
49	Iskar River	Gigen Village	KJ94	28	43.701926	24.455163
50	Iskar River	Baykal Village	KJ94	28	43.728207	24.447193
51	Danube River	640 r. km, Baykal Village	KJ94	25	43.721271	24.421942
52	Danube River	636 r. km, mouth of Iskar River	KJ94	24	43.736667	24.444881
53	Vit River	Gulyantsi Village	LJ13	27	43.634048	24.698160
54	Danube River	607 r. km, Somovit Village (port)	LJ23	25	43.690336	24.770740
55	Osam River	Muselievo	LJ23	28	43.628070	24.850584
56	Osam River	Cherkovitsa Village	LJ24	28	43.696374	24.849985
57	Danube River	599 r. km, after the inflow of Osam River	LJ24	21	43.703313	24.865325
58	Danube River	596 r. km, Nikopol Town	LJ34	21	43.711354	24.911505
59	Danube River	588 r. km, in front of Lakat Island	LJ34	22	43.712635	24.991457
60	Danube River	588 r. km, Lakat Island	LJ34	22	43.724936	24.994870
61	Danube River	563 r. km, Belene Town	LJ43	19	43.641285	25.158003
62	Danube River	564 r. km, Belene Island, marshes	LJ53	19	43.671662	25.204044
63	Danube River	554 r. km, Svishtov Town	LJ63	18	43.621100	25.361416
64	Danube River	552 r. km, downstream Svishtov Town	LJ63	18	43.616589	25.386515
65	Danube River	546 r. km, Vardim Village	LJ73	18	43.617079	25.462531
66	Danube River	536 r. km, Krivina Village	LJ83	16	43.648808	25.579632
67	Yantra River	Krivina Village	LJ83	22	43.610702	25.584005
68	Yantra River	marsh near the mouth	LJ83	22	43.632008	25.554173
69	Yantra River	Dolna Studena Village	LJ92	28	43.523179	25.710665
70	Yantra River	Tsenovo Village	LJ92	26	43.546654	25.670600
71	Danube River	527 r. km, Batin Island	LJ93	17	43.685997	25.672334
72	Danube River	522 r. km, near Gorno Ablanovo Village	LJ93	16	43.692436	25.746662
73	Rusenski Lom River	Ivanovo Village	MJ13	47	43.705965	25.977027
74	Rusenski Lom River	Bozhichen Village	MJ14	42	43.718800	25.961144
75	Danube River	497 r. km, Ruse/ upstream Ruse	MJ15	15	43.839512	25.931569
76	Rusenski Lom River	2 km south from Ruse Town	MJ15			

№	Water body	Locality/Station	UTM Code	Altitude, m	N	E
77	Rusenski Lom River	Ruse Town, Sredni kuli Quarter	MJ15	23	43.808650	25.937484
78	Rusenski Lom River	Ruse Town, Dolapite Quarter	MJ15	31	43.788010	25.925850
79	Rusenski Lom River	Ruse Town, under the railway bridge	MJ15	19	43.819343	25.936529
80	Rusenski Lom River	Ruse Town	MJ15			
81	Rusenski Lom River	mouth	MJ15	15	43.834296	25.936003
82	NA	Obraztsov Chiflik near Ruse (aerial stages)	MJ25	125	43.810999	26.033267
83	Danube River	478 r. km, Sandrovo Village	MJ26	15	43.966429	26.108657
84	Danube River	475 r. km, Aleko Island	MJ26 /MJ3 6	15	43.973725	26.124393
85	Danube River	451 r. km, Golyam Brashlyan Island, marsh	MJ47	12	44.038172	26.395564
86	Kalimok-Brushlyan	channels	MJ57	13	44.025013	26.496853
87	Danube River	432-433 r. km, Tutrakan Town	MJ67 /MJ7 7	13	44.052697	26.610797
88	Danube River	429 r. km	MJ77	12	44.058455	26.639600
89	Danube River	408 r. km, Slavianin Island	MJ77	14	44.122780	26.891058
90	Malak Preslavets Marsh		MJ88	16	44.096366	26.836295
91	Danube River	403 r. km, Popina, ponton	MJ98	11	44.130860	26.948475
92	Danube River	393 r. km, Vetren Island, marsh	NJ08	18	44.142162	27.067158
93	Srebarna Lake		NJ08	11	44.104382	27.065480
94	Danube River	381-382 r. km	NJ18	11	44.132404	27.144053
95	Danube River	382 r. km, marsh	NJ18	10	44.134440	27.136107
96	Danube River	375 r. km, Silistra Town	NJ28	10	44.125712	27.275096
97	Danube River	Bulgarian section, without specific location				
98	Yantra River	lower stretch, without specific location				
99	NA	Vidin District, without specific location (aerial stages)	FP47			
100	NA	Danubian Plain, Ruse District (aerial stages)	MJ15			

Comments on the distribution of mayflies are based on the number of their known localities/findings within the country as reported by Presolska (2014) for two periods (before and after 1990). When possible, species are classified according the IUCN categories (Vidinova 2015). The assessment of stoneflies extinction risk is conducted in accordance with the IUCN Red List Categories and Criteria (IUCN 2012) and based on the Guidelines for Using the IUCN Red List Categories and Criteria (IUCN 2017). At present, the conservation status of the Bulgarian caddis has not been evaluated and information on their distribution in Bulgaria is provided in Table 2, based on the currently available literature and observations of the authors.

RESULTS

We list a total of 104 EPT taxa— 52 mayflies, 49 caddisflies, and three stoneflies (Table 2) — from at least 100 localities in the Bulgarian section of the Lower Danube River and its neighboring territories. They comprise a total of 27 families and 47/48 genera: 13 families and 20 genera of Ephemeroptera, two families and three genera of Plecoptera, and 12 families and 24/25 genera of Trichoptera.

Table 2. List of the taxa and their records in the Bulgarian section of the Lower Danube River (D) and adjacent water basins (nonD) within the study area, together with notes on distribution (including conservation status in Bulgaria)

Taxa	Records	Distribution notes
	Ephemeroptera	
Siphlonurus (S.) aestivalis	nonD	moderately distributed
S. (S.) armatus	nonD	limited distribution, vulnerable (VU)
Ametropus fragilis	D	limited distribution, regionally extinct (EX)
Baetis (B.) buceratus	nonD	Common
B. (B.) fuscatus	D, nonD	Common
B. (B.) macani	nonD	not evaluated
B. (B.) nexus	nonD	not evaluated
B. (B.) pavidus	nonD	not evaluated
B. (B.) scambus	D, nonD	not evaluated
B. (B.) vernus	nonD	not evaluated
B. (L.) tricolor	D, nonD	not evaluated
B. (N.) gracilis	nonD	not evaluated
B. (Rh.) rhodani	nonD	widespread, very common
Centroptilum luteolum	nonD	Common
Cloeon (Cl.) dipterum	D, nonD	widespread, very common
Cl. (S.) simile	D, nonD	not evaluated
Procloeon (Pr.) bifidum	nonD	less common
Pr. (Ps.) pennulatum	nonD	not evaluated
Isonychia (I.) ignota	D, nonD	moderately distributed; fewer modern data
Oligoneuriella pallida	D, nonD	moderately distributed; no modern data
O. polonica	D	limited distribution
O. rhenana	D, nonD	widespread, common
Ecdyonurus (E.) aurantiacus	nonD	limited distribution
E. (E.) dispar	nonD	moderately distributed
E. (E.) insignis	nonD	moderately distributed
Electrogena affinis	nonD	limited distribution, endangered (EN)
E. lateralis	nonD	moderately distributed
E. mazedonica	nonD	moderately distributed
Heptagenia (D.) coerulans	D, nonD	moderately distributed; VU
H. (H.) flava	D, nonD	widespread, common
H. (H.) longicauda	nonD	moderately distributed

Taxa	Records	Distribution notes
H. (H.) sulphurea	D, nonD	moderately distributed; EN
H. (K.) fuscogrisea	D	local distribution; critically endangered (CR)
Choroterpes (Ch.) picteti	D, nonD	moderately distributed, fewer current data
Paraleptophlebia submarginata	nonD	widespread, common
Ephemera (E.) danica	nonD	widespread, common
E. (E.) lineata	nonD	moderately distributed
E. (E.) vulgata	nonD	moderately distributed, VU
Palingenia longicauda	D	limited distribution; regionally EX
Ephoron virgo	D, nonD	moderately distributed; regionally EX
Potamanthus luteus	D, nonD	widespread, common
Ephemerella ignita	nonD	widespread, very common
E. mesoleuca	nonD	limited distribution; no modern data
E. notata	nonD	moderately distributed; fewer modern data
Brachycercus harrisellus	D	limited distribution; CR
Cercobrachys minutus	D	limited distribution; CR
Caenis horaria	D, nonD	limited distribution
C. luctuosa	nonD	moderately distributed
C. macrura	D, nonD	widespread, common
C. pseudorivulorum	D, nonD	moderately distributed
C. rivulorum	nonD	limited distribution, no modern data
C. robusta	D, nonD	moderately distributed
Plecoptera		
Taeniopteryx nebulosa	D	possibly extinct (PE)
Oemopteryx loewii	D	EX
Marthamea vitripennis	D	EX
Trichoptera		
Psychomyia pusilla	nonD	very common
Ecnomus tenellus	D?, nonD	common
Neureclipsis bimaculata	D, nonD?	common
Cyrnus trimaculatus	nonD	very common
Hydropsyche contubernalis	D, nonD	very common
H. modesta	nonD	very common
H. bulgaromanorum	D, nonD	very common
H. bulbifera	nonD	very common
H. incognita	nonD	very common
H. angustipennis	nonD	very common
Hydropsyche spp.	D, nonD	NA
Cheumatopsyche lepida	nonD	common
Rhyacophila fasciata	nonD	very common
Glossosoma sp.	nonD	NA
Agraylea sexmaculata	D	common
A. multipunctata	nonD	not evaluated
Oxyethira sp.	nonD	NA
Hydroptila spp.	D, nonD	NA
Agrypnia varia	nonD	common

Table 2. (Continued)

Taxa	Records	Distribution notes
Limnephilus flavicornis	nonD	common
L. flavospinosus	D?, nonD	common
L. decipiens	D, nonD	common
L. lunatus	?nonD	very common
L. vittatus	?nonD	very common
Limnephilus sp.	nonD	NA
Potamophylax pallidus	nonD	common
Halesus digitatus	nonD	common
Micropterna nycterobia	nonD	very common
Goera pilosa	nonD	common
?Sericostoma sp./Oecismus monedula?	nonD	NA
Athripsodes cinereus	D, nonD	Rare
A. longispinosus	nonD	Rare
Athripsodes sp.	D	NA
Ceraclea alboguttata	D, nonD	Rare
C. fulva	D	very rare
C. senilis	D	very rare
C. annulicornis	D	very rare
C. aurea	D, nonD	very rare
C. riparia	D	very rare
C. dissimilis	nonD	moderately distributed
Leptocerus tineiformis	nonD	very rare
cf. Triaenodes bicolor/Ylodes conspersus	nonD	rare/ very rare
Ylodes kawraiskii	nonD	rare
Y. simulans	nonD	rare
Oecetis ochracea	D, nonD	common
O. notata	D	very rare
O. furva	D, ?nonD	rare
Setodes punctatus	D	very rare
Mystacides azurea	D, ?nonD	common

Order Ephemeroptera

Family Siphlonuridae Ulmer, 1920

Genus *Siphlonurus* Eaton, 1868

Subgenus *Siphlonurus* Eaton, 1868

Siphlonurus (*S.*) *aestivalis* (Eaton, 1903)

Localities: *3* - 30.4.2000 (Presolska 2014).

Distribution: Several localities in Bulgaria, some of them at higher altitude (to about 1000 m a.s.l.). In riparian vegetation in river spills and swamps (Russev & Vidinova 1994). Found in Europe, with the exception of the British Isles, Iceland, and the Iberian Peninsula (Puthz 1978).

Siphlonurus (*S.*) *armatus* (Eaton, 1870)

Localities: *14* - 30.4.2000 (Presolska 2014).

Distribution: Less common in Bulgaria. Its ecological preferences are almost the same as *S. aestivalis* (Russev & Vidinova 1994). In Europe recorded in Great Britain, the countries of Northern and Eastern Continental Europe (including the Baltic region), south to the Balkans (Puthz 1978, 1980).

Family Ametropodidae Bengtsson, 1913

Genus *Ametropus* Albarda, 1878

Ametropus fragilis Albarda, 1878

Localities: *27* - 08.6.1963, 13.4.1964; *34* - 14.10.1958; *75* - 19.10.1959, 19.4.1954; *97* (Russev 1968, Russev & Vidinova 1994).

Distribution: Larvae sand-dwelling in potamal sections of large rivers with shifting sandy bottom (Bauernfeind & Soldan 2012). At present, the species is hard to find in Europe due to the pollution in potamal sections of larger rivers (Russev & Vidinova 1994).

Family Baetidae Leach, 1815

Genus *Baetis* Leach, 1815

Subgenus *Baetis* Leach, 1815

Baetis (*B.*) *buceratus* Eaton, 1870

Localities: *77* (Russev et al. 1987); *98* (Janeva & Russev 1997); *53* - 10.11.2011 (Kerakova 2015).

Unpublished material: *7* - 22.7.2014; *53* - 08.8.2015; *55* - 09.8.2015.

Distribution: Large lowland rivers, sometimes dominant in β-mesosaprobic conditions of organic pollution (Bauernfeind & Soldan 2012). Frequent in Bulgaria, very often with rich populations. Widely distributed in the West Palearctic (Bauernfeind & Soldan 2012).

Baetis (*B.*) *fuscatus* (Linnaeus, 1761)

Localities: *76* - 05.10.1958 (Russev 1960, as *B. bioculatus*); *97* (Russev 1968, as *B. bioculatus*); *55* - 09.8.1971, 30.5.1972, 30.8.1972 (Russev 1977); *58* - 1988 (Russev & Uzunov 1990, 1991); *98* (Janeva & Russev 1997); *53* – 10.11.2011, 08.2012, 03.11.2012 (Kerakova 2015).

Unpublished material: *1* - 21.7.2014, 04.7.2015; *7* -22.7.2014; *11* - 21.7.2014; *12* - 04.7.2015; *19* - 03.7.2015; *26* - 24.7.2014; *53* - 16.9.2014, 08.8.2015; *55* - 09.8.2015; *56* - 17.9.2014.

Distribution: From rhithral to potamal of streams and rivers with stony bottom, preferably in lotic sections with submerged aquatic plants (Bauernfeind & Soldan 2012).

No actual data on its distribution in Bulgarian freshwaters. Found in the Palearctic and widely distributed in almost all of Europe, including some Mediterranean islands (Bauernfeind & Soldan 2012).

Baetis (*B.*) *macani* Kimmins, 1957

Localities: *98* (Janeva & Russev 1997).

Unpublished material: *56* - 17.9.2014.

Distribution: Typical lowland species. Larvae occur in all types of aquatic habitats, frequently in still water and usually on aquatic vegetation (Bauernfeind & Soldan 2012). No actual data on the species distribution in Bulgarian freshwaters. Holarctic species; recorded in Northern Europe, including Fennoscandia (Bauernfeind & Soldan 2012).

Baetis (*B.*) *nexus* Navás, 1918

Localities (unpublished material): *12* - 04.7.2015; *19* - 03.7.2015; *36* – 25.7.2014; *81* - 12.8.2015.

Distribution: Lowland waters of varying size, mostly in rivers, backwaters, and artificial water channels with rather low current velocity. Prefers places with submerged aquatic vegetation (Bauernfeind & Soldan 2012). No actual data on its distribution in Bulgarian freshwaters, but it seems to be not so common. Probably widely distributed in the West Palearctic, without clear area limits (Bauernfeind & Soldan 2012).

Baetis (*B.*) *pavidus* Grandi, 1949

Localities: 53 (Janeva 1987).

Distribution: Termophilic and rheophilic species (Jacob 2003). Larvae prefer stony habitats of medium-sized streams and brooks (Bauernfeind & Soldan 2012). No actual data on its distribution in Bulgarian freshwaters. In Europe: Atlanto-Mediterranean – Iberian Peninsula (records from Spain and Portugal), southern France, Italy and North Africa (Bauernfeind & Soldan 2012).

Baetis (*B.*) *scambus* Eaton, 1870

Localities: *76* - 5.10.1959 (Russev 1960); *66* - 5.6.1959 (Russev 1962); *97* (Russev 1966, Russev 1968); *55* - 09.8.1971 (Russev 1977); *79* - 05.10.1958, 25.10.1969 (Russev et al. 1987).

Unpublished material: *1* - 04.7.2015.

Distribution: Exclusively in lotic habitats, only as an exception in artificial water courses. Substratum preferences similar to *B. fuscatus* (Bauernfeind & Soldan 2012). No actual data on its distribution in Bulgarian freshwaters. West Palearctic species (Haybach 1998, Jacob 2003); reported in almost all European countries (Belfiore & Thomas 2017).

Baetis (*B.*) *vernus* Curtis, 1834

Localities: *55* - 09.8.1971, 30.5.1972, 30.8.1972 (Russev 1977).

Distribution: Larvae with a very broad ecological range (including substrate preferences), usually prefer the rhithral of smaller streams, but can also be found in larger lowland rivers, mountain brooks, or still waters, and artificial water bodies. No actual data on its distribution in Bulgarian freshwaters. Transpalearctic species; found throughout

Europe (including the British Islands). Not recorded on the Mediterranean islands (Bauernfeind & Soldan 2012).

Subgenus *Labiobaetis* Novikova & Kluge, 1987

Baetis (*L.*) *tricolor* Tshernova, 1928

Localities: *98* (Janeva & Russev 1997).

Unpublished material: *55* - 09.8.2015; *75* - 05.9.2014.

Distribution: Potamal of large lowland rivers, preferring low current velocity and rich, submerged vegetation, and roots (Bauernfeind & Soldan 2012). No actual data on its distribution in Bulgarian freshwaters. Found in Central and Eastern Europe (Belfiore & Thomas 2017).

Subgenus *Nigrobaetis* Novikova & Kluge, 1987

Baetis (*N.*) *gracilis* Bogoescu & Tabacaru, 1957

Localities: 53 (Janeva 1987).

Distribution: Meta- and hyporhithral of smaller colline and submontane rivers or in larger lowland rivers. Syntopic with *B. vernus*, *B. buceratus*, *Oligoneuriella rhenana*, *Ecdyonurus dispar*, and *Heptagenia coerulans*. No actual data on its distribution in Bulgarian freshwaters. Palearctic; also known from Romania and Poland (Belfiore & Thomas 2017).

Subgenus *Rhodobaetis* Jacob, 2003

Baetis (*Rh.*) *rhodani* (Pictet, 1843)

Localities: *55* - 30.5.1972, 30.8.1972 (Russev 1977); *98* (Janeva & Russev 1997).

Unpublished material: *7* - 22.7.2014; *10* - 21.7.2014; *12* - 04.7.2015; *19* - 03.7.2015; *81* - 12.8.2015.

Distribution: Lotic erosional habitats, from crenal to the potamal of rivers. Also found in pools and oligotrophic lakes, including artificial watercourses (Bauernfeind & Soldan 2012). The most eurythermic and eurytopic species within the genus (Sartori & Landolt 1999). One of the most common and ubiquitous species in Bulgaria. Recorded from all over Europe (Bauernfeind & Soldan 2012).

Genus *Centroptilum* Eaton, 1869

Centroptilum luteolum (Müller, 1776)

Localities: *13* - 22.7.2014 (unpublished material)

Distribution: Widely distributed in Europe. Common in suitable habitats in Bulgarian freshwaters. Rare and in small numbers in large lowland rivers. Larvae prefer smaller to very small streams and brooks where they live among submerged vegetation and in pools on fine sediments (pelal and psammal substrata) (Bauernfeind & Soldan 2012). Predominantly in lentic habitats with very low current velocity (Sartori & Landolt 1999).

Genus *Cloeon* Leach, 1815

Subgenus *Cloeon* Leach, 1815

Cloeon (*Cl.*) *dipterum* (Linnaeus, 1761)

Localities: *97* (Russev 1966, 1968, Russev & Kaneva-Abadjieva 1973); *4*, *44*, *62*, *71*, *84*, *85* (Russev & Kaneva-Abadjieva, 1973); *55* - 04.7.1967 (Russev 1977); *73* - 20.7.1962 (Russev et al. 1987); *49* (Janeva & Russev 1989); *43* – 04.1969 (Janeva 1991); *98* (Janeva & Russev 1997); *20* - 12.11.2010, 15.5.2011; *21* - 25.9.2010, 31.5.2012; *37* - 01.6.2012, 24.7.2012; *44* - 01.12.12; *90* - 08.6.2011, 16.7.2011, 09.9.2011, 14.5.2012; 07.8.2012; 19.11.2012; *93* - 04.9.2010, 20.11.2010, 17.5.2012 (Kazakov 2017).

Unpublished material: *6* - 21.7.2014, 05.7.2015; *31* - 01.8.2015; *34* - 18.8.2015; *51* - 19.8.2015; *93* - 22.7.2014, 25.7.2015.

Distribution: Found in almost all types of aquatic habitats. Eudominant in still waters but also found in running waters, especially in large slow-flowing rivers and backwaters (Bauernfeind & Soldan 2012). Very common in Bulgarian standing waters, especially in the presence of macrophytes. Records from all over Europe, excluding Italy and the surrounding islands (Belfiore & Thomas 2017).

Subgenus *Similicloeon* Kluge & Novikova, 1992

Cloeon (*S.*) *simile* Eaton, 1870

Localities: *80* - 28.9.1951 (Russev 1957); *97* (Russev 1966; Russev & Kaneva-Abadjieva 1973); *4*, *71*, *84*, *85* (Russev & Kaneva-Abadjieva 1973); *55* - 04.7.1967 (Russev 1977).

Distribution: Larvae predominantly in still waters with aquatic vegetation and with preferences for relatively cold oligotrophic lakes and fishponds at higher altitudes. Ecological ranges comparatively narrower; however, the species lives syntopically with C. *dipterum*, but always in considerably lower densities. No actual data on the distribution in Bulgarian freshwaters. Transpalearctic (Bauernfeind & Soldan 2012).

Genus *Procloeon* Bengtsson, 1915

Subgenus *Procloeon* Bengtsson, 1915

Procloeon (*Pr.*) *bifidum* (Bengtsson, 1912)

Localities: *26* - 24.7.2014 (unpublished material)

Distribution: Fairly numerous in metarhithral to potamal of middle-sized streams and (preferably) larger rivers. Rare in artificial water bodies. Larvae inhabit submerged vegetation as well as gravel, stone, and coarse sediments (Bauernfeind & Soldan 2012). Relatively rare in Bulgaria, despite the lack of data on its actual distribution in the country. Transpalearctic; widely distributed in Europe (Bauernfeind & Soldan 2012).

Subgenus *Pseudocentroptilum* Bogoescu, 1947

Procloeon (*Ps.*) *pennulatum* (Eaton, 1870)

Localities: *49* (Janeva & Russev 1989).

Distribution: Larvae predominantly in rhithralic sections of smaller brooks, prefer stony habitats with very slow current and partly coarse sand. Frequently also found in potamalic sections of larger rivers (Bauernfeind et al. 1995). No actual data on its distribution in Bulgarian freshwaters. Records from almost all of Continental Europe, including Great Britain (Belfiore & Thomas 2017).

Family Isonychiidae Burks, 1953

Genus *Isonychia* Eaton, 1871

Subgenus *Isonychia* Eaton, 1871

Isonychia (I.) ignota (Walker, 1853)

Localities: *97* (Russev 1968); *98* (Janeva & Russev 1997); *35* - 21.6.1961; *66* - 05.6.1960; *69* - 04.8.1964 (Russev & Vidinova 1994, Presolska 2014).

Distribution: Larvae in larger lowlands rivers, preferably in places with submerged plants and strong current (Bauernfeind & Soldan 2012). In Bulgaria, found in several large rivers in sections with lower water velocity and/or rich in macrophytes (Russev & Vidinova 1994). Transpalearctic species; recorded in almost all of Continental Europe (Bauernfeind & Soldan 2012).

Family Oligoneuriidae Ulmer, 1914

Genus *Oligoneuriella* Ulmer, 1924

Oligoneuriella pallida (Hagen, 1855)

Localities: *97* (Russev 1968, as *O. mikulskii*); *49* - 1969 (Gardanov et al. 1971, as *O.mikulskii*); *98* (Janeva & Russev 1997); *35* - 21.6.1961; *49* - 20.6.1961, 27.5.1969, 29.7.1969; *53* -20.6.1961; *55* - 10.8.1983; *66* - 05.5.1960, 08.6.1961, 07.6.1981; *67* - 05.8.1964; *69* - 05.8.1964; *87* - 07.6.1958 (Russev & Vidinova 1994, Presolska 2014).

Distribution: Larvae in mesopotamalic sections of larger rivers at lower altitudes. Stony bottom is usually preferred, but larvae also frequent on submerged or even swimming tree trunks, branches, and roots (Russev & Janeva 1983). In Bulgaria, found at several localities in bigger Danube River tributaries, Tundzha, Kamchia, Maritsa, and Struma Rivers (Presolska 2014). Also found in Southern-Central Europe, east to the Amur River Basin, south to Bulgaria and Macedonia (Bauernfeind & Soldan 2012).

Oligoneuriella polonica Mol, 1984

Localities: *66* - 05.6.1960 (Russev & Vidinova 1994).

Distribution: Larvae seem to prefer metapotamal of large lowland rivers, pebbles and smaller stones, but also from coarse shifting-sand habitats (Russev & Vidinova 1994). Found at a few localities along Vit and Yantra Rivers in Bulgaria (Russev & Vidinova 1994, Presolska 2014). Also in Poland and Hungary (Bauernfeind & Soldan 2012).

Oligoneuriella rhenana (Imhoff, 1852)

Localities: *48* - 31.5.1950 (Russev 1957); *97* (Russev 1959, 1962, 1966, 1968); *32* (Russev & Janeva 1986); *13* - 22.6.1961; *53* - 28.5.1978; *58* - 13.9.1965; *67* - 05.8.1964; *69* - 05.8.1964; *70* - 05.8.1964 (Russev & Vidinova 1994, Presolska 2014); *98* (Janeva & Russev 1997).

Unpublished material: *26* - 03.7.2015.

Distribution: Larvae under stones in the middle of riverbeds, where flow velocity and oxygen content are always higher. Can also be found on drifting tree trunks and branches (Russev & Vidinova 1994). In Bulgaria, found in almost all major river areas, mainly in hyporhithron (the white barbel zone). Recorded throughout Europe, with the exception of Great Britain, Scandinavia, the Pyrenees, the Alps, and the Caucasus (Puthz 1978).

Family Heptageniidae Needham, 1901

Genus *Ecdyonurus* Eaton, 1868

Subgenus *Ecdyonurus* Eaton, 1868

Ecdyonurus (*E.*) *aurantiacus* (Burmeister, 1839)

Localities: *25* - 12.8.2009 (Presolska 2014).

Distribution: Potamobiont, prefers epipotamalic sections of streams and rivers below 500 m a.s.l., with stony bottom and moderate to low current velocity. Moderately sensitive to organic pollution but larvae can survive in β-mesosaprobic conditions (Bauernfeind & Soldan 2012). In Bulgaria, found in some small- and medium-sized rivers of the Bulgarian Danube River and Kamchia River watersheds, and several localities in the south (Presolska 2014). Western, South-Western, and Eastern Europe, Danube River watershed in Germany and Austria, Baltic, North-Western Russia, Balkans, and Greek islands (Belfiore & Thomas 2017).

Ecdyonurus (*E.*) *dispar* (Curtis, 1834)

Localities: *35* - July 1985 (Janeva 1991).

Distribution: Larvae in rivers with stony bottom, up to approximately 600 m a.s.l., in metarhithral to epipotamal of larger rivers (Bauernfeind & Soldan 2012). In Bulgaria, found mainly in Danube River tributaries and along the Struma River, as well as from several localities in Kamchia, Veleka, Arda, and Maritsa Rivers (Presolska 2014). In Europe, recorded throughout the continent, with the exception of Fennoscandia and Italy. South-eastern boundaries not clear (Bauernfeind & Soldan 2012).

Ecdyonurus (*E.*) *insignis* (Eaton, 1870)

Localities: *98* (Janeva & Russev 1997); *8* - 22.6.1961 (Presolska 2014).

Unpublished material: *9* - 05.7.2015; *26* - 24.7.2014; *55* - 09.8.2015.

Distribution: Hyporhithral to metapotamal of rivers with stony bottom and occasionally in the shore zone of oligotrophic lakes. Highest densities in lowlands up to 300-400 m a.s.l. (Bauernfeind & Soldan 2012). Moderately spread in Bulgaria, predominantly in lowland sections of larger and deeper rivers (Presolska 2014). Recorded in Europe, including the British Islands, but not in Scandinavia (Bauernfeind & Soldan 2012).

Genus *Electrogena* Zurwerra & Tomka, 1985

Electrogena affinis (Eaton, 1887)

Localities: *32* - 10.10.1992; *35* - 27.6.1970; *50* - 24.7.1964; *70* - 05.8.1964 (Vidinova & Russev 1997, Presolska 2014).

Distribution: Larvae preferably in metarhithral to metapotamal of larger lowland rivers below 500 m a.s.l. Usually among submerged roots or logs in rather slow current (Bauernfeind & Soldan 2012). In Bulgaria, found at several localities in potamal of rivers

(Vidinova & Russev 1997, Presolska 2014). Recorded in most of Europe, excluding the Iberian Peninsula, Apennines, West Balkans, Norway, and Finland (Belfiore & Thomas 2017).

Electrogena lateralis (Curtis, 1834)

Localities: *98* (Janeva & Russev 1997); *49* - 18.7.1965; *70* - 05.8.1964 (Presolska 2014).

Distribution: Larvae in rivulets and smaller rivers with stony bottom (Bauernfeind & Soldan 2012). In Bulgaria, from many localities, mainly in hyporhithral of rivers, where it inhabits stones, decaying trunks, and branches, especially with a muddy base (Vidinova & Russev 1997). West Palearctic species; recorded in most European countries (Belfiore & Thomas 2017).

Electrogena mazedonica (Ikonomov, 1954)

Localities: *35* - 21.6.1961 (Vidinova & Russev 1997).

Unpublished material: *12* - 04.7.2015.

Distribution: Insufficiently known habitat preference. Larvae in smaller rivers (Bauernfeind & Soldan 2012). More widely spread in rhithral but also in potamal of some Bulgarian rivers from the Black Sea and Aegean Sea Basins, more rarely in Danube River tributaries. Also found in the Republic of Macedonia, Northern Greece, and Montenegro (Bauernfeind & Soldan 2012).

Genus *Heptagenia* Walsh, 1863

Subgenus *Dacnogenia* Kluge, 1988

Heptagenia (D.) coerulans Rostock, 1878

Localities: *54* - 15.9.1955 (Russev 1957); *97* (Russev 1959, 1962, 1966, 1968); *34* - 14.10.1958; *49* - 29.7.1969; *55* - 09.8.1971; *66* - 05.6.1960; *69* - 04.8.1964; *70* - 04.8.1964; *87* - 07.6.1958 (Vidinova & Russev 1997, Presolska 2014); *98* (Janeva & Russev 1997); *53* - 3.11.2012 (Kerakova 2015).

Distribution: Strict rheobiont, in larger lowland rivers up to ca. 500 m a.s.l. Preferred microhabitats are larger stones and boulders in strong current (Bauernfeind & Soldan 2012). Typical indicator of ß-mesosaprobity (Vidinova & Russev 1997). In Bulgaria, found along main stretches of bigger Danube River tributaries and the Struma River, as well as sporadically in Mesta, Maritsa, Tundzha, and Kamchia Rivers (Presolska 2014). West Palearctic but not recorded in Fennoscandia and the British Islands (Bauernfeind & Soldan 2012).

Subgenus *Heptagenia* Walsh, 1863

Heptagenia (H.) flava Rostock, 1878

Localities: *80* - 20.8.1950, 14.9.1954 (Russev 1957); 97 (Russev 1959, 1962, 1966, 1968); *49* - October 1968 (Gardanov et al., 1971); *55* - 09.8.1971, 08.8.1971 (Russev 1977); *32* (Russev & Janeva 1986); *79* - 20.8.1950, 14.9.1953, 10.6.1956, 05.10.1958, 16.4.1961 (Russev et al. 1987); *98* (Janeva & Russev 1997); *8* - 23.10.1986; *11* - 06.7.1971; *17* - 07.7.1971; *32* - 28.6.1984; *35* - 21.6.1961, 21.6.1970; *43* - 20.6.1961; *47* - 01.6.1987;

49 - 18.7.1965, 19.4.1967, 29.7.1969, 11.10.1969; *50* - 24.7.1964; *51* - 06.6.1970; *54* - 09.10.1964, 10.7.1971; *55* - 12.8.1964, 09.8.1971, 17.4.1972; *56* - 16.9.1955, 08.8.1971, 11.10.1992; *58* - 10.6.1959, 12.5.1971; *63* - 05.10.1956; *67* - 05.8.1964; *69* - 04.8.1964; *70* - 05.8.1964; *77* - 14.9.1953; *79* - 20.8.1950, 14.9.1953, 10.6.1956, 05.10.1958; *83* - 13.6.1955; *94* - 13.6.1963; *95* - 12.6.1956, 07.7.1973 (Vidinova & Russev 1997, Presolska 2014).

Unpublished material: *53* - 16.9.2014, 08.8.2015.

Distribution: Eurybiont with comparatively high ecological plasticity. Widely spread in potamal of lowland rivers, usually on submerged logs, including larger Bulgarian rivers (Vidinova & Russev 1997). Palearctic species from Continental Europe not found in Fennoscandia, the British Islands, and the Iberian Peninsula (Bauernfeind & Soldan 2012).

Heptagenia (*H.*) *longicauda* (Stephens, 1836)

Localities: *10* - 23.10.1986; *53* - 17.4.1972; *79* - 14.9.1953, 16.4.1961 (Vidinova & Russev 1997, Presolska 2014).

Unpublished material: *53* - 08.8.2015; *74* - 13.11.2011(leg. T. Trichkova)

Distribution: Larvae in potamal of larger rivers in lowlands, preferably on submerged logs and plants, and exceptionally on stones (Bauernfeind & Soldan 2012). In Bulgaria, there have been many findings in the main course of the Struma and Mesta Rivers and also several localities in bigger Bulgarian rivers (Presolska 2014). Recorded in continental Europe (including the British Islands), but not on the West and South Balkans, in North Fennoscandia, and east of Poland and Romania (Belfiore & Thomas 2017).

Heptagenia (*H.*) *sulphurea* (Müller, 1776)

Localities: *97* (Russev 1959, 1962, 1966, 1968); *32* (Russev & Janeva 1986); *77* - 14.9.1953, *79*- 14.9.1953, 10.6.1956, 16.4.1961(Russev et al. 1987); *98* (Janeva & Russev 1997); *28* - 12.6.1958, 12.10.1959, 18.4.1967; *29* - 19.9.1965; *34* - 10.10.1964; *35* - 21.6.1970; *41* - 09.7.1968; *42* - 25.10.1959; *46* - 15.10.1964; *51* - 06.6.1970; *54* - 10.7.1971; *58* - 15.10.1970; *66* - 12.6.1958, 05.6.1960; *88* - 12.6.1963; *91* - 24.10.1959; *96* - 07.7.1973; (Vidinova & Russev 1997, Presolska 2014); *53* – 10.11.2011 (Kerakova 2015).

Distribution: Larvae in metarhithral to potamal of lowland rivers, often on submerged logs, but also on coarse gravel or boulders, rarely on macrophytes (Bauernfeind & Soldan 2012). In Bulgaria, it is found predominantly in the Danube River and at the mouths of its tributaries, as well as at sporadic localities in the Struma and Tundzha Rivers (Presolska 2014). Transpalearctic; with records from almost all of Europe (Bauernfeind & Soldan 2012).

Subgenus *Kageronia* Matsumura, 1931

Heptagenia (*K.*) *fuscogrisea* (Retzius, 1783)

Localities: *28* - 18.4.1967 (Vidinova & Russev 1997, Presolska 2014).

Distribution: Mostly in lowland lakes and ponds, occasionally in slow-flowing rivers up to ca. 600 m a.s.l. exclusively on macrophytic vegetation, preferably sedges

(Bauernfeind & Soldan 2012). Very rarely found in Bulgaria and known only from a few localities (Vidinova & Russev 1997, Vidinova 2015, Presolska 2014). Palearctic species, recorded in Northern, Eastern, and Central Europe (Belfiore & Thomas 2017).

Family Leptophlebiidae Banks, 1900

Genus *Choroterpes* Eaton, 1881

Subgenus *Choroterpes* Eaton, 1881

Choroterpes (*Ch.*) *picteti* (Eaton, 1871)

Localities: *97* (Russev 1959, 1962, 1966, 1968); *98* (Janeva & Russev 1997); *13* - 14.7.1965; *54* - 15.9.1955; *63* - 23.5.1957; *69* - 04.8.1964; *70* - 05.8.1964 (Vidinova & Russev 2009).

Distribution: Larvae in different types of running waters with stony or silty bottom, mainly from epi- to mesopotamal with medium current velocity (Bauernfeind & Soldan 2012). In Bulgaria, it is found predominantly in the Danube River and Black Sea tributaries and several locations within the Arda River watershed (Presolska 2014). It is recorded in Continental Europe, with the south-eastern border insufficiently known (Bauernfeind & Soldan 2012).

Paraleptophlebia submarginata (Stephens, 1836)

Localities: *79* - 05.10.1958 (Russev et al. 1987); *7* - 01.4.1987; *8* - 27.4.1986, *10* - 27.4.1986; 23.10.1986 (Vidinova & Russev 2009, Presolska 2014).

Distribution: Larvae in streams and rivers from hypocrenal to epipotamal. Preferred microhabitats are leaf litter and submerged roots, but equally found on woody debris, among macrophytic vegetation, or on silty bottom (Bauernfeind & Soldan 2012). Widespread in Bulgaria (Presolska 2014). West Palearctic (Bauernfeind & Soldan 2012).

Family Ephemeridae Latreille, 1810

Genus *Ephemera* Linnaeus, 1758

Subgenus *Ephemera* Linnaeus, 1758

Ephemera (*E.*) *danica* Müller, 1764

Localities: *35* - 21.6.1961 (Vidinova & Russev 1997).

Distribution: Larvae preferably on sandy substratum in rhithral of streams and rivers, as well as shore zones of oligotrophic lakes (Bauernfeind & Soldan 2012). One of the most common mayflies in Bulgaria (Presolska 2014). Western Palearctic, not recorded east of the Ural Mts. (Bauernfeind & Soldan 2012).

Ephemera (*E.*) *lineata* Eaton, 1870

Localities: *8* - 27.4.1986, 23.10.1986; *10* - 23.10.1986, 09.10.1992 (Vidinova & Russev 1997, Presolska 2014).

Unpublished material: *9* - 05.7.2015.

Distribution: Larvae usually in epipotamal of larger colline rivers (Bauernfeind & Soldan 2012). In Bulgaria, it is also found in hyporhithral and is relatively rare (Vidinova

& Russev 1997, Presolska 2014). West Palearctic species with no records from Italy and the Mediterranean islands (Bauernfeind & Soldan 2012).

Ephemera (E.) vulgata Linnaeus, 1758

Localities: *3* - 30.4.2000; *5* - 28.4.1986 (Vidinova & Russev 1997, Presolska 2014).

Unpublished material: *10* - 21.7.2014.

Distribution: Larvae mostly in ponds and lakes, less frequently in potamal of slow-flowing rivers. Found in sandy substrates, but more frequently on muddy bottom mixed with sand (pellal, argyllal) (Vidinova & Russev 1997, Bauernfeind & Soldan 2012). Rarely recorded in Bulgaria. Palearctic; with records from most of Europe (Bauernfeind & Soldan 2012).

Family Palingeniidae Albarda, 1888

Genus *Palingenia* Burmeister, 1839

Palingenia longicauda (Olivier, 1791)

Localities: *97* (Russev 1959, 1962, 1966, 1968); *11* - 12.10.1958; *27* - 09.10.1956, 14.7.1957, 14.9.1957; 15.4.1958, 10.10.1958, 13.10.1964; *34* - 13.4.1964; *38* - 14.4.1964, 15.10.1964; *42* - 07.10.1956, 15.10.1958, 09.10.1959; *45* - 26.6.1960; *46* - 06.4.1961, 06.11.1961; *59* -13.7.1964;*61* - 17.4.1964; *64* - 04.4.1961, 14.7.1964, 19.10.1964; *66* - 19.10.1958; *72* - 19.10.1958; *87* - 22.10.1958, 03.6.1959; *94* - 27.9.1956, 02.6.1959; 14.6.1960, 02.9.1960, 19.4.1961, 23.10.1964 (Russev 1966, 1987); *23*, *89*, *92* (Russev & Kaneva-Abadjieva 1973, Russev 1987).

Distribution: In metapotamal of large lowland rivers with well-oxygenated and fast-flowing waters. Larvae require specially defined type of substratum (argyllal), with high amount of clay and fine silt. Highly sensitive to changes of abiotic factors and disappear rapidly from rivers with regulated banks or sections with organic pollution (Bauernfeind & Soldan 2012). The species was found at several locations in the Danube River in Bulgaria until the 1970s (Russev 1987). In Europe, there are historic records of the species from many larger rivers in Central and South-Eastern Europe. At present, it is found only in the Danube and Tisza Rivers refuge in Hungary, Slovakia, and Ukraine (Bauernfeind & Soldan 2012).

Family Polymitarcyidae Banks, 1900

Genus *Ephoron* Williamson, 1802

Ephoron virgo (Olivier, 1791)

Localities: *17* - 28.9.1954 (Russev 1957, as *Polymitarcys virgo)*; *97* (Russev 1957, 1959, 1962, 1966, 1968 – as *Polymitarcys virgo*); *2* - 29.6.1960; *11* - 16.6.1959; *15* - 29.6.1960; *16* - 07.7.1971; *27* -13.6.1959, 29.6.1960; *28* -14.9.1957, 12.10.1959; *30* - 27.6.1960; *35* - 21.6.1970; *39* - 27.6.1960; *40* - 27.6.1960; *45* - 26.6.1960; *50* - 24.8.1964; *51* - 25.6.1960; *52* - 24.6.1960; *57* - 10.7.1968; *58* - 12.9.1957; *59* - 13.7.1964; *63* - 10.6.1959; *65* - 13.4.1955, 12.9.1955; *66* - 12.9.1955, 09.6.1958; *75* - 27.10.1959,

31.8.1960; *87* - 21.9.1955, 07.6.1958, 03.6.1959, 21.6.1960; *94* - 02.6.1959, 14.6.1960 (Vidinova & Russev 1997); *98* (Janeva & Russev 1997).

Distribution: Larvae in potamal of larger lowland or medium-sized to large rivers, under boulders and large stones or burrow in cobble or sand substrata (Bauernfeind & Soldan 2012). Typical for clay ground mixed with sand. Also in the *Corophium* mud formed over the gravel of the Danube River (Vidinova & Russev 1997). In the past, considered to be the most widely-spread species along and across the Bulgarian sector of the Danube River; from several locations in the Black Sea tributaries (Presolska 2014). Last recorded in south-eastern Bulgaria in Popovska River near Dobrich Village and Sredetska River in Debelt Village (June 1987) (Vidinova 2015). Palearctic species; not found on the British Islands, Mediterranean islands, and in Fennoscandia (Bauernfeind & Soldan 2012).

Family Potamanthidae Albarda, 1888

Genus *Potamanthus* Pictet, 1843

Potamanthus luteus (Linnaeus, 1767)

Localities: *97* (Russev 1959, 1962, 1966, 1968); *8* - 27.4.1986, 23.10.1986; *10* - 27.4.1986, 11.6.1986, 13.5.1987; *13* - 22.10.1986; *27* - 07.4.1961, 08.6.1963; *53* - 20.6.1961, 04.7.1967, 28.5.1978; *56* - 29.5.1962; *60* - 13.7.1964; *63* - 03.6.1960, 03.6.1963, 10.6.1963; *66* - 12.6.1958; *68* - 29.5.1962; *70* - 17.5.1966, 20.4.1967; *88* - 12.6.1963 (Vidinova & Russev 1997, Presolska 2014); *98* (Janeva & Russev 1997).

Unpublished material: *26* - 03.7.2015.

Distribution: Larvae prefer epipotamal but also in meta- and hypopotamal with bigger stones over a muddy base. Found at many localities in main courses of major Bulgarian rivers. Palearctic species; with records from most European countries, including the British Islands, Spain, Northern Italy, and European Turkey (Vidinova & Russev 1997, Bauernfeind & Soldan 2012, Presolska 2014).

Family Ephemerellidae Klapálek, 1909

Genus *Ephemerella* Walsh, 1863

Ephemerella ignita (Poda, 1761)

Localities: *80* - 20.8.1950 (Russev 1957); *97* (Russev 1959, 1962, 1966, 1968); *32* - 21.6.1961, 28.6.1984 (Russev & Janeva 1986); *79* - 20.8.1950 (Russev et al. 1987); *10* - 20.10.1965, 01.6.1987, 09.10.1992; *13* - 22.6.1961, 17.8.1978, 24.5.1998; *33* - 11.6.1986; *35* - 21.6.1961 (Vidinova & Russev 1997, Presolska 2014).

Unpublished material: *7* - 05.7.2015; *9* - 05.7.2015; *10* - 21.7.2014; *12* - 04.7.2015; *19* - 03.7.2015; *26* - 03.7.2015; *31* - 01.8.2015; *53* - 08.8.2015.

Distribution: Predominantly in rhithral of medium-sized to large rivers, especially in meta- and hyporhithral, with very wide ecological range; larvae in almost all types of running waters (Bauernfeind & Soldan 2012). The most common mayfly in Bulgaria

(Presolska 2014). Transpalearctic species; in Europe with records from all over the continent (Belfiore & Thomas 2017).

Ephemerella mesoleuca (Brauer, 1857)

Localities: *49* -20.6.1961, 27.5.1969 (Gardanov et al. 1971, Presolska 2014).

Distribution: Rare, mostly restricted to metapotamal of large rivers, but larvae also found in small colline rivers (Balkans and Iberian Peninsula). Prefers submerged aquatic vegetation but can be found in similar habitats with particulate organic matter, gravel, and sandy substrata. In Bulgaria, there have been sporadic findings in several bigger rivers (Presolska 2014). Mediterranean species, known from the Iberian Peninsula and France, and Central Europe, Poland, Russia, and the Balkans (Bauernfeind & Soldan 2012).

Ephemerella notata Eaton, 1887

Localities: *49* -1969 (Gardanov et al. 1971);

Distribution: In meta- and hyporhithral but also in epi- and metapotamal (Vidinova & Russev 1997). Larvae usually prefer gravel bottom and places with relatively high current velocity, but not directly in the streamline (Bauernfeind & Soldan 2012). In Bulgaria, the species has been found in the north-west tributaries of the Danube River, but more commonly in big rivers flowing into the Aegean Sea – Struma, Mesta, Maritsa, and Tundzha Rivers (Presolska 2014). Recorded from England to Ukraine and Russia, north to Denmark and the Baltic, and south to France and Central and Eastern Balkans. Not found in Scandinavia, the Iberian Peninsula, the Apennines, and the West Balkans (Bauernfeind & Soldan 2012).

Family Caenidae Newman, 1853

Genus *Brachycercus* Curtis, 1834

Brachycercus harrisellus Curtis, 1834

Localities: *97* (Russev 1959, 1962, 1966, 1968); *34* - 14.10.1958 (Vidinova 2015).

Distribution: Larvae in a large spectrum of habitats, from large lowland rivers to streams in colline areas, including canals and fishpond outflows. Prefer silty or muddy microhabitats with slow current. Widely distributed all over Europe, but rare in the Mediterranean (Bauernfeind & Soldan 2012). Few historical records in Bulgaria from the Danube, Osam, and Maritsa Rivers (Presolska 2014).

Genus *Cercobrachys* Soldan, 1986

Cercobrachys minutus (Tshernova, 1952)

Localities: *97* (Russev 1968, as *Brachycercus minutus*); *59* - 13.7.1964; *64* - 19.10.1958; (Presolska 2014).

Distribution: Larvae live mostly in large lowland rivers and mostly collected from shifting-sand habitats, as well as from submerged riparian vegetation or from gravel and stony habitats, usually in small number. In Bulgaria, it is found only in the Danube and Tundzha Rivers (Presolska 2014). Transpalearctic species that is probably restricted to Central and Eastern Europe (Bauernfeind & Soldan 2012).

Genus *Caenis* Stephens, 1836

Caenis horaria (Linnaeus, 1758)

Localities: *93* - 17.8.1943 (Russev 1960); *97* (Russev 1966, 1968); *62* (Russev & Kaneva-Abadjieva 1973).

Distribution: Larvae typically in still water, predominantly lakes, and potamal of rivers. Lentic sections with stony bottom with a layer of detritus are usually preferred. Palearctic species; widely distributed all over Europe but no records from the Mediterranean islands and Greece (Bauernfeind & Soldan 2012). Comparatively rare in Bulgaria, more common in the southern part of the country, with recent findings exclusively from reservoirs, including in the Danube River Basin (Presolska 2014).

Caenis luctuosa (Burmeister, 1839)

Localities: *80* - 14.9.1953 (Russev 1957, as *Caenis moesta*); *79* - 14.9.1953 (Russev et al. 1987); *8* - 23.10.1986, *10* - 27.4.1986; 09.10.1992; *26* - 09.10.1992; *24* - 02.6.1973; *35* -10.10.1992 (Presolska 2014).

Distribution: Larvae typically in still waters, predominantly lakes and hyporhithral and epipotamal of rivers, tolerant to eutrophication, and prefer stony bottom with a layer of detritus (Bauernfeind & Soldan 2012). In Bulgaria, the species is found in some Danube River tributaries, the main course of the Struma River, and sporadic localities near the Arda and Veleka Rivers (Presolska 2014). Recorded in almost all of Europe except for the West and South Balkans (Belfiore & Thomas 2017).

Caenis macrura Stephens, 1835

Localities: *97* (Russev 1968); *60* (Russev & Kaneva-Abadjieva 1973); *73* - 20.7.1062 (Russev et al. 1987); *98* (Janeva & Russev 1997); *8* - 22.6.1961, 27.4.1986, 23.10.1986; *10* - 27.4.1986, 09.10.1992; *13* - 22.10.1986, 21.7.2009; *19* - 11.6.1986, 27.6.1986, 22.10.1986; *26* - 21.6.1961, 28.6.1974, 25.6.1978, 09.10.1992, 21.8.2009; *30* - 05.6.1960; *32* - 21.6.1961, 28.6.1984, 22.10.1986; *33* - 16.8.1978, 11.6.1986, 21.7.2009; *35* - 21.6.1961, 04.7.1983, 10.10.1992; *49* - 17.8.2004; *50* - 10.10.1964, 24.10.1964; *52* - 10.7.1968; *53* - 04.7.1967; *55* - 17.4.1972; *63* - 03.6.1962; *66* - 12.6.1958; *70* - 17.5.1966, 28.7.1967 (Presolska 2014); *53* – 31.7.2011, 10.11.2011, 08.2012, 03.11.2012 (Kerakova 2015).

Unpublished material: *12* - 04.7.2015; *13* - 22.7.2014; *81* - 12.8.2015.

Distribution: Rheophilic, in hyporhithral and epipotamal of rivers, especially with gravel and stony bottoms, as well as shore zones of lakes (Bauernfeind & Soldan 2012). The most common mayfly in Bulgaria (Presolska 2014). Palaearctic; widely distributed all over Europe (Belfiore & Thomas 2017).

Caenis pseudorivulorum Keffermüller, 1960

Localities: *53* - 22.9.2009 (Presolska 2014); 31.7.2011, 08.2012, 03.11.2012 (Kerakova 2015).

Unpublished material: *32* - 24.7.2014, *34* - 18.8.2015; *53* - 16.9.2014, 08.8.2015; *55* - 09.8.2015; *56* - 17.9.2014; *81* - 12.8.2015.

Distribution: Larvae in potamal of large lowland rivers below 300 m a.s.l.; sometimes syntopic with C. *macrura* (Bauernfeind & Soldan 2012). In Bulgaria, from larger rivers in the Aegean Basin, less common in tributaries (Ogosta and Iskar Rivers) of the Danube River (Presolska 2014). Palaearctic species; widely distributed all over Europe (Belfiore & Thomas 2017).

Caenis rivulorum Eaton, 1884

Localities: *53* - 14.4.1974 (Janeva 1987).

Distribution: Larvae are typically found in rhithral and epipotamal of brooks and rivers, as well as in still waters and lakes between 200-500 m a.s.l.; stony bottom and layer of detritus are usually preferred (Bauernfeind & Soldan 2012). Rare in Bulgaria, with records only up to 1990 m a.s.l. (Presolska 2014). Palaearctic species; widely distributed all over Europe but no records exist from Italy, the Mediterranean islands, and Greece (Bauernfeind & Soldan 2012).

Caenis robusta Eaton, 1884

Localities: *97* (Russev 1959, 1962, 1966, 1968, Russev & Kaneva-Abadjieva 1973); *62* (Russev & Kaneva-Abadjieva 1973); *73* - 20.7.1962 (Russev et al. 1987) *98* (Janeva & Russev, 1997); *86* - 15.8.2009, 08.5.2010; *62* - 19.11.2009 (Vidinova et al. 2011); *27* - 18.9.1965; *63* - 03.6.1960; *83* - 13.6.1955; *95* - 12.6.1956; *96* - 12.6.1956 (Presolska 2014); *93* - 22.4.2012, 23.4.2012, 07.8.2012 (Kerakova 2015); *20* - 12.11.2010, 15.5.2011; *21* - 25.9.2010, 16.5.2011, 04.8.2011; *22* - 06.6.2011; *37* - 01.6.2012, 24.7.2012; *44* - 01.12.2012; *90* - 07.6.2011, 16.7.2011, 09.9.2011, 14.5.2012, 07.8.2012, 19.11.2012; *93* - 17.5.2012, 04.9.2010, 20.11.2010 (Kazakov 2017).

Unpublished material: *6* - 21.7.2014; *93* - 22.7.2014, 25.7.2015.

Figure 2. *Taeniopteryx nebulosa*, imago, collection of the Institute of Biodiversity and Ecosystem Research, Bulgarian Academy of Sciences.

Distribution: Larvae in various types of lakes, potamal of rivers, and frequently artificial channels, tolerating β-mesosaprobic conditions. In Bulgaria, the species is found in potamal of main river courses, also in marshes, and lakes (Presolska 2014). Palaearctic species; widely distributed all over Europe (Bauernfeind & Soldan 2012).

Order Plecoptera

Suborder ARCTOPERLARIA

Family Taeniopterygidae Klapalek, 1905

Genus *Taeniopteryx* Pictet, 1841

Taeniopteryx nebulosa (Linnaeus, 1758) - February red stonefly (Figure 2)

Localities: *63* -16.3.1955 (Russev 1962, 1966, 1979, Tyufekchieva et al. 2013).

Distribution: Common Holarctic rheo - to limnophil. Usual in slow-flowing streams and lentic zones, rarely in submontane or montane zones. Prefers pelal, psammal, macrophytes, and fine particulate organic matter. Extinct in the Danube River.

Genus *Oemopteryx* Klapalek, 1902

Oemopteryx loewii (Albarda, 1889)

Localities: *87* -18.3.1955 (Russev 1962, 1966, 1979, Tyufekchieva et al. 2013).

Distribution: Potamal winter stonefly, typical of large European rivers. Extinct in the Danube River and most likely extinct in the whole species distribution area.

Family Perlidae Latreille, 1802

Genus *Marthamea* Klapálek, 1907

Marthamea vitripennis (Burmeister, 1839) - Figure 3.

Localities: *65* -1958 (Russev 1959, 1962, 1966, 1979).

Figure 3. *Marthamea vitripennis*, larvae, collection of the Institute of Biodiversity and Ecosystem Research, Bulgarian Academy of Sciences.

Distribution: European species, found in lithorheophilic coenosis of streams. Prefers zones with moderate to high current. Extinct in the Danube River.

Order Trichoptera Kirby, 1813

Suborder ANNULIPALPIA Martynov, 1924

Superfamily HYDROPSYCHOIDEA Curtis, 1835

Family Psychomyiidae Walker, 1852

Genus *Psychomyia* Latreille, 1829

Psychomyia pusilla (Fabricius, 1781)

Localities: *25* - 09.10.1992 (Russev et al. 1994).

Unpublished material: *54* - 09.8.2015, 1 La.

Distribution: Typical of stony sections of potamal of all Bulgarian rivers except for the Danube River. Also common in (hypo) rhithral of lowland rivers, sometimes in crenal. Typical in plains, sometimes reaching 1000-1100 m a.s.l., with broad ecological spectrum, including more polluted (β-mesosaprobic) waters. West Palearctic species; always very abundant and common in Europe (Kumanski 1988).

Family Ecnomidae Ulmer, 1903

Genus *Ecnomus* Rambur, 1842

Ecnomus tenellus Rambur, 1842

Localities: *97* (Russev 1979). Typical along the Danube River (Kumanski 1975a), in nearby swamps and lakes, e.g., Malak Preslavets and near Nikopol Town; GP24/34, LJ83, MJ15, MJ77/87/88 without specific location (Kumanski 1985); *6* - 09.10.1992 (Russev et al. 1994); *93* (Varadinova et al. 2012).

Distribution: Limno- and potamobiont, typical of lowland standing and slow-flowing waters, eurythermic. When caught, almost always abundant. Palearctic species (excluding the northernmost regions), reported in the Indo-Malayan zoogeographic region, including Pakistan, India, and Ceylon (Kumanski 1985).

Family Polycentropodidae Ulmer, 1903

Genus *Neureclipsis* McLachlan, 1864

Neureclipsis bimaculata (Linnaeus, 1758)

Localities: *97* (Kumanski 1975a, Russev 1979, 1998). Common in the Danube River but without specific locations. Likely records LJ13/14, LJ23/24, LJ82, MJ77/87/88 (Kumanski 1985).

Distribution: Typical potamobiont, common in big lowland rivers and outflows of numerous lakes in temperate and nordic regions of the Holarctic. In Bulgaria, it is common along the Danube River and rarer inside the country. Holarctic species (Kumanski 1985).

Genus *Cyrnus* Stephens, 1836

Cyrnus trimaculatus (Curtis, 1834)

Localities: *93* (Varadinova et al. 2012).

Distribution: Larvae in stagnant or slow-flowing waters, with wide ecological spectrum. Limnobiont but sometimes also found in rhithral. Widely distributed in Bulgaria from 0 to 1400 m a.s.l. West Palearctic species (Kumanski 1985).

Family Hydropsychidae Curtis, 1835

Genus *Hydropsyche* Pictet, 1834

Hydropsyche contubernalis McLachlan 1865

Localities: Danube River, near Silistra (likely *96*). More numerous are the non-Danubian localities belonging to the hyporhytron, epipotamon, and metapotamon; sometimes from mountain streams (metarhitron) (Kumanski & Botosaneanu 1974); 97 (Russev 1979, Kumanski 1985).

Distribution: Very wide distribution and ecological spectrum. Facultative potamobiont, found in bigger lowland rivers and smaller streams in premontane regions. Tolerates pollution up to α-mesosaprobic. Common in plains at 600 – 700 m and as an exception up to 1600 m a.s.l. Recorded in Europe (excluding westernmost parts), Asia Minor, and North Afghanistan (Kumanski 1985).

Hydropsyche modesta Navas, 1925

Localities: *32* - 10.10.1982 (Russev & Janeva 1986, Russev et al. 1994); *26* - 09.10.1992; *35* - 10.10.1992; *78* - 12.10.92 (Russev et al. 1994).

Distribution: One of the most common caddis in Bulgaria. Typical potamobiont with wide ecological spectrum, from bigger and calmer rivers to very small lowland brooklets. Larvae can tolerate significant pollution (α-mesosaprobic conditions). Found in all bigger Bulgarian rivers (except for the Danube River). Lowland species, usually up to 500 – 700 m a.s.l. Recorded in Southern Europe, Austria, Romania, and Asia Minor (Kumanski 1985).

Hydropsyche bulgaromanorum Malicky, 1977

Localities: as *guttata* Pict. in Klapálek 1894: Danube River (cited in Kumanski & Botosaneanu 1974);

Along the riverbank from Iron Gate to the delta (Kumanski & Botosaneanu 1974); *97* (Russev 1979, Russev 1998); Danube River, without specific location, likely records: FP45/55, GP24/34, KJ94, LJ04, LJ24/34, LJ73/83, NJ18/28 (Kumanski 1985); *17* (Russev & Uzunov 1990); *40* - 03-25.3.1988 (Russev & Uzunov 1991); *8* - 23.10.1986; *10* - 27.4, and 23.10.1986; *13* - 22.10.1986 (Russev et al. 1994).

Distribution: Typical potamobiont, very abundant in metapotamal of big rivers but frequent also in epipotamal of smaller lowland rivers. Typical lowland species, but found (rarely) in areas up to 900 m a.s.l. (mountain foothills). Especially abundant in big rivers and in the Danube River, where the species can be present in huge numbers. Tolerant of lower concentrations of dissolved oxygen and polluted waters, reaching β-mesosaprobic conditions (Kumanski 1985). In Bulgaria, it is especially common and abundant in rivers in the north (hypopotamal of the Danube River and its tributaries, Kamchiya River) and to a lesser degree in some of the southern Bulgarian rivers (e.g., Maritsa, Tundzha). Not established in Struma nor in Mesta Rivers (Kumanski 2007). Recorded in Central and South-Eastern Europe, from England to the west and till Finland to the north. Understudied areal as close to *ornatula* and *guttata* and likely confused with them before being described

(Kumanski 1985). One of the most eurytopic and most widely-distributed taxa of Hydropsychidae in most of Europe and Anatolia (Kumanski 2007).

Hydropsyche bulbifera McLachlan, 1878

Localities: *32* - 10.10.1982, *33* - 11.6.1986 and 22.10.1986 (Russev & Janeva 1986, Russev et al. 1994); *10* - 07.4.1986 (as cf.?) and 09.10.1992; *13* - 22.10.1986 and 09.10.1992; *19* - 22.10.1986 (as cf.?); *5*, *26* - 09.10.1992; *35* - 10.10.1992; *78* - 12.10.1992 (Russev et al. 1994).

Distribution: One of the most commonly and evenly-distributed species of the genus. Prefers smaller lowland rivers (potamal), but also found in bigger rivers (excluding the Danube River), and hyporhithral in premontane regions from 0 to 900 m a.s.l., with broad ecological spectrum (present in waters with β-mesosaprobic conditions). West Palearctic species, also recorded in Iran and Iraq (Kumanski 1985).

Hydropsyche incognita Pitsch, 1993

Localities: as *pellucidula* (Curtis, 1834): *53* (Janeva 1987); *8* - 27.04 and 23.10.1986, *10* - 27.04 and 23.10.1986 and 09.10.1992 (all as cf.?, Russev et al. 1994).

Distribution: The most common species of the genus in Bulgaria (Kumanski 2007). Likely having the widest ecological spectrum of the family and maybe the whole order. Facultative potamobiont, in all big and smaller lowland rivers, excluding the Bulgarian section of the Danube River (as *pellucidula*, Kumanski 1985). Larvae with preferences for moderate- to fast-flowing waters and less tolerant of lower oxygen concentrations and organic pollution unlike other species of the genus; insufficiently known areal (Pîrvu et al. 2015). Species recorded in most of Europe, excluding Great Britain, Ireland, and its northern parts, reaching the Carpathians and Bulgaria to the south-east (Kumanski 2007, Graf et al. 2008).

Hydropsyche angustipennis (Curtis, 1834)

Localities: Without a specific location, MJ13/14 (Kumanski 1985).

Distribution: Widely distributed and typical of small lowland streams and rivers. Rarely found in areas up to 1100 m a.s.l.. Evenly distributed throughout the country, often abundant. Recorded in Europe (Kumanski 1985).

Hydropsyche spp.

Localities: The genus has been recorded from numerous locations in the study area: *77*- 14.9.1953 and 16.4.1961; *79* - 14.9.1953, 26.09 and 06.7.1954 (Russev et al. 1987); *55* - 08-09.8.1971, 17.04 and 30.8.1972 (Russev 1977); spring and summer before 1969 (Janeva 1991); *guttata* group: *33* - 28.6.1984 (Russev & Janeva 1986, Russev et al. 1994); *instabilis* group: *35* - July 1985 (Janeva 1991); 97 (Russev 1998).

Unpublished material: *10* - 21.7.2014, 17 La and 05.7.2015, 40 La; *13* - 22.7.2014 and 04.7.2015, 31 La; *34* - 02.9.2014, 33 La; *81* - 18.9.2014, 63 La and 12.8.2015, 46 La; *7* - 05.7.2015, 86 La; *9* - 05.7.2015, 40 La; *11* - 18.8.2015, 9 La; *31* and *32* - 01.8.2015, 29 La; *36* - 02.8.2015, 93 La; *50* - 19.9.2015, 2 La; *55* - 09.8.2015, 10 La.

Genus *Cheumatopsyche* Wallengren, 1834

Cheumatopsyche lepida (Pictet 1834)

Localities: *8* - 23.10.1986; *10* - 27.04 and 23.10.1986; *13* - 22.10.1986 (Russev et al. 1994).

Distribution: Epipotamobiont, prefers stony-sandy substrata and moderate current speed. In Bulgaria, very widely distributed in all big (excluding the Danube River) and smaller lowland and premontane rivers up to ca. 1000 m a.s.l.. West Palearctic species; recorded in most of Europe, Asia Minor, the Caucasus, Iran, and Turkmenistan (Kumanski 1985).

Suborder "SPICIPALPIA"

Superfamily RHYACOPHILOIDEA Stephens, 1836

Family Rhyacophilidae Stephens, 1836

Genus *Rhyacophila* Pictet, 1834

Rhyacophila fasciata Hagen, 1859

Localities: *8* - 27.4.1986; *10* - 27.04 and 23.10.1986; *19* - 11.6.1986 (Russev et al. 1994).

Unpublished material: *9* - 05.7.2015, 7 La.

Distribution: Relatively eurythermic rhithrobiont, found from the Black Sea coast up to 1300 m a.s.l. in all Bulgarian mountain areas but never in large numbers. Recorded in Europe, Asia Minor, and Lebanon (Kumanski 1985).

Superfamily HYDROPTILOIDEA Stephens, 1836

Family Glossosomatidae Wallengren, 1891

Genus *Glossosoma* Curtis, 1834

Glossosoma sp.

Unpublished material: *9*, *10* - 05.7.2015, 1 La each.

Family Hydroptilidae Stephens, 1836

Genus *Agraylea* Curtis, 1834

Agraylea sexmaculata Curtis, 1834

Localities: *51* - 30.7.1969 (Kumanski 1979).

Distribution: Common in standing or slow-flowing waters rich with vegetation. Eurytopic limnophilous species, usually in plains but sometimes up to 1500 m a.s.l. (Kumanski 1985).

Agraylea multipunctata Curtis, 1834

Localities: 97 (Russev 1979); *6* - 09.10.1992 (Russev et al. 1994).

Distribution: Potamo- and limnobiont, common in waters rich with vegetation (Kumanski 1985). Its actual distribution in the country has not been evaluated, but likely to be rare.

Genus *Oxyethira* Eaton, 1873

Oxyethira sp.

montane and premontane regions, and forest streams in the plains. Can be found at elevations between 100 - 2500 m a.s.l. South-West Palearctic species (Kumanski 1988).

Family Goeridae Ulmer, 1903

Genus *Goera* Stephens, 1829

Goera pilosa (Fabricius, 1775)

Localities: Danubian Plain, without specific location (Kumanski 1975b); likely in FP74/85 and LJ13 (Kumanski 1988); *35* - July 1985 (Janeva 1991).

Unpublished material: *31* - 24.7.2014.

Distribution: Very wide ecological spectrum; larvae in calm sections of rhithral, potamal in bigger lowland rivers, and even limnal. Very common in Bulgaria, especially in slower mountain streams at lower altitudes and in epipotamal of almost all bigger rivers. One of the few caddis that tolerates β-mesosaprobic conditions. Found at elevations between 0 and 1000 m a.s.l. Recorded throughout all of Europe (excluding Iceland), Asia Minor, the Caucasus, and Ural (Kumanski 1988).

Superfamily SERICOSTOMATOIDEA Stephens, 1836

Family Sericostomatidae Stephens, 1836

?Sericostoma sp./*Oecismus monedula*?

Localities: *19* - 11.6.1986 (Russev et al. 1994).

Superfamily LEPTOCEROIDEA Leach, 1815

Family Leptoceridae Stephens, 1836

Genus *Athripsodes* Billberg, 1820

Athripsodes cinereus (Curtis, 1834)

Localities: *26* - originally *as Leptocerus cinereus* (Curt.) in Botosaneanu & Sykora (1963) as cited here and *87* - 02 July (Kumanski 1976); 97 (Russev 1979).

Distribution: Limno- and potamobiont. Very rare in Bulgaria, with only single specimens found at a few lowland localities, mostly in slow-flowing waters at 0 - 600 m a.s.l. West Palearctic species; also recorded in Central Asia (Kumanski 1988).

Athripsodes longispinosus (Martynov, 1909)

Localities: *18* - 20.6.1980 (Kumanski 1983).

Distribution: Rhithro-, potamo-, and limnobiont with wide distribution in Bulgaria. Lowland species that can be found at 0 - 400 m a.s.l. Recorded on the Eastern and South-Eastern Balkans, in Asia Minor, the Caucasus, and North Iran (Kumanski 1988).

Athripsodes sp.

Localities: *62* - August 1967; 97 (Russev 1979).

Genus *Ceraclea* Stephens, 1829

Ceraclea alboguttata (Hagen, 1860)

Localities: as *Ceraclea* (*Ceraclea*) *alboguttata*: *18* - 20.6.1980 (Kumanski 1983); *96* (Kumanski 1988).

Distribution: Potamo- and limnobiont. In Bulgaria, relatively rare, only found in rivers at 0 - 300 m a.s.l.. Recorded in Europe (excluding the Pyrenean Peninsula and Iceland) and Western Siberia (Kumanski 1988).

Ceraclea fulva (Rambur, 1842)

Localities: – as *Ceraclea (Ceraclea) fulva: 96* - 08.7.1973.

Distribution: Limno- and potamobiont. One of the rarest species of the family most likely as the southern border of its areal passes through Bulgaria. Found in plains and common inhabitant of lakes, swamps, and slow-flowing lowland rivers to the north. Recorded in Europe (excluding the Pyrenean, Apennine, and most of the Balkan Peninsulas and Iceland) and Western Siberia (Kumanski 1988).

Ceraclea senilis (Burmeister, 1839)

Localities: as *Athripsodes senilis: 96* - 08.7.1973 (Kumanski 1976, 1983); 97 (Russev 1979); as *Ceraclea (Ceraclea) senilis:* not mentioned from the Danube River (Kumanski 1988).

Distribution: as *Ceraclea (Ceraclea) senilis:* Locally distributed and rare lowland species in Bulgaria as the southern border of its areal passes through Bulgaria. Palearctic species; found in Southern Asia (Kumanski 1976), as well as in Europe, and Western Siberia (Kumanski 1983).

Ceraclea annulicornis (Stephens, 1836)

Localities: as *Athripsodes annulicornis* (Steph.): *97* (Kumanski 1976, Russev 1979); as *Ceraclea* (*Athripsodina*) *annulicornis* La: *97* (Kumanski 1983); without specific location: FP85 (Kumanski 1988).

Distribution: as *Ceraclea* (*Athripsodina*) *annulicornis*: With broad ecological spectrum, larvae found in potamal and various standing waters. Very rare in Bulgaria as the southern border of its areal passes through Bulgaria (similar to *C. fulva* and *C. senilis*). Currently based on larva from the Danube River and needs confirmation. Lowland species, found in Temperate and North Holarctic; recorded in Europe, excluding Iceland, the Pyrenean and Apennine Peninsulas, and most of the Balkans (Kumanski 1988).

Ceraclea aurea (Pictet, 1834)

Localities: as *Ceraclea* (*Athripsodina*) *aurea: 50* - 30.7.1969; *96* - 30.06-08.7.1973 (Kumanski & Malicky 1976); 97 (Russev 1979).

Distribution: as *Ceraclea* (*Athripsodina*) *aurea:* Potamobiont from plains. Very rare in Bulgaria and found at elevations between 0 and 600 m a.s.l. Species distribution is understudied. Recorded in Europe (excluding the Pyrenean and Scandinavian Peninsulas, British Isles, and Iceland) and Asia Minor (Kumanski 1988).

Ceraclea riparia (Albarda, 1874)

Localities: as *Ceraclea* (*Athripsodina*) *riparia*: *96* - 30.06 08.7.1973 (Kumanski & Malicky 1976); as *Athripsodes riparius*: 97 (Kumanski 1976, Russev 1979).

Distribution: as *Ceraclea (Athripsodina) riparia*: Typical potamobiont, found in smaller and bigger plain rivers, as well as bigger rivers. Very rare in Bulgaria and recorded in three river areas: Veleka and Fakyiska (in Strandzha Mts.) Rivers on numerous occasions, and the Danube River at Silistra. Found between 0 and 200 m a.s.l.. Recorded in Central Europe and the Balkan Peninsula (Kumanski 1988).

Ceraclea dissimilis (Stephens, 1836)

Localities: as *Ceraclea (Athripsodina) dissimilis*: *18* (Kumanski 1988).

Distribution: With broad ecological spectrum, found in streams, rivers, and lowland stagnant waters. It is common for Bulgaria in epipotamal and evenly distributed throughout the country at 0 to 600 m a.s.l.. Recorded in almost all of Europe, excluding Iceland and the southernmost parts of the Pyrenean and Balkan Peninsulas (Kumanski 1988).

Genus *Leptocerus* Leach, 1815

Leptocerus tineiformis Curtis, 1834

Localities: *18* - 20.6.1980, Europe (Kumanski 1983); *93* - (Varadinova et al. 2012).

Unpublished material: *93* - 25.7.2015, 3 La (cf., poorly preserved)

Distribution: In richly vegetated slow-flowing sections of potamal or various standing water bodies in plains. Very rare in Bulgaria and found at 0 - 300 m a.s.l. Recorded in Europe, excluding the Pyrenean Peninsula, Iceland, and Fennoscandia (Kumanski 1988).

cf. *Triaenodes bicolor* (Curtis, 1834)/*Ylodes conspersus* (Rambur, 1842)

Unpublished material: *36* - 02.8.15, 1 Pu; *93* - 25.7.2015, 4 La (cf., poorly preserved). Further material is needed to confirm identity of either species.

Distribution: Both species limnobionts and rare for Bulgaria (Kumanski 1988).

Genus *Ylodes* Milne, 1934

Ylodes kawraiskii Martynov, 1909

Localities: Nechinska bara, likely *18* (Kumanski 1988).

Distribution: Potamo- and rhithrobiont. The most common of the genus in Bulgaria. Larvae prefer slow-flowing waters near inundated riverside vegetation at 0 - about 800 m a.s.l.. Recorded in Central Europe, the Balkan Peninsula, the Caucasus, Iran, and Afghanistan (Kumanski 1988).

Ylodes simulans (Tjeder, 1929)

Localities: as *Triaenodes simulans*: *18* - 20.6.1980

Distribution: Potamo- and rhithrobiont; in Europe, excluding the westernmost and southern parts (Kumanski 1983). Similar to *Y. kawraiskii* but less common. Found at four localities in Bulgaria: a tributary of the Arda River near Krumovgrad; the Danubian Plain (Nechinska bara, right tributary of the Lom River near Smirnenski Reservoir); and Struma River (Skakavitsa Station in Zemen Gorge and Stara Kresna Station in Kresna Gorge), between 0 and about 400 m a.s.l., always with single specimens. It is likely that the southern border of its areal passes through Bulgaria. Recorded in Central and Northern Europe (excluding Iceland), reaching the Northern Balkans and Bulgaria to the south (Kumanski 1988).

Genus *Oecetis* McLachlan, 1877

Oecetis ochracea (Curtis, 1825)

Localities: Danubian Plain without specific location (Kumanski 1971); *82* (Kumanski 1976); 97 (Russev 1979); without specific location, likely in FP63, KJ94, LJ23/24, MJ67/77, NJ18 (Kumanski 1988).

Distribution: Broad ecological spectrum, in various standing and slow-flowing water bodies, coastal or plain inner bodies, including swamps, lakes, reservoirs, oxbow lakes, richly vegetated slow-flowing river sections. The most common and abundant species of the family in Bulgaria, from more than 25 localities at 0 - 600 m a.s.l.. Holarctic species (Kumanski 1988).

Oecetis furva (Rambur, 1842)

Localities: 97 (Russev 1979); along the Danube River without specific location, likely records LJ24/34 and LJ63/73 (Kumanski 1988).

Distribution: Ecologically similar to and often found with *O. ochracea* and *O. lacustris*, but much rarer than the former and more common than the latter. One of the earliest species to emerge in spring. Palearctic species; recorded in Europe but not known from the Pyrenean Peninsula and Iceland (Kumanski 1988).

Oecetis notata (Rambur, 1842)

Localities: Danube River (Botosaneanu & Sykora 1963), without specific location, likely in LJ24 (Kumanski 1988); 97 (Russev 1979).

Distribution: Facultative lowland limnobiont, prefers oxbow lakes or swamps along rivers at 0 - 200 m a.s.l.. Very rare for Bulgaria, found in swampy habitats near the rivers or very slow-flowing sections, always with single specimens. Recorded in Europe (excluding Iceland) and Siberia (Kumanski 1988).

Genus *Setodes* Rambur, 1842

Setodes punctatus (Fabricius, 1793)

Localities: 97 (Russev 1979); Danube River: likely in GP05, NJ18 (Kumanski 1988).

Distribution: Often in the Danube River, with Palearctic distribution. Oligo-β-mesosaprobic and β-mesosaprobic (Kumanski 1974). Potamobiont with preferences for plains. Very rare in Bulgaria, found in the Danube River on several occasions. Recorded in Europe (excluding Iceland and the Scandinavian Peninsula), Iran, and Siberia (until Primorsky Krai) (Kumanski 1988).

Genus *Mystacides* Berthold, 1827

Mystacides azurea (Linnaeus 1761)

Localities: Danube River on several occasions; likely in LJ83 (Kumanski 1988);

Unpublished material: *13* - 04.7.2015, 3 La (cf.).

Distribution: Wider ecological spectrum. Widely distributed in Bulgaria (about 20 localities) and usually very abundant when found, typically between 0 and 1000 m a.s.l.; Palearctic species (Kumanski 1988).

DISCUSSION

Based on the current review, a total of 104 EPT taxa have been recorded from more than 100 localities in the Lower Danube River region in Bulgaria. In this work, new localities or data are reported for 19 species of Ephemeroptera and six taxa of Trichoptera. Four mayflies (*Baetis nexus*, *Centroptilum luteolum*, *Procloeon bifidum*, and, only for the Danube River, *Caenis pseudorivulorum*), two/three genera (*Glossosoma*, *Halesus* and ?*Triaenodes*), and two species of caddisflies (*Halesus digitatus* and ?*Triaenodes bicolor*/*Ylodes conspersus*) are newly recorded for the region. We have established that EPT taxa richness is higher in the adjacent territories as compared to the main river: 55 taxa are known only from the non-Danubian localities, while only 17 are listed as typical for the main river and 49 have been recorded in both the main river and its adjacent territories.

A negative tendency of the EPT taxa richness of the Lower Danube River is apparent when comparing the studies on the benthic fauna over the last 50 years. The first review on benthic macroinvertebrates from the whole Danube River and the Bulgarian section summarized all data by the end of the 1970s (Russev 1979). It presents 51 taxa of Ephemeroptera, 50 of Plecoptera, and 77 taxa of Trichoptera from the whole river, of which 24 mayflies, three stoneflies, and 18 caddis from the Lower Danube River (Bulgarian stretch) and two mayflies and one caddis from non-Danubian localities on the territory of Bulgaria. Later, Russev (1998) established an increase of EPT taxa for the whole Danube River and a decrease in their diversity for the lower section of the river. In 2013, Graf et al. (2014) registered even less EPT taxa: 31 taxa of Ephemeroptera, one of Plecoptera, and 42 of Trichoptera from the whole Danube River. Higher number of EPT taxa has been restricted to the Upper Danube River, with Trichoptera showing the highest abundances within these sensitive groups, while taxa richness of all three orders decreases notably in the lower stretch of the river (Graf et al. 2014).

According to the current review on EPT taxa from the Bulgarian section of the Danube River and its tributary, 17 taxa have been found only in the Danube River. Most of them are rare, while, currently, seven mayflies and three stoneflies are classified as extinct, possibly extinct or endangered to a different degree. Tyufekchieva et al. (2019) assessed *Oemopteryx loewii* and *Marthamea vitripennis* as extinct and *Taeniopteryx nebulosa* as possibly extinct not only from the study area but also in Bulgaria. Mass emergence of *O. loewii* has been reported as early as 1775 from Regensburg by Schäffer (by Graf et al. 2016). Most likely this stonefly is globally extinct: the last reliable finding was reported by Russev (1962) from the Bulgarian section of the Danube River in 1955. On the one hand, this directly affects the biodiversity in the study area, while on the other, it could trigger indirect changes in the whole ecosystem. The disappearance from the river of the above-mentioned species has likely propagated along the food web of the Lower Danube River. For instance, Russev (1968) pointed out that the biomass of the mayfly *Palingenia longicauda* (currently extinct) had a significant role in fish nutrition and represented approximately 40% of the food of the Danubian bottom fish, especially for the sterlet (*Acipenser ruthenus*).

As repeatedly discussed in the papers of Zwick (1992, 2004), Fochetti & Tierno de Figueroa (2006), Kovacs & Muranyi (2008), Bojkova et al. (2011), Tyufekchieva et al. (2013), and Tyufekchieva et al. (2019), stoneflies in Europe are most affected in the plain river types, i.e., the Bulgarian section of the Danube River. Significant reduction of fauna of Plecoptera in the Danube River demonstrates the deterioration of its habitat during the last 60 years. We could speculate that the same applies for mayflies and caddisflies. The above changes in communities are likely owing to the sensitivity of most of the species from the three orders to alternations in the environments they inhabit (Eaton & Lenat 1991, Barbour et al. 1996, Wallace et al. 1996). EPT communities are greatly affected by the bottom substratum structure, interactive flow velocity, water level fluctuations, and discharge changes, as well as pollution (see Sandin & Johnson 2000, Evtimova & Donohue 2014, Jerves-Cobo et al. 2017, Savić et al. 2017).

Among the negative factors causing the reduction of mayfly, stonefly, and caddisfly diversity in bigger lowland rivers, in particular the Bulgarian section of the Danube River and its tributaries, are: pollution with waste waters, extraction of inert materials from the river bed causing degradation of habitats, destruction and dike building along the banks of the Danube River, and hydrotechnical constructions. In addition, intensive shipping along the Danube River triggers short-term fluctuations of the water level in the coastal areas, which could negatively impact most of the hydrobionts (Russev 1967). Anthropogenic factors have exerted strong negative influence even on the emblematic "Danubian mermaid" (*P. longicauda*), leading to its rapid extinction from the Lower Danube at the end of 1970s (Russev 1987). An association between benthic invertebrates and substrata has been previously demonstrated in Bulgarian rivers (Uzunov & Kovachev 2002) and for the Lower Danube River (Russev 1967). For many species, e.g., *Heptagenia* (*K.*) *fuscogrisea*, the main threat for their survival is destruction of their natural habitats caused by the cleaning of river banks overgrown with vegetation, and river regulation (Vidinova 2015).

Overall, we have found a decreasing trend of diversity of EPT taxa within the study area, which is more pronounced for the Danube River as compared to the adjacent territories, owing to the increasing pressure on the Lower Danube River. Various threats within the region could further impact distribution and richness of Ephemeroptera, Plecoptera, and Trichoptera in the adjacent areas, including deforestation, extraction of inert materials, and pollution from various sources. In order to protect and preserve these sensitive insects, it is important to protect the environments that are needed for the development of their aquatic stages.

Acknowledgments

A part of the unpublished material was collected within the following projects: 1. Mapping and Identification of the Conservation Status of Natural Habitats and Species - Phase I, Lot 2: "Mapping and Determining Conservation Status of Fishes"- C-30-43/21.03.2011; Consortium "Natura-Bulgaria"; 2. "Intercalibration of the methods for analysis of biological quality elements (BQE) for the types of surface waters on the territory of Bulgaria, corresponding to common European types in the Geographical intercalibration groups", realized/conducted by consortium DIKON-UBA.

References

Barber-James H. M., Gattolliat J. L., Sartori M. & Hubbard M. D. 2008. Global diversity of mayflies (Ephemeroptera, Insecta) in freshwater. *Hydrobiologia*, 595: 339–350.

Barbour M. T., Gerritsen J., Griffith G. E., Frydenborg R., Mccarron E., White J. S. & Bastian M. L. 1996. A Framework for Biological Criteria for Florida Streams Using Benthic Macroinvertebrates. *Journal of the North American Benthological Society* 15: 185-211.

Bauernfeind E. & Soldán T. 2012. *The Mayflies in Europe (Ephemeroptera).* Apollo Books, Ollerup, Denmark, pp. 781, ISBN 978-87-88757-45-3.

Bauernfeind E., Weichselbaumer P. & Moog O. 1995. Ephemeroptera (Eintagsfliegen). In: Moog O. (ed.) *Fauna Aquatica Austriaca Katalog für Autökologischen Einsrufung Aquatischer Organismen Österreichs*. Teil IliA. Arteninventar für Österreich. Bundes ministerium für Land und Forstwirtschaft, Wien. [Ephemeroptera (Mayflies). In: Moog O. (ed.) *Austrian Aquatic Fauna, Catalog of the Autecological Classification of Aquatic Organisms in Austria*. Part IliA. Species inventory for Austria. Federal Ministry of Agriculture and Forestry, Vienna] (In German).

Belfiore C. & Thomas A. 2017. Ephemeroptera, Mayflies. Fauna Europaea version 2017.06, https://fauna-eu.org.

Bojkova J., Soldán T., Špaček J. & Straka M. 2011. Distribution of stoneflies of the family Taeniopterygidae (Plecoptera) in the Czech Republic: earlier data, new records and recent distributional changes. *Časopsis Slezskeho Muzea Opava* [*Journal of the Silesian Museum Opava*] 60: 239–258.

Botosaneanu L. & Sykora J. 1963. Nouvelle contribution à la connaissance des trichoptères de Bulgarie [New contribution to the knowledge of caddisflies of Bulgaria]. *Acta faunistica entomologica Musei Nationalis Pragae* 9: 121-42.

Burmeister H. C. 1839. *Plecoptera. Handbuch der Entomologie*, Berlin [*Plecoptera. Handbook of Entomology*, Berlin] 2 (2): 863–881.

DeWalt R. E., Neu-Becker U. & Stueber G. 2018. Plecoptera Species File Online. Version 5.0/5.0. 04/04/2018. http://Plecoptera.SpeciesFile.org.

Eaton L. E. & Lenat D. R. 1991. Comparison of a rapid bioassessment method with North Carolina's qualitative macroinvertebrate collection method. *Journal of the North American Benthological Society* 10: 335-338.

Edmunds G. F. J. & McCafferty W. P. 1988. The Mayfly Subimago. *Annual Review of Entomology* 33: 509–529.

Evtimova V. 2013. *Water level fluctuations and their effects on lake ecology*. PhD thesis, Trinity College Dublin, University of Dublin, Dublin, Ireland, 210 p.

Evtimova V. &. Donohue I. 2014. Quantifying ecological responses to amplified water level fluctuations in standing waters: an experimental approach. *Journal of Applied Ecology* 51: 1282-1291.

Fochetti R. J. & Tierno de Figueroa M. 2006. Notes on diversity and conservation of the European fauna of Plecoptera (Insecta). *Journal of Natural History* 40 (41–43): 2361–2369.

Gardanov T., Ivanov K., Russev B., Petchinov D., Petrov P., Semov V. & Kabaktchieva E. 1971. Investigation on the self-purification processes of Iskar and Ossam Rivers and their river valleys. Papers of SIWKST, VII, 2, 17-37 (In Bulgarian with English summary).

Graf W., Csányi B., Leitner P., Paunović M., Huber T., Szekeres J., Nagy C. & Borza P. (2014). Joint Danube Survey 3. Full report on: *Macroinvertebrates. Version 1, International Commission for the Protection of the Danube River*, 46 p.

Graf W., Leitner P., Haidvogl G., Birk S., van Geest G., Buijse T. & Pletterbauer F. 2016. MARS Deliverable 5-C: Report on legacy and tipping points in large rivers, MARS project - Funded by the European Union within the 7th Framework Programme, Grant Agreement 603378.

Graf W., Murphy J., Dahl J., Zamora-Munoz C. & López-Rodríguez M. J. 2008. *Distribution and ecological preferences of European freshwater organisms.* Volume 1. Trichoptera. Pensoft Publishing, 388 p.

Haybach A. 1998. Die Eintagsfliegen (Insecta: Ephemeroptera) von Rheinland Pfalz. Zoogeographie, Faunistik, Ökologie, Taxonomie und Nomenklatur. Unter besonderer Berücksichtigung der Familie Heptageniidae und unter Einbeziehung der übrigen aus Deutschland bekannten Arten.- Dissertation Johannes-Gutenberg-Universität Mainz [The mayflies (Insecta: Ephemeroptera) of Rhineland Palatinate. Zoogeography, faunistics, ecology, taxonomy and nomenclature. With special consideration of the family Heptageniidae and including the other species known from Germany.- PhD Thesis, Johannes-Gutenberg University Mainz], 417 p.

Holzenthal R. W. 2011. Animal biodiversity: an outline of higher-level classification and survey of taxonomic richness: Order Trichoptera Kirby, 1813. *Zootaxa* 3148: 209–211.

IUCN 2012. *IUCN Red List Categories and Criteria.* Version 3.1. 2nd Edition, IUCN Species Survival Commission. IUCN, Gland, Switzerland & Cambridge, UK.

IUCN Standards and Petitions Subcommittee 2017. Guidelines for Using the IUCN Red List Categories and Criteria.Version 13. Prepared by the Standards and Petitions Subcommittee.

Jacob U. 2003. *Baetis* Leach 1815, sensu stricto oder sensu lato. Ein Beitrag zum Gattungskonzept auf der Grundlage von Artengruppen mit Bestimmungs schlüsseln. [*Baetis* Leach 1815, sensu stricto or sensu lato. A contribution to the genus concept based on species groups with identification keys]. *Lauterbornia* 47: 59-129.

Janeva I. 1987. The zoobenthos of the river Vit. I. Composition, structure and dynamics. *Hydrobiology* 31: 37-64 (In Bulgarian with English summary).

Janeva I. 1991. Saprobiological state of the river Ogosta in various periods of study. *Hydrobiology* 36: 32-48 (In Bulgarian, Russian and English summaries).

Janeva I. & Russev B. 1989. Saprobiological state of the Iskar river in the initial years following the putting into operation of the Sofia purifying station. *Hydrobiology* 34: 3-19 (In Bulgarian, English summary).

Janeva I. & Russev B. 1997. Veränderungen der Arten zusammensetzung und Güteklasse des bulgarischen Donauzuflusses Jantra nach dem Makrozoobenthon [Alterations of the Bulgarian Danube tributary Jantra River according to the macrozoobenthon]. *Lauterbornia* 31: 1-16 (In German with English summary).

Jerves-Cobo R., Everaert G., Iñiguez-Vela X., Córdova-Vela G., Díaz-Granda C., Cisneros F., Nopens I. & Goethals P. L. 2017. A methodology to model environmental preferences of EPT taxa in the Machangara River basin (Ecuador). *Water* 9(3): 195, 1-31.

Kazakov S. 2017. Structural and functional parameters of hydro-zoocenoses from Danube River wetlands. PhD thesis, IBER-BAS, Sofia, 180 p. (In Bulgarian with English summary).

Kerakova M. 2015. Trophic structure of macrozoobenthos in native and influenced freshwater ecosystems. PhD thesis, IBER-BAS, Sofia (In Bulgarian with English summary).

Kovács T. & Muranyi D. 2008. Taeniopterygidae Klapalek, 1905 species in Hungary (Plecoptera). *Folia Historico-Naturalia Musei Matraensis* 32: 103–113.

Kumanski K. 1968. Beitrag zur Erforschung der Trichopteren Bulgariens [Contribution to the exploration of Bulgarian Trichoptera] (I). *Faunistische Abhandlungen, Dresden* [*Faunistic treatises, Dresden*] 2: 109–115.

Kumanski K. 1971. Beitrag zur Untersuchung der Koecherfliegen (Trichoptera) Bulgariens [Contribution to the investigation of Bulgarian Trichoptera] (III). *Bulletin de l'Institut de zoologie et musée* [*Bulletin of the Institute of Zoology and Museum*] 33: 99–109 (In Bulgarian with Russian and German summaries).

Kumanski K. 1974. Bezüglich der Möglichkeiten für Ausnützung einiger Trichopteren-Larven als limnosaprobe Bioindikatoren. [Regarding the possibilities for exploitation of some Trichoptera larvae as limnosaprobe bioindicators] *Bulletin de l'Institut de zoologie et musée* [*Bulletin of the Institute of Zoology and Museum*] 39: 207–211.

Kumanski K. 1975a. New trichopterological data from Bulgaria (Insecta, Trichoptera) *Acta zoologica bulgarica* 2: 58–69 (In Bulgarian with Russian and English summaries).

Kumanski K. 1975b. Études sur la faune et l'écologie des trichoptères (Trichoptera) du Rhodope In: *La faune des Rhodopes. Matériaux* (Ed. by G. Peshev) [Studies on the fauna and ecology of Trichoptera (Trichoptera) of Rhodope In: *The Rhodope fauna.* Materials (Ed. By G. Peshev)] 65–91. Academie bulgare des sciences [Bulgarian Academy of Sciences, Sofia (In Bulgarian with Russian and French summaries).

Kumanski K. 1976. *Faunistic, taxonomic, zoogeographic and ecological studies on the order of caddisflies (Trichoptera, Insecta) from Bulgaria.* Institute of Zoology, Bulgarian Academy of Sciences, 297 p. (In Bulgarian).

Kumanski K. 1979. The family Hydroptilidae (Trichoptera) in Bulgaria. *Acta zoologica bulgarica* 13: 3–20.

Kumanski K. 1983. Contribution to the studies on Bulgarian caddisflies (Trichoptera). *Acta zoologica bulgarica* 21: 83–88 (In Bulgarian with Russian and English summaries).

Kumanski K. 1985. Fauna Bulgarica Volume 15: Trichoptera, Annulipalpia. In: M. Josifov (ed) *Aedibus Academiae Scientiarum Bulgaricae*, Sofia, 243 p. (In Bulgarian).

Kumanski K. 1988. Fauna Bulgarica Volume 19: Trichoptera, Integripalpia. In: M. Josifov (ed). *Bulgarian Academy of Sciences*, Sofia, 353 p. (In Bulgarian).

Kumanski K. 2007. Second addition to volumes 15 (Trichoptera, Annulipalpia) and 19 (Trichoptera, Integripalpia) of the Series Fauna of Bulgarica. *Historia naturalis bulgarica* 18: 81-94.

Kumanski K. & Botosaneanu L. 1974. Les Hydropsyche (Trichoptera) du groupe de guttata en Bulgarie et en Roumanie [Hydropsyche (Trichoptera) of the guttata group in Bulgaria and Romania]. *Acta Musei Macedonici scientiarum naturalium* 14: 25–43.

Kumanski K. & Malicky H. 1976. Beiträge zur Kenntnis der bulgarischen Köcherfliegen (Trichoptera) [Contributions to the knowledge of the Bulgarian caddisfly (Trichoptera)] *Bulletin Entomologique de Pologne* [*Entomological Bulletin of Poland*] 46: 95–126.

Lenat D. R. 1988. Water quality assessment of streams using a qualitative collection method for benthic macroinvertebrates. *Journal of the North American Benthological Society* 7: 222-233.

Linnaeus C. 1758. Neuroptera. In: Linnaeus C. *Systema naturae per regna tria naturae, secundum classes, ordines, genera, species, cum characteribus, differentiis, synonymis, locis* [*Natural system for the three kingdoms in nature, by classes, orders, genera, species, with the characteristics, differences, synonyms, localities*]. Editio decima, reformata [Tenth edition, edited], 10. ed., Holmiae, 543–552.

Malm T., Johanson K. A. & Wahlberg N. 2013. The evolutionary history of Trichoptera (Insecta): A case of successful adaptation to life in freshwater. *Systematic Entomology* 38: 459–473.

Michev T. 1999. UTM & MGRS Directory of Bulgaria. A Manuscript.

Morse J. C. (ed.) 2018. Trichoptera World Checklist. http://entweb.clemson.edu/database/trichopt/index.htm [Accessed 14.12.2018].

Mouro L. D., Zatoń M., Fernandes A. C. S. & Waichel B. L. 2015. Larval cases of caddisfly (Insecta: Trichoptera) affinity in Early Permian marine environments of Gondwana. *Nature* 6: doi:10.1038/srep19215.

Murányi D. 2008. Taxonomical problems and zoogeographical investigation of the stonefly (Plecoptera) fauna of the Carpathian Basin and the Balkans. PhD Thesis, Eötvös Loránd Tudományegyetem, Állatrendszertani és Ökológiai Tanszék, Biológia Doktori Iskola, 149 p. (In Hungarian).

Pîrvu M., Zaharia C., Satmari A. & Parvulescu L. 2015. Spatial ecology of *Hydropsyche incognita* (Trichoptera: Hydropsychidae) in the Carpathians. *European Journal of Entomology* 112(1): 106-113.

Presolska Y. 2014. Composition, distribution and ecology of Ephemeroptera (Insecta) in Bulgaria. PhD thesis, IBER-BAS, 367 p. (In Bulgarian with English summary).

Puthz V. 1978. Ephemeroptera. In: Illies, J. (ed.) *Limnofauna Europaea*. 2nd ed. Gustav Fischer, Stuttgart, 256-263.

Puthz V. 1980. Ergebnisse der Albanien-Expedition 1961 des Deutschen Entomologischen Institutes [Results of the Albanian Expedition 1961 of the German Entomological Institute]. *Beiträge zur Entomologie* [*Contributions to Entomology*] 30: 343-355 Berlin (In German with Russian and English summaries).

Rasnitsyn A. P. & Quicke L. J. (eds.) 2002. *History of Insects*. Kluwer Academic Publishers, Dodrecht, Boston, London, 2002, XII, 517 p.

Rohdendorf B. & Ranitsyn A. 1980. Historical development of class Insecta. *Trudy Paleontologicheskova Instituta* [*Proceedings of the Paleontological Institute of Academy of Sciences of USSR*] 175: 269 (in Russian).

Russev B. 1957. Beitrag zur Kenntnis der Eintagsfliegen (Ephemeroptera) Bulgariens [Contribution to the knowledge of mayflies (Ephemeroptera) in Bulgaria]. *Mitteilungen aus dem Zoologischen Institut* [*Proceedings of the Zoological Institute*] 6: 558-563 (In Bulgarian with German summary).

Russev B. 1959. Beitrag zur Erforschung des Makrobenthos der Donau am bulgarischen Ufer [Contribution to the study of the macrozoobenthos along to the Bulgarian shore of Danube River]. *Comptes Rendus, Academie Bulgare des Sciences* [*Reports, Bulgarian Academy of Sciences*] 12(4): 345–348.

Russev B. 1960. Neue Eintagsfliegen für die Fauna Bulgariens [New mayflies for the fauna of Bulgaria]. *Beiträge zur Entomologie* [*Contributions to Entomology*] *10* (7/8): 697-705 (In German with Bulgarian and Russian summary).

Russev B. 1962. Die Insektenfauna der Donau am bulgarischen Ufer [The insect fauna along to the Bulgarian shore of Danube River]. *Bull. Test Stat. Freshwater Pisciculture, Plovdiv* [*Proceedings of the experimental station in freshwater fisheries Agricultural Academy, Plovdiv*] 1: 115-128 (In Bulgarian with German summary).

Russev B. 1966. Das Zoobenthos der Donau zwischen dem 845 und 375 Flusskilometer I. Zusammensetzung, Verteilung und Ökologie [The zoobenthos of the Danube River between the 845 and 375 km river. I. Composition, distribution and ecology]. *Mitt. Zool. Inst. Mus.* [*Proceedings of the Zoological Institute and Museum*] 20: 55-131 (In Bulgarian with German summary).

Russev B. 1967. Das Zoobenthos der Donau zwischen dem 845. und 375. Flusskilometer. II. Biozoenologie und Dynamik [The zoobenthos of the Danube between the 845th and 375th river kilometer. II. Biocenology and Dynamics]. *Mitt. zool. Inst. Mus.* [*Proceedings of the Zoological Institute and Museum*] 3: 33-78 (In Bulgarian with Russian and German summary).

Russev B. 1968. Ökologische Untersuchungen über die Ephemeropteren-larven der Donau vor dem bulgarischen Ufer [Ecological studies on the Ephemeroptera larvae of the Danube River along the Bulgarian shore]. *Limnol. Berichte X. Jubiläumstag. Donauforsch* [*Limnological proceedings of the X. Anniversary IAD*] 295-303 (In German).

Russev B. 1977. Die Verunreinigung und Selbstreinigung des Ossam nach den strukturellen Aenderungen seiner Benthosfauna [The contamination and self-purification of the Ossam River according the structural changes of its benthos fauna]. *Hydrobiology* 6: 3-22 (in Bulgarian with Russian and German summaries).

Russev B. 1979. Gegenwärtige Kenntnisse über die Artenzusammensetzung des Zoobenthos der Donau [Current knowledge about species composition of zoobenthos of the Danube River]. Limnol. Ber. XIX. Jubiläumstagung Donauforschung, Sofia [Limnological proceedings of the XIX Anniversary IAD, Sofia], 306–339 (In German).

Russev B. 1987. Ecology, life history and distribution of *Palingenia longicauda* (Olivier) (Ephemeroptera). *Tijdschrift voor Entomologie* [*Journal of Entomology*] 130: 109-127.

Russev B. 1998. Das Makrozoobenthos der Donau - Dynamik der Veränderungen durch anthropogenen Einfluß. In: (E. Kusel-Fetzmann Ed.) *Plankton und Benthos der Donau, Ergebnisse der Donauforschung.* 4, IAD der SIL, Wien [The macrozoobenthos of the Danube River - dynamics of changes due to anthropogenic influence. In: (E. Kusel-Fetzmann Ed.) *Plankton and benthos of the Danube River, results of the Danube research.* 4, IAD - SIL, Vienna] 257-364 (In German).

Russev B. & Janeva I. 1983. The Significance of Mayflies (Ephemeroptera, Insecta) as Structural Constituents of Benthic Zoocoenoses of the Maritsa River. *Hydrobiology* 19: 14-24.

Russev B. & Janeva I. 1986. Hydrobiological review of the right tributary of the Danube - the river Cibrica. *Hydrobiology* 28: 36-45 (In Bulgarian with Russian and English summary).

Russev B. & Kaneva-Abadjieva V. 1973. Sur la faune du benthos des Marais de quelques Oles bulgares du Danube [On the benthos fauna of the swamps of some Bulgarian Danube islands]. *Hidrobiologia, Bucuresti* 14: 255-268 (In French with German summary).

Russev B. & Uzunov J. 1990. Das Zoobenthos und der saprobiologische Zustand der Donau während der internationalen Expedition im März 1988. Ergebnisse der Donauexpedition 1988, Eigenverl. IAD, Wien [The zoobenthos and the saprobiological state of the Danube River during the international expedition in March 1988. Results of the Danube River Expedition 1988, Self-publishing IAD, Vienna] 209-220.

Russev B. & Uzunov J. 1991. Zoobenthos und saprobiologischer Zustand der Donau während der im März 1988 durchgeführten internationalen Expedition [The zoobenthos and the saprobiological state of the Danube River during the international expedition]. *Hydrobiology* 37: 34-40 (In German with Russian summary).

Russev B. & Vidinova Y. 1994. Verbreitung und Ökologie der Vertreter einiger Familien der Ordnung Ephemeroptera (Insecta) in Bulgarien [Distribution and ecology of the representatives of some families of order Ephemeroptera (Insecta) in Bulgaria]. *Lauterbornia* 19: 107-113 (In German).

Russev B., Nikolova M. & Janeva I. 1987. Tendencies in the changes of the hydrobiological state of the Rusenski Lom river valley. *Hydrobiology* 31: 65-82 (In Bulgarian with English summary).

Russev B., Janeva I., Decheva R. & Karapetkova M. 1994. 3. Hydrobiology. In: Russev B.1994 *Limnology of Bulgarian Danube tributaries*. Publ. House "Knizen Tigar", Sofia, 130-233 (in Bulgarian with English summary).

Sartori M. & Landolt P. 1999. Atlas de distribution des Ephemeres de Suisse (Insecta, Ephemeroptera) [Atlas of the distribution of the mayflies of Switzerland (Insecta, Ephemeroptera)]. *Fauna Helvelica* 3: 214 p. (In French and German).

Sandin L. & Johnson R. K. 2000. The statistical power of selected indicator metrics using macroinvertebrates for assessing acidification and eutrophication of running waters. *Hydrobiologia* 422-423: 233-243.

Savić A., Dmitrović D. & Pešić V. 2017. Ephemeroptera, Plecoptera, and Trichoptera assemblages of karst springs in relations to some environmental factors: a case study in central Bosnia and Herzegovina. *Turkish Journal of Zoology* 41(1): 119-129.

Sinitschenkova N. D. 1985. The Jurassic stoneflies of South Siberia and adjoining territories (Perlida = Plecoptera). In: Rasnitsyn A.P. (ed.) *Yurskie nasekomye Sibiri i Mongolii* [The Jurassic Insects of Siberia and Mongolia]. Trudy Paleontol. Inst. Akad.

Nauk SSSR [Proceedings of the Paleontological Institute, Academy of Science of USSR], *211*. Nauka, Moscow: 148–171 (In Russian).

Sinitschenkova N. D. 1995. Paleontology of Stoneflies. In: Landolt P. & Sartori M. eds. *Ephemeroptera & Plecoptera: Biology- Ecology- Systematics*. MTL, Fribourg, Lachat SA, 561-565.

Tyufekchieva V., Kalcheva H., Vidinova Y., Yaneva I., Stoyanova T. & Ljubomirov T. 2013. Distribution and ecology of Taeniopterygidae (Insecta: Plecoptera) in Bulgaria. *Acta zoologica bulgarica* 65 (1): 89–100.

Tyufekchieva V., Evtimova V. & Murányi D. 2019. First Checklist of Stoneflies (Insecta: Plecoptera) of Bulgaria, with Application of the IUCN Red List Criteria at National Level. *Acta zoologica bulgarica 71* (In press).

Uzunov Y. & Kovachev S. 2002. *Hydrobiologia*. Pensoft, Sofia-Moscow, 342 pp. (in Bulgarian).

Varadinova E., Borisova P., Pehlivanov L. & Uzunov Y. 2012. Macroinvertebrate communities of the Srebarna Lake Biosphere Reserve: Species Diversity, Abundance and Modelling of the Ecological Status. In: Uzunov, Y., B. Georgiev E. Varadinoiva N. Ivanova L. Pehlivanov V. Vasilev (Eds.) *Ecosystems of the Biosphere Reserve Srebarna Lake*. Sofia, "Professor Marin Drinov" Academic Publishing House, 218 p.

Vidinova Y. 2015. Ephemeroptera. In: Golemanski et al. (Eds.) *Red Data Book of Republic of Bulgaria*. Vol. 2. Animals. BAS & MEW, S. http://e-ecodb.bas.bg/rdb/bg/ vol2/.

Vidinova Y. & Russev B. 1997. Distribution and ecology of the representatives of some Ephmeropteran families in Bulgaria.- In: Landolt P. & Sartori M. (Eds.) *Ephemeroptera & Plecoptera: Biology- Ecology- Systematics, MTL, Fribourg*, 139–146.

Vidinova Y. & Russev B. 2009. Distribution and ecology of Leptophlebiidae species (Ephemeroptera, Insecta) in Bulgaria. *Lauterbornia* 67: 151–158.

Vidinova Y., Tyufekchieva V. & Pehlivanov L. 2011. First data on benthic macroinvertebrate communities in marshes on the Belene Island (Lower Danube, Bulgarian stretch) after their renewed flooding. *Natura Montenegrina* 10(3): 227–236.

Vidinova Y., Tyufekchieva V. & Yaneva I. 2014. In memoriam: Prof. Dr. Boris Russev (29.04.1929-26.11.2002). Distinguished Bulgarian Hydrobiologist and Investigator of the Danube River. *Acta zoologica bulgarica* Supplement 7: 223–226.

Wallace J. B., Grubaugh J. W. & Whiles M. R. 1996. Biotic Indices and Stream Ecosystem Processes: Results from an Experimental Study. *Ecological Applications* 6: 140–151.

Zhang Z. Q. 2011. Animal biodiversity: An outline of higher-level classification and survey of taxonomic richness (Addenda 2013). *Zootaxa* 3148: 7–12.

Zwick P. 1992. Stream habitat fragmentation—a threat to biodiversity. *Biodiversity and Conservation* 1: 80–97.

Zwick P. 2004. Key to the West Palearctic genera of stoneflies (Plecoptera) in the larval stage. *Limnologica* 34: 315–348. www.icpdr.org/main/activities-projects/joint-danube- survey-1.

In: Biodiversity of the Bulgarian-Romanian Section ISBN: 978-1-53615-663-8
Editor: Shurulinkov, Hubenov et al.

Chapter 4

DRAGONFLIES AND DAMSELFLIES (ODONATA: INSECTA) FROM THE BULGARIAN STRETCH OF THE DANUBE RIVER AND ADJACENT TERRITORIES

Yordan Kutsarov
Kalimok-Brushlen Ltd., Tutrakan, Bulgaria

ABSTRACT

This study intends to collect and present all available information on seasonal and habitat distribution of Odonata (dragonflies and damselflies) in the Bulgarian stretch of the Danube River between chainages 845 (at the Timok River estuary) and 375 (town of Silistra) and adjoining wetlands,(up to 20 km from the river bank) included in 109 MGRS UTM 10x10 km quadrants. To date, 52 species of Odonata have been recorded in the study area, 51 of which were identified in previous publications, including by this study's author, who recorded 46 of the 51 species. From the total number of species, four have not been confirmed as present around the Danube River, one has been removed because of change in taxonomy, and one new species has been recorded in the study area.

Keywords: dragonflies, damselflies, Bulgaria, ecology, distribution, Danube River

INTRODUCTION

The Danube River is the largest river in the European Union and second largest in Europe. It forms 470 km of Bulgaria's northern border. The Danube River and its tributaries, varying from medium-sized rivers like Iskar and Yantra, to small streams, together with diverse riverside wetlands, comprise Bulgaria's largest diversity of water habitats that provide grounds for development of abundant Odonata fauna.

This paper summarises all available information on dragonflies and damselflies (Odonata) along the Bulgarian side of the Danube River and is intended to give an overview of species diversity, geographic distribution, and behavioural features (flight periods). It is hoped that with the present work one gets a better awareness of these insects and their habitats, and also an awareness of the importance of conserving them.

Materials and Methods

Study Area

The study area covered the entire 470 km stretch of the Danube River in Bulgaria, located between the town of Silistra (chainage 375) and the estuary of the Timok River (chainage 845). A MGRS 10x10 km grid was used for the purpose of mapping and statistics. The study area included all grid squares, 109 in total, which encompassed parts of the Bulgarian territory of the Danube River and a 20 km strip in-between the banks of the Danube River and the inland territory of Bulgaria (Figure 1). The total area covered was approximately 7 400 km^2 and did not include the Romanian territory of the studied squares.

Data Origin

All data were taken from the National Odonatological Database (NOD) created in the MS Access and divided into the following three categories on the basis of the collection method:

1. Data published until late 1995, with most publications comprising data about the macrozoobenthos of the Danube and its adjoining areas, and concerning findings of dragonfly and damselfly larvae (Beschovski 1965, Beschovski 1967, Russev 1962, Russev 1966). The majority of these data are presented in Fauna of Bulgaria, volume 20, Odonata (Beschovski 1994) and have been extracted and entered into the NOD by Dr. Milen Marinov with the following baseline information: date and place of observation, numbers (if any), author and publication, and UTM Code for 10x10 km., MGRS squares.

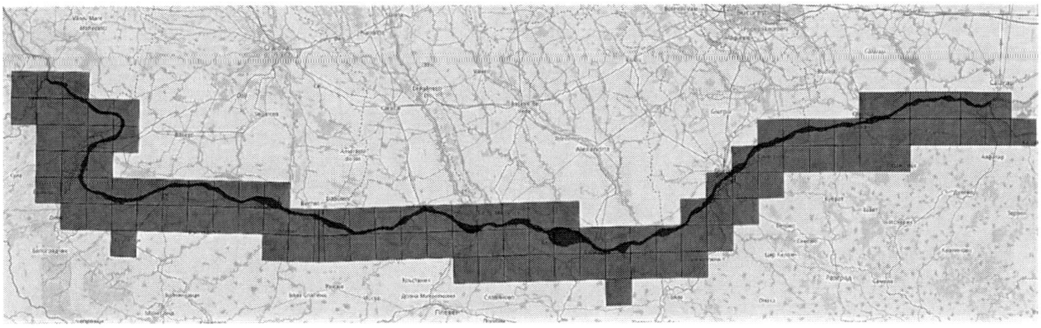

Figure 1. Study area along the lower Danube with overlaid MGRS 10x10 km.

2. Data collected during field studies by Dr. Milen Marinov between 1996 and late 2007, most of which have also been published. These data have been entered into the NOD by Dr. Milen Marinov with the following baseline information: date and place of observation, numbers (if any), author and publication, and UTM Code for 10x10 km MGRS square.
3. Data collected by the author from 2008 onwards, mainly based on field studies. A small amount of data was also collected by photographers and naturalists utilizing photographs for species identification. Data from research of odonatofauna in the Persina Nature Park and in the Natura 2000 protected areas are included in separate studies, while the remaining relevant data are published for the first time in this study and cover. Data is available for 62 MGRS squares or 57% of the studied area (Kutsarov 2012).

Collecting of Field Data and Quantity of Data

Being relatively large in size, dragonflies and damselflies of Bulgarian fauna (72 species overall) can be identified relatively easily and quickly in the field in the wild without being caught (with some exceptions). Most of the field data for the study was collected in this manner. A small portion of odonates requiring more precise study for identification, predominantly females, were caught in an entomological net and, following identification, released at the same location. In addition, a small number of exuvia were collected for laboratory identification; several larvae were caught by means of ichthyoplankton bottom meshes during WWF (World Wildlife Fund – an international environmental non-governmental organisation) organised trips for sturgeon research in the Danube River. After identification, the larvae were returned to the river.

In the past, data were recorded in a notebook and then transferred into the NOD. Starting in 2016, all field work data are being entered into an Android mobile application, ObsMapp, and stored on the Observation.org site for extraction into Microsoft Excel at the end of each year. Data utilized in this study was recorded in the NOD, with some additional data coming from Odonata sightings shared on Observation.org. Where possible, photographic data obtained by naturalists and/or photographers are added into the NOD (Plant 2005).

Scope of the Observations (Figure 2)

Observations were carried out in 65% of the MGRS squares (71 out of 109) in the study area.

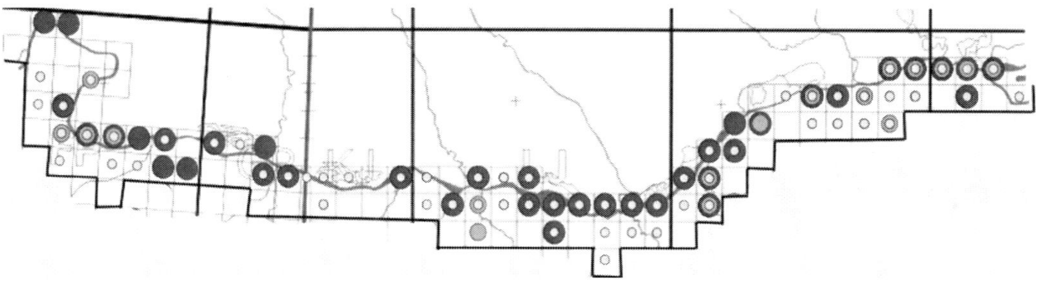

Legend:
Small circle – data collected after 2008.
Medium-sized circle – data collected between 1996 and 2007.
Large circle – data collected before 1995.

Figure 2. Geographical scope and period of the observations.

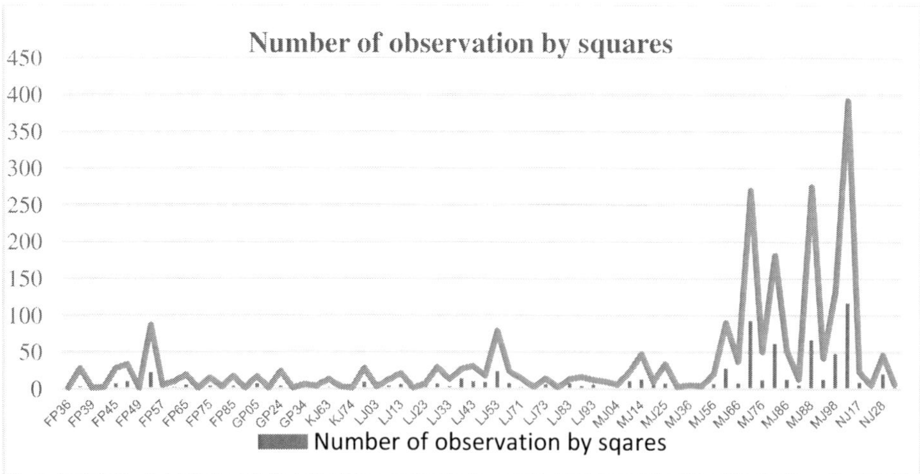

Figure 3. Number of observations by 10x10 km grid.

1. The data available until 1995 cover 39 squares or 36% of the study area.
2. The data between 1996 and 2007 cover 17 squares or 16% of the study area.
3. As of 2008, the data cover 62 squares or 57% of the study area.

Most of the observations were carried out in the eastern part of the river (Figure 3), where the wetlands are better preserved and visited much more often by entomologists.

The reduced field-work outings during the spring and autumn have left the distribution and numbers of spring and autumn species insufficiently assessed (Figure 4).

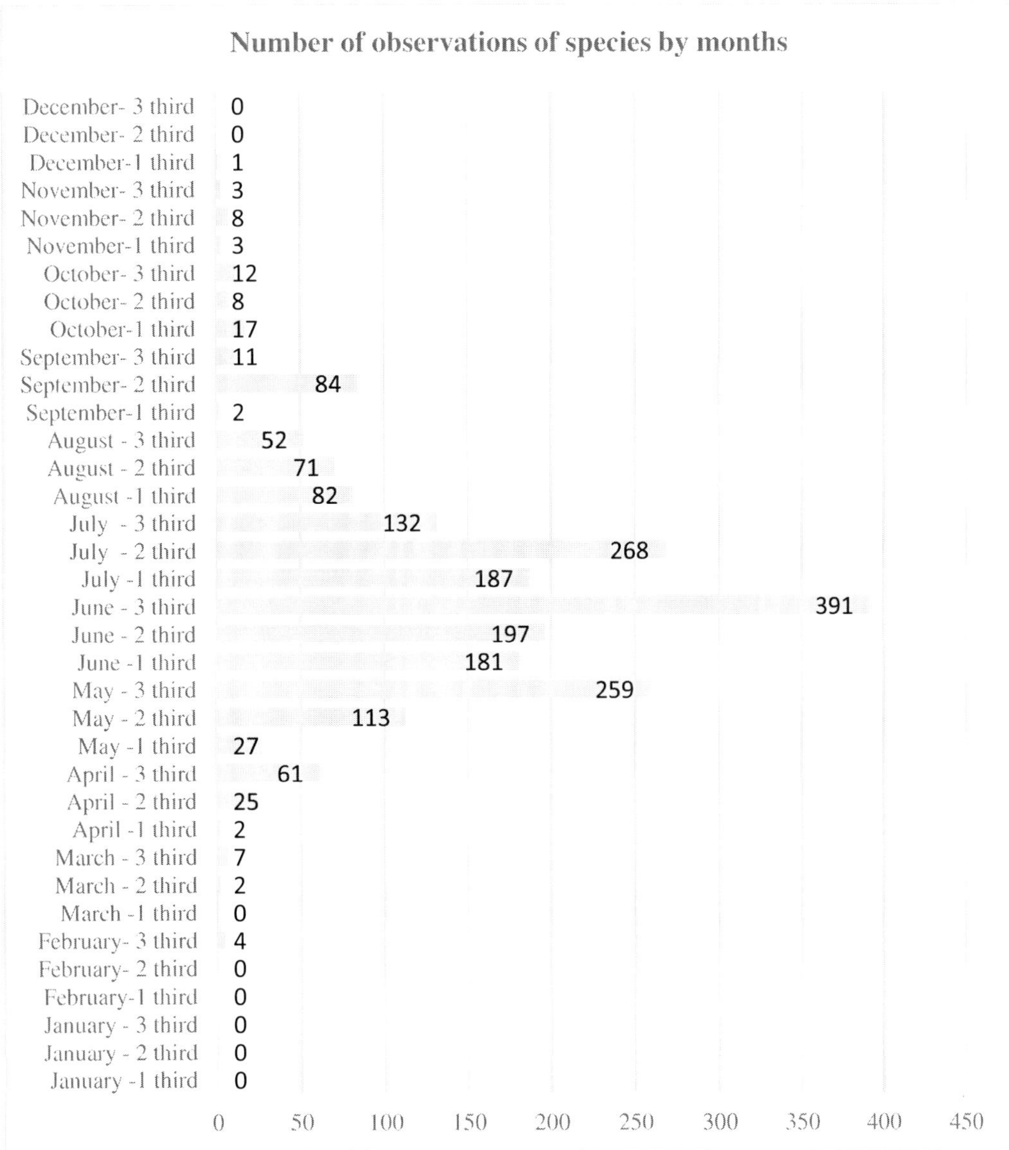

Figure 4. Number of observations by months.

Table 1. Flight periods of the Odonata species, recorded along the Bulgarian section of the Danube River. Legend: Black squares represent the peaks in numbers, with more than 30% of all observations of the species during the year; dark grey squares represent periods with 20% to 30% of the annual observations; grey squares representperiods with 10% to 20% of the annual observations; and light grey represents periods with less than 10% of the observations

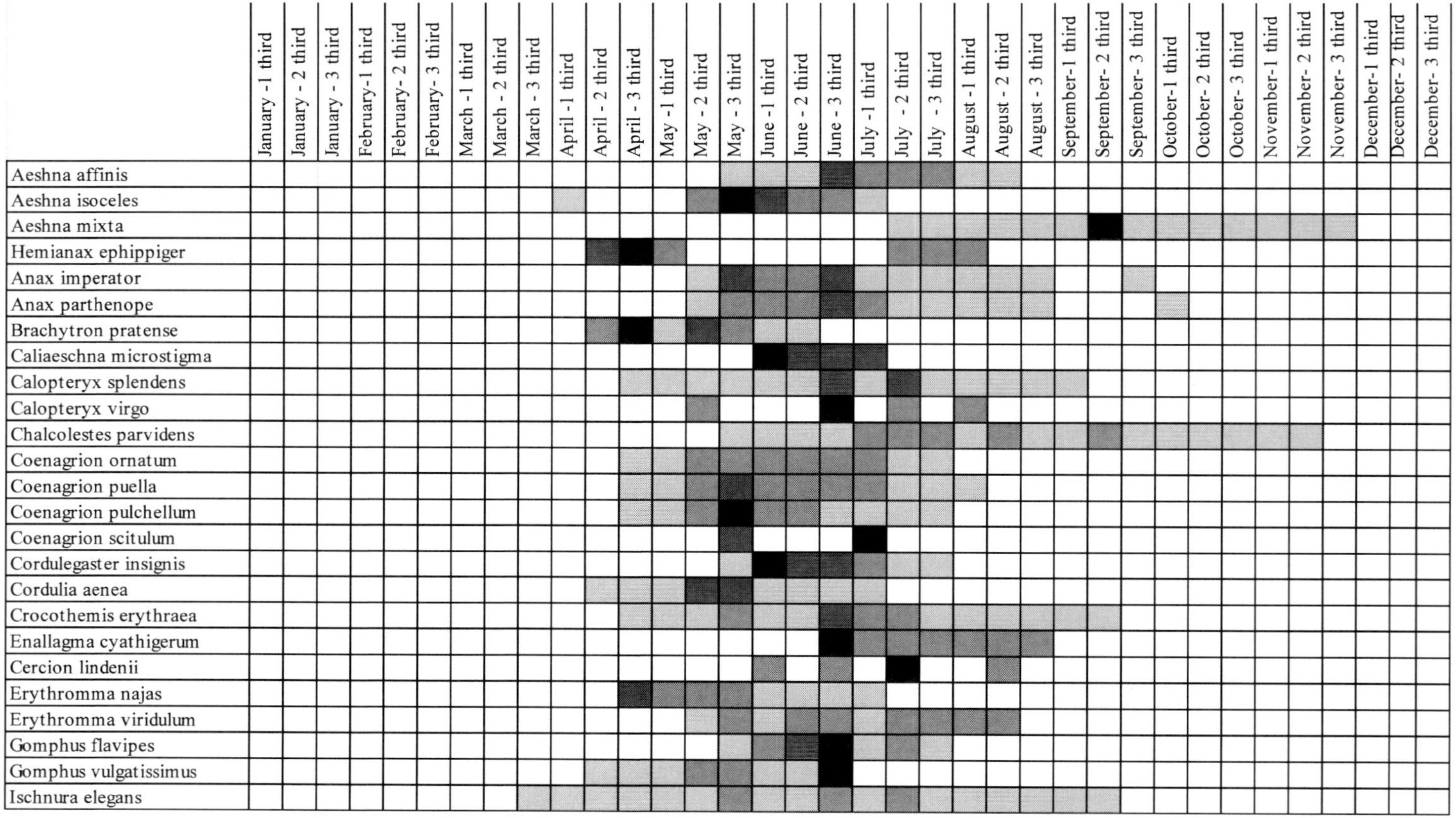

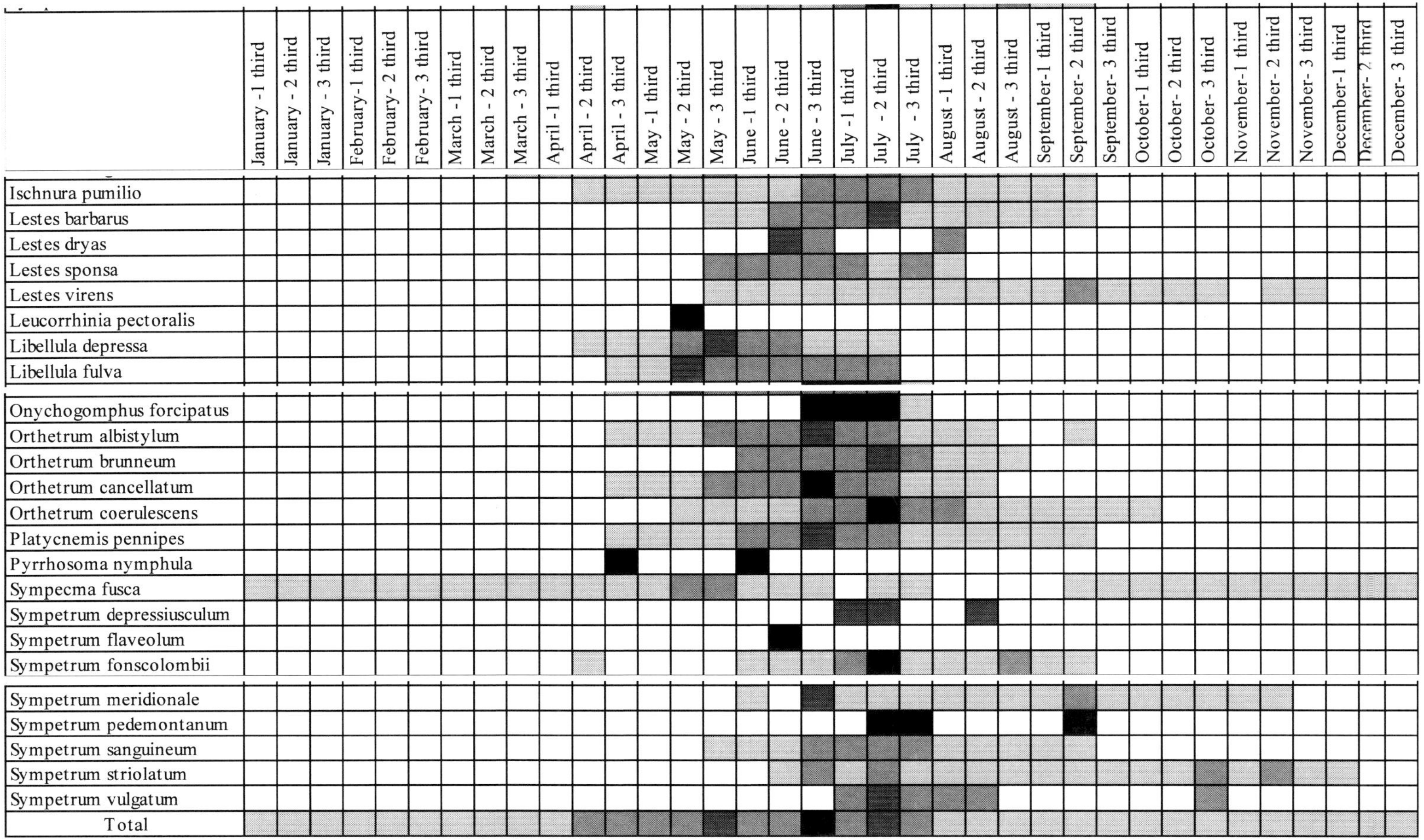
January -1 third
January - 2 third
January - 3 third
February-1 third
February- 2 third
February- 3 third
March -1 third
March - 2 third
March - 3 third
April -1 third
April - 2 third
April - 3 third
May -1 third
May - 2 third
May - 3 third
June -1 third
June - 2 third
June - 3 third
July -1 third
July - 2 third
July - 3 third
August -1 third
August - 2 third
August - 3 third
September-1 third
September- 2 third
September- 3 third
October-1 third
October- 2 third
October- 3 third
November-1 third
November- 2 third
November- 3 third
December-1 third
December- 2 third
December- 3 third
Ischnura pumilio
Lestes barbarus
Lestes dryas
Lestes sponsa
Lestes virens
Leucorrhinia pectoralis
Libellula depressa
Libellula fulva
Onychogomphus forcipatus
Orthetrum albistylum
Orthetrum brunneum
Orthetrum cancellatum
Orthetrum coerulescens
Platycnemis pennipes
Pyrrhosoma nymphula
Sympecma fusca
Sympetrum depressiusculum
Sympetrum flaveolum
Sympetrum fonscolombii
Sympetrum meridionale
Sympetrum pedemontanum
Sympetrum sanguineum
Sympetrum striolatum
Sympetrum vulgatum
Total

Results and Discussion

Flight Period

Most dragonfly and damselfly species have a clear peak in imago numbers immediately after their emergence, occurring concurrently within a few days, when the environmental parameters required by the species occur. A small number of individuals may emerge either earlier or later for a variety of reasons. This phase is followed by a gradual reduction in numbers due to predation or ageing of individuals. Usually, this reduction occurs in the two months following emergence. Seasonal weather patterns can then affect the flight period of species, causing it to occur earlier or later, as well as last longer. A few species, such as *Ischnura elegans*, start their flight period in the early spring and their imagoes fly throughout the summer. The flight periods of the species *Sympecma fusca* and *Hemianax ephippiger,* described in the ecological notes section below are more unusual.

Table 1 indicates that the month of June is the most intense flight period for dragonflies and damselflies around the Danube.

Order Odonata

Suborder Zygoptera (Damselflies)

***Chalcolestes parvidens* (Artobolevskij, 1929)**

Distribution. The species is present in suitable habitats located at lower elevations throughout Bulgaria. It inhabits willow trees along the Danube River, where temporary water bodies often linger after the high spring-time water recedes. Typical habitats include areas with willow growth around the banks of the Srebarna and the Malak Preslavets lakes (Figure 5). In the Kalimok area, the species also occurs along the channels. Another typical habitat is the small and slow flowing Studena River (Svishtov District) with its willow grove-lined banks. Also, the species *Chalcolestes viridis* (designated with a question mark on the map) has been reported in the Vidin area. Both species are very difficult to identify and, until recently had been regarded as sub-species of the same species (Marinov 1999). To date, *Chalcolestes viridis* has only been confirmed with certainty in South Eastern Bulgaria and must be confirmed along the Danube River before it is included in the regional fauna.

Ecological notes. The larvae inhabit mainly bodies of flowing or standing waters. Females lay their eggs in the soft bark of the surrounding tree and shrub vegetation (mainly willows). The adults keep close to the shady branches of the trees and shrubs. This species does not occur along rapidly-flowing rivers. It is a late-summer and autumn species, which is the reason for its poorly-studied distribution along the Danube River.

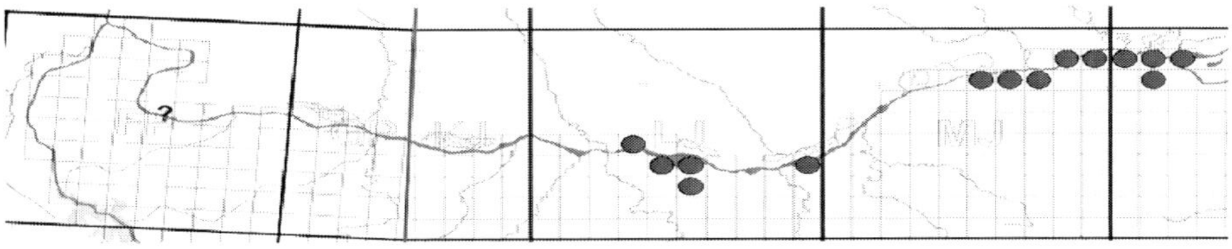

Figure 5. Distribution of *Chalcolestes parvidens* in the lower Danube area.

Lestes barbarus **(Fabricius, 1798)**

Distribution. The species is found mainly in the shallow parts of large wetlands and in the dead arms of larger rivers (Figure 6). It is also present in temporary bodies of brackish water in the remains of the Karaboaz Bog.

Ecological notes. The larvae inhabit sunny, shallow and mostly temporary water bodies with lush water-side vegetation. The species is also capable of living in brackish waters. The females often lay their eggs in the vegetation after the water body has already dried up. The larvae hatch in the spring when water returns. The species colonizes new water bodies very quickly.

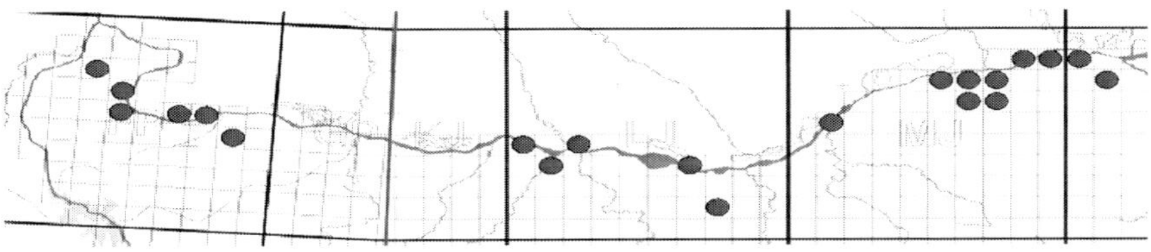

Figure 6. Distribution of *Lestes barbarus* in the lower Danube area.

Lestes dryas **(Kirby, 1890)**

Distribution. The species is mainly found in the shallows of still intact or restored wetlands in the central and eastern parts of the study area (Figure 7).

Ecological notes. The larvae of this species inhabit overgrown bodies of standing water. They are present in or around water bodies that dry up in the summer.

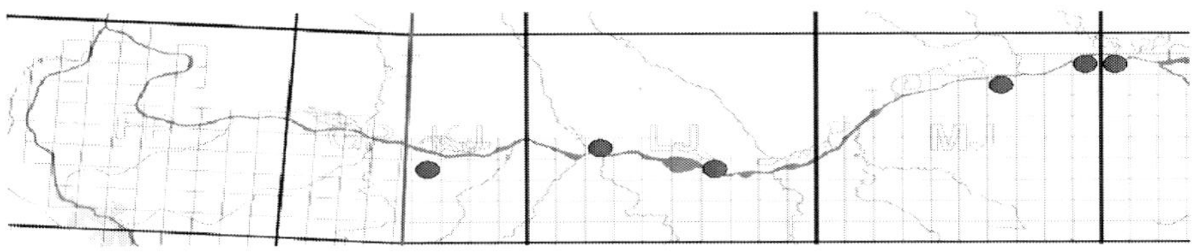

Figure 7. Distribution of *Lestes dryas* in the lower Danube area.

***Lestes sponsa* (Hansemann, 1823)**

Distribution. The species is widespread in larger wetlands along the Danube River (Figure 8).

Ecological notes. The larvae of this species inhabit a very wide range of overgrown bodies of standing water. They occur in permanent or intermittent bodies of alkaline, acidic and brackish water and can be found at altitudes between 0 and 2,500 meters above sea level (MASL).

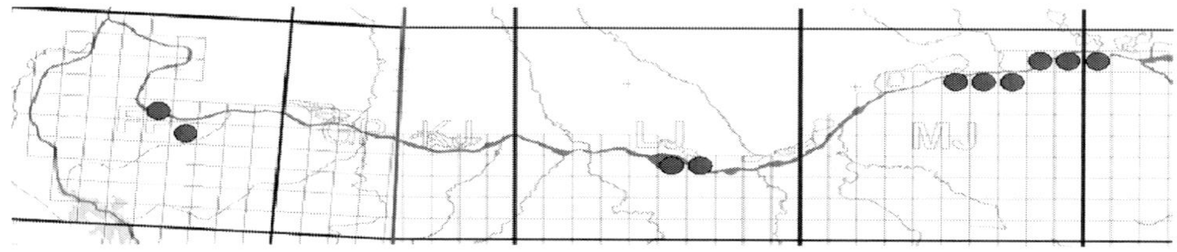

Figure 8. Distribution of *Lestes sponsa* in the lower Danube area.

***Lestes virens* (Charpentier, 1825)**

Distribution. The species occurs in larger, intact wetlands along the Danube River (Figure 9).

Ecological notes. The larvae inhabit overgrown bodies of standing water surrounded by shrubs and trees, which protect them against the wind. Unlike other species of the genus *Lestes, L. virens,* it flies throughout the summer, until the late autumn.

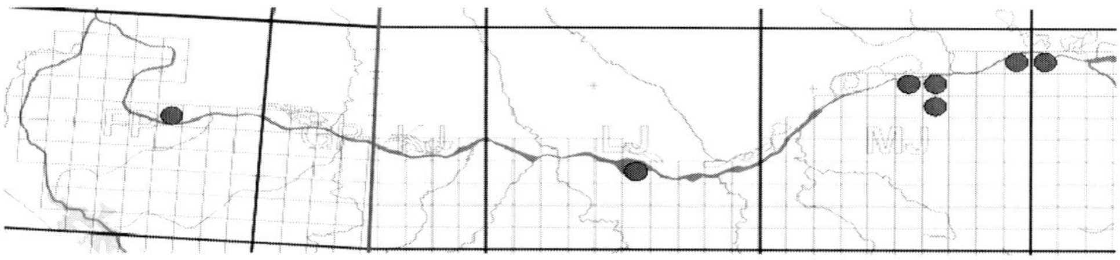

Figure 9. Distribution of *Lestes virens* in the lower Danube area.

***Sympecma fusca* (Vander Linden, 1820)**

Distribution. The species is widespread and occurs in large numbers in wetlands along the Danube River (Figure 10).

*Ecological notes.*The larvae of the species inhabit a broad range of standing or slow running waters with overgrown banks. It is important for the species to have dead vegetation that floats on the water on which to lay its eggs. This is the only species that winters as an imago and therefore has a different life cycle. After mating and laying their eggs in the early spring (March), the overwintering adults die. The new generation hatches in July and overwinters.

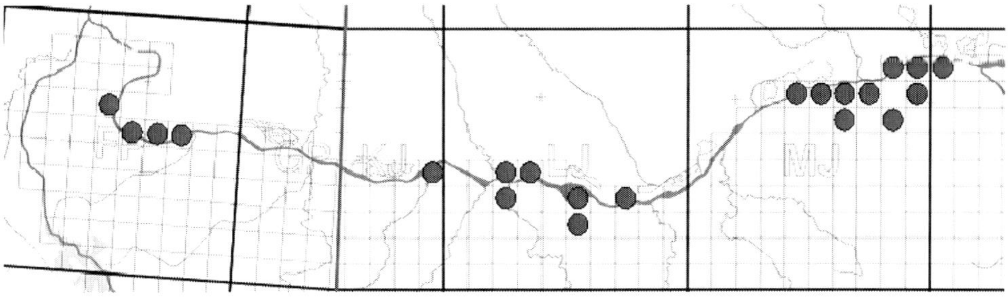

Figure 10. Distribution of *Sympecma fusca* in the lower Danube area.

Calopteryx splendens (Harris, 1782)

Distribution. The species is typical of the middle and lower stretches of rivers in Bulgaria, including the right tributaries of the Danube River (Figure 11). In the study area, it also occurs in the Danube River itself and in some smaller rivers such as the Krapinets River. The species is not present in the smallest brooks where species such as *Cordulegaster insignis*, *Caliaeschna microstigma* and *Coenagrion ornatum* are present.

Ecological notes. The larvae of the species inhabit bodies of running water. They can be found in bankside areas with inundated vegetation that is well-lit by the sun.

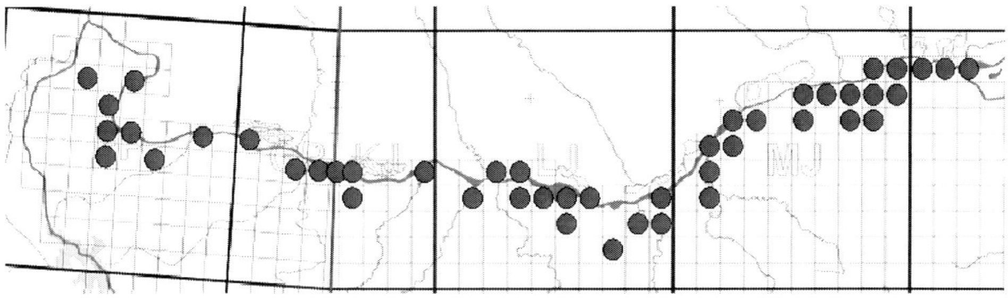

Figure 11. Distribution of *Calopteryx splendens* in the lower Danube area.

Calopteryx virgo (Linnaeus, 1758)

Distribution. The adults of the species often inhabit the upper river stretches at altitudes above 400-500 MASL and can be found in lower altitudes in South-Eastern Bulgaria and the Roussenski Lom River system flowing into the Danube River. In the study area, smaller populations were found in the Ludogorie area (near the Krapinets River) and in Vidin District, where the species was observed on the banks of the Danube River and in small, local shady brooks flowing into the Danube River (Figure 12). Reports of larvae being found in the Danube River itself are most likely the result of inaccurate identification.

Ecological notes. Larvae inhabit bodies of running water; bankside, inundated growth of sedge family (Cyperaceae) representatives is suitable for their development. Adults of the species typically stay close to floating water vegetation in shady river areas.

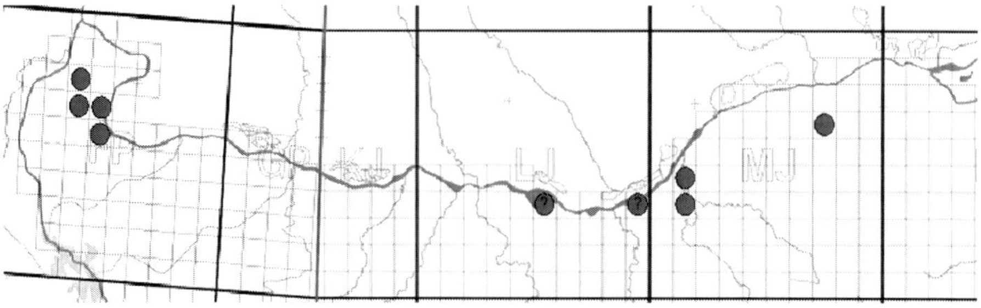

Figure 12. Distribution of *Calopteryx virgo* in the lower Danube area.

Platycnemis pennipes (Pallas, 1771)

Distribution. The species is widespread and occurs in large numbers along the Danube River (Figure 13).

Ecological notes. The larvae of this species inhabit almost all possible types of water bodies, both with running and with standing water. Flying throughout the summer, this species is the most widespread and abundant along the banks of the Danube River.

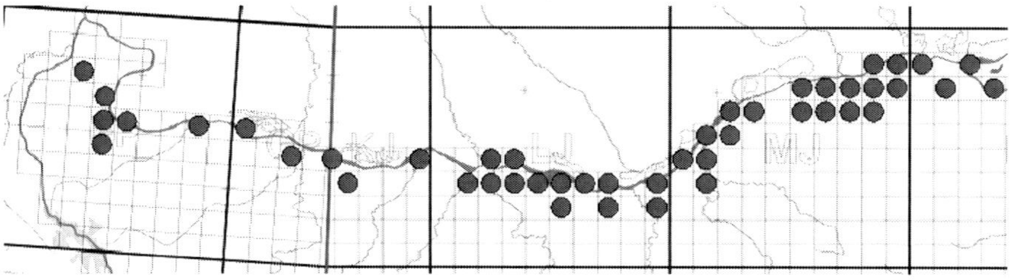

Figure 13. Distribution of *Platycnemis pennipes* in the lower Danube area.

Coenagrion ornatum (Selys, 1850)

Distribution. The species is widespread in suitable habitats along the Danube River (Figure 14).

Ecological notes. It is the only species of the *Coenagrion* genus in Bulgaria that inhabits running water. It prefers small and slow brooks rich in floating vegetation and sunlit surfaces; it also occurs in larger rivers such as the Roussenski Lom River near the "Smesite" locality to the west of Nissovo Village.

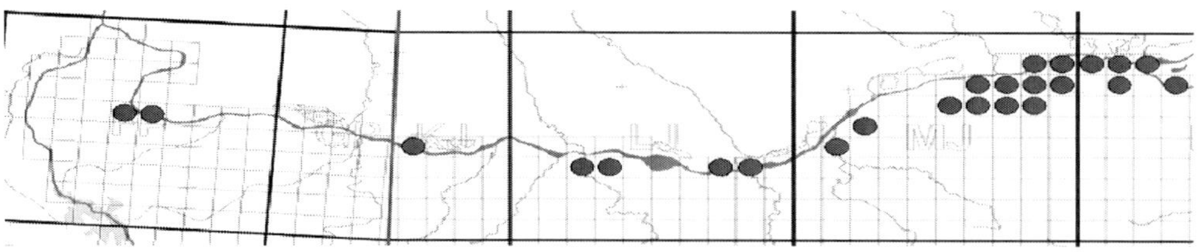

Figure 14. Distribution of *Coenagrion ornatum* in the lower Danube area.

Coenagrion puella **(Linnaeus, 1758)**

Distribution. The species is widespread and occurs in large numbers in wetlands along the Danube River during early summer (Figure 15).

Ecological notes. It inhabits standing waters rich with floating vegetation on which it lays its eggs. The species is less frequent in running waters with floating vegetation.

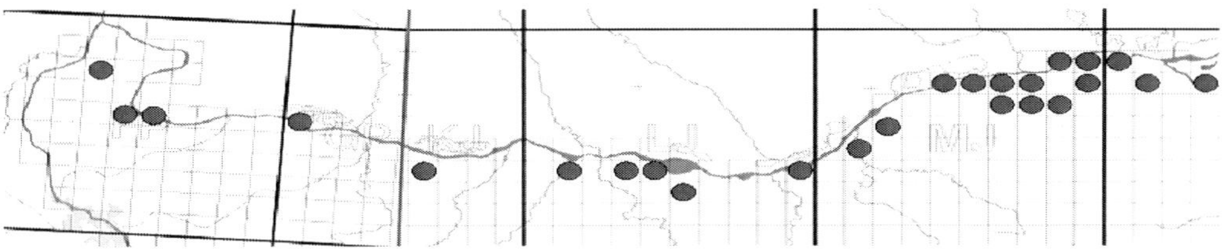

Figure 15. Distribution of *Coenagrion puella* in the lower Danube area.

Coenagrion pulchellum **(Vander Linden, 1825)**

Distribution. The species is widespread and occurs in large numbers in wetlands along the Danube River during early summer (Figure 16).

Ecological notes. The species is found in similar habitats to *Coenagrion puella*.

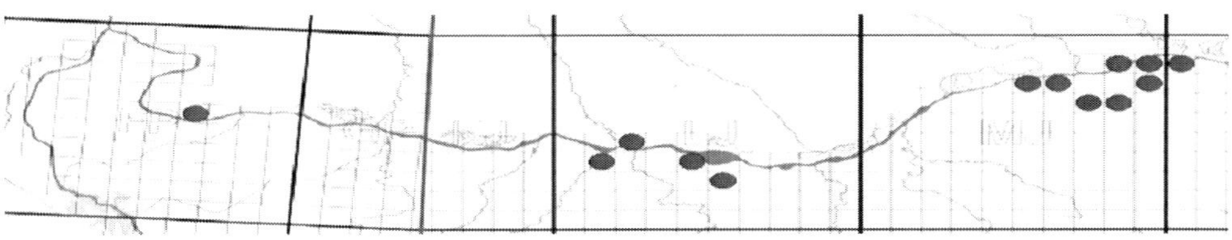

Figure 16. Distribution of *Coenagrion pulchellum* in the lower Danube area.

Coenagrion scitulum **(Rambur, 1842)**

Distribution. The species is found in small dam lakes in the Dobroudzha and Ludogorie regions (NE-Bulgaria) (Figure 17). It also occurs in fish ponds along the Cherni Lom River south of the study area.

Ecological notes. It is a southern species which has recently colonized the Lower Danube region. It prefers small water dams with inundated aquatic vegetation.

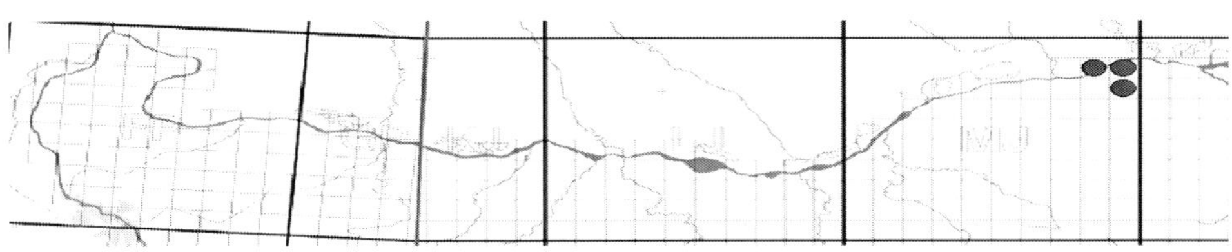

Figure 17. Distribution of *Coenagrion scitulum* in the lower Danube area.

***Enallagma cyathigerum* (Charpentier, 1840)**

Distribution. The species has been observed in widely diverse wetlands, mostly in the eastern part of the study area, and the shallows of the Danube River, small dams, marshes (Garvan Marsh) and lakes (Rabisha) (Figure 18).

Ecological notes. The species prefers standing and running water, preferably without any fish populations.

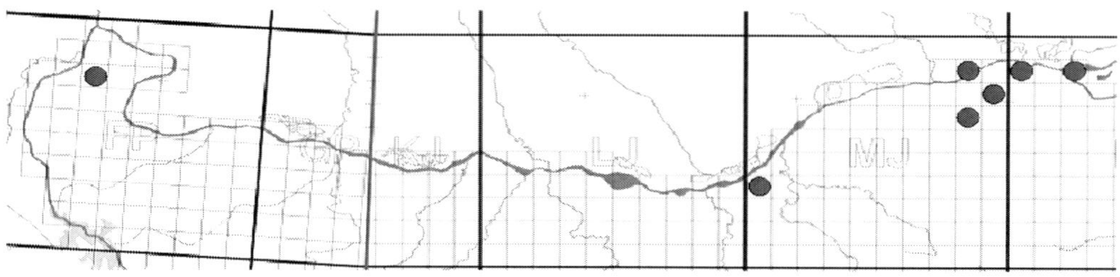

Figure 18. Distribution of *Enallagma cyathigerum* in the lower Danube area.

***Cercion lindenii* (Selys, 1840)**

Distribution. The species has been recorded in mid-sized rivers (Roussenski Lom and Yantra) and, less often, in brooks, marshes and lakes (Srebarna) (Figure 19).

Ecological notes. Unlike the two species of *Erythromma* genus, this species is frequent in the shallow places of medium-sized rivers and channels overgrown with inundated aquatic vegetation, as well as marshes and dam lakes.

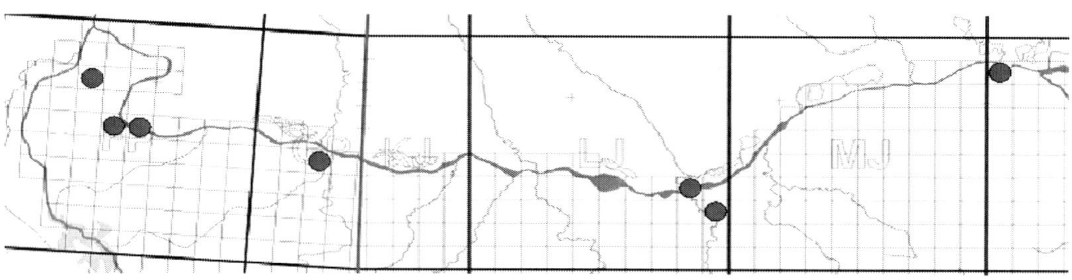

Figure 19. Distribution of *Cercion lindenii* in the lower Danube area.

***Erythromma najas* (Hansemann, 1823)**

Distribution. The species occurs in large, intact, natural marshes in the eastern part of the study area and in the marshes on Persin Island (Figure 20).

Ecological notes. It prefers standing sunlit waters that have large leaves of aquatic vegetation floating on the surface (water lilies). This spring-time species' preference to stay close to the aquatic vegetation in the water body interior makes collecting data on it difficult.

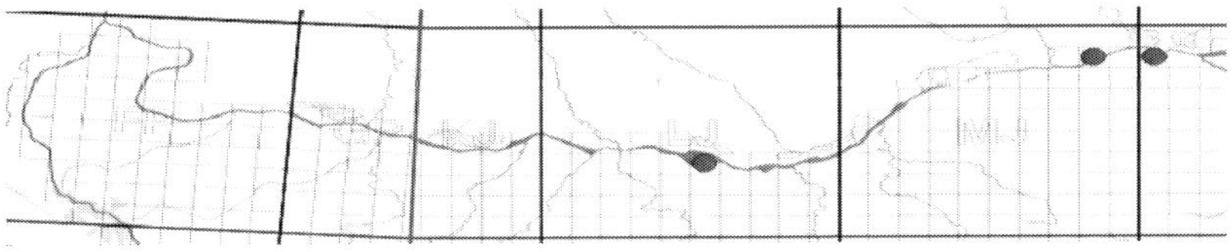

Figure 20. Distribution of *Erythromma najas* in the lower Danube area.

Erythromma viridulum (Charpentier, 1840)

Distribution. The species is observed in widely diverse standing waters and, rarely, in floodplain shallows. It is more common in the central and eastern parts of the study area (Figure 21).

Ecological notes. It likes plants with smaller floating leaves (*Lemna minor, Potamogeton perfoliatus, Elodea* sp., etc.) which makes the species far more widespread. It is a summer species that rarely comes to the banks. In very hot weather, it leaves the aquatic vegetation to hide in the shadows on the banks of the Danube River.

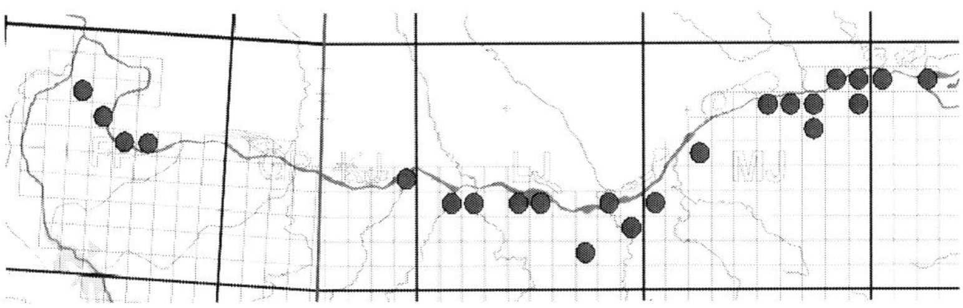

Figure 21. Distribution of *Erythromma viridulum* in the lower Danube area.

Ischnura elegans (Vander Linden, 1820)

Distribution. The species occurs everywhere in the study area, from the shallows around the Danube River to small dam lakes (Figure 22). The only place the species is not present is in fast flowing rivers and brooks.

Ecological notes. The species inhabits a large diversity of standing and slow flowing waters. It is one of the most widely-distributed species in the study area and throughout Bulgaria and Europe. It flies from early spring until late autumn.

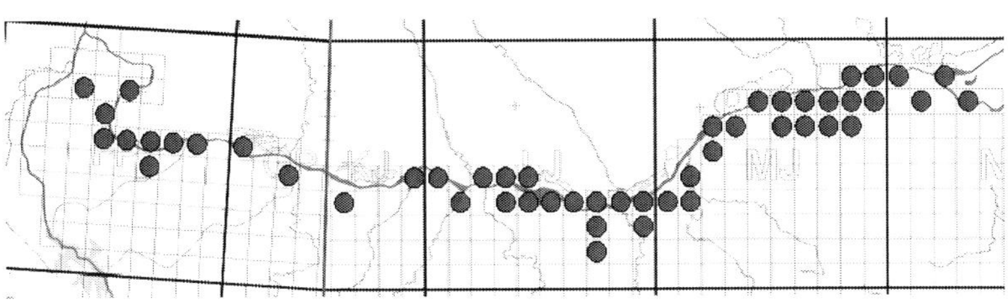

Figure 22. Distribution of *Ischnura elegans* in the lower Danube area.

Ischnura pumilio (Charpentier, 1825)

Distribution. The species is widespread in the study area (Figure 23).

Ecological notes. This species prefers slow flowing waters overgrown with herbaceous vegetation. It occurs most often in areas where brooks and rivers enter small dam lakes; it also occurs in the shallows of rivers overgrown with low grass.

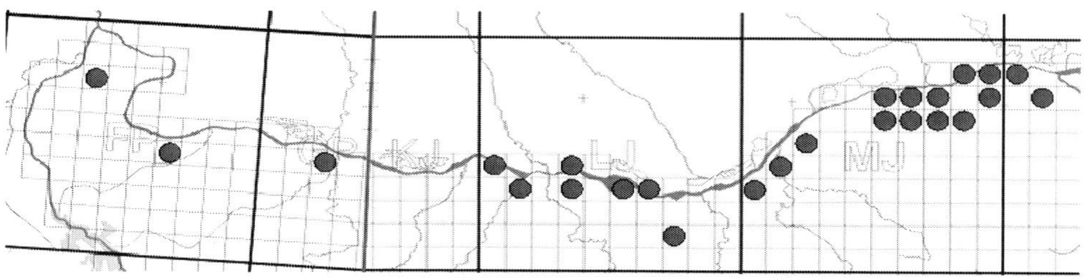

Figure 23. Distribution of *Ischnura pumilio* in the lower Danube area.

Pyrrhosoma nymphula (Sulzer, 1776)

Distribution. In the study area, the species had only been recorded in Archar River; however, it was recently found in a karst spring near the town of Tutrakan (Figure 24), also frequented by *Cordulegaster insignis*. Along the southern boundary of the area, the species has been found in a brook near the town of Borovo and in a water fountain near the town of Alfatar.

Ecological notes. The species inhabits both flowing and standing waters, but prefers shady and permanent streams in Eastern Europe. In Bulgaria, it is predominantly mountainous but has recently been found in karst springs in Dobroudzha, where its flight period is much earlier than the mountain damselfly population.

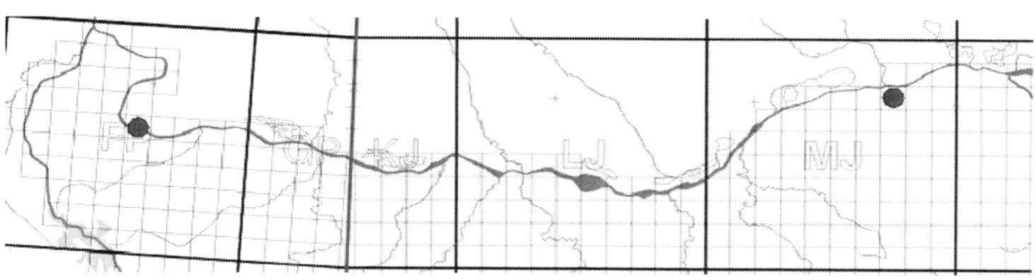

Figure 24. Distribution of *Pyrrhosoma nymphula* in the lower Danube area.

Anisoptera (Dragonflies)

Aeshna affinis (VanderLinden, 1820)

Distribution. The species is present everywhere in the study area (Figure 25).

Ecological notes. It breeds in shallow standing bodies of water overgrown with vegetation that are well-lit by the sun and protected from the wind. The water bodies may dry up during the summer. This summer species starts flying before the high heat and before the water bodies dry up.

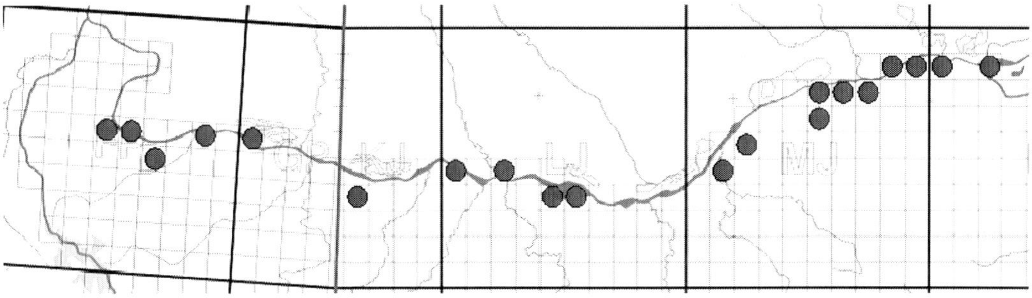

Figure 25. Distribution of *Aeshna affinis* in the lower Danube area.

Aeshna isoceles (Muller, 1767)

Distribution. The species is mainly present in large and deep water bodies with reeds along their banks. In the study area, it was only found in several dam lakes and marshes in the eastern part of the Danube (Figure 26).

Ecological notes. It prefers standing water surrounded by wide reed or cattail belts. It flies over long distances and can be found far from its regular habitats.

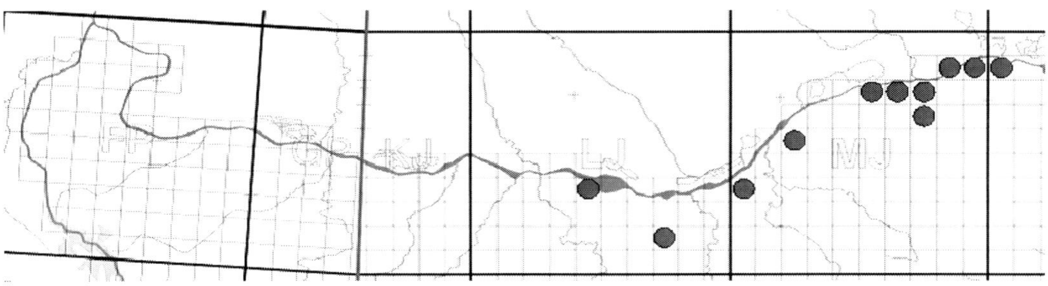

Figure 26. Distribution of *Aeshna isoceles* in the lower Danube area.

Aeshna mixta (Latreille, 1805)

Distribution. The species is found in larger water bodies with internal reed beds. The collected data do not always reflect the species' breeding areas. It was rarely found in the western and central parts of the study river section (Figure 27).

Ecological notes. It prefers large, non-drying bodies of standing water that are overgrown with thick reeds or cattail. This autumn species gathers in flocks and migrates over large distances.

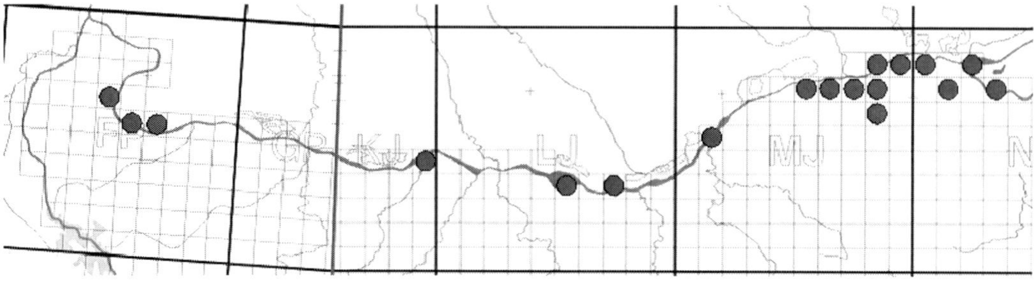

Figure 27. Distribution of *Aeshna mixta* in the lower Danube area.

Hemianax ehippiger (Burmeister, 1839)

Distribution. The species has mostly been spotted around the wetlands in the eastern part of the study area (Marinov 2001b) (Figure 28). Hatching of an imago was observed for the first time in Bulgaria in July 2015 in the Pozharevo Marsh. Massive egg-laying was established in April 2018 in the Kalimok-Brashlen Protected Area. The species' early appearance was the reason for the small number of observations.

Ecological notes. The species frequently uses temporary water bodies. It has two generations per year; the first generation develops in the south (Africa), hatches early in the year and moves north in a mass migration, arriving to the study area at the beginning of April. It breeds and lays eggs in temporary water bodies or in small dam lakes. The larvae develop quickly (2-3 months) and the new generation emerges in early July, before the water bodies dry up completely. The adults then return to the south in order to breed and lay their eggs. One autumn breeding has been observed in Southern Bulgaria to date. The species does not reach the Danube River every year during its migration.

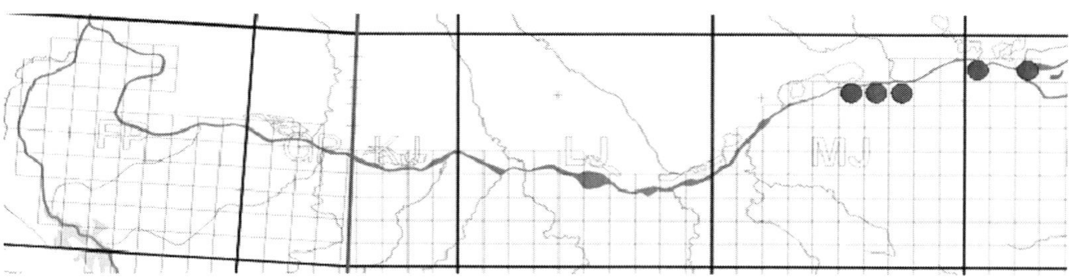

Figure 28. Distribution of *Hemianax ephippiger* in the lower Danube area.

***Anax imperator* (Leach, 1815)**

Distribution. The species is very sparsely distributed in all types of natural wetlands and in dam lakes (Figure 29).

Ecological notes. It prefers sunny standing water with water-edge vegetation and very rarely occurs in wide flood plains with slow running water.

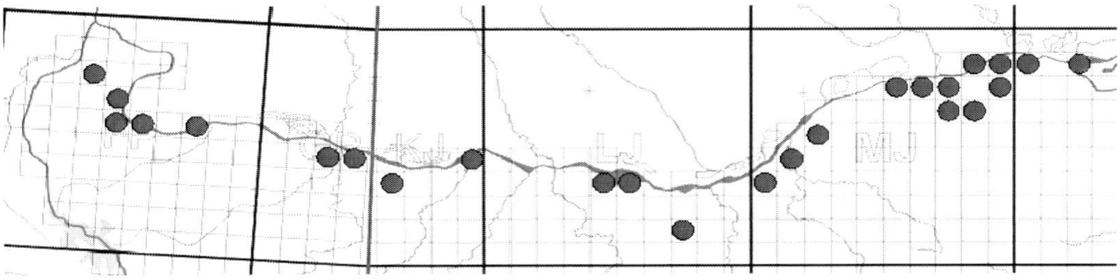

Figure 29. Distribution of *Anax imperator* in the lower Danube area.

***Anax parthenope* (Selys, 1839)**

Distribution. The species is found in water bodies of natural appearance, almost entirely in the eastern part of the study area and one locality in Vidin District (Figure 30).

Ecological notes. It prefers the same environment as the previous species but shows a preference for bankside (cattail and reed) and floating vegetation.

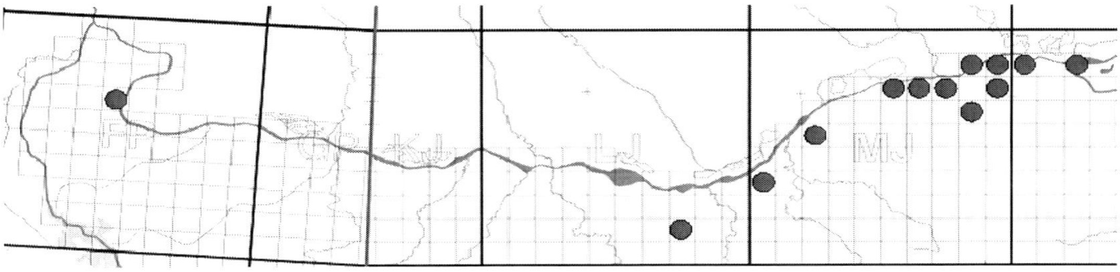

Figure 30. Distribution of *Anax parthenope* in the lower Danube area.

***Brachytron pratense* (Muller, 1764)**

Distribution. The species is found in the shallows of Srebarna Lake, the Kalimok Marshes, and the Kaykusha Marsh (Figure 31).

Ecological notes. It prefers the shallows in bodies of standing water overgrown with cattail and cock's foot grass (*Dactylis glomerata*), and with small supernatant ponds.

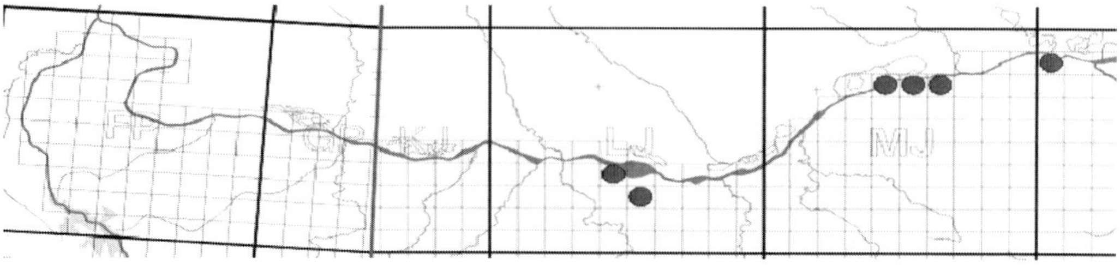

Figure 31. Distribution of *Brachytron pratense* in the lower Danube area.

Caliaeschna microstigma (Schneider, 1845)

Distribution. The species is typically found in Southern Bulgaria and, during the last few years, along the Danube River (Figure 32). In the study area, the species occurs in small shady brooks, alongside *Cordulegaster insignis*. It has also been found along the Cherni Lom River in the vicinity of the study area.

Ecological notes. It inhabits brooks and small rivers heavily-shaded by woods (Marinov 2000a).

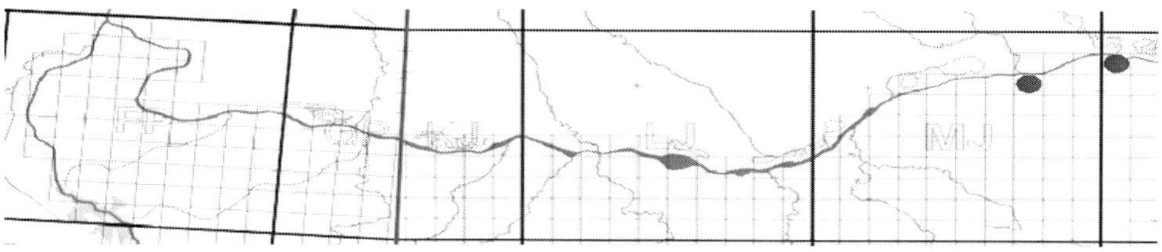

Figure 32. Distribution of *Caliaeschna microstigma* in the lower Danube area.

Gomphus flavipes (Charpentier, 1825)

Distribution. The species is found in the Danube River and the lowermost stretches of its large tributaries (Figure 33). It has been recorded only once in the standing waters of Rabisha Lake near the southwestern border of the study area, where a suitable sandy bottom is present.

Ecological notes. It typically inhabits large rivers with sandy bottoms.

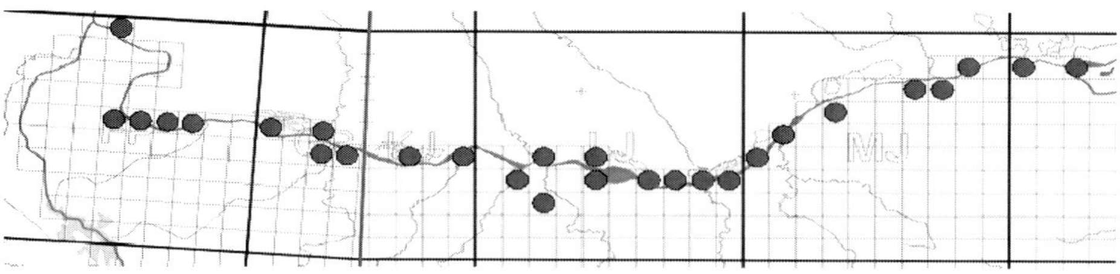

Figure 33. Distribution of *Gomphus flavipes* in the lower Danube area.

Gomphus vulgatissimus (Linnaeus, 1758)

Distribution. In the study area, the species has been found in the Danube River, its large tributaries, and in small rivers such as the Krapinets River (Figure 34).

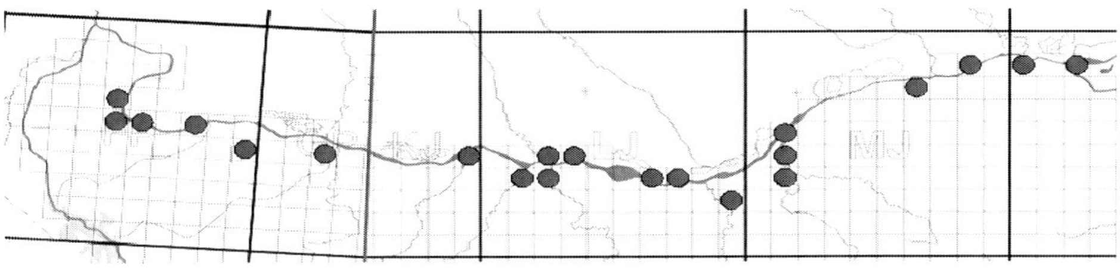

Figure 34. Distribution of *Gomphus vulgatissimus* in the lower Danube area.

Ecological notes. The species inhabits very diverse flowing waters, from the Danube River to mountain brooks.

Onychogomphus forcipatus (Linnaeus, 1758)

Distribution. The species is ubiquitous along the tributaries of the Danube River west of the Yantra River and in the Danube River to the west of the town of Nikopol (Figure 35). Its complete absence to the east of the Yantra River and north of the rivers flowing into the Black Sea should be noted.

Ecological notes. The species inhabits slow flowing rivers of varying size with sandy or predominantly sandy bottoms.

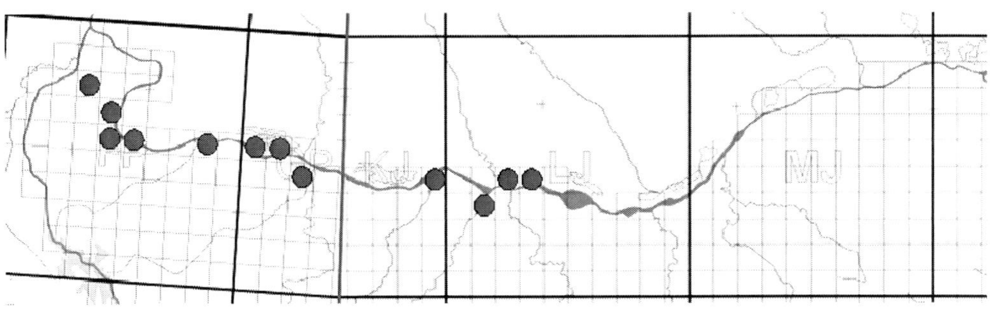

Figure 35. Distribution of *Onychogomphus forcipatus* in the lower Danube area.

Ophiogomphus cecilia (Geoffroy in Fourcroy, 1785)

Distribution. The species is known in the study area from a larva along the lower stretch of the Iskar River, which is a very suitable habitat for the species (Figure 36). A non-confirmed report of a sighting of the species exists for the Cherni Lom River, close to south of the study area.

Ecological notes. It inhabits mainly mid-size rivers and, less often, brooks with preserved natural water regimens. It is a rare species that is difficult to observe because of its small numbers.

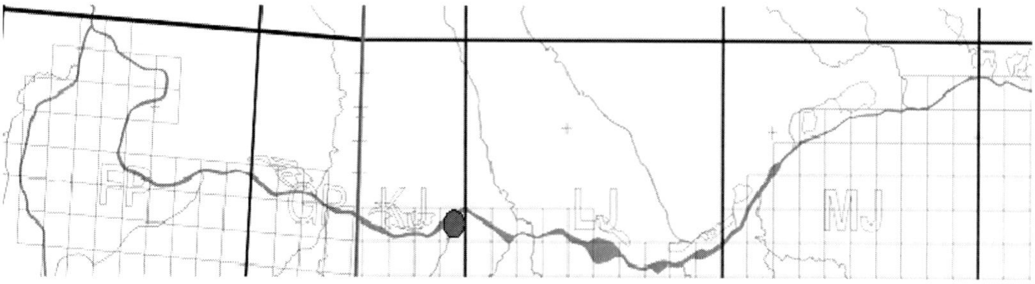

Figure 36. Distribution of *Ophiogomphus cecilia* in the lower Danube area.

Cordulegaster insignis (Figure 37a) (Schneider, 1845)

Distribution. The species is sparsely distributed in small streams flowing into the Danube River (Marinov 2007) (Figure 37).

Ecological notes. It inhabits shady brooks and springs, frequently some tens of meters long, with grassy banks on which the larva emerges. These water bodies never dry up and are so small that, in many cases, no other dragonfly or damselfly species are present.

Figure 37a. *Cordulegaster insigis.* Photo: Yordan Kutsarov.

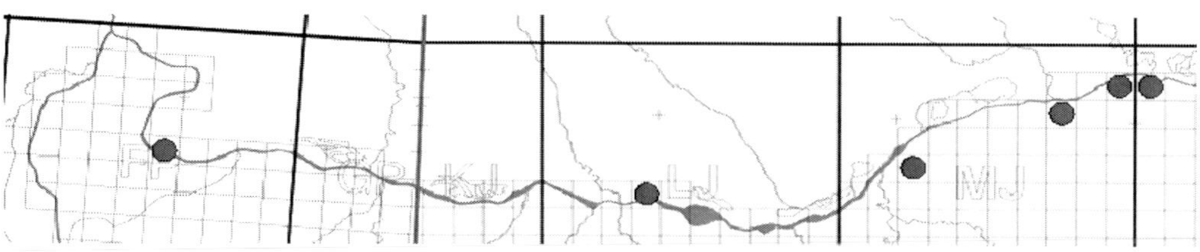

Figure 37b. Distribution of Cordulegaster insignis in the lower Danube area.

Cordulia aenea **(Linnaeus, 1758)**

Distribution. The species is ubiquitous in Srebarna Lake and Malak Preslavets Lake, where water bodies with intact natural habitats are present (Figure 38). The species appeared in the Kalimok-Brashlen Marshland six years after the project to restore the marshland was completed.

Ecological notes. It prefers mid-sized and large wetlands with standing water deeper than 1 meter and well-developed vegetation.

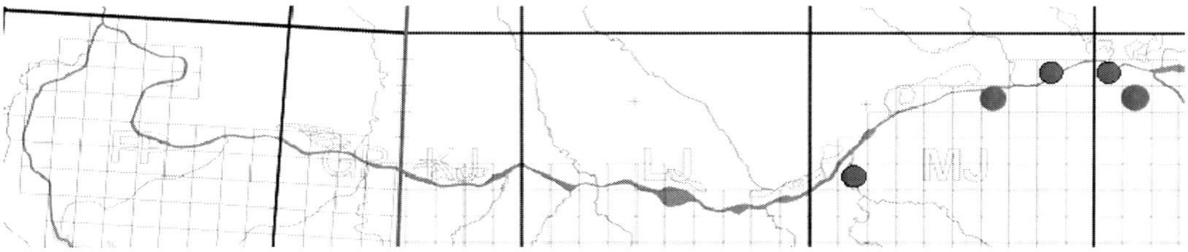

Figure 38. Distribution of *Cordulia aenea* in the lower Danube area.

Crocothemis erythraea **(Brullé, 1832)**

Distribution. The species is widespread and numerous throughout the study area (Figure 39).

Ecological notes. It is a cosmopolitan species that inhabits very diverse bodies of standing and slow flowing water.

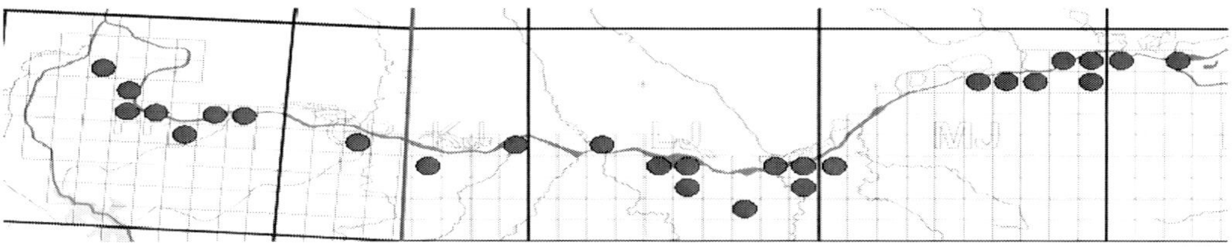

Figure 39. Distribution of *Crocothemis erythraea* in the lower Danube area.

Leucorrhinia pectoralis **(Charpentier, 1825)**

Distribution. The species was only observed once in the study area-in Srebarna Lake in 1965 (Figure 40). The lake subsequently suffered from shallowing and silting, which may explain the species' disappearance.

Ecological notes. It occurs in shallow water overgrown with low water vegetation. This northern species has been recorded in only four locations in Bulgaria. Being a spring-time species, it flies before the season of more intensive entomological studies.

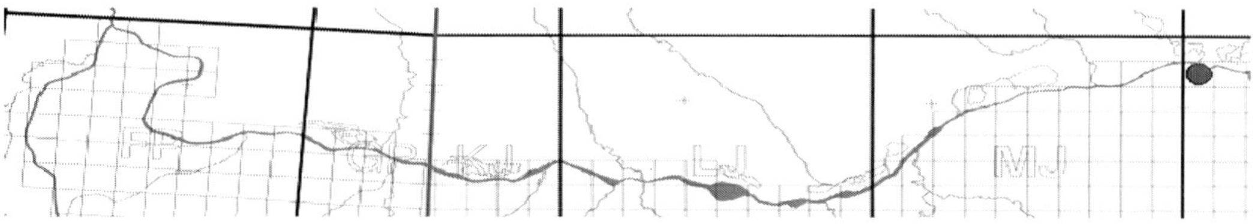

Figure 40. Distribution of *Leucorrhinia pectoralis* in the lower Danube area.

Libellula depressa (Linnaeus, 1758)

Distribution. The species is widespread in the study area (Figure 41).

Ecological notes. It is a cosmopolitan species that inhabits all types of water bodies, from slow flowing rivers to temporary puddles.

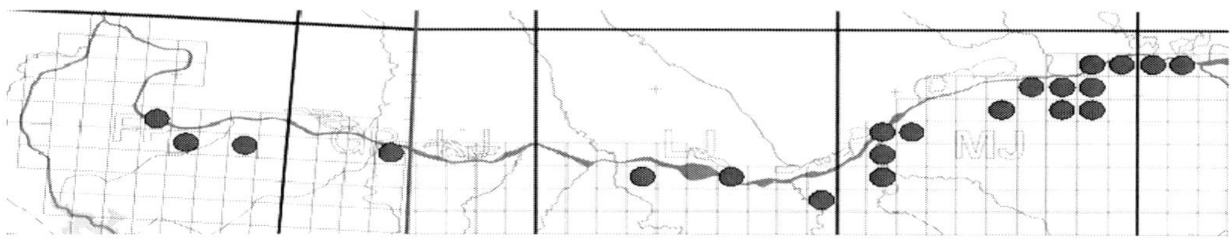

Figure 41. Distribution of *Libellula depressa* in the lower Danube area.

Libellula fulva (Muller, 1764)

Distribution. The species is found in the wetlands along the Danube River, without the central parts of the study area between Lom River mouth and Roussenski Lom River system (Figure 42). This is a migratory species that can be found outside of its breeding areas.

Ecological notes. It inhabits marshes and other standing waters, but can also be found in slow flowing waters. Known only to inhabit Southern Bulgaria until recently, the species has been colonizing the Danubian wetlands on a massive scale.

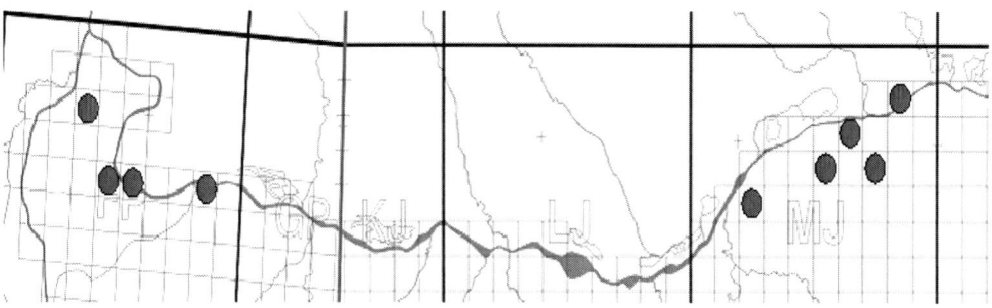

Figure 42. Distribution of *Libellula fulva* in the lower Danube area.

Libellula quadrimaculata (Linnaeus, 1758)

Distribution. The species was reported to exist in the Danube River from larvae found near the town of Oryahovo (1944) and in the Obraztsov Chiflik research institute near the town of Rousse (1961) (Russev 1962) (Figure 43). Both water bodies have undergone negative transformations and the species has, most likely, disappeared.

Ecological notes. In Bulgaria, the species inhabits bodies of standing water at higher elevations, but has also been found around the town of Varna and in the Eastern Rhodopes.

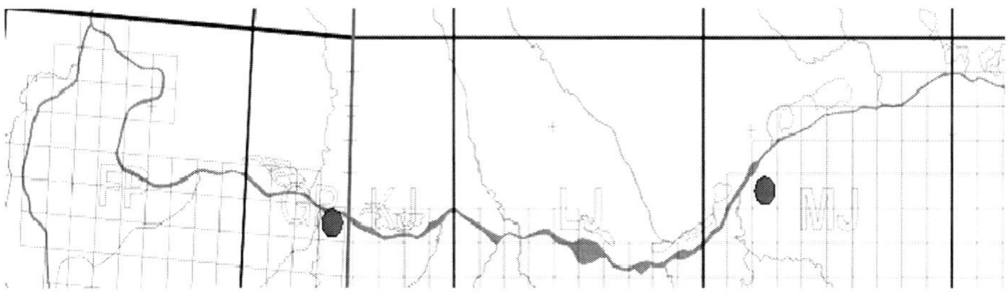

Figure 43. Distribution of *Libellula quadrimaculata* in the lower Danube area.

Orthetrum albistylum (Selys, 1848)

Distribution. The species is found throughout the entire study area (Figure 44).

Ecological notes. It inhabits standing waters and very often occurs in artificial water bodies (large and small dam lakes); it also occurs in the shallows of the Danube River, where it keeps close to river side vegetation.

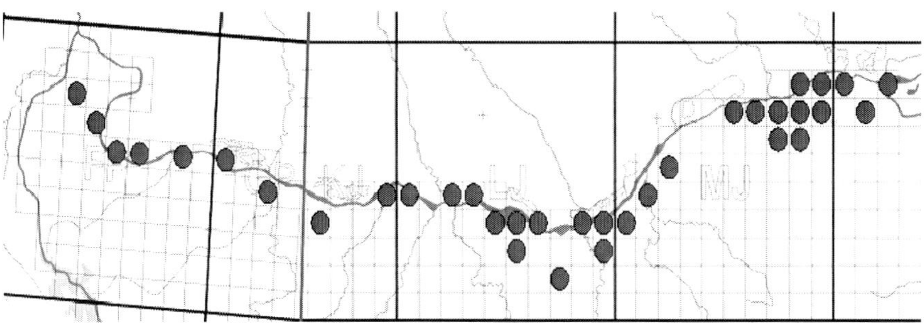

Figure 44. Distribution of *Orthetrum albistylum* in the lower Danube area.

Orthetrum brunneum (Fonscolombe, 1837)

Distribution. The species has been observed in widely diverse water bodies: the shallows of the Danube River, small dams, marshes (Garvan), and lakes (Rabisha) (Figure 45).

Ecological notes. It is most frequently found in small, sunlit and heavily-overgrown brooks with herbaceous vegetation hiding the water surface.

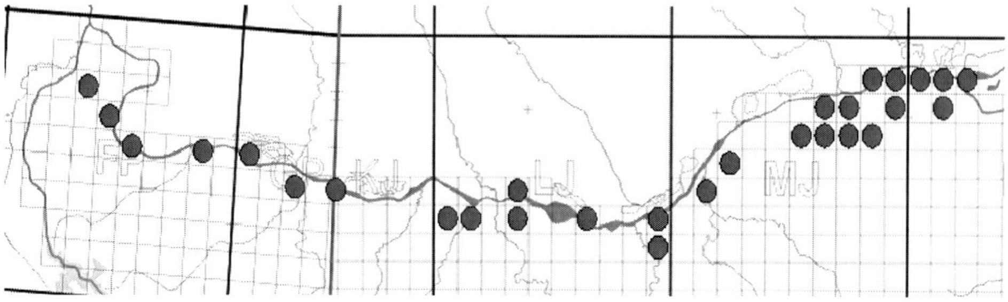

Figure 45. Distribution of *Orthetrum brunneum* in the lower Danube area.

Orthetrum cancellatum (Linnaeus, 1758)

Distribution. The species occurs throughout the entire study area but is less common in the western part of the Danube River (Figure 46).

Ecological notes. It shares the same habitats as *Orthetrum albistylum*, but, unlike the latter, it lands on bare ground around the water body.

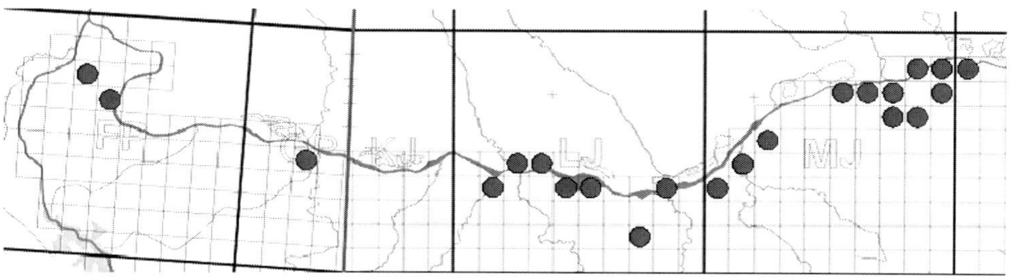

Figure 46. Distribution of *Orthetrum cancellatum* in the lower Danube area.

Orthetrum coerulescens (Fabricius, 1798)

Distribution. The species is found throughout the study area, with the exception of the area between Belene Town and Roussenski Lom River mouth (Figure 47).

Ecological notes. It shares the same environment as *Orthetrum brunneum*, but prefers the stretches of streams where the water is not completely covered by vegetation and is warmed by the sun (Marinov 2001a).

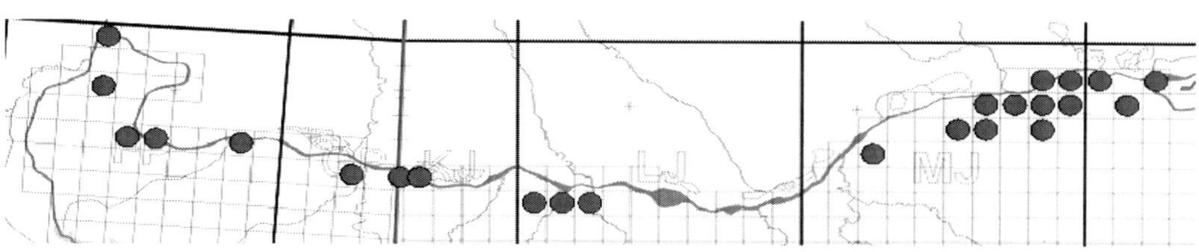

Figure 47. Distribution of *Orthetrum coerulescens* in the lower Danube area.

Sympetrum depressiusculum **(Selys, 1841)**

Distribution. The species was observed three times in Srebarna Lake and in Garvan Marsh in 2018 (Figure 48).

Ecological notes. It inhabits areas which dry up in the autumn and become heavily overgrown by herbaceous vegetation during the spring snow melt. Rice paddies are a typical habitat of the species.

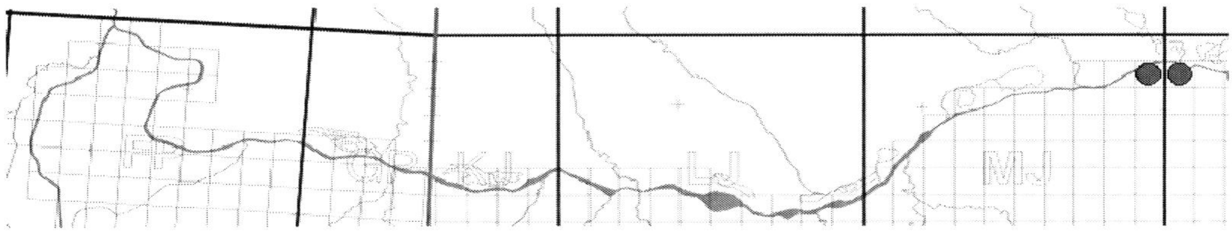

Figure 48. Distribution of *Sympetrum depressiusculum* in the lower Danube area.

Sympetrum flaveolum **(Linnaeus, 1758)**

Distribution. The species was reported in Srebarna Lake, but was most likely misidentified according to its larva and exuvia (Figure 49).

Ecological notes. In Bulgaria, the species occurs in high-mountain lakes such as Lagera Lake in the town of Smolyan and Bezbog Lake in the Pirin Moutnains; finding it along the Danube River is unlikely.

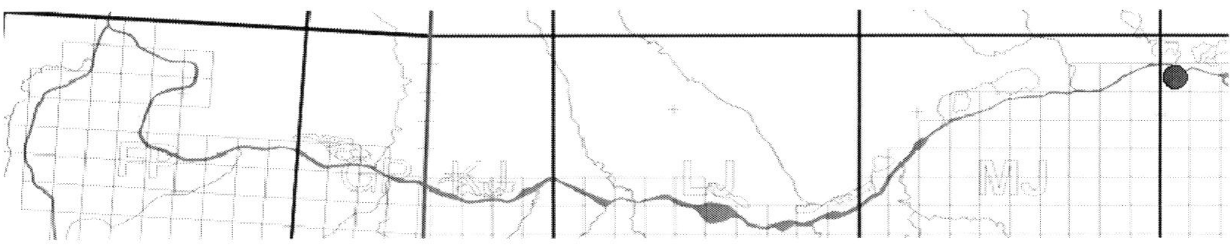

Figure 49. Distribution of *Sympetrum flaveolum* in the lower Danube area.

Sympetrum fonscolombii **(Selys, 1840)**

Distribution. The species occurs in small dams and marshes along the Danube River, more common in the easternmost parts of the study area (Figure 50).

Ecological notes. It prefers sunlit bodies of standing water and banks overgrown by bulrush.

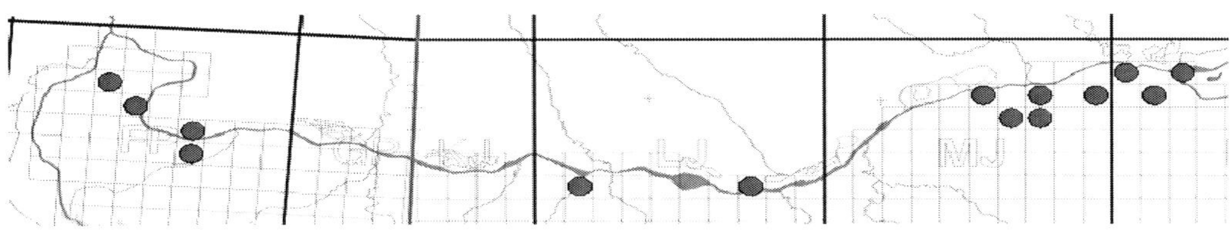

Figure 50. Distribution of *Sympetrum fonscolombii* in the lower Danube area.

***Sympetrum meridionale* (Selys, 1841)**

Distribution. The species is widespread in the study area (Figure 51).

Ecological notes. It prefers thickly overgrown and sunlit standing waters that may dry up during the summer. It flies from the spring until late autumn (November).

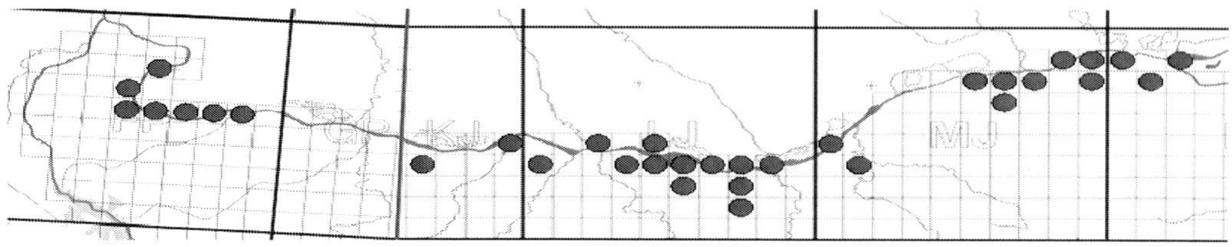

Figure 51. Distribution of *Sympetrum meridionale* in the lower Danube area.

***Sympetrum pedemontanum* (O. F. Müller in Allioni, 1766)**

Distribution. The species was found around the town of Silistra in the past, as well as Pozharevo Marsh a few years ago (Figure 52).

Ecological notes. It is most often found in sunlit artificial channels, ditches, runnels, etc., with grassy banks and slow moving water. It is less frequent in standing waters.

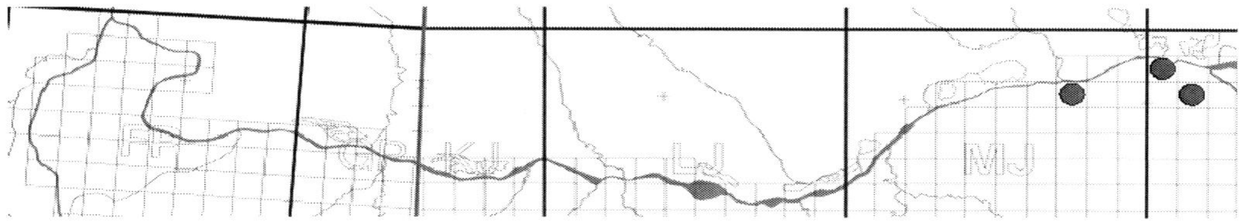

Figure 52. Distribution of *Sympetrum pedemontanum* in the Lower Danube area.

***Sympetrum sanguineum* (Muller, 1764)**

Distribution. The species is widespread in the study area (Figure 53).

Ecological notes. It lives in standing waters overgrown with vegetation.

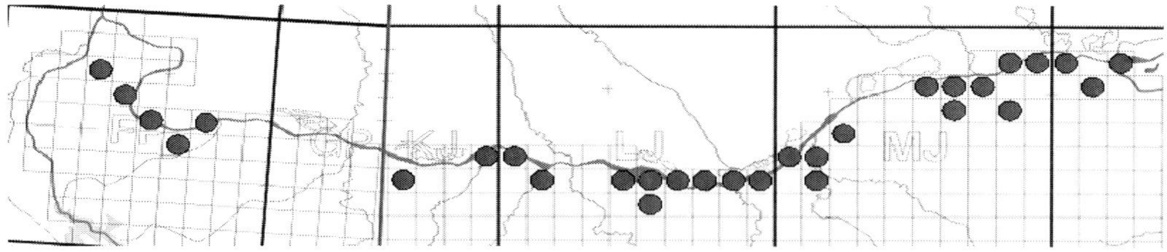

Figure 53. Distribution of *Sympetrum sanguineum* in the lower Danube area.

***Sympetrum striolatum* (Charpentier, 1840)**

Distribution. The species is common in the easternmost and western parts of the study area, while rarely being seen in the central areas (Figure 54).

Ecological notes. It prefers standing waters with abundant vegetation and avoids intermittent water bodies. It is a late autumn species, which means its numbers are most likely underestimated in this study due to the small number of observations that took place during that period.

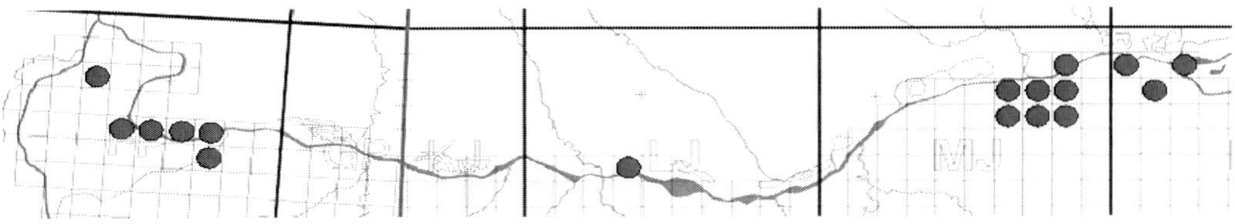

Figure 54. Distribution of *Sympetrum striolatum* in the lower Danube area.

***Sympetrum vulgatum* (Linnaeus, 1758)**

Distribution. The species is found in a small number of marshes along the Danube River (Figure 55).

Ecological notes. It resembles, and is thus difficult to differentiate from, *Sympetrum striolatum*. As a northern species that is rarely found in Bulgaria, and a late autumn species, its numbers are most likely underestimated in this study due to the limited number of observations during that period.

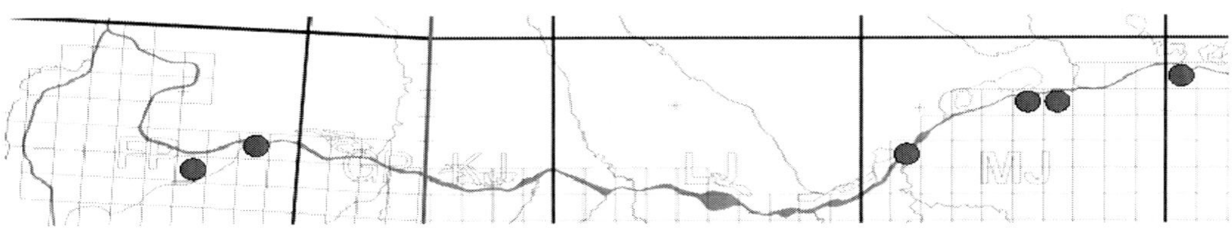

Figure 55. Distribution of *Sympetrum vulgatum* in the lower Danube area.

CONCLUSION

Conservation status, threats and protection. Endangered and protected Odonata species.

The European Habitat Directive includes 16 dragonfly and damselfly species; the European Red Data Book includes 15 that are categorized as critically endangered, endangered or vulnerable, however, the two lists are highly discrepant. Understandably, the first list includes species from north-western Europe and the second includes species from the Iberian and the Balkan peninsulas. The problem is that the species on the first list are granted legal protection and the same species are included in Bulgaria's Biodiversity Protection Act (BPA).

Six of the species listed in the European Habitats Directive occur in Bulgaria and only four of those are listed in the BPA. Six of the species included in Europe's Red Data Book also occur in Bulgaria, and none of them is included in either aforementioned list. Six other species are included as Nearly Threatened (NT) in the European Red Data Book, of which two are also listed in the EEC 92/43 Habitat Directive and are thus protected.

Four of the species in the European Habitats Directive are present in the study area, and three of them are also listed in the BPA. Two species in the European Red Data Book are also present in the study area, and two more are listed as Nearly Threatened (NT). Of these four, only one species is listed in the BPA.

Sympetrum depressiusculum is in the most critical condition, its population already meeting the Critically Endangered (CR) species criteria on the national level. The Garvan Marsh will be its second confirmed location in Bulgaria, if the presence of the species is established in the Marsh. The most endangered species are presented in Table 2.

Table 2. Odonata species of conservation importance

SPECIES	European regional Red List Category	EU27 regional Red List Category	NATURA 2000	Biodiversity Protection Act
Cordulegaster insignis	EN [B2ab(i,ii,iii,iv,v)]	EN [B2ab(i,ii,iii,iv,v)]		
Sympetrum depressiusculum	VU [A2c+3c+4c]	VU [A2c+3c+4c]		
Coenagrion ornatum	NT	NT	II	2
Caliaeschna microstigma	NT	NT		
Leucorrhinia pectoralis	LC	LC	II + IV	2
Ophiogomphus cecilia	LC	LC	II + IV	2
Gomphus flavipes	LC	LC	IV	
		II - Annex 2 of Habitats Directive		
Endangered (EN)	Near threatened (NT)	IV - Annex 4 of Habitats Directive		
Vulnerable (VU)	Least Concern (LC)	2 - Annex 2 of Law on Biological Diversity		

Dragonflies and damselflies spend their lives between aquatic and land habitats. Living most of the time in water, the species, most of which are also attached to specific microhabitats, are exceptionally sensitive to changes in the aquatic ecosystems. Changes in the land habitats around the water bodies can also threaten, to a lesser extent, the dragonfly and damselfly species.

The following are the main threats to the dragonflies and damselflies according to Kalkman (2018):

- Water management – straightening of river beds, building of dam lakes etc. complete habitat alteration and degradation – affects almost all dragonfly and damselfly species. River species such as *Coenagrion ornatum* disappear when small dam lakes are built, and the straightened river beds accelerate the water flows and change the bottom substrate, affecting almost all river species.

- Urban water pollution – Very few species manage to live in rivers affected by serious water pollution. Pollution released from a dairy farm in the Dragash Voivoda village in the Persina Nature Park has destroyed the only habitat of the protected species *Coenagrion ornatum*.
- Drainage of the marshes along the Danube River, which has been implemented since the 20-th century along the entire Bulgarian-Romanian section of Danube River, has caused the disappearance of many species there.
- Agricultural pollution – Fertilizers can change the pH of the water and insecticides destroy adult dragonflies and damselflies. Antiquated mosquito control measures around the Danube River are a particularly serious problem.
- Tourism – This threat is still minimal along the Danube River, unlike in other parts of Bulgaria. Building touristic infrastructure in the Kainardzha locality has almost completely destroyed the habitat of *Cordulegaster insignis,* which now inhabits the 10 remaining meters of the natural brook.
- Agriculture – The disappearance of rice paddies in Europe is considered the main factor for the decreasing population of *Sympetrum depressiusculum*.
- Fire – Winter-time burning of reeds can cause the complete destruction of wintering *Sympecma fusca* and of the imago of the species.
- Felling of natural tree vegetation – *Chalcolestes parvidens* lays its eggs in the soft bark of willow trees; felling or replacing the willows with hybrid poplars during the forestry activities is the cause of its disappearance.
- Grazing – Most of the dragonfly and damselfly species are attached to a particular type of aquatic habitat. For example, the species of the genus *Lestes* lay their eggs in the tissue of various bulrush species, which are afftected by grazing.
- Fish stocking – The larvae of many dragonfly and damselfly species are subject to predation by fish. Introduction of new species or increasing the numbers of predators in waterbodies causes the disappearance of some dragonfly species.
- Invasive species – It has yet to be established what damage has been caused by the small invasive fish, Chinese sleeper (*Perccottus glenii*), a very aggressive predator that has reproduced on a massive scale in Srebarna Lake during recent years.
- Climate warming – New species occur (*Libellula fulva, Coenagrion scitulum*) and some of the species found in the past have disappeared (*Leucorrhnia pectoralis*).

The protected species of dragonflies and damselflies must obtain adequate legal protection and their habitats must be protected as well. This will enable legal measures to prevent the threats listed in the previous points. Table 3 shows that almost all the species along the Danube River have some protection in the Natura 2000 protected zones. The hottest points for protection of threatened dragonfly and damselfly species along the Danube River in need of particular attention are:

Table 3. Distribution of dragonflies and damselflies among the Natura 2000 protected zones

PROTECTED AREAS / SPECIES ;	*Aeshna affinis*	*Aeshna isoceles*	*Aeshna mixta*	*Hemianax ephippiger*	*Anax imperator*	*Anax parthenope*	*Brachytron pratense*	*Caliaeschna microstigma*	*Calopteryx splendens*	*Calopteryx virgo*	*Chalcolestes parvidens*	*Coenagrion ornatum*	*Coenagrion puella*	*Coenagrion pulchellum*	*Coenagrion scitulum*	*Cordulegaster insignis*	*Cordulia aenea*	*Crocothemis erythraea*	*Enallagma cyathigerum*	*Cercion lindenii*	*Erythromma najas*	*Erythromma viridulum*	*Gomphus flavipes*	*Gomphus vulgatissimus*	*Ischnura elegans*	*Ischnura pumilio*
Archar			V																				V		V	
Batin		V			V	V			V	?	V		V					V	V			V	V		V	V
Boblata	V								V			V	V	V											V	
Vidbol									V																	
Park of Vidin																										
Vojnitsa										V																
Vartopski dol	V				V				V			V	V			V		V		V		V		V	V	
Delejna																										
Dolno Linevo																										
Zlatia																								V		
Kalimok-Bruchlen	V	V	V	V	V	V	V		V		V	V	V	V			V	V				V	V	V	V	V
Karaboaz	V		V		V				V				V	V				V				V	V	V	V	V
Kozloduj	V																								V	
Lomovete	V		V		V				V	V		V	V				V						V	V	V	V
Ludogorie																										
Ludogorie-Srebarna		V	V		V				V		V	V	V				V					V			V	V
Ludogorie-Boblata	V				V				V	V		V	V	V								V			V	V
Makresh																										
Marten-Riachovo									V														V		V	
Marsh Mominbrod																										
Nikopolsko plato												V										V		V	V	V
Novo selo																							V			
Obnova - Karaman dol																										
Orizishteto																										
Orsoia	V		V						V					V				V				V	V		V	V
Ostrov	V				V				V			V	V					V					V		V	
Bliznatsi island									V	V													V		V	
Vardim island			V									V											V	V	V	
Kutotvo island																										
Kozloduj islands																							V			
Tchajka island									V		V															
Ostrovska step - Vadim																										
Persina	V	V	V		V		V		V	?	V	V	V	V		V		V			V	V	V	V	V	V
Pozharevo - Garvan	V	V	V	V	V	V		V	V		V	V	V	V	V	V	V	V	V		V	V	V	V	V	V

PROTECTED AREAS / SPECIES ;	*Lestes barbarus*	*Lestes dryas*	*Lestes sponsa*	*Lestes virens*	*Leucorrhinia pectoralis*	*Libellula depressa*	*Libellula fulva*	*Libellula quadrimaculata*	*Onychogomphus forcipatus*	*Ophiogomphus cecilia*	*Orthetrum albistylum*	*Orthetrum brunneum*	*Orthetrum cancellatum*	*Orthetrum coerulescens*	*Platycnemis pennipes*	*Pyrrhosoma nymphula*	*Sympecma fusca*	*Sympetrum depressiusculum*	*Sympetrum flaveolum*	*Sympetrum fonscolombii*	*Sympetrum meridionale*	*Sympetrum pedemontanum*	*Sympetrum sanguineum*	*Sympetrum striolatum*	*Sympetrum vulgatum*	*Number of species in each Natura 2000 zone*
Archar							V					V			V		V				V					8
Batin											V				V					V	V		V			18
Boblata						V					V	V	V	V	V		V						V	V		15
Vidbol												V			V											3
Park of Vidin																										0
Vojnitsa																										1
Vartopski dol	V		V			V	V		V		V	V		V	V	V	V				V		V	V		25
Delejna																										0
Dolno Linevo																										0
Zlatia																										1
Kalimok-Bruchlen	V	V	V	V		V	V				V	V	V	V	V		V			V	V		V	V	V	36
Karaboaz	V								V	V	V		V	V	V		V			V	V		V			23
Kozloduj																										2
Lomovete	V					V					V	V	V		V						V		V		V	21
Ludogorie																										0
Ludogorie-Srebarna	V					V					V	V		V	V		V			V	V	V	V	V		23
Ludogorie-Boblata	V			V		V	V				V	V	V	V	V		V			V	V		V	V		24
Makresh																										0
Marten-Riachovo															V											4
Marsh Mominbrod																										0
Nikopolsko plato						V								V	V						V					9
Novo selo																										1
Obnova - Karaman dol																										0
Orizishteto																										0
Orsoia	V		V	V		V					V	V			V		V			V	V		V	V	V	22
Ostrov		V										V		V	V						V		V			14
Bliznatsi island									V						V						V					7
Vardim island															V		V				V		V			9
Kutotvo island																										0
Kozloduj islands									V																	2
Tchajka island																										2
Ostrovska step - Vadim																										0
Persina	V	V	V	V		V			V		V	V	V		V		V				V		V	V		33
Pozharevo - Garvan	V	V	V	V		V	V				V	V	V	V	V	V	V	?		V	V	V	V	V		42

SPECIES / PROTECTED AREAS ;	*Aeshna affinis*	*Aeshna isoceles*	*Aeshna mixta*	*Hemianax ephippiger*	*Anax imperator*	*Anax parthenope*	*Brachytron pratense*	*Caliaeschna microstigma*	*Calopteryx splendens*	*Calopteryx virgo*	*Chalcolestes parvidens*	*Coenagrion ornatum*	*Coenagrion puella*	*Coenagrion pulchellum*	*Coenagrion scitulum*	*Cordulegaster insignis*	*Cordulia aenea*	*Crocothemis erythraea*	*Enallagma cyathigerum*	*Cercion lindenii*	*Erythromma najas*	*Erythromma viridulum*	*Gomphus flavipes*	*Gomphus vulgatissimus*	*Ischnura elegans*	*Ischnura pumilio*	*Lestes barbarus*	*Lestes dryas*	*Lestes sponsa*	*Lestes virens*	*Leucorrhinia pectoralis*	*Libellula depressa*	*Libellula fulva*	*Libellula quadrimaculata*	*Onychogomphus forcipatus*	*Ophiogomphus cecilia*	*Orthetrum albistylum*	*Orthetrum brunneum*	*Orthetrum cancellatum*	*Orthetrum coerulescens*	*Platycnemis pennipes*	*Pyrrhosoma nymphula*	*Sympecma fusca*	*Sympetrum depressiusculum*	*Sympetrum flaveolum*	*Sympetrum fonscolombii*	*Sympetrum meridionale*	*Sympetrum pedemontanum*	*Sympetrum sanguineum*	*Sympetrum striolatum*	*Sympetrum vulgatum*	*Number of species in each Natura 2000 zone*
Rabovo																																																				0
Vit river	V								V					V									V			V									V			V		V	V										V	10
Iskar river																																																				0
Lom river	V								V									V						V	V								V		V		V	V		V	V						V			V		13
Ogosta river					V				V									V		V			V	V	V	V									V		V	V	V	V	V											14
Skut river									V																V													V		V	V											5
Yantra river									V			V						V		V		V	V	V	V							V					V	V	V		V						V					14
Svishtov forest																																																				0
Srebarna	V	V	V	V	V	V	V	V	V		V	V	V	V			V	V	V	V	V	V	V	V	V	V	V	V	V	V	V	V					V	V	V	V	V		V	?	?	V	V	V			V	41
Studena river		V			V	V			V									V				V			V		V										V		V								V					11
Timok																																								V												1
Hursovska river																																																				0
Tzar Petrovo																																																				0
Tzibar	V								V				V										V		V										V		V	V			V											9
number of Natura 2000 sites for each species	#	7	#	3	#	5	3	2	#	6	7	#	#	9	1	3	5	#	3	4	3	#	#	#	#	#	#	5	6	6	1	#	6	0	9	1	#	#	#	#	#	2	#	2	1	8	#	3	#	9	5	

1. Pozharevo-Garvan Protected Area with 42 species of which 4 are threatened.
2. Srebarna Protected Area with 41 species of which 4 are threatened (Beschovski 1998, Marinov 2000b).
3. Kalimok-Brashlen Protected Area with 36 species of which 2 are threatened.
4. Persina Protected Area with 33 species of which 3 are threatened (*C. ornatum* has already disappeared).
5. Vartopski Dol Protected Area with 25 species of which 2 are threatened.

ACKNOWLEDGMENTS

I would like to express my sincere gratitude to some researchers-amateurs who have constantly been providing information without which this paper would not have been possible. Special thanks go to the WWF Bulgaria Danube-Carpathian program, which has given me the chance to participate in trips along the Danube River to collect new information, especially in the westernmost parts of the study area. I also thank the Persina Nature Park Directorate for the opportunity to carry out a full year of research in the park, the first of its kind along the Danube River, and for its efforts in making Persina Nature Park one of the hotspots for the protection of dragonflies and damselflies along the Danube River.

REFERENCES

Beschovski V. 1965. Odonata (dragonflies) from the Bulgairan riverside of the Danube and from certain wetlands in Northern Bulgaria. *Bull. Inst. Zool. Mus.* 18: 159-168.

Beschovski V. 1967. Ecological survey of the larvae of Odonata in the Bulgarian rivers. *Bull. Inst. Zool. Mus.* 24: 5-20.

Beschovski V. 1994. Fauna of Bulgaria, Insecta, Odonata. Bulgarian Academy of Sciences, vol. 23, 372 p.

Beschovski V. & Marinov M. 1998. Order Odonata. In: Michev T., Georgiev B., Petrova A. & Stoyneva M. (eds.) 1998. *Biodiversity of the Srebarna Biosphere reserve.* Ministry of Environment and Waters, Pensoft, 73 p.

Kutsarov Y., Bechev D., Kostadinov R. & Marinov M. 2012. The Bulgarian Odonata database – current status, organisation and a case study new entries. *ZooNotes* 33: 1-25.

Kalkman V. J., Boudot J.P., Bernard R., De Knijf G. & Suhling F. 2018. Diversity and conservation of European dragonflies and damselflies (Odonata). *Hydrobiologia.* https://doi.org/10.1007/s10750-017-3495-6.

Marinov M. 1999. Chalcolestes parvidens (Artobolevski) and Somatochlora meridionalis Nielsen in Bulgaria (Zygoptera: Lestidae, Anisoptera: Corduliidae). *Notulae Odonatologicae* 5(3): 31-33.

Marinov M. 2000a. An atypical Caliaeschna microstigma (Schneider) habitat in Bulgaria (Anisoptera: Aeshnidae). *Notulae Odonatologice* 5(6): 83.

Marinov M. 2000b. A contribution to the fauna of Srebarna biosphere reserve. *Acta zoologica bulgarica* 52 (2): 37-42.

Marinov M. 2001a. The Orthetrum coerulescens complex in Bulgaria (Odonata: Libellulidae). *International Journal of Odonatology* 4 (1): 35-40.

Marinov M. 2001b. Review of Hemianax ephippiger (Burm.) records from Bulgaria (Anisoptera: Aeshnidae). *Notulae Odonatologice* 5 (8): 105-106.

Marinov M., Grebe B. & Kutsarov Y. 2007. Cordulegaster insignis (Schneider, 1845) in Bulgaria with notes on its Ecological notes. In: Tyagi B.K. (ed.): *Odonata: Biology of Dragonflies*. Scientific Publishers (India), Jodhpur, pp. 51-61.

Plant C., Sands R. & Fasham M. 2005. Dragonflies and damselflies. In: Hill D., Fasham M., Tucker G., Shewry M. & Shaw P. (eds.) *Handbook of Biodiversity Methods: Survey, Evaluation and Monitoring,* Cambridge: Cambridge University Press, pp. 322-327.doi:10.1017/CBO9780511542084.017

Russev B. 1962. The insect fauna of Danube River in front of the Bulgarian shore. *Bull. Research freshwater-fisheries station, Plovdiv* 1: 115-128.

Russev B. 1966. Zoobenthos of Danube River between 845 and 375 river kilometer. I. Composition, distribution and ecology. *Bull. Inst. Zool. Mus.* 20: 55-131.

In: Biodiversity of the Bulgarian-Romanian Section
Editor: Shurulinkov, Hubenov et al.
ISBN: 978-1-53615-663-8

Chapter 5

SPECIES COMPOSITION AND DISTRIBUTION OF NEMATOCERA (DIPTERA: INSECTA) FROM THE BULGARIAN DANUBE RIVER BANK

Zdravko Hubenov
National Museum of Natural History,
Bulgarian Academy of Sciences, Sofia, Bulgaria

ABSTRACT

The Nematocera suborder of the Danube River, the Danubian islands and the nearby parts of the Bulgarian Danube River bank are scrutinized. A total of 153 species from 14 families have been established. Three families are better studied – Culicidae (20 species – 13.1%), Ceratopogonidae (36 species – 23.5%) and Chironomidae (63 species – 41.2%). The established species belong to 52 areographical categories. The fauna can be divided into 2 main groups: 1) species with Mediterranean type of distribution (12 species – 7.8%) – more thermophilic and distributed mainly in the southern parts of the Palaearctic. Four species of southern type, distributed in the Palaearctic and beyond it, can be formally related to this group as well; 2) species with Palaearctic and Eurosiberian type of distribution (141 species – 92.2%) – more cold-resistant and widely distributed in the Palaearctic. Sixty-four species of northern type, distributed in the Palaearctic and beyond it, can be formally related to this group as well. Of the established taxa, 14 are pests on the forestry or agriculture, and 13 species have a human or veterinary medical significance.

Keywords: Diptera, Nematocera, Bulgarian Danubian bank, faunistic composition, zoogeography

ABBREVIATIONS

[ ] – names and synonyms under which the species are reported from Bulgaria;
● – species without precise locality;
♦ – species, presented everywhere, without precise locality;
▲ – species of human or veterinary medical significance;
■ – enemy for forestry or agriculture;
♠ – lack of data;
♣ – outdated information;
* – literary sources, containing data for the Danube River, Danubian islands or nearby territories;
E – Eastern Danubian bank;
M – Middle Danubian bank;
W – Western Danubian bank.

Areographical and Zoogeographical Categories

? – possible category;
cee – Central and East European;
csena – Central and South European-North African;
cset – Central (Middle) and South European-Turanian;
des – Disjunct Eurosiberian;
dp – Disjunct Palaearctic;
e – European;
ean – European-Anatolian;
ee – East European;
ena – European-North African;
esanca – Eurosiberian-Anatolian-Central Asian;
esca – Eurosiberian-Central Asian;
ewca – European-West Central Asian;
h – Holarctic;
hn – Holarctic-Neotropical;
hno – Holarctic-Neotropical-Oriental;
ho – Holarctic-Oriental;
hoa – Holarctic-Oriental-Australian;
hoes – Holoeurosiberian;
hpta – Holarctic-Paleotropical-Australian;
hptn – Holarctic-Paleotropical-Neotropical;
mca – Mediterranean-Central Asian;

mwca – Mediterranean-West Central Asian;
nmi – North Mediterranean-Iranian;
nmsfe – North Mediterranean and South Far East;
pa – Palaearctic-Australian;
pat – Palaearctic-Afrotropical;
po – Palaearctic-Oriental;
poa – Palaearctic-Oriental-Australian;
ppt – Palaearctic-Paleotropical;
ppta – Palaearctic-Paleotropical-Australian;
ptm – Paleotropical-Mediterranean;
sena – South European-North African;
sk – Semicosmopolitan;
sp – South Palaearctic;
spo – South Palaearctic-Oriental;
swpo – Southwest Palaearctic-Oriental;
tes – Transeurosiberian;
tp – Transpalaearctic;
wces – West and Central Eurosiberian;
wcp – West and Central Palaearctic;
wes – West Eurosiberian;
wp – West Palaearctic;
wpo – West Palaearctic-Oriental.

INTRODUCTION

The studies on Diptera of the Bulgarian Danube River bank began 113 years ago. The first data on the two-winged fauna from the territory along the Danube River were reported by Kovachev (1905), Manolov (1909a) and Nedelkov (1909, 1912). A number of works have been published by other authors (Joakimoff 1899; Theobald 1907, 1910; Drenowsky 1922, 1929a, 1929b; Drenowsky & Enderlein 1923; Buresch 1924, 1926, 1938; Enderlein 1924; Drensky 1926; Drensky & Drensky 1928; Czerný 1930; Drensky 1931a, 1931b, 1932, 1942; Szilady 1934; Zilahi 1934; Thienemann 1936; Lackschewitz 1940a). In some works, there are data on the Danube River, the Danubian islands and the nearby areas. After the World War II, studies on the Bulgarian Nematocera were conducted by Kozarov (1949), Strenzke (1951a, 1951b), Bozhkov (1953, 1957, 1958, 1959, 1961, 1962, 1965, 1966a, 1966b, 1967, 1974a, 1974b, 1991), Kozarov & Bozhkov (1953); Drensky (1955), Stefanov (1955, 1956), Dimov (1957), Marinov (1957), Guéorguiev & Beron (1962) Naidenov (1962), Bozhkov & Christova (1965), Bozhkov et al. (1969), Michailova (1982, 1985, 1989, 1994, 1996, 2006), Krzemiński (1984), Bechev (1986b, 1994, 1996a, 1997,

2000, 2002a, 2004b, 2006b, 2006c, 2010), Dimitrova (1989), Skuhravá et al. (1991, 1992), Beron (1994, 2015, 2016), Popov (1999), Beschovski (2006), Bobeva et al. (2013, 2014, 2015), Bobeva (2014), Adler & Crosskey (2018). Some of these works refer to the territory around the Danube River as well. The hydrobiological research of the Bulgarian waters was of great merit for the study of the water Diptera (Valkanov 1936, 1941a, 1941b, 1948, 1954, 1957a; Arndt 1943; Caspers 1951a, 1951b, 1951c, 1952, 1957; Strenzke 1951; Cvetkov 1955a, 1955b, 1957, 1958, 1962; Angelov 1956; Dimitrov 1957, 1960a, 1960b, 1962a, 1962b, 1962c, 1962d, 1963a, 1963b, 1966, 1969, 1970, 1972; Russev 1959, 1962, 1963, 1966a, 1966b; Zaschev & Angelov 1959; Mihailova-Neikova 1961; Kovachev 1975, 1976, 1990; Janeva 1987, 1989, 1991; Russev et al. 1987, 1991, 1994; Stoichev 1994, 1996, 1998, 2000a, 2000b, 2001a, 2001b, 2002, 2004; Janeva & Russev 1997; Kovachev et al. 1999; Janeva et al. 2001; Stoichev & Cernev 2001; Varadinova et al. 2012, 2013; Stoyanova et al. 2013; Dashinov 2017). In the applied entomological literature there are numerous data for the two-winged insects that are plant pests or have medical significance (Malkov 1902a, 1902b, 1902c, 1903, 1906b, 1907; Manolov 1907; Hitilov 1912a, 1912b; Chichkoff & Konsuloff 1914; Konsuloff 1921a, 1921b, 1922a, 1922b, 1922c 1923a, 1923b; Konsuloff & Paspalev 1924, 1925; Markov 1923, 1925a, 1925b, 1929, 1937; Tschorbadjiev 1925a, 1925b, 1925e, 1926, 1927, 1929, 1932, 1939; Markov & Morov 1929; Vasilev 1931; Lazarov 1935, 1939, 1949; Slivenski 1935, 1940, 1946; Drensky 1939, 1949, 1950, 1958; Petkoff 1939; Kitanov 1943; Nikolova 1943, 1949a, 1962; Avlavidov 1947, 1948, 1958, 1959, 1961, 1970; Zhelev 1948; Drensky 1949, 1950, 1958; Ganov 1949; Boychev 1950a, 1950b; Paspalev 1950, 1951; Hristova 1954; Buresch & Lazarov 1956; Popoff 1956; Kovchazov 1958, 1961, 1976; Popoff & Nikolova 1958; Zamfirov 1958, 1961; 1962a, 1962b, 1962c, 1963; Hristova, 1959; Kovachevski et al. 1949, 1959; Atanassov & Christova 1960; Dimtchev et al. 1960, 1962; Kalaydzhiev et al. 1960; Atanassov & Petrov 1961; Atanassov et al. 1962; Grigorov 1962, 1972, 1976; Choleva 1964; Harizanov 1964; Sengalevich 1964; Donchev 1968; Christova & Todorova 1969; Christova et al. 1971; Minarzh & Christova 1971; Zlatanov 1971; Christova & Dorovski 1972; Dorovski 1976; Christova 1980; Harizanov et al. 1996; Mikov 2005, 2008a, 2008b, 2011; Kutsarov 2006; Mikov et al. 2011, 2013; Agushev 2012, 2014, 2015; Nedelchev 2013; Agushev & Bileva 2014).

The aim of this paper is to scrutinize the fauna, distribution and zoogeography of the suborder Nematocera from the Danube River sector of Bulgaria and to present a bibliography of the established species on the whole territory of the country.

INVESTIGATED REGIONS, MATERIALS AND METHODS

The Bulgarian Danube River bank (the coastline between the Danube River and the Danubian Plane) begins at the mouth of the Timok River (845 river km) and ends at Silistra

Town (374 river km). It is 471 km long, from 250 m to 15 km wide, up to 100 m high and has an area of about 2000 km^2. The Vit and Yantra rivers divide it into the Western, Middle and Eastern parts (Michailov et al. 1966; Michailov & Popov 1978; Kiradzhiev 2013). The bank is the widest in the Vidin, Chernopol, Belene and Brashlen (Pobrezhie) lowlands. It is the narrowest in the villages of Vetren, Somovit, Dolno Ryahovo and Malak Preslavets. The steep slopes cover 171 km (36% of the total length of the bank). The lowlands along the river cover about 293 km (62% of the length of the bank). These include the flooded terrace and parts of the overflowing terraces. The parts of the river valleys at the river mouths cover about 6.3 km (1.3%) of the bank. The area of the lowlands along the Danube River is about 1300 km^2. Most of them, before the construction of the dikes, were swamps. Now they are drained and are converted into cultivable lands. Of more than 20 marshes with an area of about 100 km^2 in the past, nowadays Srebarna and Garvan (Leshtava) with 5 km^2 area are left. The Danubian islands also belong to the floodplains along the Danube River. The total area of the Bulgarian islands (about 84) is about 115 km^2 (Michailov & Popov 1978; Kiradzhiev 2013). Along the Danubian bank, one flooded and some overflowing terraces are outlined. The flooded terrace defines the basic lines of the bank. It comprises 56.1% of the bank's length and covers an area of 732.5 km^2. The altitude of the flooded terrace at Timok River is 35 m, and at Silistra Town – 16 m. From the Bulgarian side, 13 rivers (Topolovets, Voynishka, Vidbol, Archar, Skomlya, Lom, Tsibrica, Ogosta, Iskar, Vit, Osam, Yantra and Roussenski Lom) are flowing into the Danube River. Of their total runoff, 26% appertain to the Iskar River and 20% to the Yantra River (Hristova 2012). The water levels of the Danube River are characterized by a spring maximum and an autumn minimum (Mladenov 1978). The climate is continental with temperatures from -30^0 to +42^0C. The winds are oriented along the riverbed. The low rainfall defines the Danubian coast as one of the dry parts of the country (Sabev & Stanev 1959; Tichkov 1982; Velev 1990, 2002; Stanev 1991). The vegetation belongs to the xerothermic oak forests belt. Various interzonal communities are presented. Under human impact, the natural vegetation is largely destroyed (Stojanov 1948, 1966; Velchev et al. 1982, 1989; Bondev 1991, 1997, 2002; Velchev 1997, 2002). The coastal zone of the Danube River belongs to the Danubian zoogeographical region and has an European and Eurosiberian faunistic character (Georgiev 1980, 1982, 1997, 2002).

The material has been collected after 1900 and is collected from 46 localities (Table 1). The main part of it is stored in the National Museum of Natural History, Sofia and the Institute of Biodiversity and Ecosystem Research. All species reported from the Danube River, the Danubian islands and the surroundings of the river (taxa of human and veterinary medical significance, forest and agricultural pests, and also species listed in various foreign publications) are included. Many collectors did not give accurate localities and indicated only the Danube River or the Danubian bank. For a number of widespread and numerous species the authors did not give the localities and mentioned they occur everywhere.

Table 1. Localities of Nematocera (Diptera: Insecta) from the Bulgarian Danubian bank

Localities	GPS Navigation (°N, °E)	UTM Kode	Species
along the whole Bulgarian Danubian bank (♦)		FP39-NJ28	36
Archar Village	43°48'45.03", 22°55'06.45"	FP55	2
Aydemir Village	44°06'49.54", 27°10'40.70"	NJ19	1
Belene Town	43°39'10.38", 25°07'38.83"	LJ43	2
Belene Town, surroundings	43°39'02.53", 25°17'34.41"	LJ63	3
Botevo Village	43°50'23.22", 22°51'37.75"	FP45	1
Danubian River bank (●)		FP39-NJ28	30
Dolni Tsibar Village	43°49'01.09", 23°31'06.95"	GP05	3
Gamzovo Village	44°05'50.07", 22°45'39.07"	FP48	2
Iskar River near Baykal Village	43°42'08.24", 24°27'19.95"	KJ94	2
Iskar River mouth	43°43'54.43", 24°26'28.93"	KJ94	1
Kalimok (Biological Experimental Station)	44°00'41.34", 26°26'21.56"	MJ57	32
Karamanovo Village	43°33'37.83", 25°31'42.90"	LJ82	3
Kozloduy Town, surroundings	43°46'36.03", 23°44'25.05"	GP25	4
Krivina Village, near the Danubian River bank	43°38'42.88", 25°34'42.18"	LJ83	4
Kutovo Village	44°00'49.53", 22°58'25.61"	FP57	1
Lom Town, near the Danubian River bank	43°37'15.53", 25°21'32.53"	FP75	6
Lom Town, surroundings	43°49'17.00", 23°11'13.10"	FP75	9
Lom River, before Lom Town	43°46'55.41", 23°14'03.63"	FP75	3
Malak Preslavets Village	44°05'54.96", 26°50'01.52"	MJ88	1
Mechka Village, surroundings	43°42'22.23", 25°47'08.33"	MJ04	5
Nikopol Town	43°42'21.42", 24°53'43.37"	LJ36	11
Nova Cherna Village	44°01'04.58", 26°27'22.88"	MJ57	3
Novo Selo Village	44°09'49.13", 22°47'09.74"	FP49	3
Ogosta River mouth	43°44'43.54", 23°53'01.89"	GP34	2
Ogosta River near Glozhene Village	43°41'31.89", 23°49'33.83"	GP24	4
Oresh Village	43°34'44.45", 25°14'21.86"	LJ52	2
Oryahovo Town	43°44'19.96", 23°57'41.35"	GP34	5
Ostrov Village	43°40'57.57", 24°08'58.92"	KJ63-KJ73	5
Popina Village	44°07'40.24", 26°56'56.55"	MJ98	1
Rousse Town, surroundings	43°50'42.70", 25°56'35.49"	MJ15	32
Ryahovo Village	43°59'41.21", 26°14'43.28"	MJ37	2
Silistra Town, surroundings	44°07'25.22", 27°16'03.97"	NJ28	24
Somovit Village	43°41'21.04", 24°46'26.81"	LJ23-LJ24	7
Srebarna Lake	44°06'24.64", 27°04'22.52"	NJ08	17
Sredna Kula (suburb of Rousse Town)	43°48'34.10", 25°56'10.22"	MJ15	1
Stanevo Village, near the Danubian River bank	43°50'39.30", 23°26'38.07"	FP95	1
Svishtov Town, surroundings	43°37'15.53", 25°21'32.53"	LJ63	8
Tsar Simeonovo Village	43°51'09.96", 22°50'35.19"	FP45	1
Tutrakan Town	44°02'51.63", 26°36'08.38"	MJ67-J77	7
Vardim Village	43°36'50.79", 25°27'54.93"	LJ72-LJ73	1
Vetren Village	44°08'24.52", 27°01'53.32"	NJ08	1
Vidin Town, near the Danubian River bank	43°59'31.33", 22°53'11.99"	FP57	22
Vidin Town, surroundings	43°58'36.73", 22°52'19.10"	FP57	12
Vit River near Gulyantsi Town	43°38'02.45", 24°41'53.19"	LJ13	4
Vrav Village	44°11'43.78", 22°44'12.86"	FP39	2

The processing of literary data has reviewed all data that refer to the Bulgarian Diptera. Recent electronic issues have been used to update the names of the taxa. The classification of the dipterans is based on the works of Soós & Pap (1984-1993), Wood & Borkend (1989), Papp & Darvas (1997, 2000a, 2000b), Nartshuk (2003), Ziegler (2003), Oosterbroek (2006) and Yeates et al. (2007).

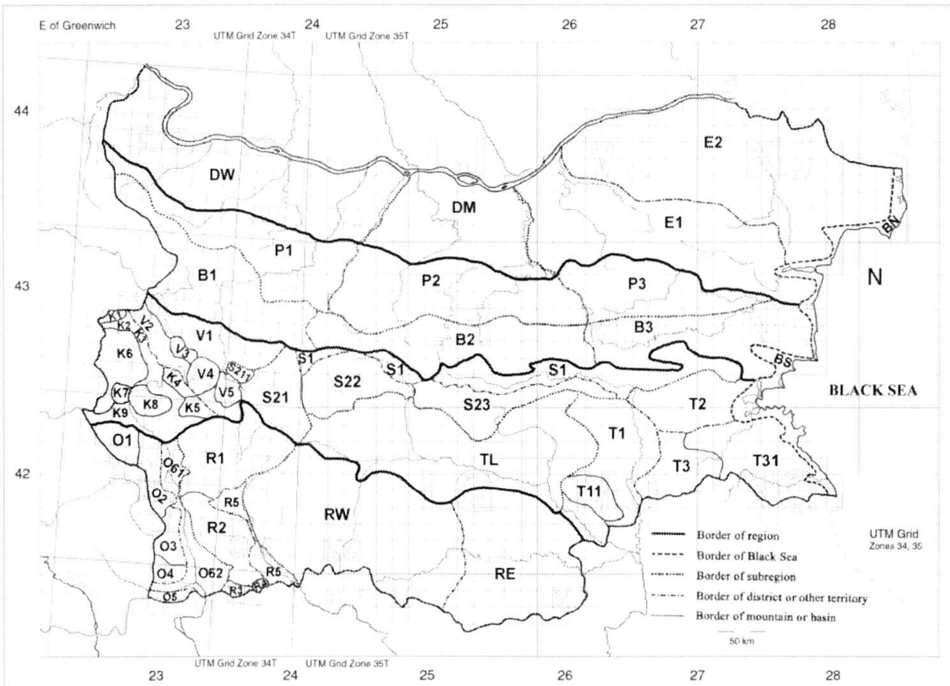

Figure 1. Natural geographical territorial units of Bulgaria.

Abbreviations used: **B** – Black Sea Coast; **B1** – Western Stara Planina (Balkan) Mts.; **B2** – Middle Stara Planina (Balkan) Mts.; **B3** – Eastern Stara Planina (Balkan) Mts.; **BN** – Northern Black Sea Coast; **BS** – Southern Black Sea Coast; **D** – Danubian Plain; **DM** – Middle Danubian Plain; **DW** – Western Danubian Plain; **E1** – Popovo-Provadiya district; E2 – Loudogorie-Dobroudzha district; **P1** – Western Predbalkan; **K1** – Rouy Mt.; **K2** – Trun Basin (Znepole); **K3** – Strazha-Cherna Gora-Rudini Mts.; **K4** – Golo Burdo Mt.; **K5** – Verila Mts.; **K6** – Kraishte; **K7** – Zemenska Planina Mt.; **K8** – Konyavska Planina Mt.; **K9** – Kyustendil Basin; **O1** – Osogovska Planina Mts.; **O2** – Vlahina Planina Mts.; **O3** – Maleshevska Planina Mts.; **O4** – Ograzhden Mts.; **O5** – Belasitsa Mts.; **O6** – Srednostroumska Valley; **O61** – Boboshevo-Simitli Valley; **O62** – Kroupnik-Sandanski-Petrich Valley; **P2** – Middle Predbalkan; **P3** – Eastern Predbalkan; **R** – Rila-Rhodope Massif; **R1** – Rila Mts.; **R2** – Pirin Mts.; **R3** – Slavyanka Mt.; **R4** – Sturgach Mt.; **R5** – Mesta Valley; **RE** – Eastern Rhodope Mts.; **RO** – Osogovo-Belasitsa group; **RP** – Rila-Pirin group; **RR** – Rhodope Mts.; **RW** – Western Rhodope Mts.; **S** – Stara Planina Range System; **S1** – Podbalkan Basins; **S2** – Sredna Gora Mts.; **S21** – Ihtimanska Sredna Gora Mts.; **S211** – Lozenska Planina Mt.; **S22** – Sushtinska Sredna Gora Mts.; **S23** – Surnena Sredna Gora Mts.; **SB** – Stara Planina (Balkan) Mts.; **SP** – Predbalkan; **T** – Transitional Region; **T1** – Sakar-Toundzha district; **T11** – Sakar Mt.; **T2** – Bakadzhik-Bourgas district; **T3** – Strandzha-Dervent district; **T31** – Strandzha Mts.; **TK** – Kraishte-Konyavo district; **TL** – Thracian Lowland; **TS** – Srednogorie-Podbalkan subregion; **TT** – Toundzha-Strandzha subregion; **TV** – Vitosha district; **V1** – Sofia Basin; **V2** – Zavalska-Viskyar Mts.; **V3** – Lyulin Mt.; **V4** – Vitosha Mt.; **V5** – Plana Mts.

Weaknesses in the literature data which limit the obtaining of an equivalent information when comparing territories and systematic groups include: different levels of study of the individual taxa; insufficient research of many families in the corresponding areas; a lack of exact localities for the part of the recorded species; insufficient data for the localities and vertical distribution; existence of rich synonymy; outdated data; a lack of generalized investigations; significant differences in the number of taxa in the separate areas; unexplored territories; prolonged periods of data accumulation for most regions; predominance in recent years of the ecological studies versus those of the fauna; priority research on groups of economic and medical significance[1].

The areographical categorization of the species is based on the available literature and recent electronic issues. Zoogeographical analysis for the species categorization was used. This method allows obtaining data information about species complexes with different zoogeographical character based on the published data regarding species distribution and results of the faunistic research. These complexes contain zoogeographical information about the taxonomic groups which, combined with the origin of the ranges, determines the zoogeographical character of the fauna. The distribution of the species according to the zoogeographical categories is scrutinized. The classification of the areas is based on the works of Geptner (1936), Darlington (1957), Kryzhanovsky (1965, 1976, 2002), de Lattin (1967), Müller (1974, 1980), Udvardi (1975), Crosskey & White (1977), Malicky et al. (1983), Gorodkov (1984), Grehan (1988, 1993), Vigna Taglianti et al. (1999)[2], Procheş & Ramdhani (2012), Holt et al. (2013) and Ficetola et al. (2017).

For each species are given: 1) a recent scientific name and synonyms under which it was reported from Bulgaria; 2) distribution on the investigated territory; 3) areographical characteristics; 4) literature for Bulgaria, as an asterisk indicates the literary sources containing data for the Danube River, the Danubian islands or nearby areas; 5) distribution throughout the country – Figure 1 (Hubenov 1997). More than one areographical category is given for some species. This is related to the differences in the distribution of the taxa presented in the separate literary sources. For introduced taxa, the natural range is first indicated, followed by the secondary (anthropogenic) range. Unfortunately, in most publications is not specified whether the range is natural or anthropogenic one. The zoogeographical analysis only includes the first arealographical categories. Before the

---

[1] These weaknesses lead to 6 problems: 1) Continuous supplementation of an existing historical list of the fauna. As a result, species diversity in a given area is higher than in reality. 2) Incomparability of data in terms of time periods. Data comparisons between two areas very often cover different periods as it is not possible to study all taxonomic groups and territories simultaneously. 3) Inaccurate zoogeographical characteristic of some taxa due to the lack of information on their vertical distribution. 4) Incomplete reporting of anthropogenic influences, successional and landscape changes on the communities in the separate regions. Thus, a number of well-studied areas in the past have already been significantly changed. 5) Prioritization of research in areas under monitoring or environmental protection legislation. 6) Workload of the specialists with environmental, medical and other applied investigations that do not allow them to devote time to faunistic research.

[2] The inversion of the nomenclature of the ranges and the border between the West and East Palaearctic along the Ural Mts. of Vigna Taglianti et al. (1999) was not followed. The traditional nomenclature and the border along the Yenisey River was used.

following categories there is a question mark "?" (for differences between the literary sourccs) or a lcttcr "i" (for introduction data).

Results and Discussion

A total of 153 species of suborder Nematocera, belonging to 14 families, has been reported from the Bulgarian Danube River bank so far (about 10% of the all 1540 species found in Bulgaria). The families Chironomidae (63 or 41.8%), Ceratopogonidae (36 or 23.5%) and Culicidae (20 or 13.1%) comprise the greatest number of species. These families were investigated in the studied territory. The remaining families contain from 1 to 11 species. They are poorly investigated that is evident from the review of the known species composition in Bulgaria (Table 2). Large groups, such as Limoniidae, Mycetophilidae and Cecidomyiidae (with 220 to 260 species) are represented by several species. At the same time, 12 families of Nematocera, known for the Bulgarian fauna, are not reported from the studied area. The dipterans are a highly mobile group and after further exploration of the studied territory, can be expected to reach 40-50% of the species composition of the most families found in the country. This is evident in the better studied families Culicidae and Ceratopogonidae, which are represented by about 40-50% of the species found in Bulgaria.

Only the best studied family – Culicidae, is evenly distributed in the different parts of the studied territory. In the remaining families the taxa distribution is connected with the exploration of the corresponding region and is not systematic. This is evident when comparing the established species in regard to the localities (Table 1). Four areas of detailed research (over 20 species found) are outlined. First are the surroundings of Rousse and Vidin, followed by Silistra – towns with scientific centers where research has begun at the earliest. The big number of species in the Kalimok area is related to the studies on the family Ceratopogonide conducted in the Biological experimental station Kalimok of the Bulgarian Academy of Sciences. Thus, most species (122 species – 79.7%) have been found in the eastern part of the Danube River sector (Table 2). The middle part of the Danubian river bank where five of the families have not been reported, is most poorly studied.

In comparison with Pirin, Rila, Vitosha and Vrachanska Planina Mountains (Hubenov 2015, 2016, 2018, 2019) it is seen that in the surroundings of the Danube River, a doubly smaller number of species of sudorder Nematocera have been established. It should be kept in mind that Vitosha and Rila are the most studied Bulgarian mountains and in Pirin Mt. and Vrachanska Planina Mts. there are some well-studied big families. With regard to the variety of habitats, the mountains significantly exceed the Danube River sector. At the same time, it is evident that the well-studied families Culicidae and Ceratopogonidae are poorly represented in the mountains.

Of the established taxa, 14 are pests on the forestry or agriculture, and 13 species have a human or veterinary medical significance.

Table 2. Nematocera from the Bulgarian sector of the Danube River

Families	Species in Bulgaria	Species from the Bulgarian Danubian bank			
		Total	Western Danubian bank	Middle Danubian bank	Eastern Danubian bank
Tipulidae	89	4 (4.5%)	3 (2.6%)	4 (5.5%)	3 (2.5%)
Limoniidae	221	2 (0.9%)	2 (1.7%)		
Pediciidae	25				
Cylindrotomidae	1				
Blephariceridae	6				
Bibionidae	13	2 (15.4%)	1 (0.9%)	1 (1.4%)	2 (1.6%)
Hesperinidae	1				
Mycetophilidae	255	2 (0.8%	2 (1.7%)		
Ditomyiidae	2				
Bolitophilidae	10				
Diadocidiidae	3				
Keroplatidae	48	1 (2.1%)	1 (0.9%)		
Sciaridae	84	1 (1.2%)	1 (0.9%)	1 (1.4%)	1 (0.8%)
Cecidomyiidae	260	11 (4.2%)	10 (8.8%)	10 (13.9%)	10 (8.2%)
Psychodidae	12	3 (25.0%)	2 (1.7%)	1 (1.4%)	1 (0.8%)
Trichoceridae	7				
Anisopodidae	3				
Scatopsidae	4	1 (25.0%)	1 (0.9%)		
Ptychopteridae	5				
Dixidae	1				
Chaoboridae	1	1 (100%)			1 (0.8%)
Culicidae	47	20 (42.6%)	19 (16.7%)	17 (23.6%)	19 (15.6%)
Thaumaleidae	1				
Simuliidae	71	6 (8.5%)	4 (3.5%)	2 (2.8%)	4 (3.3%)
Ceratopogonidae	66	36 (54.5%)	18 (15.8%)		35 (28.7%)
Chironomidae	305	63 (20.7%)	50 (43.9%)	36 (50.0%)	46 (37.7%)
Total	1541	153 (9.9%)	114 (74.5%)	72 (47.1%)	122 (79.7%)

NEMATOCERA

Tipulomorpha

Tipulidae

***Nephrotoma scalaris* (Meigen, 1818)** [*Pachyrhina*] – M; Svishtov, LJ63, N43°37'15.53", E25°21'32.53"; wp, ? wpo; Nedelkov *1912; Szilady 1934; Oosterbroek & Theowald 1992; Popov 1999; Oosterbroek 2017; B1, DM, TL; 50-320 m.

***Nigrotipula nigra* (Linnaeus, 1758)** – W, M, E; ♦; FP39-NJ28; tp, ? h; Nedelkov 1912; Szilady *1934; Oosterbroek & Theowald 1992; Popov 1999; Oosterbroek 2017; TL; 206 m.

***Tipula* (*Tipula*) *oleracea* Linnaeus, 1758** – W, M, E; ♦, ■; FP39-NJ28; ena (hn, i); Malkov 1907, Nedelkov 1912, Tschorbadjiev *1932, Nikolova 1949a, 1962; Drensky 1955; Buresch & Lazarov 1956; Popov 1956; Popov & Nikolova *1958; Grigorov 1972, 1976; Oosterbroek & Theowald 1992; Popov 1999; Oosterbroek 2017; DE, V1, V4, TL; 150-1000 m.

***Tipula* (*Tipula*) *paludosa* Meigen, 1830** – W, M, E; ●, ■; wp, ? tp (h, i); Nikolova 1949, 1962; Drensky 1955; Popov & Nikolova *1958; Grigorov *1972, *1976; Oosterbroek & Theowald 1992; Popov 1999; Oosterbroek 2017; E2, V1, B; 0-550 m.

Limoniidae

***Erioptera* (*Mesocyphona*) *bivittata* (Loew, 1873)** [*Molophilus*, *Ormosia*] – W; Lom, FP75, N43°49’56.70”, E23°13’43.91”; tp; Szilady *1934; Lackschewitz 1940a; Krzemiński *1984; Savchenko et al. 1992; Oosterbroek 2017; DW, TL, R1; 30-2389 m.

***Rhabdomastix* (*Rhabdomastix*) *subparva* Stary, 1971** [*R. schistacea* (Schummel, 1829)] – W; Lom, FP75, N43°49’56.70”, E23°13’43.91”; e; Szilady *1934; Starý 2004a; Mendl 1986; Krzemiński & Starý 1989; Savchenko et al. 1992; Oosterbroek 2017; DW, B, T31, R2; 30-550 m.

Bibionomorpha

Bibionidae

***Bibio hortulanus* (Linnaeus, 1758)** – W, M, E; ♦; ■; Vidin, FP57, N43°59’31.33”, E22°53’11.99”; Rousse, MJ15, N43°50’42.70”, E25°56’35.49”; wp; Joakimoff 1899; Kovachev *1905; Nedelkov 1912; Szilády *1934; Drensky 1955; Popov & Nikolova 1958; Grigorov 1972; DW, E1, E2, V1, TL, R1; 20-1200 m.

***Bibio marci* (Linnaeus, 1758)** – E; Rousse, MJ15, N43°50’42.70”, E25°56’35.49”; ena; Kovachev *1905; Nedelkov 1912; Drensky 1928; Popov & Nikolova 1958; Nikolova & Natskova 1965; DM, E1, E2, V1, K9, R1, TL; 25-600 m.

Mycetophilidae (Fungivoridae)

***Mycetophila mitis* (Johannsen, 1912)** – W; Kozloduy, GP25, N43°46’36.03”, E23°44’25.05”; h; Bechev *1996a, 1997, 2000, 2002a, *2010; DW; 60 m.

***Platurocypta punctum* (Stannius, 1831)** – W; Kozloduy, GP25, N43°46’36.03”, E23°44’25.05”; h; Bechev *1994, 1997, 2000, 2002a, 2006b, *2010; DW, B1, B2, RW; 60-1500 m.

Keroplatidae (+ Macroceridae)

***Orfelia nemoralis* (Meigen, 1818)** – W; Kozloduy, GP25, N43°46’36.03”, E23°44’25.05”; h, ? e; Bechev *1986b, 2000, 2002a, 2004b, 2006b, 2006c; DW, E1, E2, TL, O62, RW, BN, BS; 10-1400 m.

Sciaridae (Lycoriidae)

***Bradysia fenestralis* (Zetterstedt, 1838)** [*Sciara*, *Neosciara*] – W, M, E; ♦; ■; ♠; ? e; Choleva 1964; Grigorov *1972.

Cecidomyiidae

***Dasineura mali* (Kieffer, 1904)** – W, M, E; ♦; ■; hoes, i, hna; Grigorov 1962, *1972; Naidenov 1962; Skuhravá et al. 1991, 1992; B1, V1, K9, TL, BN; 0-1000 m.

***Dasineura napi* (Loew, 1850)** [*D. brassicae* (Tavares, 1904)] – W, M, E; ♦; ■; ♠; e; Malkov 1903; Buresch & Lazarov 1956; Skuhravá et al. 1991, 1992.

***Dasineura pyri* (Bouché, 1847)** [*Perrisia*] – W, M, E; ♦; ■; e, i, ha; Malkov 1902; Tschorbadjiew 1939; Lazarov 1949; Buresch & Lazarov 1956; Popoff *1956; Kovachevski et al. *1959; Naidenov 1962; Harizanov 1964; Grigorov 1972; Skuhravá et al. 1991, 1992; Beschovski 2006; E1, E2, P1, V1, K9, TL, O61, R1, RW, BN, BS; 10-1000 m.

***Dryomyia circinans* (Giraud, 1861)** – W, M, E; ♦; ■; nmsfe, dp; Drensky 1955; Naidenov 1962, 1963; Zlatanov *1971; Dimitrova 1987, 1989; Skuhravá et al. 1991, 1992; DW, E2, B3, K9, V4, V5, S211, T31, R2, BN, BS; 10-1000 m.

***Dryomyia lichtensteinii* (F. Löw, 1878)** – ♦; ■; ♠; sena; Zlatanov *1971; Skuhravá et al. 1991, 1992; 10-1000 m.

***Mayetiola destructor* (Say, 1817)** – W, M, E; ♦; ■; wp, i, ha; Malkov 1902b; Hitilov 1912a, 1912b; Drenowsky 1922; Drenowsky & Enderlein 1923; Tschorbadjiew 1925a, 1925c, 1926, 1927, 1929, 1932 1939; Petkoff 1939; Lazarov 1935, 1939; Buresch & Lazarov 1956; Lyubenov 1956, 1960; Popoff *1956; Zamfirov 1958, 1961; 1962a, 1962b, 1962c, 1963; Makarov 1959; Kovachevski et al. *1959; Naidenov 1963; Grigorov *1972, *1976; Skuhravá et al. 1991, 1992; Harizanov et al. *1996; DW, DM, E1, P1, V1, S1, T2, TL, BN; 0-1000 m.

***Contarinia medicaginis* Kieffer, 1895** – W, M, E; ♦; ■; ? h; Hristova, 1954; Popoff *1956; Hristova 1959; Kovachevski et al. 1959; Naidenov 1962; Donchev 1968; Grigorov *1972, 1976; Skuhravá et al. 1991, 1992; Harizanov et al. 1996; V1, TL, BN; 10-600 m.

***Contarinia nasturtii* (Kieffer, 1888)** [*C. torquens* Meijere 1906] – W, M, E; ♦; ■; ean, i, h; Nikolova 1943; Buresch & Lazarov 1956; Popoff 1956; Popoff & Nikolova *1958; Kovachevski et al. 1959; Naidenov 1963; Grigorov 1972, 1976; Dimitrova 1989; Skuhravá et al. 1991, 1992; Beschovski 2006; DW, DM, P1, P2, B1, V1, V4, TL, R1, R2, RW; 100-1550 m.

***Contarinia pyrivora* (Riley, 1886)** [*Diplosis*] – W, M, E; ♦; ■; e, i, ha; Malkov 1902c, 1906b; Vasilev 1931; Zhelev 1948; Kovachevski et al. *1959; Lazarov 1949; Buresch & Lazarov 1956; Popoff 1956; Naidenov 1962; Sengalevich 1964; Grigorov *1972, 1976; Skuhravá et al. 1991, 1992; V1, TL, RE; 150-650 m.

***Haplodiplosis marginata* (von Roser, 1840)** [*H. equestris* Wagner, 1871] – W, M, E; ♦; ■; e, Popoff 1956; Makarov 1959; Kovachevski et al. *1959; Zamfirov 1961; Grigorov *1972; Skuhravá et al. 1991, 1992; Harizanov et al. 1996; DW, P1, V1; 50-650 m.

***Sitodiplosis mosellana* (Gehin, 1857)** [*Stenodiplosis*]– W, M, E; ♦; ■; tp, i, h; Popoff *1956; Makarov 1959; Zamfirov 1961; Grigorov 1972; Skuhravá et al. 1991, 1992; DW; 100-150 m.

Psychodidae (+ Phlebotomidae)

***Phlebotomus* (*Phlebotomus*) *papatasi* (Scopoli, 1786)** – W, E; Vidin, FP57, 43°58'36.73", 22°52'19.10"; Rousse, MJ15, 43°50'42.70", 25°56'35.49"; ▲; ppt; Chichkoff & Konsuloff *1914; Drensky 1926, 1931a, 1942, 1955; Drensky & Drensky *1928; Kitanov 1943; Ganov 1949; Boychev 1950; DM, E1, E2, S1, TL,T1, O5, O62, RW, RE, BN, BS; 0-1400 m.

? ***Phlebotomus* (*Adlerius*) *chinensis* Newstead, 1916** – W; Vidin, FP57, 43°58'36.73", 22°52'19.10"; ▲; ptm; Boychev *1950; DW; 30-40 m.

***Psychoda albipennis* Zetterstedt, 1850** – M; Svishtov, LJ63, N43°37'15.53", E25°21'32.53"; e, ? k; Nedelkov *1909, 1912; DM, V1; 20-600 m.

Scatopsidae

***Coboldia fuscipes* (Meigen, 1830)** – W; Lom, FP75, N43°49'56.70", E23°13'43.91; hpta, sk; Szilády 1934; DW; 30-40 m.

Culicomorpha

Chaoboridae

***Chaoborus* (*Chaoborus*) *crystallinus* (De Geer, 1776)** [*Corethra plumicornis* (Fabricius, 1787)] – E; Rousse, MJ15, 43°50'42.70", 25°56'35.49"; Srebarna Lake, NJ08, 44°06'24.64", 27°04'22.52"; Silistra, NJ28, 44°07'25.22", 27°16'03.97"; h; Chichkoff & Konsuloff *1914; Stoichev 1998; Kovachev et al. 1999; Uzunov et al. *2001; Varadinova et al. *2012; E1, E2, V1, BN; 0-600 m.

Culicidae

***Anopheles* (*Anopheles*) *atroparvus* van Thiel, 1927** [*A. maculipennis* var. *atroparvus* van Thiel, 1927] – W, M, E; ♦; FP39-NJ28; Novo Selo, FP49, 44°09'49.13", 22°47'09.74"; Gamzovo, FP48, 44°05'50.07", 22°45'39.07"; Vrav, FP39, 44°11'43.78", 22°44'12.86"; Vidin, FP57, 43°58'36.73", 22°52'19.10"; Lom, FP75, 43°37'15.53", 25°21'32.53"; Dolni Tsibar, GP05, 43°49'01.09", 23°31'06.95"; Oryahovo, GP34, 43°44'19.96", 23°57'41.35"; Ostrov, KJ63-KJ73, 43°40'57.57", 24°08'58.92"; Nikopol, LJ36, 43°42'21.42", 24°53'43.37"; Oresh, LJ52, 43°34'44.45", 25°14'21.86"; Svishtov, LJ63, 43°37'15.53", 25°21'32.53"; Rousse, MJ15,

43°50'42.70", 25°56'35.49"; ▲; ewca; Markov 1937; Drensky 1939, 1949, 1950, 1958; Slivenski 1946; Avlavidov 1947, 1970; Boychev *1950b; Dimtchev et al. *1960, *1962; Bozhkov 1966a, *1991; Christova 1980; Minář 1990; Mikov 2005; Kutsarov *2006; DW, DM, E1, E2, BN; 0-200 m.

***Anopheles* (*Anopheles*) *claviger* (Meigen, 1804)** [*A. bifurcatus* Linnaeus, 1758] – W, M, E; •; wcp; Chichkoff & Konsuloff 1914; Konsuloff 1921a, 1922a, 1922b; Markov 1925a, 1925b, 1929, 1937; Drenowsky 1929a; Drensky 1931a, 1931b, 1932; Slivenski 1935, 1946; Avlavidov 1947, 1959; Kozarov 1949; Drensky 1949, 1950, 1958; Paspalev 1950, 1951; Bozhkov 1961, 1967, 1974, *1991; Dimtchev et al. *1962; Bozhkov et al. 1969; Christova & Todorova 1969; Christova et al. 1971; Mikov 2005, 2008a, 2008b; Beschovski 2006; Agushev 2014; Agushev & Bileva 2014; DM, P3, B1, B2, B3, V1, V4, K9, TL, O61, O62; T2; T31, R5, RW, RE, BN, BS; 0-1300 m.

***Anopheles* (*Anopheles*) *hyrcanus* (Pallas, 1771)** [*A. hyrcanus* var. *pseudopictus* Grassi, 1900; *Myzorhynchus sinensis* Wiedemann, 1828; *M. pseudopictus* (Grassi, 1900)] – W, M, E; ♦; FP39-NJ28; Srebarna Lake, NJ08, 44°06'24.64", 27°04'22.52"; Silistra, NJ28, 44°07'25.22", 27°16'03.97"; spo; Chichkoff & Konsuloff *1914; Konsuloff 1921a, 1921b, 1922a, 1922c; Markov 1925a, 1929, 1937; Drenowsky 1929a; Drensky 1931a; Slivenski *1935, 1946; Drensky 1949, 1950, 1958; Boychev *1950b; Dimtchev et al. *1962; Dorovski 1976; Bozhkov *1991; Mikov 2005; Kutsarov *2006; Agushev 2014; Agushev & Bileva 2014; DW, DM, E1, E2, P1, O62, TL; 10-200 m.

***Anopheles* (*Anopheles*) *maculipennis* Meigen, 1818** – W, M, E; ♦; FP39-NJ28; Aydemir, NJ19, 44°06'49.54", 27°10'40.70"; Belene, LJ43, 43°39'10.38", 25°07'38.83"; Dolni Tsibar, GP05, 43°49'01.09", 23°31'06.95"; Gamzovo, FP48, 44°05'50.07", 22°45'39.07"; Karamanovo, LJ82, 43°33'37.83", 25°31'42.90"; Lom, FP75, 43°37'15.53", 25°21'32.53"; Nikopol, LJ36, 43°42'21.42", 24°53'43.37"; Novo Selo, FP49, 44°09'49.13", 22°47'09.74"; Oryahovo, GP34, 43°44'19.96", 23°57'41.35"; Ostrov, KJ63-KJ73, 43°40'57.57", 24°08'58.92"; Rousse, MJ15, 43°50'42.70", 25°56'35.49"; Ryahovo, MJ37, 43°59'41.21", 26°14'43.28"; Silistra, NJ28, 44°07'25.22", 27°16'03.97"; Somovit, LJ23-LJ24, 43°41'21.04", 24°46'26.81; Srebarna Lake, NJ08, 44°06'24.64", 27°04'22.52"; Svishtov, LJ63, 43°37'15.53", 25°21'32.53; Tutrakan, MJ67-MJ77, 44°02'51.63", 26°36'08.38"; Vidin, FP57, 43°58'36.73", 22°52'19.10"; ▲; wcp; Manolov 1907, *1909a, 1909b; Theobald 1910; Nedelkov 1912; Chichkoff & Konsuloff *1914; Konsuloff 1921a, 1922a, 1922b, 1922c, 1923a; Markov 1925a, 1925b, 1929, 1937; Drenowsky 1929a; Markov & Morov 1929; Drensky 1931a; Drensky 1939, *1949, 1950, *1958; Arndt 1943; Slivenski 1946; Avlavidov 1947, 1948, 1958, 1959, 1961; Kozarov 1949; Boychev *1950b; Paspalev 1950, 1951; Bozhkov 1953, *1957, 1961, 1962, 1965, 1966b, *1991; Kozarov & Bozhkov 1953; Stefanov 1955, 1956; Dimov *1957; Marinov 1957; Kovchazov *1958, *1961, 1976; Atanassov & Christova 1960; Atanassov & Petrov 1961; Dimtchev et al. *1962; Atanassov et al. *1962; Mikov 2005, 2008a, 2011;

Beschovski 2006; Kutsarov 2006; Mikov et al. 2011; Agushev 2012, 2014, 2015; Agushev & Bileva 2014; DW, DM, E1, E2, B1, B2, B3, K9, V1, V4, S1, TL, T1, T2, T3, O62, T31, R1, R5, RW, RE, BN, BS; 0-2190 m.

***Anopheles* (*Anopheles*) *melanoon* Hackett, 1934** [*A. maculipennis* var. *melanoon* Hackett, 1934] – W, E; •; Danubian bank; Karamanovo, LJ82, 43°33'37.83", 25°31'42.90; Nova Cherna, MJ57, 44°01'04.58", 26°27'22.88"; Tsar Simeonovo, FP45, 43°51'09.96", 22°50'35.19"; nmi; Markov 1937; Drensky 1939, 1949, 1950, *1958; Paspalev 1950, 1951; Dimchev et al. *1962; Dorovski 1976; Christova 1980; Mikov 2005, 2008a; DM, E1, E2, O62, RE, BS; 20-250 m.

***Anopheles* (*Anopheles*) *messeae* Falleroni, 1926** [*A. maculipennis* var. *messeae* Falleroni, 1926] – W, M, E; Archar, FP55, 43°48'45.03", 22°55'06.45"; Belene, LJ43, 43°39'10.38", 25°07'38.83"; Dolni Tsibar, GP05, 43°49'01.09", 23°31'06.95"; Karamanovo, LJ82, 43°33'37.83", 25°31'42.90"; Kozloduy, GP25, 43°46'36.03", 23°44'25.05"; Lom, FP75, 43°37'15.53", 25°21'32.53"; Malak Preslavets, MJ88, 44°05'54.96", 26°50'01.52"; Nikopol, LJ36, 43°42'21.42", 24°53'43.37"; Nova Cherna, MJ57, 44°01'04.58", 26°27'22.88"; Novo Selo, FP49, 44°09'49.13", 22°47'09.74"; Oresh, LJ52, 43°34'44.45", 25°14'21.86"; Oryahovo, GP34, 43°44'19.96", 23°57'41.35"; Ostrov, KJ63-KJ73, 43°40'57.57", 24°08'58.92"; Popina, MJ98, 44°07'40.24", 26°56'56.55"; Rousse, MJ15, 43°50'42.70", 25°56'35.49"; Ryahovo, MJ37, 43°59'41.21", 26°14'43.28"; Silistra, NJ28, 44°07'25.22", 27°16'03.97"; Somovit, LJ23-LJ24, 43°41'21.04", 24°46'26.81"; Srebarna Lake, NJ08, 44°06'24.64", 27°04'22.52"; Sredna Kula, MJ15, 43°48'34.10", 25°56'10.22"; Svishtov, LJ63, 43°37'15.53", 25°21'32.53"; Tutrakan, MJ67-MJ77, 44°02'51.63", 26°36'08.38"; Vardim, LJ72-LJ73, 43°36'50.79", 25°27'54.93"; Vidin, FP57, 43°58'36.73", 22°52'19.10"; Vrav, FP39, 44°11'43.78", 22°44'12.86"; ▲; esanca; Konsuloff 1921b; Markov 1937; Drensky 1939, 1949, *1950, *1958; Slivenski *1946; Boychev *1950b; Paspalev 1950, 1951; Bozhkov 1953, 1961, 1991; Stefanov 1956; Dimchev et al. *1962; Mikov 2005, 2008; Kutsarov 2006; Agushev 2014, 2015; Agushev & Bileva 2014; DW, DM, E1, E2, V1, TL, T2, O62, RW, RE, BS; 0-800 m.

***Anopheles* (*Cellia*) *superpictus* Grassi, 1899** [*A.* (*Myzomyia*) *superpictus* Grassi, 1899] – W; Archar, FP55, 43°48'45.03", 22°55'06.45"; ▲; swpo; Konsuloff 1921a, 1922a, 1923; Markov 1925a, 1925b, 1929, 1937, 1950; Drenowsky 1929; Markov & Morov *1929; Slivenski 1935, 1940, 1946; Avlavidov 1947, 1948; Paspalev 1950, 1951; Drensky 1949, 1950, 1958; Kozarov 1949; Boychev 1950b; Stefanov 1955; Bozhkov 1961, *1991; Christova 1980; Minář 1990; Mikov 2005; Beschovski 2006; Agushev 2014, 2015; Agushev & Bileva 2014; DW, P1, B3, S1, TL, T1, T2, T3, O61, O62, R5, RE; 100-500 m.

***Culiseta* (*Culiseta*) *annulata* (Schrank, 1776)** [*Theobaldia*] – W, M, E; ♦; FP39-NJ28; ▲; wp; Nedelkov 1909, 1912; Theobald 1910; Chichkoff & Konsuloff 1914; Konsuloff 1922b; Drenowsky *1929a, 1929b; Bozhkov 1958, 1959, 1961, 1962, 1965, 1967,

1991; Bozhkov et al. 1969; Christova & Todorova 1969; Christova et al. 1971; Christova & Dorovski 1972; Dorovski 1976; Beschovski 2006; Kutsarov *2006; Mikov 2011; Agushev 2012, 2015; DW, DM, E1, E2, B1, V1, S22, TL, O62, R1, R5, RW, BN, BS; 0-1400 m.

***Coquillettidia* (*Coquillettidia*) *richiardii* (Ficalbi, 1889)** [*Mansonia*] – W, M, E; ♦; FP39-NJ28; Nikopol, LJ36, 43°42'21.42", 24°53'43.37"; Silistra, NJ28, 44°07'25.22", 27°16'03.97"; Somovit, LJ23-LJ24, 43°41'21.04", 24°46'26.81"; Srebarna Lake, NJ08, 44°06'24.64", 27°04'22.52"; Vidin, FP57, 43°59'31.33", 22°53'11.99"; ▲; wp; Chichkoff & Konsuloff *1914; Bozhkov & Christova 1965; Bozhkov et al. 1969; Dorovski 1976; Bozhkov *1991; Kutsarov *2006; Mikov et al. *2011; DW, DM, E1, E2, BN, BS; 0-250 m.

***Aedes* (*Aedes*) *cinereus* Meigen, 1818** – W, M, E; ♦; FP39-NJ28; Rousse, MJ15, 43°50'42.70", 25°56'35.49"; Vidin, FP57, 43°58'36.73", 22°52'19.10"; h; Manolov 1907; Theobald 1907, 1910; Chichkoff & Konsuloff *1914; Bozhkov 1961, 1991; Bozhkov & Christova 1965; Dorovski 1976; Christova & Dorovski 1972; Kutsarov *2006; Mikov 2011; Agushev 2014, 2015; Agushev & Bileva 2014; DW, DM, E1, E2, BS; 0-100 m.

***Aedes* (*Aedimorphus*) *vexans* (Meigen, 1830)** – W, M, E; ♦; FP39-NJ28; Nikopol, LJ36, 43°42'21.42", 24°53'43.37"; Silistra, NJ28, 44°07'25.22", 27°16'03.97"; Tutrakan, MJ67-MJ77, 44°02'51.63", 26°36'08.38"; Vidin, FP57, 43°58'36.73", 22°52'19.10"; hpta; Chichkoff & Konsuloff *1914; Drenowsky 1929a; Bozhkov 1962, 1965, 1967, 1991; Bozhkov & Christova 1965; Christova & Todorova 1969; Bozhkov et al. 1969; Christova et al. 1971; Dorovski 1976; Kutsarov *2006; Mikov et al. 2011; Agushev & Bileva 2014; Agushev 2015; DW, DM, E1, E2, B1, P1, V1, TL, O62, BN, BS; 0-600 m.

***Aedes* (*Ochlerotatus*) *annulipes* (Meigen, 1830)** [*Ochlerotatus*] – W, M, E; ♦; FP39-NJ28; ean; Nedelkov 1912; Chichkoff & Konsuloff 1914; Bozhkov et al. 1969; Christova et al. 1971; Bozhkov 1991; Kutsarov *2006; DW, DM, E1, E2, V1, BN, BS; 0-700 m.

***Aedes* (*Ochlerotatus*) *cantans* (Meigen, 1818)** [*Ochlerotatus*] – W, M, E; ♦; FP39-NJ28; ▲; h; Drenowsky 1929a; Bozhkov et al. 1969; Christova et al. 1971; Bozhkov 1991; Cutsarov *2006; Mikov 2011; DW, DM, E1, E2, O62, BN, BS; 0-150 m.

***Aedes* (*Ochlerotatus*) *communis* (De Geer, 1776)** – W, M, E; ♦; FP39-NJ28; Nikopol, LJ36, 43°42'21.42", 24°53'43.37"; Rousse, MJ15, 43°50'42.70", 25°56'35.49"; Tutrakan, MJ67-MJ77, 44°02'51.63", 26°36'08.38"; Vidin, FP57, 43°58'36.73", 22°52'19.10"; ▲; h; Theobald 1907, 1910; Chichkoff & Konsuloff *1914; Bozhkov 1959, 1961, 1966b, 1991; Bozhkov & Christova 1965; Cutsarov *2006; DW, DM, E1, E2, V4, R1, BS; 0-1450 m.

***Aedes* (*Ochlerotatus*) *dorsalis* (Meigen, 1830)** – M, E; Rousse, MJ15, 43°50'42.70", 25°56'35.49"; Somovit, LJ23-LJ24, 43°41'21.04", 24°46'26.81; Tutrakan, MJ67-MJ77, 44°02'51.63", 26°36'08.38"; h, ? ho; Manolov 1907; Theobald 1907, 1910;

Chichkoff & Konsuloff *1914; Drenowsky 1929a; Bozhkov 1961, *1991; Dorovski 1976; DM, E1, E2, K9, V1, O62, BS; 0-600 m.

***Aedes* (*Ochlerotatus*) *punctor* (Kirby, 1837)** – W, M, E; ♦; FP39-NJ28; h; Bozhkov 1959, 1991; Minarzh & Christova 1971; Kutsarov *2006; DW, DM, E1, E2, V4, R1; 20-1400 m.

***Culex* (*Culex*) *pipiens* Linnaeus, 1758** [*C. pipiens* var. *ciliaris* Linnaeus, 1758; *C. pipiens* var. *molestus* Forskål, 1775] – W, M, E; ♦; FP39-NJ28; ▲; sk; trogloxene; Joakimoff 1899; Kovachev *1905; Manolov 1907; Nedelkov *1909, 1912; Chichkoff & Konsuloff *1914; Konsuloff 1922b; Drenowsky *1929a; 1929b; Markov & Morov 1929; Czerný 1930; Slivenski 1935, 1946; Drensky 1942; Caspers 1951a; Bozhkov 1959, 1961, 1962, 1965, 1966b, *1991; Kalaydzhiev et al. 1960; Guéorguiev & Beron 1962; Bozhkov & Christova 1965; Russev 1966a; Bozhkov et al. 1969; Christova & Todorova 1969; Dorovski 1976; Beron 1994, 2015; Janeva & Russev 1997; Kutsarov *2006; Mikov 2011; Agushev 2012, 2014, 2015; Agushev & Bileva 2014; DW, DM, E1, E2, P2, B1, B2, B3, V1, V4, S22, S23, TL, T1, T31, O62, R1, RW, BN, BS; 0-2100 m.

***Culex* (*Barraudius*) *modestus* Ficalbi, 1890** – W, M, E; ♦; FP39-NJ28; ▲; tp, ? wcp; Christova & Todorova 1969; Bozhkov et al. 1969; Christova et al. 1971, 1972; Christova & Dorovski 1972; Dorovski 1976; Bozhkov 1991; Kutsarov *2006; Mikov 2011; DW, DM, E1, E2, V1, TL; BN, BS; 0-600 m.

***Culex* (*Neoculex*) *territans* Walker, 1856** [*C. sergentii* Theobald, 1903; *C. apicalis* Adams, 1903] – W, M, E; ♦; FP39-NJ28; h; Chichkoff & Konsuloff 1914; Drenowsky 1929a; 1929b; Arndt 1943; Bozhkov 1961, 1962, 1965, 1967, 1991; Russev 1966a; Bozhkov et al. 1969; Christova et al. 1971, 1972; Dorovski 1976; Beschovski 2006; Kutsarov *2006; Agushev 2014, 2015; Agushev & Bileva 2014; DW, DM, E2, P1, V1, TL; T3, T31, O62, R5, BN, BS; 0-600 m.

***Culex* (*Maillotia*) *hortensis* Ficalbi, 1889** – W, E; Rousse, MJ15, 43°50'42.70", 25°56'35.49"; Vidin, FP57, 43°59'31.33", 22°53'11.99"; swpo; Chichkoff & Konsuloff *1914; Drenowsky 1929a; 1929b; Bozhkov 1959, 1961, 1965, 1966b, 1967, 1991; Christova & Todorova 1969; Bozhkov et al. 1969; Christova et al. 1971, 1972; Dorovski 1976; Beschovski 2006; Kutsarov *2006; DW, E1, E2, V1, S23, T31, O62, R1, RW, BN, BS; 0-1390 m.

Simuliidae

***Simulium* (*Simulium*) *argenteostriatum* Strobl, 1898** [*Cleitosimulium*; *S. alternans* Enderlein, 1921; *S. schoenbaueri* Enderlein, 1921] – W; •; csena; Enderlein 1924; Konsuloff & Paspalev 1925; Tschorbadjiew *1925e; Buresch 1938; Kovachev 1969, 1975, 1976, 1979, 1990, 2000; Janeva 1987; Russev et al. *1994; Beschovski 2006; Adler & Crosskey 2018; DW, P1, P2, B1, B2, V1, V4, R1, R2, R5, RW; 50-2100 m.

***Simulium* (*Simulium*) *colombaschense* (Scopoli, 1780)** – W, M, E; ♦; FP39-NJ28; Vidin, FP57, 43°58'36.73", 22°52'19.10"; ▲; cee; Nedelkov 1912; Konsuloff 1923b; Markov 1923; Buresch 1924; Enderlein *1924; Konsuloff & Paspalev 1924; Tschorbadjiew 1925e; Buresch 1938; Russev *1966b; Russev et al. 1994; Adler & Crosskey 2018; DW, DM, E1, E2, P1, B1, V1, R1; 30-900 m.

***Simulium* (*Simulium*) *ornatum* Meigen, 1818** [*Odagmia konsuloffi* (Enderlein, 1924); *S. pratorum* Friederichs, 1920] – W, M, E; •; hop; Enderlein, 1924; Konsuloff & Paspalev 1925; Tschorbadjiew *1925e; Buresch 1926b, 1938; Kovachev 1975, 1976, 1990; Janeva 1987, 1989; Russev et al. *1994; Janeva et al. 2001; Beschovski 2006; Stoyanova et al. 2013; Uzunov et al. 2011; Varadinova et al. 2013; Adler & Crosskey 2018. DW, DM, E1, E2, P1, P2, P3, B1, B2, V1, V4, O61, O62, R1, R2, R5, RW, RE; 50-1100 m.

***Simulium* (*Eusimulium*) *angustipes* Edwards, 1915** [*E. latizonum* (Rubtsov, 1956); *E. securiforme* (Rubtsov, 1956)] – E; Vetren, NJ08, 44°08'24.52", 27°01'53.32"; wcp; Russev *1962; Kovachev 1976, 1979, 2000; Russev et al. 1994; DW, DM, E2, P1, P2, O62, R1, R2, RW; 20-2240 m.

***Simulium* (*Wilhelmia*) *balcanicum* (Enderlein, 1924)** [*W. balcanica severinense* (Dinulescu, 1966)] – E; Krivina, LJ83, 43°38'42.88", 25°34'42.18"; ean, ? eant; Enderlein, 1924; Tschorbadjiew 1925e; Buresch 1926b, 1938; Russev 1962, *1966b; Kovachev 1975, 1976, 1990; Janeva 1987; Russev et al. 1994; Uzunov et al. 2011; Varadinova et al. 2013; Adler & Crosskey 2018; DW, DM, E1, E2, P1, P2, P3, B1, B2, V1, V4, O61, O62, R5, RR; 20-700 m.

***Simulium* (*Wilhelmia*) *equinum* (Linnaeus, 1758)** [*W. secundum* (Baranov, 1926); *W. equina ivashentzovi* Rubtsov, 1940] – W; •; Lom River, before Lom Town, FP75, 43°46'55.41", 23°14'03.63"; tp, ? po; Enderlein, 1924; Tschorbadjiew 1925e; Buresch 1938; Russev *1962; Kovachev *1969; Russev et al. 1994; Adler & Crosskey 2018; DW, DM, P1, P2, B1, V1; 20-600 m.

Ceratopogonidae (Heleidae)

***Forcipomyia* (*Forcipomyia*) *bipunctata* (Linnaeus, 1767)** – W; Lom, FP75, 43°37'15.53", 25°21'32.53"; Vidin, FP57, 43°58'36.73", 22°52'19.10"; h; Zilahi 1934; DW, V4, RW; 50-800 m.

***Culicoides* (*Beltranmyia*) *circumscriptus* Kieffer, 1918** [*C. pulcher* Zilahi-Sebess, 1934] – W, E; Vidin, FP57, 43°58'36.73", 22°52'19.10"; Kalimok, MJ57, 44°00'41.34", 26°26'21.56"; ppt; Zilahi 1934; Bobeva et al. *2013, *2014; Nedelchev *2013; Bobeva *2014; Pudar et al. 2018; DW, E2, P2, TL, T1, T2, O61, RE; 20-370 m.

***Culicoides* (*Beltranmyia*) *salinarius* Kieffer, 1914** – W, E; Vidin, FP57, 43°58'36.73", 22°52'19.10"; Kalimok, MJ57, 44°00'41.34", 26°26'21.56"; tp; Bobeva et al. *2013; Nedelchev *2013; Pudar et al. 2018; DW, DM, E2, P1, P2, K9, TV, V1, TL, T1, T2, O61, RW, RE, BN, BS; 0-1000 m.

***Culicoides* (*Culicoides*) *fagineus* Edwards, 1939** – W, E; Vidin, FP57, 43°58'36.73", 22°52'19.10"; Rousse, MJ15, 43°50'42.70", 25°56'35.49"; Kalimok, MJ57, 44°00'41.34", 26°26'21.56"; wcp; Bobeva et al. *2013; Nedelchev *2013; Pudar et al. 2018; DW, E1, E2, P1, E2, B1, K9, TV, V1, T1, T2, O61, RW, RE, BS; 20-1000 m.

***Culicoides* (*Culicoides*) *flavipulicaris* Dhafarov, 1964** – E; Nova Cherna, MJ57, 44°01'04.58", 26°27'22.88"; dp, ? nmca; Bobeva et al. *2013; Nedelchev 2013; Pudar et al. 2018; E2, K9, PV, V1; 20-700 m.

***Culicoides* (*Culicoides*) *grisescens* Edwards, 1939** – E; Kalimok, MJ57, 44°00'41.34", 26°26'21.56"; tp; Bobeva et al. *2013; Nedelchev 2013; Pudar et al. 2018; E2, P2, TV, V1, T1, T2, RE, BS; 0-600 m.

***Culicoides* (*Culicoides*) *impunctatus* Goetghebuer, 1920** – W, E; Vidin, FP57, 43°58'36.73", 22°52'19.10"; Kalimok, MJ57, 44°00'41.34", 26°26'21.56"; wces; Bobeva et al. 2013; Nedelchev 2013; Pudar et al. 2018; DW, E2, P1, B1, TV, V1, O61, BN, BS; 0-700 m.

***Culicoides* (*Culicoides*) *newsteadi* Austen, 1921** [*C. halophilus* Kieffer, 1924] – W, E; Vidin, FP57, 43°58'36.73", 22°52'19.10"; Kalimok, MJ57, 44°00'41.34", 26°26'21.56"; Silistra, NJ28, 44°07'25.22", 27°16'03.97"; wp; Bobeva et al. *2013; Nedelchev *2013; Bobeva *2014; Pudar et al. 2018; DW, DM, E1, E2, P1, P2, B1, K9, TV, V1, S1, TL, T1, T2, O61, RE, BN, BS; 0-700 m.

***Culicoides* (*Culicoides*) *pulicaris* (Linnaeus, 1758)** [*Ceratopogon*] – W, E; Vidin, FP57, 43°58'36.73", 22°52'19.10"; Rousse, MJ15, 43°50'42.70", 25°56'35.49"; Kalimok, MJ57, 44°00'41.34", 26°26'21.56"; Silistra, NJ28, 44°07'25.22", 27°16'03.97"; pat; Nedelkov 1912; Bobeva et al. *2013; Nedelchev *2013; Pudar et al. 2018; DW, DM, E1, E2, P1, P2, B1, K9, TV, V1, S1, T1, T2, TL, RW, RE, BN, BS; 0-1000 m.

***Culicoides* (*Culicoides*) *punctatus* (Meigen, 1804)** – W, E; Vidin, FP57, 43°58'36.73", 22°52'19.10"; Rousse, MJ15, 43°50'42.70", 25°56'35.49"; Kalimok, MJ57, 44°00'41.34", 26°26'21.56"; Silistra, NJ28, 44°07'25.22", 27°16'03.97"; pat; Bobeva et al. *2013; Nedelchev *2013; Bobeva *2014; Pudar et al. 2018; DW, DM, E1, E2, P1, P2, B1, K9, TV, V1, S1, TL, T1, T2, O61, RE, RW, BN, BS; 0-1000 m.

***Culicoides* (*Avaritia*) *dewulfi* Goetghebuer, 1936** – E; Kalimok, MJ57, 44°00'41.34", 26°26'21.56"; dp; Bobeva et al. *2013; Nedelchev 2013; Pudar et al. 2018; E2, P2; 20-250 m.

***Culicoides* (*Avaritia*) *obsoletus* (Meigen, 1818)** – W, E; Vidin, FP57, 43°58'36.73", 22°52'19.10"; Rousse, MJ15, 43°50'42.70", 25°56'35.49"; Kalimok, MJ57, 44°00'41.34", 26°26'21.56"; Silistra, NJ28, 44°07'25.22", 27°16'03.97"; h; Zilahi 1934; Bobeva et al. *2013; Nedelchev *2013; Bobeva *2014; Pudar et al. 2018; DW, DM, E1, E2, P1, P2, B1, K9, V1, S1, TV, TL, T1, T2, RW, RE, BN, BS; 0-1000 m.

***Culicoides* (*Avaritia*) *scoticus* Downes & Kettle, 1952** – E; Kalimok, MJ57, 44°00'41.34", 26°26'21.56"; wcp; Bobeva et al. 2013; Nedelchev 2013; Pudar et al. 2018; E2, RE; 20-250 m.

Culicoides **(*Silvaticulicoides*) *fascipennis* (Staeger, 1839)** – W, E; Vidin, FP57, 43°58'36.73", 22°52'19.10"; Rousse, MJ15, 43°50'42.70", 25°56'35.49"; Kalimok, MJ57, 44°00'41.34", 26°26'21.56"; Silistra, NJ28, 44°07'25.22", 27°16'03.97"; tp; Bobeva et al. 2013; Nedelchev 2013; Pudar et al. 2018; DW, DM, E1, E2, P1, P2, B1, E2, K9, V1, S1, TL, T1, T2, O61, RW, RE, BN, BS; 0-1000 m.

Culicoides **(*Silvaticulicoides*) *pallidicornis* Kieffer, 1919** [*C. pallidicornis* var. *bruneoscutellatus* Zilahi-Sebess, 1934] – E; Rousse, MJ15, 43°50'42.70", 25°56'35.49"; Kalimok, MJ57, 44°00'41.34", 26°26'21.56"; h; Zilahi 1934; Bobeva et al. *2013; Nedelchev *2013; Pudar et al. 2018; E1, E2, P2, K9, TV, V1, TL, T1, T2, O61, R1, RE, BN, BS; 0-1400 m.

Culicoides **(*Silvaticulicoides*) *subfasciipennis* Kieffer, 1919** – E; Kalimok, MJ57, 44°00'41.34", 26°26'21.56"; Silistra, NJ28, 44°07'25.22", 27°16'03.97"; wcp; Bobeva et al. *2013; Nedelchev *2013; Pudar et al. 2018; DM, E1, E2, P1, P2, B1, K9, TV, V1, TL, T1, T2, O61, RE, BN, BS; 0-700 m.

Culicoides **(*Sensiculicoides*) *alazanicus* Dzhafarov, 1961** – E; ean; Bobeva 2014; Bobeva et al. 2014, 2015; Pudar et al. 2018. E2; 20 m.

Culicoides **(*Sensiculicoides*) *festivipennis* Kieffer, 1914** – E; Rousse, MJ15, 43°50'42.70", 25°56'35.49"; Kalimok, MJ57, 44°00'41.34", 26°26'21.56"; Silistra, NJ28, 44°07'25.22", 27°16'03.97"; hop; Bobeva et al. *2013, 2014; Nedelchev *2013; Bobeva *2014; Pudar et al. 2018; DM, E1, E2, P1, P2, B1, K9, V1, TL, T1, T2, O61, R1, BS; 0-600 m.

Culicoides **(*Sensiculicoides*) *gejgelensis* Dzhafarov, 1964** – E; Kalimok, MJ57, 44°00'41.34", 26°26'21.56"; mca; Bobeva et al. *2013; Nedelchev 2013; Pudar et al. 2018; E2, K9, TV, V1, S1, O61, RW, BS, RE; 0-1000 m.

Culicoides **(*Sensiculicoides*) *griseidorsum* Kieffer, 1918** – E; Kalimok, MJ57, 44°00'41.34", 26°26'21.56"; ena; Bobeva *2014; Szadziewski et al. 2016; E2; 20 m.

Culicoides **(*Sensiculicoides*) *kurensis* Dzhafarov, 1960** – W, E; Vidin, FP57, 43°58'36.73", 22°52'19.10"; Rousse, MJ15, 43°50'42.70", 25°56'35.49"; Kalimok, MJ57, 44°00'41.34", 26°26'21.56"; hom; Bobeva et al. *2013; Nedelchev *2013; Pudar et al. 2018; DW, P1, E1, E2, B1, V1, RW, RE; 30-1000 m.

Culicoides **(*Sensiculicoides*) *odiatus* Austen, 1921** – W, E; Vidin, FP57, 43°58'36.73", 22°52'19.10"; Rousse, MJ15, 43°50'42.70", 25°56'35.49"; Kalimok, MJ57, 44°00'41.34", 26°26'21.56"; Silistra, NJ28, 44°07'25.22", 27°16'03.97"; wcp; Bobeva et al. *2013; Nedelchev *2013; Pudar et al. 2018; DW, DM, E1, E2, P1, P2, B1, K9, V1, T1, T2, O61, RE, BS; 0-600 m.

Culicoides **(*Sensiculicoides*) *pictipennis* (Staeger, 1839)** [*C. arcuatus* (Winnertz, 1852)] – W, E; Vidin, FP57, 43°58'36.73", 22°52'19.10"; Kalimok, MJ57, 44°00'41.34", 26°26'21.56"; wcp; Zilahi 1934; Bobeva et al. *2013; Nedelchev *2013; Bobeva 2014; Pudar et al. 2018; DW, E2, P1, P2, B1, TV, V1, T1, T2, O61, R1, RW, RE; 30-2900 m.

***Culicoides* (*Sensiculicoides*) *simulator* Edwards, 1939** – E; Rousse, MJ15, 43°50'42.70", 25°56'35.49"; Kalimok, MJ57, 44°00'41.34", 26°26'21.56"; Silistra, NJ28, 44°07'25.22", 27°16'03.97"; wcp; Bobeva et al. *2013; Nedelchev *2013; Pudar et al. 2018; DM, E1, E2, V1, S1, TL, T1, T2, BN, BS; 0-600 m.

***Culicoides* (*Oecacta*) *longipennis* Khalaf, 1957** – E; mwca; Kalimok, MJ57, 44°00'41.34", 26°26'21.56"; Bobeva et al. *2013; Nedelchev 2013; Pudar et al. 2018; E2, K9, V1, TV, S1, O61, RW, RE; 20-1000 m.

***Culicoides* (*Oecacta*) *vexans* (Staeger, 1839)** – E; Kalimok, MJ57, 44°00'41.34", 26°26'21.56"; Silistra, NJ28, 44°07'25.22", 27°16'03.97"; esca; Bobeva et al. *2013; Nedelchev *2013; Pudar et al. 2018. E1, E2, P2, TV, V1, S1, TL, T1, T2, RE, BN, BS; 0-700 m.

***Culicoides* (*Monoculicoides*) *nubeculosus* (Meigen, 1830)** – W, E; Vidin, FP57, 43°58'36.73", 22°52'19.10"; Kalimok, MJ57, 44°00'41.34", 26°26'21.56"; des, ? tes; Bobeva et al. *2013; Nedelchev *2013; Bobeva *2014; Pudar et al. 2018; DW, E1, E2, P1, P2, K9, V1, S1, O61, RE, BS; 0-600 m.

***Culicoides* (*Monoculicoides*) *parroti* Kieffer, 1922** – E; Kalimok, MJ57, 44°00'41.34", 26°26'21.56"; Silistra, NJ28, 44°07'25.22", 27°16'03.97"; wp; Bobeva et al. *2013; Nedelchev *2013; Pudar et al. 2018; DM, E2, P1, P2, TL, RE, BN, BS; 0-300 m.

***Culicoides* (*Monoculicoides*) *puncticollis* (Becker, 1903)** [C. *impressus* Kieffer, 1918] – W, E; Vidin, FP57, 43°58'36.73", 22°52'19.10"; Rousse, MJ15, 43°50'42.70", 25°56'35.49"; Kalimok, MJ57, 44°00'41.34", 26°26'21.56"; sp; Zilahi 1934; Bobeva et al. *2013; Nedelchev *2013; *Bobeva 2014; Pudar et al. 2018; DW, E1, E2, P1, K9, TV, V1, S1, TL, T1, T2, O61, RW, RE, BN, BS; 0-1000 m.

***Culicoides* (*Monoculicoides*) *riethi* Kieffer, 1914** – E; Rousse, MJ15, 43°50'42.70", 25°56'35.49"; Kalimok, MJ57, 44°00'41.34", 26°26'21.56"; Silistra, NJ28, 44°07'25.22", 27°16'03.97"; po, ? ho; Bobeva et al. *2013; Nedelchev *2013; Bobeva *2014; Pudar et al. 2018; DM, E1, E2, P1, P2, B1, K9, V1, S1, TL, T1, T2, O61, RE, BN, BS; 0-600 m.

***Culicoides* (*Monoculicoides*) *stigma* (Meigen, 1818)** – E; Kalimok, MJ57, 44°00'41.34", 26°26'21.56"; Silistra, NJ28, 44°07'25.22", 27°16'03.97"; wp; Bobeva et al. *2013; Nedelchev *2013; Pudar et al. 2018; E2, P1, P2, B1, V1, TL, T1, T2, RW, RE, BS; 0-1000 m.

***Culicoides* (*Wirthomyia*) *reconditus* Campbell & Pelham-Clinton, 1960** – W, E; Vidin, FP57, 43°58'36.73", 22°52'19.10"; Kalimok, MJ57, 44°00'41.34", 26°26'21.56"; e; Bobeva et al. *2013; Nedelchev *2013; Pudar et al. 2018; DW, E2, P1, P2, B1, S1, TL, RE, BS; 0-450 m.

***Culicoides* (*Pontoculicoides*) *saevus* Kieffer, 1922** [*C. drenskii* Zilahi-Sebess, 1934] – E; Kalimok, MJ57, 44°00'41.34", 26°26'21.56"; wcp; Zilahi 1934; Bobeva et al. *2013; Nedelchev 2013; Pudar et al. 2018; E2, B1, V1; 20-600 m.

***Culicoides* (*Pontoculicoides*) *sejfadinei* Dzhafarov, 1958** – E; Kalimok, MJ57, 44°00'41.34", 26°26'21.56"; wcp; Bobeva et al. *2013; Nedelchev 2013; Pudar et al. 2018; E2, K9, TV, V1, S1, T1, T2, O61, RE, RW; 20-1000 m.

***Culicoides* (*Pontoculicoides*) *tauricus* Gutsevich, 1959** – W, E; Vidin, FP57, 43°58'36.73", 22°52'19.10"; Kalimok, MJ57, 44°00'41.34", 26°26'21.56"; ? cset; Bobeva et al. *2013; Nedelchev *2013; Pudar et al. 2018; DW, DM, E1, E2, P1, B1; 20-400 m.

***Culicoides* (*Remmia*) *schultzei* (Enderlein), 1908** – W, E; Vidin, FP57, 43°58'36.73", 22°52'19.10"; Rousse, MJ15, 43°50'42.70", 25°56'35.49"; Silistra, NJ28, 44°07'25.22", 27°16'03.97"; ppta; Bobeva et al. *2013; Nedelchev *2013; Pudar et al. 2018; DW, DM, E1, E2, P1, P2, B1, TL, T1, T2; 20-450 m.

Chironomidae

***Tanypus* (*Tanypus*) *punctipennis* Meigen, 1818** [*Pelopia*; *Protenthes punctipennis* var. *ferrugineus* Kieffer, 1918] – W; ♦; Vidin, FP57, 43°58'36.73", 22°52'19.10"; Lom, FP75, 43°37'15.53", 25°21'32.53"; hno; Zilahi *1934; Cvetkov 1955a, 1955b, 1957, 1962; Valkanov 1957a; Dimitrov 1957, 1960a, 1960b, 1962a, 1962b, 1962c, 1962d, *1963a, 1966, 1970, 1972; Mihailova-Neikova 1961; Russev 1966a, *1966b; Janeva 1987; Russev et al. 1994; Stoichev 1994, 1996; DW, DM, E1, E2, P2, V1, V5, S211, S1, TL, RW, RE, BN, BS; 0-900 m.

***Procladius* (*Holotanypus*) *choreus* (Meigen, 1804)** – W, M, E; •; ho; Russev et al. 1994; Stoichev *1994, 1996; Michailova 2006; DW, DM, E1, E2, P1, P2, RW; 20-1500 m.

***Procladius* (*Holotanypus*) *ferrugineus* (Kieffer, 1918)** – W, M, E; •; tes, ? dp; Russev et al. 1994; Stoichev 1994, 1996; DW, DM, E1, E2; 20-500 m.

***Anatopynia plumipes* (Fries, 1823)** – W, M, E; •; dp, ? tp; Russev et al. 1994; Stoichev *1994, 1996, 2000a, 2002; Michailova 2006; DW, DM, E1, E2, P2, P3, S1, T1, R1, RW, RE, BN, BS; 0-2340 m.

***Apsectrotanypus trifascipennis* (Zetterstedt, 1838)** [*Anatopynia*] – W, M, E; •; wes; Dimitrov 1963a; Russev et al. 1994; Stoichev *1994, 1996; Dashinov 2017; DW, E1, E2, B2, V5, R1; 20-1150 m.

***Ablabesmyia* (*Ablabesmyia*) *longistyla* Fittkau, 1962** – W, M, E; •; dp; Stoichev *1994, 1996. DW; 30-50 m.

***Ablabesmyia* (*Ablabesmyia*) *monilis* (Linnaeus, 1758)** – W, M, E; Vidin, FP57, 43°58'36.73", 22°52'19.10"; Somovit, LJ23-LJ24, 43°41'21.04", 24°46'26.81; Nikopol, LJ36, 43°42'21.42", 24°53'43.37"; Krivina, LJ83, 43°38'42.88", 25°34'42.18"; hno; Cvetkov 1955a, 1962; Valkanov 1957a; Russev *1959, *1962, 1966a, *1966b; Dimitrov 1962b, *1963a, 1966; Michailova 1982; Russev et al. 1994; Stoichev *1994, 1996; DW, DM, E1, E2, P1, P2, V1, TL, R1, RW, RE, BN; 0-1500 m.

***Larsia curticalcar* (Kieffer, 1918)** [*Ablabesmyia*] – W, M, E; •; FP39-NJ28; Belene, LJ43, 43°39'10.38", 25°07'38.83"; Belene, surroundings, LJ63, 43°39'02.53", 25°17'34.41"; Silistra, NJ28, 44°07'25.22", 27°16'03.97"; wp; Russev *1959, *1962, 1966a, *1966b; Dimitrov 1962a, 1962d, *1963a, 1966; Janeva 1991; Russev et al. 1994; Stoichev *1994, 1996, 2002; Kenderov et al. 2012; DW, DM, E1, E2, P1, P2, S1, B1, V1, V5, S211, T1, R1, R2, R5, RW, RE, BN, BS; 20-2440 m.

***Thienemannimyia lentiginosa* (Fries, 1823)** [*Ablabesmyia*] – M; Nikopol, LJ36, 43°42'21.42", 24°53'43.37"; Somovit, LJ23-LJ24, 43°41'21.04", 24°46'26.81"; dp; Cvetkov 1955a; Valkanov 1957; Dimitrov 1963a; Russev *1966b; Janeva 1987; Russev et al. *1994; Stoichev 1994, 1996; DW, B2, P2, BN; 0-660 m.

***Prodiamesa rufovittata* Goetghebuer, 1932** – W, M, E; •; wces; Dimitrov 1960a, 1962d, 1963a, 1966; Russev et al. 1994; Stoichev *1994, 1996; DW, P1, V5, S211, RW; 50-1150 m.

***Brillia bifida* (Kieffer, 1909)** [*B. modestus* Meigen, 1830; ? *B. immaculata* Botnariuc & Cure, 1956 (? nomina dubia); ? *B. pallida* Chernovskii, 1949 (? nomina dubia)] – W, M, E; •; po, ? ho; Russev 1961; Dimitrov 1962d, 1963a, 1966; Janeva 1987, 1991; Michailova 1989; Russev et al. 1991, 1994; Stoichev *1994, 1996, 2000a; Janeva & Russev 1997; Stoichev & Chernev 2001; Uzunov et al. 2011; Varadinova et al. 2013; Dashinov 2017; DW, DM, E1, P2, V4, R1, R5, RW, BN; 0-2535 m.

***Bryophaenocladius subvernalis* (Edwards, 1929)** [*B. subvernalis* var. *bulgarica* Cure; *B. subvernalis* var. *vitae* Cure] – W; Vit River near Gulyantsi, LJ13, 43°38'02.45", 24°41'53.19"; tp, ? dp; Janeva *1987; Russev et al. 1994; DW, DM; 25-30 m.

***Camptocladius stercorarius* (De Geer, 1776)** – W; Vit River near Gulyantsi, LJ13, 43°38'02.45", 24°41'53.19"; hoa; Nedelkov 1912; Dimitrov 1962b, 1963a, 1966; Janeva *1987; Russev et al. 1994; DW, DM, P1, P2, V1, TL; 20-600 m.

***Cricotopus (Cricotopus) algarum* (Kieffer, 1911)** – W, M, E; Lom, FP75, 43°37'15.53", 25°21'32.53"; Vit River near Gulyantsi, LJ13, 43°38'02.45", 24°41'53.19"; Nikopol, LJ36, 43°42'21.42", 24°53'43.37"; Srebarna Lake, NJ08, 44°06'24.64", 27°04'22.52"; wces; Cvetkov 1955a; Valkanov 1957a; Dimitrov 1957, 1960a, 1962a, 1962b, 1962c, 1966; Russev *1959, 1961, *1966b; Janeva *1987, 1991; Michailova 1989, 1996; Russev et al. 1991, 1994; Stoichev *1994, 1996, 1998, 2000a, 2000b, 2002; Janeva & Russev 1997; Stoichev & Chernev 2001; Uzunov et al. *2001; Uzunov et al. 2011; Varadinova et al. *2012, 2013; Kenderov et al. 2012; DW, DM, E1, E2, P1, P2, B1, V4, S1, TL, R1, RW, RE, BN; 0-2394 m.

***Cricotopus (Cricotopus) annulator* Goetghebuer, 1927** [*Eukiefferiella alpestris* Goetghebuer, 1941] – W, M, E; •; h; Dimitrov 1962d; Michailova 1982; Janeva 1987; Russev et al. 1987, 1991, 1994; Stoichev *1994, 1996; Stoichev & Chernev 2001; Uzunov et al. 2011; Varadinova et al. 2013; Dashinov 2017; DW, DM, E1, E2, P1, P2, P3, V5, T31, R1, R2, R5, BS, 0-2365 m.

***Cricotopus* (*Cricotopus*) *trifascia* Edwards, 1929** – E; Srebarna Lake, NJ08, 44°06'24.64", 27°04'22.52"; ho; Janeva 1987; Michailova 1989; Russev et al. 1994; Stoichev 1994, 1996; Janeva & Russev 1997; Uzunov et al. *2001; Uzunov et al. 2011; Varadinova et al. *1912, 2013; DW, DM, E1, P1, P2, O62, V5; 25-350 m.

***Cricotopus* (*Isocladius*) *ornatus* (Meigen, 1818)** – E; Srebarna Lake, NJ08, 44°06'24.64", 27°04'22.52"; h, ? ho; Caspers 1951a; Strenzke 1951; Michailova *1989, 1996, 2006; Janeva 1991; Russev et al. 1991, 1994; Stoichev 1996, 1998; Janeva & Russev 1997; Kovachev et al. 1999; Uzunov et al. *2001; Varadinova et al. *1912; DW, DM, E2, P1, P2, V1, RW, BN, BS; 0-1550 m.

***Cricotopus* (*Isocladius*) *sylvestris* (Fabricius, 1794)** – E; Srebarna Lake, NJ08, 44°06'24.64", 27°04'22.52"; hno; Cvetkov 1955a, 1955b, 1962; Angelov 1956; Dimitrov 1957, 1960b, 1962a, 1962b, 1962c, 1962d, 1963a, 1963b, 1966; Valkanov 1957a; Belcheva 1959; Zaschev & Angelov 1959; Mihailova-Neikova 1961; Russev 1961, 1966a, 1966b; Janeva 1987, 1991; Michailova 1989, 1996, 2006; Russev et al. 1994; Stoichev *1994, 1996, 1998, 2000a, 2000b, 2001a, 2002; Janeva & Russev 1997; Kovachev et al. 1999; Stoichev & Chernev 2001; Uzunov et al. *2001; Uzunov et al. 2011; Kenderov et al. 2012; Varadinova et al. *1912; 2013. DW, DM, E1, E2, P1, P2, B1, V1, V4, V5, S1, S211, TL, T1, T31, O61, O62, R1, R2, R5, RW, RE, BN, BS; 0-2535 m.

***Cricotopus* (*Isocladius*) *tricinctus* (Meigen, 1818)** – W; Ogosta River near Glozhene, GP24, 43°41'31.89", 23°49'33.83"; h; Russev et al. 1987, 1994; Janeva *1991; Janeva & Russev 1997; Uzunov et al. 2011; Varadinova et al. 2013; DW, DM, E1, E2, R5, BN; 0-200 m.

***Eukiefferiella brevicalcar* (Kieffer, 1911)** – W, M, E; •; wp; Janeva 1987; Russev et al. 1991, 1994; Stoichev *1994, 1996; Stoichev & Chernev 2001; Dashinov 2017; DW, DM, P2, R1; 30-1960 m.

***Eukiefferiella clypeata* (Thienemann, 1919)** [? *E. clypeata* (Kieffer, 1923)] – W, M, E; •; tp; Russev et al. 1991, 1994; Stoichev *1994, 1996; Stoichev & Chernev 2001; Uzunov et al. 2011; Varadinova et al. 2013. DW, DM, R1, R5; 200-1550 m.

***Eukiefferiella gracei* (Edwards, 1929)** [*E. longicalcar* Thienemann, 1926; *E. potthasti* Lehmann, 1972] – M, E; Svishtov, LJ63, 43°37'15.53", 25°21'32.53"; Krivina, LJ83, 43°38'42.88", 25°34'42.18"; h, ? wcp; Dimitrov 1960a, 1962a, 1962d, 1963a, 1966; Russev 1966a, *1966b; Michailova 1982; Janeva 1987; Stoichev *1994, 1996; Russev et al. 1991, 1994; Janeva & Russev 1997; Uzunov et al. 2011; Kenderov et al. 2012; Varadinova et al. *2012, 2013; DW, DM, E1, E2, P1, S1, TL, R1, RW; 50-1200 m.

***Eukiefferiella lobifera* Goetghebuer, 1934** – W, M, E; •; h; Dimitrov 1963a, 1966; Stoichev *1994, 1996; Russev et al. 1991, 1994; DW, P1, B1, T31, RW, BS; 0-1200 m.

***Eukiefferiella similis* Goetghebuer, 1939** – W, M, E; ♦; FP39-NJ28; Vidin, FP57, 43°58'36.73", 22°52'19.10"; Lom, FP75, 43°37'15.53", 25°21'32.53"; Tutrakan,

MJ67-MJ77, 44°02'51.63", 26°36'08.38"; Silistra, NJ28, 44°07'25.22", 27°16'03.97"; Srebarna Lake, NJ08, 44°06'24.64", 27°04'22.52"; wcp, ? tp; Russev *1959, 1966a, *1966b; Dimitrov 1962d, *1963a, 1966; Janeva 1987; Stoichev 1994, 1996; Russev et al. 1994; Janeva & Russev 1997; Stoichev & Chernev 2001; Uzunov et al. *2001; Kenderov et al. 2012; Uzunov et al. 2011; Varadinova et al. *2012, 2013; DW, DM, E1, E2, P1, P2, B1, V4, TL, R1, RW; 20-2350 m.

***Limnophyes minimus* (Meigen, 1818)** [*L. pusillus* Eaton, 1875] – W; Vidin, FP57, 43°58'36.73", 22°52'19.10"; Oryahovo, GP34, 43°44'19.96", 23°57'41.35"; hptn, sk; Cvetkov 1955a, 1957; Valkanov 1957; Dimitrov 1962b, *1963a; Russev *1959, *1962, 1966a; Russev et al. 1994; Uzunov et al. 2011; Varadinova et al. 2013; DW, DM, E1, E2, TL, R5, BN; 0-200 m.

? *Limnophyes transcaucasicus* Tshernovskij, 1949 – M; Nikopol, LJ36, 43°42'21.42", 24°53'43.37"; ? wces; Russev 1966b; Janeva & Russev 1997; DM; 20 m.

***Metriocnemus* (*Metriocnemus*) *albolineatus* (Meigen, 1818)** [*M. atratulus* Zetterstedt, 1850] – E; Rousse, MJ15, 43°50'42.70", 25°56'35.49"; ho; Nedelkov 1912; Janeva 1987; Michailova *1989, 2006; Russev et al. 1994; DW, E1, E2, P2, V1, RW; 20-1000 m.

***Metriocnemus* (*Metriocnemus*) *eurynotus* (Holmgren, 1883)** [*M. hygropetricus* Kieffer, 1912] – W, M, E; •; ho; Arndt 1943; Dimitrov 1963a; Russev 1966b; Janeva 1987; Stoichev *1994, 1996; Russev et al. 1994; Kenderov et al. 2012; P1, P2, P3, B1, B3, S1, T1, O61, O62, R1, RE; 300-1000 m.

***Orthocladius* (*Euorthocladius*) *thienemanni* Kieffer, 1906** – W; Ogosta River near Glozhene, GP24, 43°41'31.89", 23°49'33.83"; h; Michailova 1985, 1989, 2006; Janeva 1987, 1991; Russev et al. 1987, 1991, 1994; Stoichev 1994, 1996; Janeva & Russev 1997; Uzunov et al. 2011; Varadinova et al. 2013; DW, DM, E1, P1, P2, V1, R2, RW; 30-1100 m.

***Parakiefferiella bathophila* (Kieffer, 1912)** [*Chironomus*, *Orthocladius*] – W, E; Lom, FP75, 43°37'15.53", 25°21'32.53"; Srebarna Lake, NJ08, 44°06'24.64", 27°04'22.52"; h; Russev *1966b; Janeva 1987; Russev et al. 1994; Stoichev 1994, 1996; Uzunov et al. *2001; Uzunov et al. 2011; Varadinova et al. 2012, 2013; DW, DM, E1, E2, P1, P2, S1, TL, T1, O61, O62, RW, RE; 20-700 m.

***Psectrocladius* (*Psectrocladius*) *barbimanus* (Edwards, 1929)** [*P. ishimicus* Chernovskij, 1049] – E; Srebarna Lake, NJ08, 44°06'24.64", 27°04'22.52"; h; Dimitrov 1962b, 1963a; Michailova 1989; Uzunov et al. *2001; Varadinova et al. *2012; E2, V1, TL; 20-550 m.

***Thienemanniella clavicornis* (Kieffer, 1911)** [*Metriocnemus*] – W, M, E; •; wp; Janeva 1987, 1991; Russev et al. 1991, 1994; Stoichev *1994, 1996; Janeva & Russev 1997; Uzunov et al. 2011; Varadinova et al. 2013; Dashinov 2017; DW, DM, E1, E2, P1, P2, V5, R1; 20-1950 m.

***Tvetenia calvescens* (Edwards, 1929)** [*Eukiefferiella*] – W, M, E; •; h; Dimitrov 1962a, 1962d, 1963a, 1966; Russev 1964, 1966a; Janeva 1987, 1991; Russev et al. 1991, 1994; Stoichev *1994, 1996; Stoichev & Chernev 2001; Kenderov et al. 2012; Uzunov et al. 2011; Varadinova et al. 2013; Dashinov 2017; DW, DM, E1, E2, P1, P2, P3, B1, B3, S1, TL, T1, T31, O61, O62, R1, R5, RW, BS; 20-2250 m.

***Chironomus* (*Chironomus*) *annularius* Meigen, 1818** – E; Srebarna Lake, NJ08, 44°06'24.64", 27°04'22.52"; h; Nedelkov 1912; Michailova 1989, 1996; Uzunov et al. *2001; Varadinova et al. *2012; DW, E2, V1, TL, R5, BN; 0-750 m.

***Chironomus* (*Chironomus*) *plumosus* (Linnaeus, 1758)** [*Ch. plumosus f.l. semireductus Lenz, 1924*] – W, M, E; ♦; FP39-NJ28; Ogosta River near Glozhene, GP24, 43°41'31.89", 23°49'33.83"; Iskar River near Baykal, KJ94, 43°42'08.24", 24°27'19.95"; hno; Nedelkov 1912; Thienemann 1936; Valkanov 1936, 1941b, 1957a; Arndt 1943; Caspers 1951a, 1951b; Strenzke 1951a; Cvetkov 1955a, 1955b, 1957, 1958, 1962; Dimitrov 1957, 1960a, 1960b, 1962a, 1962b, 1962c, 1963a, 1966; Belcheva 1959; Russev 1959, 1961, 1962, 1964, 1966a, *1966b; Zaschev & Angelov 1959; Mihailova-Neikova 1961; Michailova 1982, 1989, 2006; Janeva 1987; Russev et al. 1987, 1991, 1994; Stoichev 1994, 1996, 2001a; Janeva & Russev 1997; Kovachev et al. 1999; Michailova & Krastanov 2000; Uzunov et al. *2001; Uzunov et al. 2011; Kenderov et al. *2012; Varadinova et al. 2012, 2013; DW, DM, E1, E2, P1, P2, P3, B1, B2, B3, V1, V4, S1, S2, TL, T1, T31, O61, O62, R1, R2, R5, RW, RE, BN, BS; 0-2394 m.

***Chironomus* (*Chironomus*) *riparius* Meigen, 1804** [*Ch. thummi* Kieffer, 1911] – W, M, E; ♦; FP39-NJ28; hn; Nedelkov 1912; Thienemann 1936; Valkanov 1936, 1957a; Caspers 1951a; Strenzke 1951; Cvetkov 1955a; Belcheva 1959; Kolarov 1959; Russev *1959, 1961, 1964, 1966a, *1966b; Dimitrov 1960b, 1962a, 1962b, 1962d, *1963a, 1966; Michailova 1982, *1989, 1996, 2006; Russev et al. 1987, 1991; Janeva 1987, 1991; Stoichev *1994, *1996, 1998, 2000a, 2000b, 2001a, 2002; Janeva & Russev 1997; Kovachev et al. 1999; Russev et al. 1994; Stoichev & Chernev 2001; Uzunov et al. 2011; Kenderov et al. 2012; Varadinova et al. 2012, 2013; DW, DM, E1, E2, P1, P2, P3, B1, V1, V4, V5, S1, S211, TL, T1, T31, O61, O62, R1, R2, R5, RW, RE, BN, BS; 0-2535 m.

***Cryptochironomus* (*Cryptochironomus*) *defectus* (Kieffer, 1913)** – W, E; Stanevo, FP95, 43°50'39.30", 23°26'38.07"; Ostrov, KJ63-KJ73, 43°40'57.57", 24°08'58.92"; Silistra, NJ28, 44°07'25.22", 27°16'03.97"; pa; Cvetkov 1955a, 1955b, 1957, 1962; Valkanov 1957a; Dimitrov 1957, 1960a, 1960b, 1962b, 1962c, 1962d, *1963a, 1966; Mihailova-Neikova 1961; Russev *1962, *1963, 1966a, *1966b; Russev et al. 1987, 1991, 1994; Janeva 1987, 1991; Michailova 1989, 1996; Stoichev *1994, *1996, 1998, 2000a, 2000b, 2001, 2002; Janeva & Russev 1997; Kovachev et al. 1999; Stoichev & Chernev 2001; Uzunov et al. *2001; Uzunov et al. 2011; Kenderov et al. 2012;

Varadinova et al. *2012, 2013; Dashinov 2017; DW, DM, E1, E2, P1, P2, P3, B1, B2, B3, V1, V5, S1, S211, TL, T1, T31, O61, O62, R1, R2, RW, RE, BN, BS; 0-2545 m.

Dicrotendipes modestus **(Say, 1823)** [*Trichotanypus*] – W; Vidin, FP57, 43°58'36.73", 22°52'19.10"; Lom, FP75, 43°37'15.53", 25°21'32.53"; h; Zilachi 1934; DW; 30 m.

Dicrotendipes nervosus **(Staeger, 1839)** [*Limnochironomus*] – W, M, E; Rousse, MJ15, 43°50'42.70", 25°56'35.49"; Srebarna Lake, NJ08, 44°06'24.64", 27°04'22.52"; ho; Cvetkov 1955a, 1955b, 1957, 1962; Valkanov 1957a; Mihailova-Neikova 1961; Dimitrov 1962b, 1962c, 1963a, 1966; Russev *1966b; Michailova 1982, *1989, 1996; Russev et al. 1994; Stoichev 1994, 1996, 1998, 2000a, 2002; Janeva & Russev 1997; Janeva & Russev 1997; Kovachev et al. 1999; Stoichev & Chernev 2001; Uzunov et al. 2011; Kenderov et al. 2012; Varadinova et al. *2012, 2013; DN, DM, E1, E2, P2, V1, TL, O61, O62, R1, R5, RE, BN, BS; 0-2440 m.

Einfeldia pagana **(Meigen, 1838)** [*Chironomus biappendiculatus* Kruglova, 1940] – W, M, E; •; ho; Dimitrov *1963a; *Russev 1966b; Michailova 1982; Janeva 1987; Russev et al. 1994; Stoichev 1994, 1996; Dashinov 2017; DW, DM, E1, E2, P1, P2, V1, V5. TL; 20-900 m.

Endochironomus tendens **(Fabricius, 1775)** [? *E. signaticornis* (Kieffer, 1913) – nomina dubia] – E; Mechka (surroundings), MJ04, 43°42'22.23", 25°47'08.33"; tp; Cvetkov 1955a, 1955b, 1962; Valkanov 1957a; Dimitrov 1962b, 1963a; Michailova *1982, 1989, *1992, 1996, 2006; Janeva 1987; Russev et al. 1994; Stoichev 1994, 1996, 1998; Janeva & Russev 1997; Kovachev et al. 1999; Kenderov et al. 2012. DW, DM, E1, P1, P2, V1, V5, TL, O61, RW, BN, BS; 0-1550 m.

Glyptotendipes **(*Glyptotendipes*)** ***cauliginellus*** **(Kieffer, 1913)** [*G. gripekoveni* Kieffer, 1913] – W, M, E; •; Mechka (surroundings), MJ04, 43°42'22.23", 25°47'08.33"; Srebarna Lake, NJ08, 44°06'24.64", 27°04'22.52"; po; Cvetkov 1955a, 1955b, 1962; Valkanov 1957a; Dimitrov 1960b, 1962a, 1962b, 1963a, 1963b; Mihailova-Neikova 1961; Russev 1966b; Michailova *1982, 1989, 1996, 2006; Russev et al. 1994; Stoichev *1994, 1996, 1998; Janeva & Russev 1997; Kovachev et al. 1999; Uzunov et al. *2001; Kenderov et al. 2012; Varadinova et al. *2012, 2013; DW, DM, E1, E2, B1, TL, R5, RW, BN, BS; 0-1550 m.

Glyptotendipes **(*Glyptotendipes*)** ***glaucus*** **(Meigen, 1818)** – W, E; Ogosta River near Glozhene, GP24, 43°41'31.89", 23°49'33.83"; Mechka (surroundings), MJ04, 43°42'22.23", 25°47'08.33"; Srebarna Lake, NJ08, 44°06'24.64", 27°04'22.52"; tes, ? wces; Michailova *1982, 1989, 1996, 2006; Janeva *1991; Russev et al. 1994; Stoichev 1994, 1996, 1998; Janeva & Russev 1997; Kovachev et al. 1999; Uzunov et al. *2001; Kenderov et al. 2012; Varadinova et al. *2012; DW, DM, E1, TL, O61, O62, RW, BN, BS; 0-1550 m.

Glyptotendipes **(*Glyptotendipes*)** ***pallens*** **(Meigen, 1804)** [*G. polytomus* Kieffer, 1909] – E; •; Srebarna Lake, NJ08, 44°06'24.64", 27°04'22.52"; po; Cvetkov 1955a, 1962, Valkanov 1957a; Mihailova-Neikova 1961; Dimitrov 1962b, 1963a; Russev et al.

1994; Stoichev *1994, 1996; Janeva & Russev 1997; Stoichev & Chernev 2001; Uzunov et al. *2001; Varadinova et al. *2012; DW, DM, E1, E2, TL, R1, BN, BS; 0-1570 m.

***Glyptotendipes* (*Glyptotendipes*) *paripes* (Edwards, 1929)** – E; Rousse, MJ15, 43°50’42.70”, 25°56’35.49”; h; Michailova *1979, 1982, 1989. DW, E1, E2, P1, TL, O61, BS; 20-400 m.

***Microtendipes chloris* (Meigen, 1818)** – W, M, E; •; Rousse, MJ15, 43°50’42.70”, 25°56’35.49”; wpo; Dimitrov 1957, 1962d, *1963a; Russev *1959, *1962, *1966b; Janeva 1987; Russev et al. 1994; Stoichev 1994, 1996; Janeva & Russev 1997; Kenderov et al. 2012; Varadinova et al. 2013; DW, DM, E1, E2, P1, P2, V5, S211, TL, T5, RW; 20-1200 m.

***Microtendipes pedellus* (De Geer, 1776)** – W, E; Iskar River near Baykal, KJ94, 43°42’08.24”, 24°27’19.95”; Rousse, MJ15, 43°50’42.70”, 25°56’35.49”; ho; Michailova *1989, 2006; Russev et al. 1994; Stoichev 1994, 1996; Janeva & Russev 1997; Uzunov et al. 2011; Kenderov et al. *2012; Dashinov 2017; DW, DM, E1, P1, P2, V5, O61, RW; 20-1000 m.

***Parachironomus gracilior* (Kieffer, 1918)** [*P. arcuatus* Goetghebuer, 1919] – E; Rousse, MJ15, 43°50’42.70”, 25°56’35.49”; po; Michailova *1989; E1, E2, P1, BN; 0-200 m.

? *Paratendipes connectens* Lipina, 1926 [? nomen dubium] – W, M, E; •; Svishtov, LJ63, 43°37’15.53”, 25°21’32.53”; ? wces; Russev *1959, *1962, *1966b; Dimitrov *1963a; DW, DM, E1, E2; 20-30 m.

? *Paratendipes intermedius* Chernovskij, 1949 [? nomen dubium] – E; Rousse, MJ15, 43°50’42.70”, 25°56’35.49”; ee; Russev *1962, *1966b; Dimitrov *1963a; Varadinova et al. 2013; DW, DM, E1, E2, V5; 20-900 m.

***Paratendipes nudisquama* (Edwards, 1929)** [*P. transcaucasicus* Chernovskii, 1949; *Limnophyes*] – W, M, E; •; Nikopol, LJ36, 43°42’21.42”, 24°53’43.37”; hno; Dimitrov 1962b, 1962d, 1963a, 1966; Russev 1966a, *1966b; Russev et al. 1994; Stoichev *1994, 1996; Janeva & Russev 1997; Uzunov et al. 2011; Varadinova et al. 2013; DW, DM, E1, E2, P2, V5, S211, TL, T1, O61, O62, R1, R5, RW, RE; 20-1200 m.

***Polypedilum* (*Uresipedilum*) *convictum* (Walker, 1856)** – W, M, E; •; Oryahovo, GP34, 43°44’19.96”, 23°57’41.35”; Somovit, LJ23-LJ24, 43°41’21.04”, 24°46’26.81”; Krivina, LJ83, 43°38’42.88”, 25°34’42.18”; Mechka, MJ04, 43°42’22.23”, 25°47’08.33”; Rousse, MJ15, 43°50’42.70”, 25°56’35.49”; ho; Dimitrov 1957, 1960a, 1960b, 1962b, 1962c, 1963a, 1966; Russev *1959, *1962, 1966a, *1966b; Cvetkov 1962; Michailova 1982; Janeva 1987, 1991; Russev et al. 1987, 1991, 1994; Stoichev *1994, 1996; Janeva & Russev 1997; Uzunov et al. 2011; Varadinova et al. 2013; Dashinov 2017; DW, DM, E1, E2, P1, P2, V5, S1, TL, R1, R5, RW, RE; 20-1400 m.

***Polypedilum* (*Polypedilum*) *nubeculosum* (Meigen, 1804)** [*Paratendipes*] – W, M, E; •; Vit River near Gulyantsi, LJ13, 43°38’02.45”, 24°41’53.19”; Srebarna Lake, NJ08,

44°06'24.64", 27°04'22.52"; h; Caspers 1951a, 1951b; Strenzke 1951; Cvetkov 1955a, 1955b, 1957, 1962; Dimitrov 1957, 1960a, 1960b, 1962a, 1962b, 1962c, 1963a, 1966; Valkanov 1957a; Mihailova-Neikova 1961; Russev 1964, 1966a; Michailova 1982, 1989, 1996; Janeva *1987; Russev et al. 1991, 1994; Stoichev *1994, *1996, 1998, 2000a, 2001a, 2004; Janeva & Russev 1997; Kovachev et al. 1999; DW, DM, E1, E2, P2, S1, TL, R1, RW, RE, BN, BS; 0-2535 m.

***Polypedilum* (*Polypedilum*) *nubifer* (Skuse, 1889)** [*P. pharao* Kieffer, 1925; *P. aberrans* Chernovskii, 1949] – W; Lom, FP75, 43°37'15.53", 25°21'32.53"; poa; Caspers 1951a; Strenzke 1951; Cvetkov 1955a, 1962; Valkanov 1957a; Dimitrov 1960b, 1962b, 1963a, 1966; Russev 1964, 1966a, 1966b; Russev et al. 1994; Stoichev 1994, 1996; Michailova 1989, 2006; DW, DM, E1, E2, P1, P2, TL, RW, RE, BN, BS; 0-1550 m.

***Polypedilum* (*Tripodura*) *scalaenum* (Schrank, 1803)** [*P. breviantennatum* Chernovskii, 1949] – W, M; Botevo, FP45, 43°50'23.22", 22°51'37.75"; Lom, FP75, 43°37'15.53", 25°21'32.53"; Belene, surroundings, LJ63, 43°39'02.53", 25°17'34.41"; ho; Cvetkov 1955a, 1955b, 1957; Valkanov 1957a; Dimitrov 1957, 1960a, 1962b, 1962c, 1963a, 1966; Russev *1962, 1964, 1966a, *1966b; Janeva 1987, 1991; Russev et al. 1994; Stoichev 1994, 1996; Janeva & Russev 1997; Stoichev & Chernev 2001; Uzunov et al. 2011; Varadinova et al. 2013; DW, DM, E1, E2, P2, V5, S211, S1, TL, R1, R5, RW, RE, BN; 0-2003 m.

***Polypedilum* (*Tripodura*) *tetracrenatum* Hirvenoja, 1962** [*Procladius*] – W; •; Ogosta River mouth, GP34, 43°44'43.54", 23°53'01.89"; dp; Janeva *1991; Russev et al. 1994; Stoichev *1994, 1996; Janeva & Russev 1997; DW, DM, E1, E2; 20-100 m.

***Cladotanytarsus* (*Cladotanytarsus*) *mancus* (Walker, 1856)** [*Tanytarsus*] – W, M, E; ♦; Mechka (surroundings), MJ04, 43°42'22.23", 25°47'08.33"; Ostrov, KJ63-KJ73, 43°40'57.57", 24°08'58.92"; Silistra, NJ28, 44°07'25.22", 27°16'03.97"; wcp; Cvetkov 1955a, 1955b, 1957; Valkanov 1957a; Russev *1959, *1962, 1966a, *1966b; Dimitrov 1960b, 1962b, 1962c, 1962d, 1963a, *1966; Russev et al. 1994; Stoichev 1994, 1996, 2001a; Janeva & Russev 1997; DW, DM, E1, E2, P1, P2, V1, V5, S1, S211, TL, R1, RW, RE, BN; 0-1200 m.

? *Cladotanytarsus sexdentatus* (Chernovskij 1949) [*Tanytarsus*; ? nomen dubium] – W; Kutovo, FP57, 44°00'49.53", 22°58'25.61"; wes; Russev *1966b; Stoichev 1994, 1996; Janeva & Russev 1997; DW, DM, P2; 50-200 m.

***Micropsectra junci* (Meigen, 1818)** [*M. praecox* Meigen, 1818; *Microtendipes*] – W, M; Lom, FP75, 43°49'17.00", 23°11'13.10"; Svishtov, LJ63, 43°37'15.53", 25°21'32.53"; h; Dimitrov 1957, 1963a, 1966; Belcheva 1959; Russev *1959, *1962, 1966a, *1966b; Janeva 1987; Michailova 1982; Russev et al. 1994; Stoichev 1994, 1996, 2000a, 2004; Janeva & Russev 1997; DW, DM, E1, E2, P1, P2, V1, S1, R1, R5, RW; 20-2368 m.

***Paratanytarsus lauterborni* (Kieffer, 1909)** [*Tanytarsus*] – W, M, E; ♦; FP39-NJ28; Tutrakan, MJ67-MJ77; 44°02'51.63", 26°36'08.38"; h; Russev *1959, *1962, *1966b; Dimitrov 1960b, 1962a, 1962b, 1962d, *1963a, 1963b, 1966; Mihailova-Neikova 1961; Cvetkov 1962; Michailova 1982, 1989, 2006; Russev et al. 1991, 1994; Janeva & Russev 1997; DW, DM, E1, E2, P1, B1, V5, S1, S211, TL, R1, RW, BS; 0-1550 m.

***Tanytarsus excavatus* Edwards, 1929** – W; Lom River, before Lom Town, FP75, 43°46'55.41", 23°14'03.63"; e; Russev et al. *1991, 1994; DW; 30-40 m.

***Tanytarsus gregarius* Kieffer, 1909** [*T. lobatifrons* Kieffer, 1913] – W, M, E; ♦; FP39-NJ28; Iskar River mouth, KJ94, 43°43'54.43", 24°26'28.93"; h; Caspers 1951b; Valkanov 1957a; Dimitrov 1960a, 1960b, 1962b, 1962c, 1962d, *1963a, 1963b, 1966; Russev 1964, 1966a, *1966b; Janeva 1987; Russev et al. 1991, 1994; Stoichev *1994, 1996, 2001a; Janeva & Russev 1997; Stoichev & Chernev 2001; Stoichev & Danova 2003; Uzunov et al. 2011; *Kenderov et al. 2012; Varadinova et al. 2012, 1013; DW, DM, E1, E2, P1, P2, TL, T1, T31, V5, S1, S211, O61, O62, R1, R5, RW, RE, BN, BS; 0-2709 m.

***Tanytarsus usmaensis* Pagast, 1931** – W; Lom River, before Lom Town, FP75, 43°46'55.41", 23°14'03.63"; h; Janeva 1991; Russev et al. *1991, 1994. DW, DM; 30-500 m.

***Virgatanytarsus arduennensis* (Goetghebuer, 1922)** [*Tanytarsus*] – W; Ogosta River mouth, GP34, 43°44'43.54", 23°53'01.89"; ppt; Janeva *1991; Russev et al. 1991, 1994; Janeva & Russev 1997; DW, DM, P1; 30-250 m.

The zoogeographical categorization of the species is made on the basis of current data about their distribution. Thus, the dipterans are divided into 52 categories, combined into two main groups and five complexes (Table 3).

Table 3. Distribution of Nematocera from the Bulgarian sector of the Danube River

Arealographical categories	Bulgarian Danubian bank			
	Total number	Western	Middle	Eastern
Species distributed in Palaearctic and out of it	68 (44.4%)	56 (48.3%)	30 (41.1%)	50 (41.0%)
North type	64 (41.8%)	52 (44.8%)	29 (39.7%)	48 (39.3%)
Semicosmopolitan	1 (0.7%)	1 (0.9%)	1 (1.4%)	1 (0.8%)
Holarctic-Paleotropical-Neotropical	1 (0.7%)	1 (0.9%)		
Holarctic-Paleotropical-Australian	2 (1.3%)	2 (1.7%)	1 (1.4%)	1 (0.8%)
Holarctic-Neotropical-Oriental	5 (3.3%)	4 (3.4%)	3 (4.1%)	4 (3.3%)
Holarctic-Oriental-Australian	1 (0.7%)	1 (0.9%)		
Holarctic-Neotropical	1 (0.7%)	1 (0.9%)	1 (1.4%)	1 (0.8%)
Holarctic-Oriental (ho)	10 (6.5%)	8 (6.9%)	7 (9.6%)	9 (7.4%)
Palaearctic-Oriental-Australian	1 (0.7%)	1 (0.9%)		
Palaearctic-Paleotropical	3 (2.0%)	3 (2.6%)		2 (1.6%)
Palaearctic-Afrotropical	2 (1.3%)	2 (1.7%)		2 (1.6%)

Arealographical categories	Bulgarin Danubian Bank			
	Total number	Western	Middle	Eastern
Palaearctic-Oriental	4 (2.6%)	1 (0.9%)	1 (1.4%)	4 (3.3%)
Palaearctic-Australian	1 (0.7%)	1 (0.9%)		1 (0.8%)
West Palaearctic-Oriental	1 (0.7%)	1 (0.9%)	1 (1.4%)	1 (0.8%)
Holarctic (h)	30 (19.6%)	24 (20.7%)	14 (19.2%)	21 (17.2%)
South type	4 (2.6%)	4 (3.4%)	1 (1.4%)	2 (1.6%)
South Palaearctic-Oriental	1 (0.7%)	1 (0.9%)	1 (1.4%)	1 (0.8%)
Southwest Palaearctic-Oriental	2 (1.3%)	2 (1.7%)		1 (0.8%)
Paleotropical-Mediterranean	1 (0.7%)	1 (0.9%)		
Species with Palaearctic distribution	85 (55.6)	60 (51.7%)	43 (58.9%)	72 (59.0%)
Palaearctic type	51 (33.%)	33 (28.4%)	24 (32.9%)	45 (36.9%)
Holopalaearctic	2 (1.3%)	1 (0.9%)	1 (1.4%)	2 (1.6%)
Transpalaearctic	11 (7.2%)	9 (7.8%)	4 (5.5%)	8 (6.6%)
West and Central Palaearctic	13 (8.5%)	7 (6.0%)	4 (5.5%)	13 (10.7%)
West Palaearctic	12 (7.8%)	9 (7.8%)	9 (12.3%)	11 (9.0%)
Disjunct Palaearctic	6 (3.9%)	3 (2.6%)	3 (4.1%)	4 (3.3%)
South Palaearctic	1 (0.7%)	1 (0.9%)		1 (0.8%)
European-North African	3 (2.0%)	1 (0.9%)	1 (1.4%)	3 (2.5%)
Eurosiberian-Anatolian-Central Asian	1 (0.7%)	1 (0.9%)	1 (1.4%)	1 (0.8%)
Eurosiberian-Central Asian	1 (0.7%)			1 (0.8%)
European-West Central Asian	1 (0.7%)	1 (0.9%)	1 (1.4%)	1 (0.8%)
Eurosiberian type	26 (17.0%)	21 (18.1%)	17 (23.3%)	20 (16.4%)
Holoeurosiberian	1 (0.7%)	1 (0.9%)	1 (1.4%)	1 (0.8%)
Transeurosiberian	2 (1.3%)	2 (1.7%)	2 (2.7%)	1 (0.8%)
West and Central Eurosiberian	5 (3.3%)	4 (3.4%)	4 (5.5%)	4 (3.3%)
West Eurosiberian	2 (1.3%)	2 (1.7%)	1 (1.4%)	1 (0.8%)
Disjunct Eurosiberian	1 (0.7%)	1 (0.9%)		1 (0.8%)
European	9 (5.9%)	8 (6.9%)	6 (8.2%)	6 (4.9%)

Species Distributed in the Palaearctic and beyond It

This group (68 species or 44.4%) includes 18 categories, of which 15 combine species of northern type (widely distributed in the Holarctic and Palaearctic) and three – species of southern type (distributed only in the southern parts of the Palaearctic). The percentage difference between the separate parts of the studied territory is from 41.0% (Eastern Sector) to 44.8% (Western Sector). It is very likely other species of the group of northern type in the Danube River and nearby areas to be established owing to their distribution and insufficient studies of the parts of the studied territory. It is accepted that the taxa of northern type have vast areas and ecological flexibility. In the Superpala-earctic complex, the Holarctic species (30 species or 19.6%) are best presented. Of the other areographical categories, the Holarctic-Oriental (10 species or 6.5%) and the Holarctic-Neotropical-Oriental (5 species or 3.3%) forms are better presented. The species of southern type are not determinant for the zoogeographical characteristic of the fauna due to their fewness (1-4 species or 1.4-3.4%). Usually the Superpalaearctic complex is scanty presented and is not determinant for the zoogeographical characteristic of the taxa in the Bulgarian terrestrial fauna (with the exception of the coastal fauna). In highly mobile forms (such as Diptera) the complex is well presented and can exceed 20-25% (Hubenov 2015). It is best presented in the west part of the studied area and less represented in the other two areas (Table 3). In the two-winged insects significant number of synanthropic and synbovil forms with cosmopolitan or subcosmopolitan distribution occur. They have anthropogenic ranges, structured with the development of the human civilization (before the beginning of the contemporary research).

Species Distributed only in the Palaearctic, but in More than One Subregion (Palaearctic Type)

Taxa, whose areas include more than one Palaearctic subregion in latitudinal direction, belong to this complex. They are well represented in the high mobile groups and comprise about 25-30% of the species composition. A total of 51 dipteran species (33.0%) from this group, combined into 10 areographical categories, has been established in the Danubian sector (Table 3). Its character is determined by the Transpalaearctic (11 species or 7.2%), West Palaearctic (12 species or 7.8%) and West and Central Palaearctic (13 species or 8.5%). The correlation of these categories remains the same in the separate parts of the Danubian sector and varies from 5.5 to 12.3% (4 to 13 species). Six species (3.9%) – from 3 to 4 species in the different parts of the studied territory, have a longitudinal disjunction of the areas with regard to Siberia and Central Asia. Probably some of these species are presented with sparse populations and will be studied in more detail as a result of further research. Most often a latitudinal disjunction of the areas of this group lacks (Gorodkov 1984, Josifov 1988, Hubenov 2015). A significant part of the species with wide vertical distribution (about 20%) belong to the Palaearctic complex. It includes from 28.4% to 36.9% (24 to 45 species) of the species composition in the separate parts of the Danubian sector. The vast areas and wide vertical distribution of the taxa of this group are an indication of the greater ecological flexibility of its species. The Palaearctic complex (like the Superpalaearctic complex) is very well represented in the studied area. This is probably related to the incomplete study of the territory where the lack of sufficient research and the non-systematic collection of a material have led to the collection of common and widespread species. A significant part of the taxa are connected with water which also increases the number of species with large areas.

Species Distributed within one Subregion of the Palaearctic

This complex (34 species – 22.2%) combine species with Eurosiberian and Mediterranean type of distribution (17 categories). The Mediterranean-Central Asian species are also included here according to Kryzhanovsky (1965, 2002) and Lopatin (1989) who combine the Mediterranean and Central Asian subregions (Table 3). The species with Mediterranean type of distribution are accepted in a general way and include Submediterranean, Subiranian and Pontian faunistic elements that could be considered separately from the Mediterranean ones as well (Gruev & Kusmanov 1994, 1999; Gruev 1995; Gruev & Bechev 2000).

The Eurosiberian Species

The Eurosiberian species (26 species – 17.0% of 9 areographical categories) include from 16.4 to 23.3% (17 to 21 species) of the species composition in the different parts of the Danubian sector. The European (9 species or 5.9%) and the West and Central Eurosiberian (5 species or 3.3%) taxa are the most numerous. The European-Anatolian species are also well represented. The ratio of these categories is different for the separate families. Only one taxa with disjunctive range is presented (*Culicoides flavipulicaris* Dhafarov, 1964), for which also a North Mediterranean-Central Asian range is possible. The low percentage of the Eurosiberian species (in the mountains they reach 40-50%) is related to the low altitude of the studied area and the specificity of the taxa considered.

The Mediterranean Species

The Mediterranean species (8 species – 5.2% of 8 areographical categories) include from 2.7 to 5.7% (2 to 7 species) of the species composition in the different parts of the Danubian sector. Their relatively scarce populations are due to the lower ecological flexibility of the Mediterranean forms in comparison with the previous ones. Because of the big variety of these areas, the group is divided into many subgroups with different origin, distribution and ecological peculiarities of the taxa. This complexity contributes to establishing of various zoogeographical classifications for Bulgaria (Josifov 1981, 1986, 1988, 1999; Gruev 1988, 1995, 2000a, 2000b, 2000c, 2002; Heiss & Josifov 1990; Gruev & Kusmanov 1994; Hubenov 1996, 2008; Gruev & Bechev 2000; Popov 2002). The Mediterranean forms are most poorly represented in the middle part of the Danubian sector. The low percentage of the Mediterranean species, regardless of the appropriate xerothermic habitats in the studied area, is related to the insufficient research of the most families.

Conclusion

The Nematocera suborder of the Danube River, the Danubian islands and the nearby parts of the Bulgarian Danube River bank are scrutinized. A total of 153 species from 14 families have been established. Three families are better studied – Culicidae (20 species – 13.1%), Ceratopogonidae (36 species – 23.5%) and Chironomidae (63 species – 41.2%). The established species belong to 52 areographical categories. The fauna can be divided into 2 main groups: 1) species with Mediterranean type of distribution (12 species – 7.8%) – more thermophilic and distributed mainly in the southern parts of the Palaearctic. Four species of southern type, distributed in the Palaearctic and beyond it, can be formally related to this group as well; 2) species with Palaearctic and Eurosiberian type of distribution (141 species – 92.2%) – more cold-resistant and widely distributed in the Palaearctic. Sixty-four species of northern type, distributed in the Palaearctic and beyond it, can be formally related to this group as well. Endemic forms have not been found. Of the established species, 14 are pests on the forestry or agriculture, and 13 species have a human or veterinary medical significance.

Acknowledgments

The study was made within the frame of the International Association for Danube Research (IAD), supported by the DIAS network.

References

Adler P. & Crosskey R. 2018. *World Blackflies (Diptera: Simuliidae): a comprehensive revision of the taxonomic and geographical inventory*. 134 p. https://biomia.sites.clemson.edu/ pdfs/blackflyinventory.pdf.

Agushev T. 2012. Preliminary study of species composition, ecology, and health importance of family Culicidae in Plovdiv City, Bulgaria. *Science & Techynologies* 2 (1): 204-207. (In Bulgarian with English summary).

Agushev T. 2014. Preliminary results for the species composition and dynamics of mosquito family (Culicidae) in the city of Plovdiv. *Ecologica Balkanica* 5 (Special edition): 123-131.

Agushev T. 2015. Vidov sastav, ekologiya i znachenie na semeystvo komari (Culicidae) v grad Plovdiv [Species composition, ecology and importance of the mosquito family (Culicidae) in the Plovdiv Town]. Ph.D. thesis, Plovdiv, Agricultural University. 172 p. (In Bulgarian).

Agushev T. & Bileva T. 2014. Prouchvane na vidoviya sastav na komarite ot semeystvo Culicidae v grad Plovdiv [Study of the species composition of the mosquitoes of the Culicidae family in the Plovdiv Town]. *Natural & Mathematical science* 4 (3): 79-83.

Angelov A. 1956. Hydrologische und hydrobiologische Untersuchungen über den Rabischa-See [Hydrological and hydrobiological investigations on the Lake Rabischa]. *Annuaire de l'Université de Sofia, Faculté de Biologie, Géologie et Géographie* [*Annual of the Sofia University, Faculty of Biology, Geology and Geography*] 59 (1 – Zoology): 1-3 (In Bulgarian, Russian and German summaries).

Arndt W. 1943. Beiträge zur Kenntnis der Süsswasserfauna Bulgariens. (Ergebnisse eines Aufenthaltes in Bulgarien im Juli – August 1924) [Contributions to the knowledge of the freshwater fauna of Bulgaria. (Results of a stay in Bulgaria in July - August 1924)]. *Mitteilungen Königl Naturwissenschaftlichen Instituten in Sofia* (*Bulgarien*) [*Messages of the Royal scientific institutes in Sofia* (*Bulgaria*)] 16: 189-206.

Atanassof D., Avlavidov T., Kovchazov G., Hristova T. & Deanov I. 1962. Sastoyanie na chuvstvitelnostta na *An. maculipennis* Meig. i *An. sacharovi* Favre. po otnoshenie na DDT i dieldrin v nyakoi rayoni na Bulgaria [Sensibility state of *An. maculipennis* Meig. and *An. sacharovi* Favre. with regard to DDT and dieldrin in some regions of Bulgaria]. *Byuletin na NIEM* [*Bulletin of NIEM*] 3: 121-128. (In Bulgarian).

Atanassov D. & Christova T. 1960. The results from the investigation of the resistance of Anopheles maculipennis to the chloric insecticides in some region of the country. *Trudy naucho-issledovatelskogo instituta epidemologii i mikrobiologii* [*Proceedings of the Research Institute of Epidemiology and Microbiology*] 7: 279-287. (In Russian, English summary).

Atanassov D. & Petrov B. 1961. Prouchvane anofeliynostta ns nyakoi napoitelni sistemi v Narodna Republika Bulgariya [Study of the Anaphelian fauna of some irrigation systems in the Republic of Bulgaria] *Higiene* 4 (3): 51-55. (In Bulgarian).

Avlavidov T. 1947. Anofeliynata fauna v Dolniya Kamchiyski kray [The Anaphelian fauna in the Lower Kamchia region]. *Lekarski pregled* [*Medical review*] 9 (9/10): 127-129. (In Bulgarian).

Avlavidov T. 1948. Za kasnata esenna malariya prez 1947 godina [For the late autumn malaria in 1947 in northern Bulgaria]. *Lekarski pregled* [*Medical review*] 10 (9/10): 134-136. (In Bulgarian).

Avlavidov T. 1958. Po vaprosa za izkorenyavane na malariyata vav Varnenski okrag [To the question for the malaria eradication in the Varna district]. *Hygiene, Epidemology and Microbiology* 2 (2): 16-19. (In Bulgarian).

Avlavidov T. 1959. To the question for the sensibility of the anophelian mosqutos against DDT in the region of Varna. *Hygiene, Epidemology and Microbiology* 3 (5): 40-47. (In Bulgarian).

Avlavidov T. 1961. Prichini za vaznikvane na malariynite zabolyavaniya v ozdravena veche teritoriya [Causes of malaria disease rising in already recovered territory]. *Hygiene* 4 (4): 27-30. (In Bulgarian).

Avlavidov T. 1970. Fenologichni osobenosti na anofeliynite komari v severiztochniya kray na nashata strana do likvidiraneto na malariyata [Fenological characteristics of the anopheline mosquitoes in the northeastern part of Bulgaria until the malaria eradication]. *Letopisi na HEI* [*Chronicles of HEI*] 4 (26): 70-74. (In Bulgarian).

Bechev D. 1986b. New species from the Superfamily *Mycetophiloidea* (Diptera) for the Bulgarian Fauna. II. *Acta zoologica bulgarica* 32: 56-59. (in Bulgarian, English summary).

Bechev D. 1994. New species Mycetophiloidea (Diptera) to the fauna of Bulgaria. VI. *Travaux Scientifiques Universite de Plovdiv, Biologie – Animalia* [*Scientific Works of the Plovdiv University, Biology – Animalia*] 30 (6): 25-28.

Bechev D. 1996a. New species of fungus gnats (Diptera: Sciaroidea, excluding Sciaridae) to the fauna of Bulgaria. VII. *Travaux Scientifiques Universite de Plovdiv, Biologie – Animalia* [*Scientific Works University of Plovdiv, Biology – Animalia*] 32 (6): 19-21.

Bechev D. 1997. Check-list of the fungus gnats (Diptera: Sciaroidea, excluding Sciaridae) of the Balkan Peninsula. *Travaux Scientifiques Universite de Plovdiv, Biologie – Animalia* [*Scientific Works of the Plovdiv University, Biology – Animalia*] 33: 9-22.

Bechev D. 2000. Seasonal activity and vertical distribution of fungus gnats (Diptera: Sciaroidea, excluding Sciaridae) in Western and Central parts of Star Planina Ridge (Bulgaria). *Travaux Scientifiques Universite de Plovdiv, Biologie – Animalia* [*Scientific Works of the Plovdiv University, Biology – Animalia*] 36: 51-66.

Bechev D. 2002a. Check-list of the fungus gnats of the families Bolitophilidae, Diadocidiidae, Ditomyiidae, Keroplatidae and Mycetophilidae in Bulgaria (Diptera: Nematocera). *Travaux Scientifiques Universite de Plovdiv, Biologie – Animalia* [*Scientific Works of the Plovdiv University, Biology – Animalia*] 38: 81-112.

Bechev D. 2004b. New distributional data about the fungus gnats of the families Bolitophilidae, Diadocidiidae, Ditomyiidae and Keroplatidae in Bulgaria (Diptera: Nematocera: Sciaroidea). *Travaux Scientifiques Universite de Plovdiv, Biologie – Animalia* [*Scientific Works of the Plovdiv University, Biology – Animalia*] 40: 129-136.

Bechev D. 2006b. Fungus Gnats (Diptera: Bolitophilidae, Diadocidiidae, Ditomyiidae, Keroplatidae, Mycetophilidae) of the Western Rhodopes (Bulgaria). In: Beron P. (Eds.). *Biodiversity of Bulgaria.* 3. Biodiversity of Western Rhodopes (Bulgaria and Greece) I. Sofia, National Museum of Natural History & Pensoft, 701-710.

Bechev D. 2006c. The fungus gnats of the families Bolitophilidae, Diadocidiidae, Ditomyiidae and Keroplatidae (Diptera: Sciaroidea) of Bulgaria. *Scientific Studies of the University of Plovdiv, Biology, Animalia* 42: 21-83.

Bechev D. 2010. On the family Mycetophilidae (Insecta: Diptera) in Bulgaria. *ZooNotes* Supplement 1: 1-72.

Belcheva R. 1959. Prouchvaniya varhu yazovir "V. Kolarov" s ogled ribostopanskoto mu izpolzvane [Studies on the Vasil Kolarov Dam with a view to fish-breeding]. *Annuaire de l'Université de Sofia, BGGF* [*Annual of the Sofia University, BGGF*] 51(1): 302-338. (In Bulgarian).

Beron P. 1994. *Résultats des recherhes biospéléologiques en Bulgarie de 1971 à 1994 et liste des animaux cavernicoles bulgares* [*Results of the biospeleological research in Bulgaria from 1971 to 1994 and list of the Bulgarian cavernous animals*]. Tranteeva 1. Sofia, Fédération bulgare de spéléologie [Bulgarian Federation of Speleology]. 137 p.

Beron P. 2015. Cave fauna of *Bulgaria.* Sofia, East-Wst. 440 p.

Beron P. 2016. Terrestrial cave invertebrates of the Vrachanska Planina mountains. In: Bechev D. & Georgiev D. (Eds.). Faunistic diversity of Vrachanski Balkan Nature Park. *ZooNotes S*upplement 3: 185-230.

Beschovski V. 2006. Faunistic review of some Nematocera families (Diptera) establisched in the Western Rhodopes (Bulgaria). In: Beron P. (Ed.). *Biodiversity of Bulgaria.* 3. Biodiversity of Western Rhodepes (Bulgaria and Greece) 1. Sofia, National Museum of Natural History & Pensoft, 729-752.

Bobeva A. 2014. Molecular studies of the specificity of avian malaria parasites (Haemosporidia) to the vectors of the genus Culicoides Latreille, 1809 (Diptera: Ceratopogonidae). Ph.D. thesis, Sofia, Bulgarian Academy of Sciences. 81 p. (In Bulgarian).

Bobeva A., Zehtindjiev P., Bensch S. & Radrova J. 2013. A survey of biting midges of the genus *Culicoides* Latreille, 1809 (Diptera: Ceratopogonidae) in NE Bulgaria, with respect to transmission of avian haemosporidians. *Acta Parasitologica* 58 (4): 585-591. DOI: 10.2478/s11686-013-0185-z

Bobeva A., Ilieva M., Dimitrov D. & Zehtindjiev P. 2014. Degree of associations among vectors of the genus *Culicoides* (Diptera: Ceratopogonidae) and host bird species with respect to haemosporidian parasites in NE Bulgaria. *Parasitology Research* 113 (12): 4505-4511. DOI: 10.1007/s00436-014-4140-1

Bobeva A., Zehtindjiev P., Ilieva M., Dimitrov D., Mathis A. & Bensch S. 2015. Host preferences of ornithophilic biting midges of the genus *Culicoides* in Eastern Balkans with respect to transmission of haemosporidian parasites. *Medical and Veterinary Entomology* 29 (3): 290-296. DOI: 10.1111/mve.12108

Bondev I. 1991. *The vegetation of Bulgaria*. Sofia, Kliment Ohridski University Press. 184 p. (In Bulgarian).

Bondev I. 1997. Geobotanic regioning. In: Yordanova M. & Donchev D. (Ed.). *Geography of Bulgaria. Sofia*. Bulgarian Academy of Sciences, 269-305. (In Bulgarian with English summary).

Bondev I. 2002. Geobotanic regioning. In: *Geography of Bulgaria.* Sofia, ForCom, 336-352. (In Bulgarian).

Boychev D. 1950a. Prinos kam rasprostranenieto na roda Phlebotomus (Diptera) v Bulgaria [Contribution to the distribution of the genus Phlebotomus (Diptera) in Bulgaria]. *Godishnik na Selskostopanskata akademiya "G. Dimitrov". Zootehnicheski fakultet* [*Annual of the Georgi Dimitrov Agricultural Academy. Zootechnical Faculty*] 3: 101-107. (In Bulgarian).

Boychev D. 1950b. Podvidovete na *Anopheles maculipennis* po nasheto dunavsko kraybrezhie [Subspecies of *Anopheles maculipennis* on the Bulgarian Danubian coast]. *Godishnik na Selskostopanskata akademiya "G. Dimitrov". Zootehnicheski fakultet* [*Annual of the Georgi Dimitrov Agricultural Academy. Zootechnical Faculty*] 3: 109-112. (In Bulgarian).

Bozhkov (Božkov) D. 1953. Können die Larven der *Anopheles maculipennis* Eier ihrer eigenen Art aufessen ? [Can the larvae of *Anopheles maculipennis* eat eggs of their own kind?]. *Bulletin de l'Institut zoologque de l'Academie Bulgare des Sciences* [*Bulletin of the Zoological Institute of the Bulgarian Academy of Sciences*] 2: 163-188. (In Bulgarian with German summary).

Bozhkov D. 1957. Malariynite komari kato obekt za izuchvane ot uchenicheskite zoologicheski krazhotsi [The malaria mosquitoes, as a subject of study by the pupil's zoological groups]. *Priroda i znanie* [*Nature and knowledge*] 10 (2): 20-22. (In Bulgarian).

Bozhkov D. 1958. Eine für die Fauna Bulgariens neue Art blutsaugende Stechmücke [A new type of blood-sucking mosquito for the fauna of Bulgaria]. *Comptes rendus de l'Academie bulgare des Sciences* [*Reports of the Bulgarian Academy of Sciences*] 11 (6): 501-502.

Bozhkov D. 1959. Beitrag zur Erforschung der Stechmückenfauna im Rhodopen- und Rilagebirge [Contribution to the research of mosquito fauna in the Rhodope and Rila mountains]. *Bulletin de l'Institut zoologque de l'Academie Bulgare des Sciences* [*Bulletin of the Zoological Institute of the Bulgarian Academy of Sciences*] 8: 109-119. (In Bulgarian with German summary).

Bozhkov D. 1961. Mückenlarven aus dem Gebiet von Burgas und dem Strandžagebirge [Mosquito larvae from the area of Burgas and the Strandzha Mt.]. In: Paspalev G. (Ed.). *Prirodni ognishta na zaraza v Strandzha planina* [*Natural sources of infection in Strandja Mt.*]. Sofia, Bulgarian Academy of Sciences, 147-153. (In Bulgarian with German summary).

Bozhkov D. 1962. Komarni larvi ot Petrichko i Gotse Delchevsko [Mosquitoes larvae from Petrich and Gotse Delchev]. In: Paspalev G. (Ed.) *Prirodni ognishta na zaraza v Petrichko i Gotse Delchevsko.* Sofia, Bulgarian Academy of Sciences, 219-224. (In Bulgarian).

Bozhkov D. 1965. Stechmückenlarven aus Thrakien [Mosquito larvae from Thrace]. In: Paspalev G. (Ed.). *Die Fauna Thrakiens* 2. Sofia, Bulgarischen Akademie der Wissenschaften [*The fauna of Thrace* 2. Sofia, Bulgarian Academy of Sciences] 255-264. (In Bulgarian with German summary).

Bozhkov D. 1966a. Krovososushtie komary Bolgarii [Blood-sucking mosquitoes of Bulgaria]. *Entomologicheskoe obozrenie* [*Entomological review*] 45 (3): 570-574. (In Russian).

Bozhkov D. 1966b. Beitrag zum studium der Steichmücken im Vitoscha-Gebirge [Contribution to the study of the mosquitoes in the Vitosha Mt.]. *Bulletin de l'Institut de zoologie et musée* [*Bulletin of the Institute of Zoology and Museum*] 21: 65-67. (In Bulgarian with German summary).

Bozhkov D. 1967. Stechmücken aus dem westlichen Balkangebirge [Mosquitoes from the Western Stara Planina Mts.]. *Bulletin de l'Institut de zoologie et musée* [*Bulletin of the Institute of Zoology and Museum*] 25: 41-64. (In Bulgarian with German summary).

Bozhkov D. 1974a. Stechmücken aus dem mittleren und dem östlichen Balkangebirge (Mit einigen Notizen über die Stechmücken des Balkangebirges überhaupt) [Mosquitoes from the Middle and Eastern Stara Planina Mts. (with some notes about the mosquitoes of the Stara Planina Mts. at all)]. *Bulletin de l'Institut de zoologie et musée* [*Bulletin of the Institute of Zoology and Museum*] 41: 161-175. (In Bulgarian with German summary).

Bozhkov D. 1974b. Kravosmucheshtite komari na balgarskite planini [The blood-sucking mosquitoes of the Bulgarian mountains]. *Priroda* [*Nature*] 23 (3): 40-42. (In Bulgarian).

Bozhkov D. 1991. *Kravosmucheshtite komari* [*The blood-sucking mosquitoes*]. Sofia, Bulgarian Academy of Sciences. 223 p. (In Bulgarian).

Bozhkov D. & Christova T. 1965. Kravosmucheshtite komari kato prenositeli na virusi v Evropa [The blood-sucking mosquitoes as virus carriers in Europe]. *Priroda* [*Nature*] 14 (4): 71-74. (In Bulgarian).

Bozhkov D., Christova T. & Canev I. 1969. Stechmücken an der bulgarischen Schwarzmeerküste [Mosquitoes of the Bulgarian Black Sea coast]. *Bulletin de l'Institut de zoologie et musée* [*Bulletin of the Institute of Zoology and Museum*] 29: 151-166. (In Bulgarian with German summary).

Buresch I. 1924. Referate und Berichte [Presentations and reports]. *Mitteilungen der Bulgarischen Entomologischen Gesellschaft in Sofia* [*Messages of the Bulgarian Entomological Society in Sofia*] 1: 16-20. (In Bulgarian).

Buresch I. 1926b. Referate und Berichte im Jahre 1925 [Presentations and reports in 1925]. *Mitteilungen der Bulgarischen Entomologischen Gesellschaft in Sofia* [*Messages of the Bulgarian Entomological Society in Sofia*] 3: 22-27. (In Bulgarian).

Buresch I. 1938. Ausländische Literatur über die Fauna Bulgariens, Thraciens und Mazedoniens. VI [Foreign literature on the fauna of Bulgaria, Thrace and Macedonia.

VI]. *Mitteilungen aus den Königl Naturwissenschaftlichen Instituten in Sofia* [*Messages of the Royal Institute of Natural Sciences in Sofia*] 11: 182-184. (In Bulgarian).

Buresch I. & Lazarov A. 1956. Harmful insects for agriculture and forestry enterprise in Bulgaria. Bibliography and catalogue. *Trudove na Instituta po zoologiya* [*Works of the Institute of Zoology*] 5. Sofia, Bulgarian Academy of Sciences. 291 p. (In Bulgarian).

Caspers H. 1951a. Biozönotische Untersuchungen über die Strandarthropoden im bulgarischen Küstenbereich des Schwarzen Meeres [Biocenotic research on the beach arthropods in the Bulgarian coastal area of the Black Sea]. *Hydrobiologia, den Haag* [*Hydrobiology, Hague*] 3 (2): 131-193.

Caspers H. 1951b. Quantitative Untersuchungen über die Bodentierwelt des Schwarzen Meeres im bulgarischen Küstenbereich [Quantitative studies on the soil animal world of the Black Sea in the Bulgarian coastal area]. *Archiv für Hydrobiologie* [*Archive for Hydrobiology*] 45 (1-2): 1-192.

Caspers H. 1951c. Rhythmische Erscheinungen in der Fortpflanzung von *Clunio marinus* (Dipt., Chironomidae) und das Problem der lunaren Periodizität bei Organismen [Rhythmic phenomena in the reproduction of *Clunio marinus* (Dipt., Chironomidae) and the problem of the lunar periodicity in the organisms]. *Archiv für Hydrobiologie* [*Archive for Hydrobiology*] Supl. 18 (3): 415-594.

Caspers H. 1952. Untersuchung Über der bulgarischen Küste des Schwarzen Meeres [Investigation on the Bulgarian coast of the Black Sea]. *Zoologisher Anzeiger* [*Zoological journal*] 148 (5/8): 243-259.

Caspers H. 1957. Black Sea and Sea of Azov. *Treafise on Marine Ecology and Palaeoecology* 1: 801-889.

Chichkoff G. & Konsuloff, S. 1914. Izuchavaniya varhu komarite (Culicidae) v Bulgaria I tehnite larvi [Studies on the mosquitoes (Culicidae) in Bulgaria and their larvae]. *Annuaire de l'Université de Sofia. Faculté physico-mathématique* [*Annual of the Sofia University. Physico-mathematical faculty*] 8/9: 1-51. (In Bulgarian).

Choleva B. 1964. Vredni muhi po pechurkite [Harmful flies on the field mushrooms]. *Rastitelna zhashtita* [*Plant protection*] 6: (in Bulgarian).

Christova T. 1980. Epidemiologichno znachenie i geografsko rasprostranenie na anofeliynite komari v Bulgariya [Epidemiological significance and geographical distribution of the Anopheles mosquitoes in Bulgaria]. *DDD periodichen byuletin* [*DDD periodical newsletter*] 11 (1): 3-6. (In Bulgarian).

Christova T., Canev I. & Smilova D. 1971. Promeni vav vidoviya sastav na kravosmucheshtite komari ot bulgarskoto Chernomorsko kraybrezhie [Changes in the species composition of the blood-sucking mosquitoes from the Bulgarian Black Sea coast]. *Letopisi na HEI* [*Chronicles of HEI*] 5 (31): 173-183. (In Bulgarian).

Christova T. & Dorovski D. 1972. Razprostranenie na komarni larvi ot sem. Culicidae v Plevenski okrag [Distribution of mosquito larvae from the Culicidae family in Pleven district]. *Epidemiologiya, mikrobiologiya i infektsiozni bolesti* [*Epidemiology, microbiology and infectious diseases*] 9 (2): 185-192. (In Bulgarian).

Christova T. & Todorova M. 1969. Komarite v grad Sofiya i borbata sreshtu tyah: 1. Vidov sastav i razprostranenie na kravosmucheshtite komari v Sofiya [The mosquitoes in Sofia and the fight against them: 1. Species composition and distribution of the mosquitoes in Sofia]. *Letopisi na HEI* [*Chronicles of HEI*] 3 (16): 88-92. (In Bulgarian).

Crosskey R. & White G. 1977. The Afrotropical Region. A recommendet term in zoogeography. *Journal of Natural History* 11: 541-544.

Cvetkov L. 1955a. Die Chironimidenfauna der Bulgarischen Schwarzenmeerseen [The Chironimidae fauna of the Bulgarian Black Sea lakes]. *Bulletin de l'Institut zoologque de l'Academie des Sciences de Bulgarie* [*Bulletin of the Zoological Institute of the Bulgarian Academy of Sciences*] 4/5: 215-249. (In Bulgarian with German summary).

Cvetkov L. 1955b. Untersuchungen über das Futter der Fische im Beloslavsko-See [Studies on the diet of fish in the Beloslavsko Lake]. *Bulletin de l'Institut zoologque de l'Academie des Sciences de Bulgarie* [*Bulletin of the Zoological Institute of the Bulgarian Academy of Sciences*] 4/5: 329-352. (In Bulgarian with German summary).

Cvetkov L. 1957. Zoobenthos des Beloslav-Sees [Zoobenthos of the Beloslav lake]. *Bulletin de l'Institut zoologque de l'Academie des Sciences de Bulgarie* [*Bulletin of the Zoological Institute of the Bulgarian Academy of Sciences*] 6: 381-439. (In Bulgarian with German summary).

Cvetkov L. 1958. Investigations into the microbenthos of the Bulgarian Black Sea lakes. *Bulletin de l'Institut zoologique de l'Academie des sciences de Bulgarie* [*Bulletin of the Zoological Institute of the Bulgarian Academy of Sciences*] 7: 219-250. (In Bulgarian with English summary).

Cvetkov L. 1962. Sur la prolifération quantitative des Chironomidés et la productivité des bassins de pisciculture [On the quantitative proliferation of Chironomids and the productivity of the fish ponds]. *Bulletin de l'Institut de zoologie et musée* [*Bulletin of the Institute of Zoology and Museum*] 12: 173-196. (In Bulgarian with French summary).

Czerný L. 1930. Dipteren auf Schnee und Höhlen [Diptera on the snow and caves]. *Mitteilungen Königl Naturwissenschaftlichen Instituten in Sofia* (Bulgarien) [*Messages of the Royal scientific institutes in Sofia* (Bulgaria)] 3: 113-118.

Darlington P. 1957. *Zoogeography*. New York, Wiley. 675 p.

Dashinov D. 2017. New data on the Chironomidae (Diptera) fauna from the Upper Iskar River catchment (Bulgaria) – spatial distribution and ecological notes. *Ecological Engineering and Environment Protection* 9: 16-21.

Dimitrov M. 1957. Hidrologichna i hidrobiologichna harakteristika na yazovir "Al. Stamboliyski" [Hydrological and hydrobiological characteristic of the Aleksandar Stamboliiski Dam]. *Travaux de l'Institut de recherches scientifiques sur la pêche et les industries s'y rattachant* (*Varna*) [*Works of the Institute for Scientific Research on Fisheries and Related Industries* (*Varna*)] 1: 159-197. (In Bulgarian).

Dimitrov M. 1960a. Hidrologichna i hidrobiologichna harakteristika na yazovir "G. Dimitrov" [Hydrological and hydrobiological characteristic of the Georgi Dimitrov Dam]. *Bulletin de l'Institut central de recherche scientifique de pisciculture et de pêcherie* (*Varna*) [*Bulletin of the Central Scientific Research Institute for Fish Farming and Fisheries (Varna)*] 1: 87-122. (In Bulgarian with German summary).

Dimitrov M. 1960b. Bentosat na nyakoi malki yazoviri v Plovdivski okrag [The benthos of some small dams in the Plovdiv district]. *Bulletin de l'Institut central de recherche scientifique de pisciculture et de pêcherie* (*Varna*) [*Bulletin of the Central Scientific Research Institute for Fish Farming and Fisheries* (*Varna*)] 1: 123-138. (In Bulgarian).

Dimitrov M. 1962a. The benthos in the Batak Dam. *Bulletin of the test station of Freshwater Pisciculture* 1: 59-80. (In Bulgarian with English summary).

Dimitrov M. 1962b. Bentosat na ribovadnite stopanstva v Plovdivski okrag [The benthos of the fish-farms in the Plovdiv district]. *Bulletin of the test station of Freshwater Pisciculture* 1: 21-57. (In Bulgarian).

Dimitrov M. 1962c. The Benthos in the "Studen Kladenets" Dam. *Bulletin of the test station of Freshwater Pisciculture* 1: 81-94. (In Bulgarian with English summary).

Dimitrov M. 1962d. Materiali po hironomidnata fauna (larvae) na r. Iskar [Materials on the Chironomidae fauna (larvae) of the Iskar River]. *Bulletin of the test station of Freshwater Pisciculture* 1: 129-135. (In Bulgarian).

Dimitrov M. 1963a. Studies of Chironomid fauna (Larvae) in Bulgaria. *Bulletin of the test station of Freshwater Pisciculture* 2: 5-24. (In Bulgarian with English summary).

Dimitrov M. 1963b. Biologicheski osnovi i efektivnost na zelenoto torene v ribovadnite baseyni [Biological basics and efficiency of the green fertilization in the fish-farms]. *Priroda* [*Nature*] 12 (6): 74-77. (In Bulgarian).

Dimitrov M. 1966. Chironomidae larvae (Family Chironomidae, Diptera) and Oligachaeta (Order Oligochaeta) of the Maritsa River and some of its tributaries. In: *Die Fauna Thrakiens 3. Sofia, Bulgarische Akademie der Wissenschaften* [*The fauna of Thrace 3. Sofia, Bulgarian Academy of Sciences*] 307-318. (In Bulgarian, Russian and English summaries).

Dimitrov M. 1969. The benthos of the Pysachnik Dam. *Proceedings of the freshwater fishery research station* (*Plovdiv*) 6: 31-39. (In Bulgarian, Russian and English summaries).

Dimitrov M. 1970. Das Benthos des Stausees "Ivailovgrad" [The benthos of the Ivailovgrad reservoir]. *Proceedings of the freshwater fishery research station* (Plovdiv) 7: 35-43. (In Bulgarian, Russian and German summaries).

Dimitrov M. 1972. Das Benthos im Stausee "Ovtschariza" [The benthos of the Ovchtaritsa reservoir]. *Proceedings of the freshwater fishery research station* (Plovdiv) 9: 83-96. (In Bulgarian, Russian and German summaries).

Dimitrova B. 1987. A Contribution to the fauna of the subfamily Cecidomyiinae (Diptera, Cecidomyiidae) in Vitosha, Bulgaria. In: *Savremenni postizheniya na balgarskata zoologiya.* Sofia, Bulgarian Academy of Sciences, 25-29. (In Bulgarian).

Dimitrova B. 1989. Faunistic, zoogeographical and environmental studies of the subfamily Cecidomyiinae (Diptera, Cecidomyiidae) of the Vitosha Mt. Ph.D. thesis, Sofia, Bulgarian Academy of Sciences. 174 p. (In Bulgarian).

Dimov I. 1957. Zimuvane i diapauza na *Anopheles maculipennis typicus* v Sofiysko [Wintering and diapause of *Anopheles maculipennis typicus* in Sofia district]. *Higiena, Epidemologiya i Mikrobiologiya* [*Hygiene, Epidemiology and Microbiology*] 1 (2): 36-41. (In Bulgarian).

Dimtchev D., Stefanov S., Kovtchazov G. & Christova T. 1960. Sur la composition des especes de la faune des *Anopheles* et des varietes de *A. maculipennis* Meigen sur le littoral bulgare du Danube [On the composition of species of *Anopheles* fauna and varieties of *A. maculipennis* Meigen on the Bulgarian coast of the Danube]. *Folia medica* 2 (1): 25-31.

Dimtchev D., Stefanov S., Kovtchazov G. & Christova T. 1962. Prinos kam prouchvane na anofeliynata fauna po dunavskoto kraybrezhie s ogled epidemiologiyata na malariyata v tozi rayon [Contribution to the study of the Anopheles fauna along the Danubian coast in view of the epidemic of malaria in this region]. *Sbornik trudove na Visshiya meditsinski institut "I. P. Pavlov" v Plovdiv* [*Collection of papers of the Pavlov Higher Medical Institute in Plovdiv*] 16: 245-262. (In Bulgarian).

Donchev K. 1968. Vredni galitsi po lyutsernata [Harmful Cecidomyiidae on the alfalfa]. *Rastitelna zashtita* [*Plant protection*] 8: 19-21. (In Bulgarian).

Dorovski D. 1976. Prouchvaniya varhu vidoviya sastav, razprostranenieto i borbata s komarite v Plevenski okrag [Studies on the species composition, distribution and mosquito control in the Pleven district]. Ph.D. thesis, Sofia, National Center of Infectious and Parasitic Diseases. 155 p. (In Bulgarian).

Drenowsky A. 1922. Nepoznat vreden vid peperuda po zhitnite rasteniya u nas [Unknown harmful kind of butterfly on the cereal plants in Bulgaria]. *Estestvoznanie i geografiya* [*Natural Studies and Geography*] 6 (1): 44-45. (In Bulgarian).

Drenowsky A. 1929a. I. Anatomo-morfologichni i biologichni belezhki varhu malarichniya komar i negovata larva. II. Kratka sistematika na petrichkata komarna fauna [I. Anatomic-morphological and biological notes on the malarial mosquito and its larva. II. A brief systematics of the Petrich mosquito's fauna]. *Izvestiya na Direktsiyata na narodnoto zdrave 14 (priturka)* [*Notifications of the People's Health Directorate 14 (supplement)*] 2: 118-137. (In Bulgarian).

Drenowsky A. 1929b. Deynostta na petrichkata protivomalarichna opitna stantsiya prez 1928 godina [The activity of the Petrich Antimalarial Experimental Station in 1928]. *Bulgarska klinika* [*Bulgarian Clinic*] 1 (10): 745-777. (In Bulgarian).

Drenowsky A. & Enderlein G. 1923. *Novi vredni muhi za zhitnite rasteniya v Bulgariya* [*New harmful flies on the cereal plants in Bulgaria*]. Arhiv na Ministerstvoto na zemedelieto i darzhavnite imoti [Archives of the Ministry of Agriculture and State Property] 3: 1-20. (In Bulgarian).

Drensky K. 1926. Sanitary significans of the Insekts. *Mitteilungen der Bulgarischen Entomologischen Gesellschaft in Sofia* [*Reports of the Bulgarian Entomological Society in Sofia*] 3: 59-70. (In Bulgarian with English summary).

Drensky K. 1939. The varieties of *Anopheles maculipennis* Meigen and their relation to the distribution of malaria. *Mitteilungen der Bulgarischen Entomologischen Gesellschaft in Sofia* [*Reports of the Bulgarian Entomological Society in Sofia*] 10: 31-44. (In Bulgarian with English summary).

Drensky K. 1949. Malariynite komari u nas [Malaria mosquitoes in Bulgaria]. *Izvestiya na institutite po eksperimentalna meditsina, sotsialna meditsina, obshta biologiya i veterinarno-eksperimentalna meditsina* [*Notifications of the Institutes of Experimental Medicine, Social Medicine, General Biology and Veterinary-Experimental Medicine*] 1: 246-279. (In Bulgarian).

Drensky K. 1950. Malariynite komari u nas [Malaria mosquitoes in Bulgaria]. In: *II nauchen sabor na balgarskite lekari, chast II, malariya* [*II Scientific Council of Bulgarian Doctors, Part II, Malaria*]. Sofiya, 27-48. (In Bulgarian).

Drensky K. 1958. Anofeliynite komari i borbata protiv malariyata [Anopheline mosquitoes and the fight against the malaria]. *Priroda* [*Nature*] 7 (3): 43-48. (In Bulgarian).

Drensky P. 1928. Referate und Berichte im Jahre 1926-1927 [Presentations and reports in 1926-1927]. *Mitteilungen der Bulgarischen Entomologischen Gesellschaft in Sofia* [*Reports of the Bulgarian Entomological Society in Sofia*] 4: 18-21. (In Bulgarian).

Drensky P. 1931a. Kleine entomologische Mitteilungen [Small entomological messages]. *Mitteilungen der Bulgarischen Entomologischen Gesellschaft in Sofia* [*Reports of the Bulgarian Entomological Society in Sofia*] 6: 123-141. (In Bulgarian).

Drensky P. 1931b. Referate und Berichte im Jahre 1930 [Presentations and reports in 1930]. *Mitteilungen der Bulgarischen Entomologischen Gesellschaft in Sofia* [*Reports of the Bulgarian Entomological Society in Sofia*] 6: 19-23. (In Bulgarian).

Drensky P. 1932. Referate und Berichte im Jahre 1931 [Presentations and reports in 1931]. Mitteilungen der Bulgarischen Entomologischen Gesellschaft in Sofia [Reports of the Bulgarian Entomological Society in Sofia] 7: 26-28. (In Bulgarian).

Drensky P. 1942. Über die Insektenfauna des Küstengebietes nördlich von Warna [About the insect fauna of the coastal area north of Varna]. *Mitteilungen der Bulgarischen Entomologischen Gesellschaft in Sofia* [*Reports of the Bulgarian Entomological Society in Sofia*] 12: 15-44. (In Bulgarian with German summary).

Drensky P. 1955. Fliegen und Arachnoideen aus den Wäldern und den Feldschutzwaldstreifen der Süd-Dobrudscha [Flies and arachnoids from the forests and the protective field of the South Dobrudzha]. In: *Sbornik na ekspeditsiyata po polezashtitnite poyasi v Dobrudzha prez 1952 godina* [*Compilation of the expedition on the protective belts in Dobroudja in 1952*], Sofia, Bulgarian Academy of Sciences, 385-397. (In Bulgarian with German summary).

Drensky P. & Drensky K. 1928. Beitrag zur Erforschung von Genus *Phlebotomus* (Dipt.) und des Dreitägigen Fiebers in Bulgariaen [Contribution to the research of the genus *Phlebotomus* (Dipt.) and the three-day fever in Bulgaria]. *Mitteilungen der Bulgarischen Entomologischen Gesellschaft in Sofia* [*Reports of the Bulgarian Entomological Society in Sofia*] 4: 31-56. (In Bulgarian with German summary).

Enderlein G. 1924. Die Simuliden Bulgariens und ihre Schäden im Jahre 1923 [The Simulids of Bulgaria and their damage in 1923]. *Zoologischer Anzeiger* [*Zoological journal*] 61 (11/12): 280-288.

Ficetola G., Mazel F. & Thuiller W. 2017. Global determinants of zoogeographical boundaries. *Nature Ecology & Evolution* 1 (89): 1-7.

Ganov V. 1949. Klinika na papatatsievata treska u nas [The clinical picture of the Phlebotomidae fever in Bulgaria]. *Izvestiya na Instituta za eksperimentalna meditsina, sotsialna meditsina, obshta biologiya i veterinarna eksperimentalna meditsina* [*Notifications from the Institute of Experimental Medicine, Social Medicine, General Biology and Veterinary Experimental Medicine*] 1: 452-464. (In Bulgarian).

Georgiev V. 1980. Zoogeografsko rayonirane na Bulgaria [Zoogeographical regioning of Bulgaria]. *Geografiya* [*Geography*] 7: 1-4. (In Bulgarian).

Georgiev V. 1982. Zoogeographical regions on the basis of the land fauna. In: *Geography of Bulgaria* 1. Sofia, Bulgarian Academy of Sciences, 472-477. (In Bulgarian with English summary).

Georgiev V. 1997. Zoogeographical regioning. In: Yordanova M. & Donchev D. (Eds.). *Geography of Bulgaria.* Sofia, Bulgarian Academy of Sciences, 330-333. (In Bulgarian with English summary).

Georgiev V. 2002. Zoogeographical regioning. In: *Geography of Bulgaria.* Sofia, ForCom, 373-375. (In Bulgarian with English summary).

Geptner V. 1936. *Obshtaya zoogeografiya* [*General zoogeography*]. Moscow, Biomedgiz. 548 p. (In Russian).

Gorodkov K. B. 1984. Ranges types of insects of tundra and forests zones of the European Part of USSR. In: *Provisional Atlas of the insects of the European Part of USSSR.* Leningrad, Nauka, 3-20. (In Russian).

Grehan J. 1988. The natural biogeographic regions. *Revista di Biologia – Biology Forum* 81 (4): 569-575.

Grehan J. 1993. Conservation biogeography and the biodiversity crisis: a global problem in space/time. *Biodiversity Letters* 1: 134-140.

Grigorov S. 1962. Prinos kam entomofaunata na Bulgariya [Contribution to the entomofauna of Bulgaria]. *Rastitelna zashtita* [*Plant protection*] 10 (1): 48-54. (In Bulgarian).

Grigorov S. 1972. *Spetsialna entomologiya* [*Special entomology*]. Sofia, Zemizdat. 592 p. (In Bulgarian).

Grigorov S. 1976. *Spetsialna entomologiya* [*Special entomology*]. Sofia, Zemizdat. 545 p. (In Bulgarian).

Gruev B. 1988. *General biogeography*. Sofia: Nauka i iskustvo [Science and art]. 396 p. (In Bulgarian).

Gruev B. 1995. About the mediterranean faunistic complex in Bulgaria. *Annuaire de l'Université de Sofia "Kliment Ohridski", Faculté de Biologie* [*Annual of the Sofia University Kliment Ohridski, Faculty of Biology*] 86/87 (1 – Zoology): 75-82.

Gruev B. 2000a. Souteuropean peninsular endemism of Alticinae (Coleoptera: Chrysomelidae). *Travaux Scientifiques Universite de Plovdiv, Animalia* [*Scientific Works of the Plovdiv University, Animalia*] 36 (6): 35-50.

Gruev B. 2000b. About the Atlantic faunistical element in Bulgaria. *Travaux Scientifiques Universite de Plovdiv, Animalia* [*Scientific Works of the Plovdiv University, Animalia*] 36 (6): 67-72. (in Bulgarian, English summary).

Gruev B. 2000c. About the submediterranean zone of the Palaearctic realm and the submediterranean faunistic element in Bulgaria. *Travaux Scientifiques Universite de Plovdiv, Animalia* [*Scientific Works of the Plovdiv University, Animalia*] 36 (6): 73-94. (In Bulgarian with English summary).

Gruev B. 2002. A comparative study on Alticinae (Coleoptera: Chrysomelidae) in the Balkan Peninsula and Asiatic Turkey. Causes of the similarities and the differences of the fauna. *Travaux Scientifiques Universite de Plovdiv Animalia* [*Scientific Works of the Plovdiv University, Animalia*] 38 (6): 49-79.

Gruev B. & Bechev D. 2000. Zoogeographical belonging of the leaf beetle species of Lamprosomatinae, Eumolpinae, Chrysomelinae, Alticinae, Hispinae and Cassidinae (Coleoptera, Chrysomelidae) and their distribution in the biogeographical regions of Bulgaria. *Travaux Scientifiques Universite de Plovdiv, Animalia* [*Scientific Works of the Plovdiv University, Animalia*] 36 (6): 5-34. (in Bulgarian with English summary).

Gruev B. & Kusmanov B. 1994. *General biogeography*. Sofia: University K. Ohridski. 498 p. (In Bulgarian).

Gruev B. & Kusmanov B. 1999. *General biogeography*. Plovdiv: Plovdiv University. 344 p. (In Bulgarian).

Guéorguiev V. & Beron P. 1962. Essai sur la faune cavernicole de Bulgarie [Test on the cavernous fauna of Bulgaria]. *Annales de Spéléologie* [*Annals of Speleology*] 17 (2/3): 285-411.

Harizanov A. 1964. Krushova listna muha [Pear leaf fly]. *Ovoshtarstvo* [*Fruit growing*] 6: (In Bulgarian).

Harizanov A., Angelova R. & Babrikova T. 1996. *Entomology*. Sofia, Zemizdat. 416 p. (In Bulgarian).

Heiss E. & Josifov M. 1990. Vergleichende Untersuchung über Artenspektrum, Zoogeographie und Ökologie der Heteropteren-Fauna in Hochgebirgen Österreichs und Bulgariens [Comparative study on the species spectrum, zoogeography and ecology of the heteropteran fauna in high mountains in Austria and Bulgaria]. *Bericht des Naturwissenschaftlich-Medizinischen Vereins Innsbruck* [*Reports of the Scientific-Medical Association Innsbruck*] 77: 123-161.

Hitilov I. 1912a. Shvedskata i hesenskata muhi [The Swedish and Hessian flies]. *Zemedelie* [*Agriculture*] 18 (1): 3-5. (In Bulgarian).

Hitilov I. 1912b. Rolyata na navremennoto zaoravane na starnishteto za unishtozhavane na nepriyatelite na zhitnite rasteniya [The role of the ploughing on time of the stubble for destroying of the pests on the cereal plants]. *Zemedelie* [*Agriculture*] 18 (11/12): 171-173. (In Bulgarian).

Holt B., Lessard J.P., Borregaard M., Fritz S., Araújo M., Dimitrov D., Fabre P.H., Graham C., Graves G., Jønsson K., Nogués-Bravo D., Wang Z., Whittaker R., Fjeldså J. & Rahbek C. 2013. An ubdate of Wallace's zoogeographic regions of the world. *Science* 339: 74-78.

Hristova E. 1959. Lyutsernov tsvetov komar (Contarinia medicaginis Kieff.) i borbata s nego [Alfalfa blossom mosquito (Contarinia medicaginis Kieff.) and the fight against it]. *Nauchni trudove na nauchno-izsledovatelskiya institut po furazhite v Pleven* [*Scientific papers of the Pleven Institute for Feed Science*] 1: 55-57. (In Bulgarian).

Hristova K. 1954. Lyutsernov tsvetov komar – *Contarinia medicaginis* i borbata s nego [Alfalfa blossom mosquito – *Contarinia medicaginis* and the fight against it]. *Byuletin po rastitelna zashtita* [*Plant Protection Bulletin*] 3 (1): 67-69. (In Bulgarian).

Hristova N. 2012. *Hydrology of Bulgaria.* Sofia, Tip-top pres. 832 p. (In Bulgarian with English summary).

Hubenov Z. 1996. Zoogeographische Charakteristik der bulgarischen Raupenfliegen (Diptera, Tachinidae) [Zoogeographical characteristics of the Bulgarian caterpillar flies (Diptera, Tachinidae)]. *Historia Naturalis bulgarica* 6: 49-58.

Hubenov Z. 1997. Possibilities for using of a system from the really defined natural territories for the faunistic researches in Bulgaria. *Acta zoologica bulgarica* 49: 5-9.

Hubenov Z. 2008. Recent fauna of Bulgaria – Animalia: Invertebrata. *Acta zoologica bulgarica* 60 (1): 3-21.

Hubenov Z. 2015a. Areographical structure of the Bulgarian non-marine invertebrate fauna (Metazoa: Invertebrata). *Acta zoologica bulgarica* 67 (2): 203-213.

Hubenov Z. 2015b. Two-winged insects (Insecta: Diptera) of Pirin. *Historia naturalis bulgarica* 21: 215-256.

Hubenov Z. 2016. The Dipterans (Insecta: Diptera) of the Rila Mountains. *Historia naturalis bulgarica* 23: 37-99.

Hubenov Z. 2018. The Dipterans (Insecta: Diptera) of the Vitosha Mountains. *Historia naturalis bulgarica* 26: 1-66.

Hubenov Z. 2019. The Dipterans (Insecta: Diptera) of the Vrachanska Planina Mountains. In: Bechev D. & Georgiev D. (Eds.). Faunistic diversity of Vrachanski Balkan Nature Park. Part 2. *ZooNotes* Supplement 6

Janeva I. 1987. The zoobenthos of the river Vit. I. Composition, structure and dynamics of the zoocoenoses. *Hydrobiology* 31: 37-64. (In Bulgarian with English summary).

Janeva I. 1989. Changes in the saprobiological state of the Černa river and their effect on the composition and structure of the benthic organisms. *Hydrobiology* 34: 20-29. (In Bulgarian with English summary).

Janeva I. 1991. Saprobiological state of the river Ogosta in various periods of study. *Hydrobiology* 36: 32-48. (In Bulgarian with English summary).

Janeva I. & Russev B. 1997. Veränderungen der Artenzusammensetzung und Güteklasse des bulgarischen Donauzufluss Jantra nach dem Makrozoobenthon [Changes in the species composition and quality class of the Bulgarian Danube tributary Jantra after macrozoobenthon]. *Lauterbornia* 31: 1-16.

Janeva I., Vidinova Y. & Tyufekchieva V. 2001. Contemporary saprobiological characteristics of Arda River in the section of future "Gorna Arda" cascade building. *Acta zoologica bulgarica* 53 (2): 37-46.

Joakimoff D. 1899. Prinos kam faunata ot nasekomi na Rila planina [Contribution to the insect fauna of Rila Mountain]. *Periodichesko spisanie na Balgarskoto knizhovno druzhestvo* [*Periodical Journal of the Bulgarian Literary Society*] 60: 858-884. (In Bulgarian).

Josifov M. 1981. Nasekomite ot razred Heteroptera na Balkanskiya poluostrov [Insects from the order Heteroptera on the Balkan Peninsula]. Dr. Sci. Thesis, Sofia: Bulgarian Academy of Sciences (Instute of Zoology), 31-288. (In Bulgarian).

Josifov M. 1986. Verzeichnis der von der Balkanhalbinsel bekannten Heteropterenarten (Insecta, Heteroptera) [List of the Heteroptera species known from the Balkan Peninsula (Insecta, Heteroptera)]. *Faunistische Abhandlungen Staatliches Museum für Tierkunde Dresden* [*Faunistic contributions of the Dresden Zoological Museum*] 14 (6): 61-93.

Josifov M. 1988. Über den zoogeographischen Charakter der Südeuropäischen Insektenfauna unter besonderer Berücksichtigung der Heteropteren [On the zoogeographical character of the southern European insect fauna with special consideration of the Heteroptera]. *Berichte des Naturwissenschaftlich-Medizinischen Vereins in Insbruck* [*Reports of the scientific-medical association in Innsbruck*] 75: 177-184.

Josifov M. 1999. Heteropterous insects in the Sandanski-Petrich Kettle, Southwestern Bulgaria. *Historia naturalis bulgarica* 10: 35-66.

Kalaydzhiev L., Bozhkov D., Stefanov S. & Mihaylova L. 1960. Varhu izpolzvaneto na komarnite larvi kato hrana na mladi dagovi pastarvi [On the use of the mosquito larvae as a food for young rainbow trouts]. *Izvestiya na instituta po fiziologiya* [*Notifications from the Institute of Physiology*] 4: 235-239. (In Bulgarian).

Kiradzhiev S. 2013. Entsiklopedichen geografski rechnik na Balgariya [Encyclopedic Geographical Dictionary of Bulgaria]. Sofia, *Iztok-Zapad* [*East-West*]. 624 p. (In Bulgarian).

Kitanov B. 1943. Materiali po upotrebata na rasteniyata v narodnite sredstva za borba sreshtu parazitite i stopanski vrednite zhivotni [Materials on the plants use in the people means of fight against parasites and pests]. *Spisanie na zemedelskite ispitatelni instituti v Balgariya* [*Journal of the Agricultural Institutions in Bulgaria*] 13 (1/2): 29-37. (In Bulgarian).

Konsuloff S. 1921a. Die Lebensbedingungen der Anophelinen in Bulgarien im Zusammenhang mit der Malariaverbreitung [The living conditions of the anophelins in Bulgaria in connection with the spread of malaria]. *Archif für Schiffs- und Tropen-Hygiene* [*Archive for ship and tropical hygiene*] 25: 227-240.

Konsuloff S. 1921b. Orizoseyaneto i malariyata [Rice fields and malaria]. *Spisanie na zemedelskite izpitatelni instituti v Balgaria* [*Journal of the Agricultural Testing Institutes of Bulgaria*] 2 (1/2): 86-120. (In Bulgarian).

Konsuloff S. 1922. Biologichni usloviya za malariyata u nas [Biological conditions for malaria in Bulgaria]. *Annuaire de l'Université de Sofia* [*Annual of the Sofia University*] 17: 1-32. (In Bulgarian).

Konsuloff S. 1922b. Über die Doppeleatmung der Mückenlarven [About the double breathing of the mosquito larvae]. *Biologischen Zentralblatt* [*Biological Papers*] 42 (4): 188-192.

Konsuloff S. 1922c. Die Bekämpfung der Malariamücken in den Reisfeldern [Fighting the malaria mosquitoes in the rice fields]. *Zeitschrift für angewandte Entomologie* [*Journal of Applied Entomology*] 8 (2): 283-297.

Konsuloff S. 1923a. Biologichni belezhki po borbata s malariyata u nas [Biological notes on the fight against malaria in Bulgaria]. *Annuaire de l'Université de Sofia* [*Annual of the Sofia University*] 19 (1a): 139-150. (In Bulgarian with German summary).

Konsuloff S. 1923b. Zla muha [Evil flies]. *Priroda* [*Nature*] 23 (10): 150-152. (In Bulgarian).

Konsuloff S. & Paspalev G. 1924. Predvaritelni izuchvaniya varhu zlata muha [Preliminary studies on the evil fly]. *Trudove na Balgarskiya nauchen zemedelsko-stopanski institut* [*Works of the Bulgarian Scientific and Agricultural Research Institute*] 10: 3-13. (In Bulgarian).

Konsuloff, S. & Paspalev G. 1925. Izuchvaniya varhu zlata muha (Simuliidae) [Studies on the evil fly (Simuliidae)]. *Annuaire de l'Université de Sofia* [*Annual of the Sofia University*] 21: 59-66. (In Bulgarian with German summary).

Kovachev S. 1969. Beitrag zur erforschung der Simulliden-Fauna von Bulgarien (Diptera, Simullidae) [Contribution to the study of the Simullidae fauna of Bulgaria (Diptera,

Simullidae)]. *Bulletin de l'Institut de zoologique et Musée* [*Bulletin of the Zoological Institute and Museum*] 30: 191-195. (In Bulgarian with German summary).

Kovachev S. 1975. Materialien über die Simuliidenfauna (Diptera, Simuliidae) der Rhodopen [Materials on the Simuliid fauna (Diptera, Simuliidae) of the Rhodopes]. In: *La Faune des Rhodopes. Matériaux* [*The Wild Rhodopes. Materials*]. Sofia, BAN, 205-208. (In Bulgarian with German summary).

Kovachev S. 1976. Simulidae (Diptera) from the Strouma and Mesta River systems. *Hydrobiology* 3: 67-77. (In Bulgarian with English summary).

Kovachev S. 1979. Bemerkungen über die Zusammensetzung, Verbreitung und Ökologie der Simuliiden (Diptera, Simuliidae) in der Bulgarischen Donauzuflüssen [Remarks on the composition, distribution and ecology of the Simuliids (Diptera, Simuliidae) in the Bulgarian Danube tributaries]. In: *XXI Arbeitstag IAD* (*Novi Sad*) [*XXI working day IAD* (*Novi Sad*)], 392-396.

Kovachev S. 1990. Black-flies (Insecta, Diptera, Simulidae) from Vitosha. In: *Fauna of Southwestern Bulgaria.* 3. Sofia, Bulgarian Academy of Sciences, 150-153. (In Bulgarian with English summary).

Kovachev S. 2000. Qualitative structure and distribution of Black-flies (Diptera: Simuliidae) from high mountain streams of glacial origin in the Rila Mountains. In: Golemsnski V. & Naidenov W (Eds.). *Biodiversity and evolution of glacial water ecosystems in the Rila Mountains.* Sofia, Institute of Zoology, BAS, 67-74.

Kovachev S., Stoichev S. & Hainadjieva V. 1999. The zoobenthos of several lakes along the Northern Bulgarian Black Sea coast. *Lauterbornia* 35: 33-38.

Kovachev V. 1905. A Contribution to the study of the entomological fauna of Bulgaria. *Godishnik na Roussenskata gimnaziya za uchebnata godina 1904-1905* [*Annual of the Rousse High School for the academic year 1904-1905*] 3-12. (In Bulgarian).

Kovachevski I., Lazarov A., Bogdanov V., Hristov A., Balevski A. & Martinov S. 1949. *Spravochnik po zashtita na rasteniyata ot bolesti i nepriyateli* [*A guide of plant protection from diseases and pests*]. Sofia, Zemizdat. 424 p. (In Bulgarian).

Kovachevski I., Lazarov A., Bogdanov V., Hristov A., Balevski A., Nikolova V., Dodov D. & Martinov S. 1959. *Spravochnik po zashtita na rasteniyata ot bolesti i nepriyateli* [*A guide of plant protection from diseases and pests*]. Sofia, Zemizdat. 424 p. (In Bulgarian).

Kovchazov T. 1958. Rezultati ot prouchvaniyata za rezistentnost na *Anopheles maculipennis* spryamo DDT v Roussenski okrag [Results of the resistance studies of *Anopheles maculipennis* vs. DDT in Rousse District]. *Higiena, epidemiologiya i mikrobiologiya* [*Hygiene, epidemiology and microbiology*] 2 (2): 20-25. (In Bulgarian).

Kovchazov T. 1961. Otnosno izkorenyavaneto na malariyata v Roussenski okrag [On the eradication of malaria in the Rousse District]. *Epidemiologiya, microbiologiya i virusologiya* [*Epidemiology, microbiology and virology*] 4 (4): 23-26. (In Bulgarian).

Kovchazov T. 1976. Prouchvaniya varhu chuvstvitelnostta na *Anopheles maculipennis* spryamo insektitsidi vav Varnenski okrag [Studies on the sensitivity of Anopheles maculipennis towards insecticides in the Varna District]. *Letopisi na HE*I [*Chronicles of HEI*] 9 (6): 161-163. (In Bulgarian).

Kozarov G. 1949. Malarichnite komari prez zimata [Malaria mosquitoes in winter]. *Priroda i znanie* [*Nature and knowledge*] 2 (7): 8-11. (In Bulgarian).

Kozarov G. & Bozhkov D. 1953. Izuchavane varhu diapauzata i zimuvaneto na *Anopheles maculipennis maculipennis* Meig. v Bulgariya [Study on the diapause and wintering of *Anopheles maculipennis maculipennis* Meig. in Bulgaria]. *Annuaire de l'Université de Sofia* [*Annual of the Sofia University*] 47: 157-175. (In Bulgarian).

Kryzhanovsky O. 1965. *Composition and origin of the terrestrial fauna of Central Asia (based chiefly on the beetles material).* Moscow-Leningrad, Nauka. 430 p. (In Russian).

Kryzhanovsky O. 1976. On the principles of zoogeographical regionalization of land. *Zoologichesky Zhurnal* [*Zoological Journal*] 55 (7): 965-975. (In Russian with English summary).

Kryzhanovsky O. 2002. Composition and distribution of entomofauna of the globe. Moscow, *KMK*. 237 p. (In Russian).

Krzemiński W. 1984. Limoniidae of Bulgaria (Diptera, Nematocera). I. *Acta zoologica bulgarica* 24: 27-34.

Krzemiński W. & Starý J. 1989. Limoniidae of Bulgaria (Diptera, Nematocera). II. *Bulletin Entomologique de Pologne* 59 (1): 253-279.

Kutsarov Y. 2006. *Komarite (Culicidae) po porechieto na reka Dunav i savremenni metodi za borba s tyah* [*The Mosquitoes (Culicidae) along the Danube River and modern methods to control them*]. Tutrakan, Protected site "Kalimok – Brashlen". 32 p. (In Bulgarian).

Lackschewitz P. 1940a. Die paläarktischen Rhamphidiinen und Eriopterinen (Diptera) des Wiener Naturhistorischen Museums [The Palaearctic Rhamphidiinen and Eriopterinen (Diptera) of the Viennese natural history museum]. *Annalen des Naturhistorischen Museums in Wien* [*Annals of the Natural History Museum in Vienna*] 50 (1939): 1-67.

Lattin de G. 1967. Grundriss der Zoogeographie [Bases of zoogeography]. *Hochschullehrbücher für Biologie* [*College textbooks for biology*] 12. Jena, VEB Gustav Fischer Verlag. 602 p.

Lazarov A. 1935. Hesenska muha, *Mayetiola destructor* Mg. [Hessian fly, *Mayetiola destructor* Mg.]. *Zemedelie* [*Agriculture*] 39 (7): 55. (In Bulgarian).

Lazarov A. 1939. Referate u. Mitteilungen [Presentations and Messages]. *Mitteilungen der Bulgarischen Entomologischen Gesellschaft in Sofia* [*Reports of the Bulgarian Entomological Society in Sofia*] 10: 169-170. (In Bulgarian with German summary).

Lazarov A. 1949. *Vrednite nasekomi po ovoshtnite kulturi v Bulgariya i borbata s tyah* [*Harmful insects on the fruit trees in Bulgaria and the fight against them*]. Sofia, Bulgarian Academy of Sciences. 328 p. (In Bulgarian).

Lopatin I. K. 1989. *Zoogeography*. Minsk: Vysheyshaya shkola [High School]. 392 p. (in Russian).

Lyubenov Y. 1956. Znachenie na agrotehnikata v borba s zhitnite muhi [The importance of the agrotechnics in the fight with the wheat flies]. *Celskostopanska misal* [*Agricultural Thought*] 4 (6/7): 87-88. (In Bulgarian).

Lyubenov Y. 1960. Prouchvaniya varhu znachenieto na agrotehnicheskite meropriyatiya v borbata s nay-vazhnite vidove zhitni muhi (*Oscinosoma frit* L., *Chlorops pumilionis* Bjerk., *Phorbia genitalis* Schabl. и *Mayetiola destructur* Say) v Bulgaria [Studies on the importance of the agro-technical measures in the fight against the most important kinds of cereal flies (*Oscinosoma frit* L., *Chlorops pumilionis* Bjerk., *Phorbia genitalis* Schabl. and *Mayetiola destructur* Say) in Bulgaria]. *Nauchni trudove na tsentralnia nauchnoizsledovatelski institut za zshtita na rasteniyata* [*Scientific papers of the Central Research Institute for Plant Protection*] 3: 99-142. (In Bulgarian).

Makarov M. 1959. Novi vidove muhi, nepriyateli po pshenitsata v Bulgariya [New species of flies, enemies on the wheat in Bulgaria]. *Nauchni trudove VSI Georgi Dimitrov* [*Scientific Works of the High Agriculture Institute Georgi Dimitrov*] 6 (34): 365-384. (In Bulgarian).

Malicky H., Ant H., Aspöck H., de Jong R., Thaler K. & Varga Z. 1983. Argumente zur Existenz und Chorologie mitteleuropäischer (extramediterran-europäischer) Faunen-Elemente [Arguments on the existence and chorology of the Central European (extra-Mediterranean-European) faunal elements]. *Entomologia Generalis* 9: 101-119.

Malkov K. 1902a. Vredni nasekomi po ovoshtnite darveta [Insect pests of on the fruit trees]. *Oralo* [*Plowing*] 10: 147-149. (In Bulgarian).

Malkov K. 1902b. Po-vazhni bolesti i povredi po zhitnite rasteniya [More important diseases and damages on the cereals]. *Oralo* [*Plowing*] 9 (17): 262-264. (In Bulgarian).

Malkov K. 1902c. Stradat li ovoshtnite darveta, zelenchutsite i zemedelskite rasteniya ot razni bolesti i kakvo mozhe da se napravi za tyahnoto unishtozhenie [Do fruit trees, vegetables and agricultural plants suffer from various diseases and what can be done to destroy them]. *Sadovo* 5 (6/7): 66-69. (In Bulgarian).

Malkov K. 1903. *Bolesti i povredi po rapitsata i sredstva za tyahnoto unishtozhavane* [*Diseases and damages to the cole and means of destroying them*]. Plovdiv. 54 p. (In Bulgarian).

Malkov K. 1906b. Nay-rasprostranenite bolesti i povredi, poyaveni po zemedelskite rasteniya i ovoshtni darveta prez 1906 godina v Knyazhestvo Bulgariya [The most common diseases and damages occurring on the agricultural plants and fruit trees in 1906 in Bulgaria]. *Godishen otchet na Darzhavnata zemedelska opitna stantsiya v*

Sadovo za 1906 godina [*Annual Report of the State Agricultural Testing Station in Sadovo for 1906*] 4: 156-176. (In Bulgarian).

Malkov K. 1907. Edin prinos kam prouchvane vrednite nasekomi po kulturnite rasteniya u nas [A contribution to the study of harmful insects on crops in Bulgaria]. *Trudove na Darzhavnata zemedelska opitna stantsiya v Sadovo* [*Works of the State Agricultural Testing Station in Sadovo*] 2: 47-54. (In Bulgarian).

Manolov S. 1907. Malariyata i komarite v Burgas i okolnostite [Malaria and mosquitoes in Burgas and the surroundings]. *Letopis na lekarskiya sayuz v Balgariya* [*Chronicle of the medical association in Bulgaria*] 5 (3): 115-128. (In Bulgarian).

Manolov S. 1909a. Malariyata i komarite v gr. Tutrakan [Malaria and mosquitoes in the Tutrakan Town]. *Letopis na lekarskiya sayuz v Balgariya* [*Chronicle of the medical association in Bulgaria*] 7 (1): 19-20. (In Bulgarian).

Manolov S. 1909b. Po sistematikata i biologiyata na komarite v Burgasko [On the systematics and biology of mosquitoes in Burgas]. *Letopis na lekarskiya sayuz v Balgariya* [Chronicle of the medical association in Bulgaria] 7 (10): 644-648. (In Bulgarian).

Marinov T. 1957. Über die vertikale verbreitung von *Anopheles maculipennis* Mg. [About the vertical distribution of *Anopheles maculipennis* Mg.] *Bulletin de l'Institut zoologique de l'Academie des sciences de Bulgarie* [*Bulletin of the Zoological Institute of the Bulgarian Academy of Sciences*] 6: 599-601. (In Bulgarian with German summary).

Markov K. 1923. Golubashkata muha [*Simulia columbascense* (Sch.)] [The *Simulia columbascense* (Sch.)]. *Svedeniya po zemedelieto* [*Information on agriculture*] 4 (3/4): 13-18. (In Bulgarian).

Markov K. 1925a. *Malariyata i borbata s neya* [*Malaria and the fight against it*]. Sofia, Direktsiya na narodnoto zdrave, Inspektorat po malariyata [Directorate of Public Health, Malaria Inspectorate]. 52 p. (In Bulgarian).

Markov K. 1925b. *Materiali za izuchavaneto na malariyata v Bulgariya [Materials for studying the malaria in Bulgaria*]. Glavna direktsiya na narodnoto zdrave [General Directorate of Public Health] 2. Sofia, Darzhavna pechatnitsa. 95 p. (In Bulgarian).

Markov K. 1929. Epidemiologiya I borba s malariyata v Bulgariya [Epidemiology and the fight against malaria in Bulgaria]. *Sbornik na BAN 25, Prirodomatematichski klon* [*Collection of the Bulgarian Academy of Sciences 25, Natural Mathematical Department*] 11: 3-48. (In Bulgarian).

Markov K. 1937. *Malariyata v Bulgariya prez poslednite petnadeset godini (1920-1935)* [*The malaria in Bulgaria in the last fifteen years (1920-1935)*]. Sofia, Druzhestvo za narodno zdrave [Society for People's Health]. 47 p.

Markov K. & Morov T. 1929. Nyakoi novi danni varhu raprostranenieto na malarichnite komari v Bulgariya [Some new data on the distribution of the malaria mosquitoes in Bulgaria]. *Bulgarska klinika* [*Bulgarian Clinic*] 1 (9): 669-680.

Mendl H. 1986. Stelzmücken (Limoniidae) aus Bulgarien (Diptera, Nematocera) [Mosquitoes (Limoniidae) from Bulgaria (Diptera, Nematocera)]. *Spixiana* 9 (3): 275-282.

Michailov Tz., Mishev K. & Popov W. 1966. Dunavska halmista ravnina [Danubian hilly plain]. In: Beshkov A. (Ed.). *Geography of Bulgaria* 1. Sofia, Bulgarian Academy of Sciences, 57-80. (In Bulgarian).

Michailov Tz. & Popov W. 1978. Geomorphologie des Donauufers [Geomorphology of the Danube River bank]. In: Russev B. & Naidenow W. (Eds.). *Limnologie des Bulgarischen Donauabschnitts.* Sofia, Bulgarische Akademie der Wissenschaften [*Limnology of the Bulgarian Danube River section.* Sofia, Bulgarian Academy of Sciences], 7-28. (In Bulgarian with German summary).

Mihailova-Neikova M. 1961. Hydrobiological research of the Mandra Lake with regard to its importance as a fishing ground. *Annuaire de l'Université de Sofia, Faculté de Biologie, Géologie et Géographie* [*Annual of the Sofia University, Faculty of Biology, Geology and Geography*] 53 (1 – Zoology): 57-123. (In Bulgarian with English summary).

Michailova P. 1974. Zwei neue Arten der Gattung *Chironomus* (Diptera, Chironomidae) von der bulgarischen Schwarzmeerküste [Two new species of the genus Chironomus (Diptera, Chironomidae) from the Bulgarian Black Sea coast]. *Zoologische Beiträge* [*Zoological contributions*] 20 (2): 339-357.

Michailova P. 1980. A review of the genus *Clunio* Haliday, 1855 (Diptera, Chironomidae). *Zoologischer Anzeiger* [*Zoological Journal*] 5 (6): 417-432.

Michailova P. 1982. Chironomids (Diptera, Chironomidae) in fish raising pools in Bulgaria. *Acta zoologica bulgarica* 20: 51-68. (In Bulgarian with English summary).

Michailova P. 1985. Cytotaxonomic review of some species of the genus *Orthocladius* Van der Vulp (Diptera, Chironomidae). *Entomologische Abhandlungen Staatliches Museum für Tierkunde Dresden* [*Entomological contributions of the Drezden Zoological Museum*] 48 (9): 149-166.

Michailova P. 1989. The polytene chromosomes and their significance to the systematics and phylogeny of the family Chironomidae, Diptera. *Acta Zoologica Fenica* 186: 1-107.

Michailova P. 1994. *Chironomus bonus* Shilova et Dyvarsheishvili (Diptera, Chironomidae) from Bulgaria; karyotype and morphology. *Studia dipterologica* 1 (2): 186-194.

Michailova P. 1996. Cytotaxonomy of Chironomidae (Diptera) from Lake Shabla (Bulgaria): cytogenetic evidence for introgressive hybridization. *Hydrobiologia* 318: 25-42.

Michailova P. 2006. Family Chironomidae (Diptera) from the Rhodopes Mountains. In: Beron P. (Ed.). *Biodiversity of Bulgaria.* 3. Biodiversity of Western Rhodopes

(Bulgaria and Greece) 1. Sofia, National Museum of Natural History & Pensoft. 711-718.

Michailova P. & Dimitrova B. 1984. Caryological characteristics of *Acricotopus lucens* (Zett.) (Diptera, Chironomidae) from Bulgaria. *Comptes rendus de l' Académie bulgare des Sciences* [*Reports of the Bulgarian Academy of Sciences*] 37 (8): 1081-1084.

Michailova P. & Krastanov B. 2000. Cytotaxonomical differentiation oh *Chironomus plumosus* group (Diptera: Chironomidae) from fish pools near Plovdiv, Bulgaria. *Acta zoologica bulgarica* 52 (1): 2940.

Mikov O. 2005. History of investigations of malaria mosquitoes in Bulgaria. *Infectology* 42 (4): 39-47. (In Bulgarian).

Mikov O. 2008a. Biologichni, ekologichni i epidemiologichni prouchvaniya na anofeliyni komari v nyakoi regioni na Bulgariya [Biological, ecological and epidemiological studies of the anopheline mosquitoes in some regions of Bulgaria]. Ph.D. thesis, Sofia, National Center of Infectious and Parasitic Diseases. 155 p. (In Bulgarian).

Mikov O. 2008b. Determination of the season of possible malaria transmission by anopheline mosquitos in South-Western Bulgaria. *Meditsinskaya Parazitologiya i Parazitarnye Bolezni* [*Medical Parasitology and Parasitic Diseases*] 1: 31-33.

Mikov O. 2011. Razprostranenie na komarite, potentsialni prenositeli na virusa Zapaden Nil v Bulgariya [Distribution of the mosquitoes, potential carriers of the West Nile virus in Bulgaria]. 9 *National Congress on the Clinical Microbiology and Infection of the Bulgarian Association of Microbiologists*. Plovdiv, 28-30.IV.2011. In: Sbornik nauchni trudove. Plovdiv, 45-46.

Mikov O., Vuchev O. & Lalkovski N. 2011. Razprostranenie na komarite, potentsialni prenositeli na dinofilariozi po horata i zhivotnite v Bulgariya [Distribution of the mosquitoes, potential carriers of human and animal dinofilariasis in Bulgaria]. In: *Collection of papers and posters from the anniversary scientific session, 110th anniversary of NDNIVMI*. Sofia, Bulgarian Food Safety Agency, 152-156.

Mikov O., Nikolov G., Schaffner F. & Mathis A. 2013. First record and establishment of *Aedes albopictus* in Bulgaria. In: *VBORNET-EMCA Joint Meeting "Invasive mosquitoes and public health in the European context"*, Antwerp, Belgium, 28-29 November 2013 (Conference paper).

Minář J. 1990. Family Culicidae. In: Soós Á. & Papp L. (Eds.). *Catalogue of Palaearctic Diptera*. 2. Budapest, Akadémiai Kiadó, Hungarian Natural History Museum, 74-113.

Minarzh Y. & Christova T. 1971. Materialy po izucheniyu fauny moshek i komarov v Bolgarii (Diptera: Simulidae, Culicidae) [Materials on the study of the fauna of black flies and mosquitoes in Bulgaria (Diptera: Simulidae, Culicidae)]. *Trudy nauchno-issledovatelskogo instituta epidemiologii i mikrobiologii* [*Works of the Research Institute of Epidemiology and Microbiology*] 15: 245-247. (In Russian).

Mladenov N. 1978. Über die hydrologischen Verhältnisse der Donau zwischen Strom-km 845 und 375 [About the hydrological conditions of the Danube between river km 845 and 375]. In: Russev B. and Naidenow W. (Eds.). *Limnologie des Bulgarischen Donauabschnitts* [*Limnology of the Bulgarian Danube River section*]. Sofia, Bulgarische Akademie der Wissenschaften, 29-60. (In Bulgarian with German summary).

Müller P. 1974. *Aspects of zoogeography*. The Hague, W. Junk Publishers. 208 p.

Müller P. 1980. *Biogeographie*. Stuttgart, Ulmer. 414 p.

Naidenov W. 1962. Bulgarische Itonidinae (Diptera, Itonididae) und ihre Gallbildungen [Bulgarian Itonidinae (Diptera, Itonididae) and their gall formations]. *Bulletin de l'Institut de zoologie et musée* [*Bulletin of the Institute of Zoology and Museum*] 11: 141-165. (In Bulgarian with German summary).

Naidenov W. 1963. Shikalkoobrasuvane [Gallnuts formation]. *Priroda i znanie* [*Nature and knowledge*] 16 (1): 21-23. (In Bulgarian).

Nartshuk E. P. 2003. *Key to families of Diptera (Insecta) of the fauna of Russian and adjacent countries*. St. Petersburg, Proceedings of the Zoological Institute 294, Russian Academy of Sciences. 251 p. (In Russian).

Nedelchev N. 2013. *Kulikoidite v Balgaria* [*The culicoides in Bulgaria*]. Sofia, NTS. 211 p. (In Bulgarian).

Nedelkov N. 1909. Nashata entomologichna fauna [Our entomological fauna]. *Arhiv na Ministerstvoto na narodnoto prosveshtenie* [*Archives of the Ministry of National Enlightenment*] 1 (3): 83-135. (In Bulgarian).

Nedelkov N. 1912. Sixth contribution to the entomological fauna of Bulgaria. *Revue Academy Sciences de Bulgarie* [Bulgarian Academy Science Review] 2: 177-218. (In Bulgarian).

Nikolova V. 1943. Zeleviyat komar *Contarinia torquens* Meig. i borbata sreshtu nego [*Contarinia torquens* Meig. and the fight against it]. *Gradinarstvo* [*Gardening*] 24 (2): 27-29. (In Bulgarian).

Nikolova V. 1949a. Zelevata dalgonozhka kato nepriyatel na zelenchukovite kulturi u nas [*Tipula oleracea* as an enemy of the vegetable crops in Bulgaria]. *Gradinarstvo* [*Gardening*] 30 (7): 207-210. (In Bulgarian).

Nikolova V. 1962. Pochvoobrazuvashti nasekomi i tyahnoto znachenie za selskoto stopanstvo [Soil-forming insects and their importance for agriculture]. *Priroda* [*Nature*] 11 (2): 92-96. (In Bulgarian).

Nikolova V. & Natskova V. 1965. Vredni i polezni za selskoto stopanstvo kosmati komari – sem. Bibionidae (Diptera, Nematocera) [Harmful and useful for the agriculture march flies - Bibionidae (Diptera, Nematocera)]. *Priroda* [*Nature*] 14 (5): 87-80. (In Bulgarian).

Oosterbroek P. 2006. *The European families of the Diptera. Identification, diagnosis, biology*. KNNV Publishing. 209 p. DOI: 10.1163/9789004278066

Oosterbroek P. 2017. *Catalogue of the Craneflies of the World (Diptera, Tipuloidea: Pediciidae, Limoniidae, Cylindrotomidae, Tipulidae). ccw.naturalis.nl/* (Last update: 01 Nov 2017).

Oosterbroek P. & Theowald Br. 1992. Family Tipulidae. In: Soós Á., Papp L. & Oosterbroek P. (Eds.). *Catalogue of Palaearctic Diptera.* 1. Budapest, Akadémiai Kiadó, Hungarian Natural History Museum. 56-178.

Papp L. & Darvas B. (Eds.). 1997. *Contributions to a Manual of Palaearctic Diptera* 2. *Nematocera and Lower Brachycera.* Budapest, Science Herald. 592 p.

Papp L. & Darvas B. (Eds.). 2000a. *Contributions to a Manual of Palaearctic Diptera* 4: Appendix. Budapest, Science Herald. 604 p.

Papp L. & Darvas B. (Eds.). 2000b. *Contributions to a Manual of Palaearctic Diptera* 1: *General and Applied Dipterology.* Budapest, Science Herald. 978 p.

Paspalev G. 1950. Izuchavane na malariynite komari v Iztochnite Rodopi [Study of the malaria mosquitoes in the Eastern Rhodopes]. *II nauchen sabor na balgarskite lekari. II. Malariya [II scientific meeting of the Bulgarian doctors. II. Malaria]* 92-100. (In Bulgarian).

Paspalev G. 1951. Recherches sur les musquets malariens dans la région des Rhodopes orientales [Research on the malaria mosquitoes in the Eastern Rhodopes region]. *Buletin de l'Institut zoologique de l'Academie Bulgare des sciences [Bulletin of the Zoological Institute of the Bulgarian Academy of Sciences]* 1: 200-233. (In Bulgarian with French summary).

Petkoff P. 1939. *Predpazvane na posevite i gradinite ot vredni nasekomi [Protection of the crops and gardens from harmful insects].* Sofia, Izvori na znanieto [Sources of knowledge] 13. 112 p. (In Bulgarian).

Popoff W. 1956. *Zemedelska entomologia [Agricultural entomology].* Sofia, Zemizdat. 464 p. (In Bulgarian).

Popoff W. & Nikolova V. (1958). *Nepriyateli po zelenchukovite rasteniya v Balgariya i borbata s tyah [The enemies of the vegetable plants in Bulgaria and the fight against them].* Sofia, Izdatelstvo na BAN [Publishing House of the Bulgarian Academy of Sciences]. 225 p. (In Bulgarian).

Popov A. 2002. Zoogeographical analysis of Neuroptera in Bulgaria. *Acta zoologica hungarica* 48 (Suppl. 2): 271-180.

Popov S. 1999. *Species of family Tipulidae (Diptera) known to the fauna of Bulgaria till* 1998. Travaux Scientifiques Universite de Plovdiv, Biologie – Animalia [Scientific Works of the Plovdiv University, Biology – Animalia] 35 (6): 33-36. (In Bulgarian with English summary).

Procheş Ş. & Ramdhani S. 2012. The World's zoogeographical regions confirmed by cross-taxon analyses. *BioScience* 62 (3): 260-270.

Pudar D., Petrić D., Allène X., Alten B., Ayhan N., Cvetkovikj A., Garros C., Goletić T., Gunay F., Hlavackova K., Ćupina A. I., Kavran M., Lestinova T., Mathieu B., Mikov

O., Pajović I., Rakotoarivony I., Stefanovska J., Vaselek S., Zuko A. & Balenghien T. 2018. An update of the *Culicoides* (Diptera: Ceratopogonidae) checklist for the Balkans. *Parasites & Vectors* 11: 468. https://doi.org/10.1186/s13071-018-3051-x

Russev B. 1959. Beitrag zur Erforschung des Macrobenthos der Donau am bulgarischen Ufer [Contribution to the research of the macrobenthos of the Danube River on the Bulgarian shore]. *Comptes rendus de l'Academie bulgare des Sciences* [*Reports of the Bulgarian Academy of Sciences*] 12 (4): 345-348.

Russev B. 1962. Nasekomnata fauna na reka Dunav pred bulgarskiya bryag [Insect fauna of the Danube River in front of the Bulgarian coast]. *Notifications of the Experimental Freshwater Fishing Station in Plovdiv* 1: 115-128. (In Bulgarian).

Russev B. 1963. Anthropogene lithoreophile Biozönose in der Donau vor dem bulgarischen Ufer [Anthropogenic lithoreophilic biocenosis in the Danube River in front of the Bulgarian shore]. *Comptes rendus de l'Academie bulgare des Sciences* [*Reports of the Bulgarian Academy of Sciences*] 16 (4): 545-548.

Russev B. 1966a. Hydrobiologische untersuchungen der Marica. I [Hydrobiological studies of the Marica. I]. In: *Die Fauna Thrakiens* 3. Sofia, Bulgarische Akademie der Wissenschaften [*The fauna of Thrace* 3. Sofia, Bulgarian Academy of Sciences], 231-291. (In Bulgarian, Russian and German summaries).

Russev B. 1966b. Das Zoobenthos der Donau zwischen dem 845. und 357. Flusskilometer. I. Zusammensetzung, Verteilung und Ökologie [The Zoobenthos of the Danube between the 845th and 357th river kilometers. I. Composition, distribution and ecology]. *Bulletin of the Institute of Zoology and Museum* 20: 55-131. (In Bulgarian, Russian and German summaries).

Russev B., Yaneva I., Detcheva R. & Karapetkova M. 1994. Zusammensetzung der Hydrofauna [Composition of the hydrofauna]. In: Russev B. (Ed.) *Limnology of the Bulgarian Danube tributaries*. Sofia, Knizhen Tigar. pp. 130-174.

Sabev L. & Stanev S. 1959. Climatic regions of Bulgaria and their climate. *Proceedings of the Institute of Hydrology and Meteorology* 5: 29-174. (In Bulgarian).

Savchenko E. N., Oosterbroek P. & Starý J. 1992. Family Limoniidae. In: Soós Á., Papp L. & Oosterbroek P. (Eds.). *Catalogue of Palaearctic Diptera*. 1. Budapest, Akadémiai Kiadó, Hungarian Natural History Museum, 183-369.

Sengalevich G. 1964. Nepriyateli po krushovite plodove [Pear fruits pests]. *Ovoshtarstvo* [*Fruit growing*] 11 (8): 30-33. (In Bulgarian).

Skuhrava M., Skuhravy V., Dončev K. & Dimitrova B. 1991. Gall midges (Cecidomyiidae, Diptera) of Bulgaria. I. Faunistic researches in the 1978-1987 period. *Acta zoologica bulgarica* 42: 3-26.

Skuhrava M., Skuhravy V., Dončev K. & Dimitrova B. 1992. Gall midges (Cecidomyiidae, Diptera) of Bulgaria. II. Host plant relations and economic importance. *Acta zoologica bulgarica* 43: 23-42.

Slivenski M. 1935. *Prinos kam izuchvaneto na malariyata v Bulgariya* [*Contribution to the study of malaria in Bulgaria*]. Sofia, Pridvorna pechatnitsa [Court printing works]. 230 p. (In Bulgarian).

Slivenski M. 1940. *Malariyata u nas. Prinos kam iuchavaneto i* [*Malaria in Bulgaria. Contribution to its study*]. Sofia, Pridvorna pechatnitsa [Court printing works]. 250 p. (In Bulgarian).

Slivenski M. 1946. *Malariyata. Prinos kam izuchavaneto i u nas, Zapadna Trakiya (Belomorieto) i Makedoniya* [*Malaria. Contribution to its study in Bulgaria, Western Thrace (Belomorie) and Macedonia*]. Sofia, Voenno-izdatelski fond [Military-publishing fund]. 256 p. (In Bulgarian).

Soós A. & Pap L. (Eds.) 1984-1993. *Catalogue of Palaearctic Diptera* 1-13. Budapest, Hungarian Natural History Museum.

Stanev S. 1991. Climatic regions of Bulgaria. In: *The climate of Bulgaria.* Sofia, Bulgarian Academy of Sciences, 53-89. (In Bulgarian).

Starý J. 2004a. Revision of European species of the genus *Rhabdomastix* (Diptera: Limoniidae). Part 2: Subgenus *Rhabdomastix* s. str. *European Journal of Entomology* 101: 657-687.

Stefanov S. 1955. Prekratyavane na diapauzata i nachaloto na polovata aktivnost pri *Anopheles maculipennis* i *Anopheles superpictus* v Blagoevgradsko [Discontinuance of the diapause and beginninig of the sexual activity for *Anopheles maculipennis* and *Anopheles superpictus* in the Blagoevgrad district] *Annuaire de l'Université de Sofia, BGGF* [*Annual of the Sofia University, BG*GF] 48 (1): 157-175. (In Bulgarian).

Stefanov S. 1956. Sezonni izmeneniya na nyakoi morfologichni belezi na yaytsata pri *Anopheles maculipennis maculipennis* i *Anopheles maculipennis messae* v Sofia [Seasonal changes in some morphological features of the eggs in *Anopheles maculipennis maculipennis* and *Anopheles maculipennis messae* in Sofia]. *Annuaire de l'Université de Sofia, BGGF* [*Annual of the Sofia University, B*GGF] 49 (1): 161-198. (In Bulgarian).

Stoichev S. 1994. Contribution to the study of the chironomid fauna (Diptera, Chironomidae) in the Bulgarian stretch of the Danube River and its tributaries. Frequency of occurrence and dominant analysis of the species found. *Hydrobiology* 39: 91-101. (In Bulgarian with English summary).

Stoichev S. 1996. On the Chironomid fauna from Bulgarian inland waters. *Lauterbornia* 25: 117-123.

Stoichev S. 1998. The zoobenthos from the lakes Schabla-Ezerets (northern Black Sea coast of Bulgaria). In: Golemansky V., Naidenov W. (Eds.). *Biodiversity of Schabla Lake System,* Sofia, "Prof. Marin Drinov" Academic Publishing House, 91-99.

Stoichev S. 2000a. The zoobenthos from several glacial lakes in the Rila Mountains, Bulgaria. In: Golemanski V. & Naidenov W. (eds.). *Biodiversity and evolution of glacial water ecosystems in the Rila Mountains*. Sofia, Institute of Zoology, 155-162.

Stoichev S. 2000b. Zoobenthos in outflows of some glacial lakes in the Rila Mountains. In: Golemanski V. & Naidenov W. (eds.). *Biodiversity and evolution of glacial water ecosystems in the Rila Mountains*. Sofia, Institute of Zoology, 163-168.

Stoichev S. 2001a. The zoobenthos of the Koprinka reservoir, Central Bulgaria. *Lauterbornia* 40: 39-41.

Stoichev S. 2001b. Of the Chironomid Fauna from Blagoevgradska Bistritsa River (Rila Mountain, Southwest Bulgaria). *Lauterbornia* 40: 119-121.

Stoichev S. 2002. Hydrofaunistic investigation of the Urdini Ezera Glacial Lakes, Northwestern Rila Mountains, West Bulgaria. *Acta zoologica bulgarica* 54 (1): 63-68.

Stoichev S. 2004. The benthic Invertebrates in the Marichini Ezera Glacial Lakes, Eastern Rila Mountains, South-Wewst Bulgaria. *Acta zoologica bulgarica* 56 (1): 277-282.

Stoichev S. & Chernev N. 2001. On the Chironomid fauna from Blagoevgradska Bistritsa River (Rila Mountain, Southwest Bulgaria). *Lauterbornia* 40: 119-121.

Stojanov N. 1948. *Die Vegetation unserer Donauinsel und ihre Bewirtschaftung* [*The vegetation of the Bulgarian Danubian islands and its management*]. Sofia, Bulgarian Academy of Sciences. 188 p. (In Bulgarian with German summary).

Stojanov N. 1966. Rastitelna pokrivka [Plant cover]. In: Beshkov A. (Ed.). *Geography of Bulgaria* 1. Sofia, Bulgarian Academy of Sciences, 447-482. (In Bulgarian).

Stoyanova T., Traykov I., Bogoev V., Yaneva I., Vidinova Y., Tyufekchieva V. & Kenderov L. 2013. Composition of the macrozoobenthos in semi-mountainous river in South-Western Bulgaria. *Natura Montenegrina* 12 (3-4): 803-811.

Strenzke K. 1951a. Chironomiden von der bulgarischen Küste des Schwarzen Meeres [Chironomids from the Bulgarian coast of the Black Sea]. *Archiv für Hydrobiologie* [*Archive of the Hydrobiology*] Supl. 18 (4): 678-691.

Strenzke K. 1951b. Eine südosteuropäische Chironomiden-Gattung *Halliella* (Diptera) [A southeast European chironomid genus Halliella (Diptera)]. *Deutsche Zoologische Zeitschrift* [German Zoological Journal] 1 (1): 15-23.

Szadziewski R., Filatov S. & Dominiak P. 2016. A redescription of *Culicoides griseidorsum* Kieffer, 1918, with comments on subgeneric position of some European taxa (Diptera: Ceratopogonidae). *Zootaxa* 4107 (3): 413-422. http://www.mapress.com /j/zt/

Theobald F. 1907. *A monograph of the Culicidae or mosquitoes*. 4. London. 639 p.

Theobald F. 1910. *A monograph of the Culicidae or mosquitoes*. 5. London. 646 p.

Thienemann A. 1936. Haffmücken und andere Salzwasserchironomiden [Damselflies and other saltwater chironomids]. *Kieler Meeresforschungen* [*Kieler Sea Research*] 1: 167-178.

Tichkov H. 1982. Climatic partition. In: *Geography of Bulgaria. Physical geography*. Sofia, Bulgarian Academy of Sciences, 240-247. (In Bulgarian with English summary).

Tschorbadjiew P. 1925a. Konstatirani zhivotinski nepriyateli po kulturnite rastenia v Bulgariya prez 1924 g. [Established animal pests on the cultivated plants in Bulgaria in 1924]. *Svedeniya po zemedelieto* [*Information of agriculture*] 6 (4/6): 4-19. (In Bulgarian).

Tschorbadjiew, P. 1925b. *Otchet na entomologicheskata sektsiya pri Zemedelskiya izpitatelen institut v Sofiya za 1924 g.* [*Report of the Entomological Section at the Agricultural Testing Institute in Sofia in 1924*]. Sofia, 167-175. (In Bulgarian).

Tschorbadjiew P. 1925c. Rezultati ot edna anketa po zlata muha v Bulgaria. [Results of a poll on the evil fly in Bulgaria]. *Svedeniya po zemedelieto* [*Information of agriculture*] 6 (9/19): 3-16. (In Bulgarian).

Tschorbadjiew P. 1926. Nepriyateli po kulturnite rasteniya v Bulgariya prez 1925 g. [Pests on the cultivated plants in Bulgaria in 1925]. *Svedeniya po zemedelieto* [*Information of agriculture*] 7 (2): 20-68. (In Bulgarian).

Tschorbadjiew P. 1927. *Nepriyateli po kulturnite rasteniya v Bulgariya prez 1925 g.* [*Pests on the cultivated plants in Bulgaria in 1925*] Godishen otchet na Darzhavnata zemedelska opita stantsiya v Sofia za 1925 g. [Annual Reports of the State Agricultural Expertise Station in Sofia for 1925] Sofia, 207-256. (In Bulgarian).

Tschorbadjiew P. 1929. Nepriyateli po kulturnite rasteniya v Bulgariya prez 1927 g. [Pests on the cultivated plants in Bulgaria in 1927]. *Svedeniya po zemedelieto* [*Information on agriculture*] 10 (3/4): 3-59. (In Bulgarian).

Tschorbadjiew P. 1932. Nepriyateli po kulturnite rasteniya v Bulgariya prez 1928 i 1929 godini [Pests on the cultivated plants in Bulgaria in 1928 and 1929]. *Svedeniya po zemedelieto* [*Information of agriculture*] 13 (3/4): 3-48. (In Bulgarian).

Tschorbadjiew P. 1939. Materialien über die schädlichen Insekten und anderen Feinde der Kulturpflanzen in Bulgarien [Materials about the harmful insects and other enemies of the crops in Bulgaria]. *Mitteilungen der Bulgarischen Entomologischen Gesellschaft in Sofia* [*Messages of the Bulgarian Entomological Society in Sofia*] 10: 55-72. (In Bulgarian with German summary).

Udvardi M. 1975. *A classification of the Biogeographical Provinces of the World.* Morges, IUCN Paper 18. 49 p.

Uzunov Y., Tsankova V., Todorov I. & Varadinova E. 2001. The macrozoobenthic fauna of the Biospher reserve Srebarna Lake in North-Eastern Bulgaria. *Lauterbornia* 40: 43-51.

Uzunov Y., Varadinova E., Yaneva I., Stoichev S., Vidinova Y. & Kumanski K. (2011). Long-term changes of the bottom Invertebrate fauna of the Mesta River in Southwestern Bulgaria. *Annuaire de l'Université de Sofia "St. Kliment Ohridski", Faculte de Biologie* [*Annual of the Sofia University St. Kliment Ohridski, Faculty of Biology*] 99 (1 – Zoology): 33-52.

Valkanov A. 1936. Notizen über die Brackwässer Bulgariens. II. Versuch einer hydrographischen und biologischen Erforschung derselben [Notes on the brackish waters of Bulgaria. II. Attempt of the hydrographical and biological research]. *Annual of the Sofia University. Physics-Mathematics Faculty* 32 (3): 209-341. (In Bulgarian with German summary).

Valkanov A. 1941a. Über das Auffinden von anabiotischen Dipteren in Bulgarien [About finding anabiotic dipterans in Bulgaria]. – *Annual of the Sofia University St. Kliment Ohridski, Physics-Mathematics Faculty* 37 (3): 201-205. (In Bulgarian with German summary).

Valkanov A. 1941b. Nashite dobrudzhanski ezera [Bulgarian Dobrudzha Lakes]. *Ribarski pregled* [*Fishing review*] 11 (6): 6-9. (In Bulgarian).

Valkanov A. 1948. *Thalasomyia frauenfeldi* Schiner vom Schwarzen Meer [*Thalasomyia frauenfeldi* Schiner from the Black Sea]. *Works of the biological marine station in Varna* 14: 103-112. (In Bulgarian with German summary).

Valkanov A. 1954. Betrag zur Kenntnis unserer Schwarzmeerfauna [Contribution to the knowledge of the Bulgarian Black Sea fauna]. *Works of the biological marine station in Varna* 18: 49-53. (In Bulgarian with German summary).

Valkanov A. 1957. Katalog unserer Schwarzmeerfauna [Catalog of the Bulgarian Black Sea fauna]. *Works of the biological marine station in Varna* 19: 1-62. (In Bulgarian with German summary).

Varadinova E., Borisova P., Pehlivanov L. & Uzunov Y. 2012. Macroinvertebrate communities of the Srebarna Lake Biosphere Reserve: species diversity, abundance and modelling of the ecological status. In: Uzunov Y. Georgiev B., Varadinova E., Ivanova N., Pehlivanov L. & Vasilev V. (Eds.). *Ecosystems of the Biosphere Reserve Srebarna Lake.* Sofia, Professor Marin Drinov Academic Publishing House, 93-113.

Varadinova E., Kerakova M., Alexandrova M., Soufi R., Stoichev S., Vidinova Y., Tyufekchieva V. & Uzunov Y. 2013. Botton Invertebrate communites (the macrozoobenthos): Key biological quality element for ecological classification of the Mesta River basin. In: Uzunov Y., Pehlivanov L., Georgiev B. & Varadinova E. (Eds.). *Mesta River: Biological quality elements and ecological status.* Sofia, Professor Marin Drinov Academic Publishing House, 61-96.

Vasilev P. 1931. Krushoviyat plodov komar, *Contarinia pirivora* [Pear fruit mosquito, *Contarinia pirivora*]. *Lozarski pregled* [*Vineyard review*] 16 (4/5): 112-113. (In Bulgarian).

Velchev V. 1997. Features and regularities of distribution of contemporary vegetation. In: Yordanova M. & Donchev D. (Eds.). *Geography of Bulgaria.* Sofia, Bulgarian Academy of Sciences, 265-269. (In Bulgarian with English summary).

Velchev V. 2002. Characteristic features and regularities in the distribution of the present-day vegetation. In: Copralev I. et al. (Eds.). *Geography of Bulgaria.* Sofia, ForKom, 321-324. (In Bulgarian with English summary).

Velchev V., Ganchev S. & Bondev, I. 1982. Rastitelni poyasi [Vegetation belts]. In: Galabov Zh. (Ed.). *Geography of Bulgaria. Physical geography.* Sofia, Bulgarian Academy of Sciences, 439-443. (In Bulgarian with English summary).

Velchev V. Bondev I., Kochev H., Russakova V., Vassilev P., Meshinev T., Nikolov V., Georgiev N. & Valchev V. 1989. Vegetation. In: Mishev K. (Ed.). *Natural and economic potential of the mountains in Bulgaria.* Sofia, Bulgarian Academy of Sciences, 273-337. (In Bulgarian with English summary).

Velev S. 1990. *Klimatat na Balgaria [The climate of Bulgaria*]. Sofia, Narodna prosveta [National education]. 179 p.

Velev S. 2002. Climatic regioning. In: Copralev I. (Ed.). *Geography of Bulgaria.* Sofia, ForKom, 155-156. (in Bulgarian with English summary).

Vigna Taglianti A., Audisio P., Biondi M., Bologna M., Carpaneto G., Biase A., Fattorini S., Piattella E. & Sindaco R. 1999. A proposal for a chorotype classification of the Near East fauna, in the framework of the Western Palaearctic Region. *Biogeographia* 20: 31-59.

Wood D. M. & Borkent A. 1989. Phylogeny and classification of the Nematocera. In: McAlpine J. F. & Wood D. M. (Eds.). *Manual of Nearctic Diptera* 3. Ottawa, Research Branch Agriculture Canada, 1333-1370.

Yeates D., Wiegmann B., Courtney G., Meier R., Lambkin Ch. & Pape T. 2007. Phylogeny and systematics of Diptera: Two decades of progress and prospects. *Zootaxa* 1668: 565-590.

Zamfirov Ts. 1958. Nay-blagopriyatnite srokove za seitba na pshenitsa i oves v rayona na Instituta v Knezha, kato sredstvo za borba s zhitnite muhi [Most favorable sowing times for wheat and oats in the Knezha Institute region as a means of fight against wheat flies]. *Nauchni trudove na kompleksniya selskostopsnski naucho-izsledovatelski institut v Knezha* [*Scientific works of the Complex Agricultural Research Institute in Knezha*] 1: 163-172. (In Bulgarian).

Zamfirov Ts. 1961. Vidov sastav i nyakoi biologichni osobenosti na zhitnite muhi v rayona na Instituta v Kneha [Species composition and some biological features of the wheat flies in the region of the Knezha Institute]. *Izvestiya na instituta po tsarevitsata* [*Corn Institute Notifications*] 1: 199-225. (In Bulgarian).

Zamfirov Ts. 1962. Prouchvaniya varhu vazmozhnostite za borba s zhitnite muhi chrez predposevno tretirane na semenata na zhitnite kuturi s insektitsidi [Studied on the possibilities of fighting against wheat flies by treating the wheat seeds with insecticides before sowing]. *Izvestiya na instituta po tsarevitsata* [*Corn Institute Notifications*] 3: 181-192. (In Bulgarian).

Zamfirov Ts. 1962b. Vazmozhnosti za borba s zhitnite muhi pri pshenitsata *Oscinosoma frit* L., *Chlorops pumilionis* Bjerk. and *Mayetiola destructor* Say chrez naprashvane na poseva s insektitsidi [Possibilities for controlling wheat flies *Oscinosoma frit* L., *Chlorops pumilionis* Bjerk. and *Mayetiola destructor* Say in wheat by sowing treatment with insecticides]. *Izvestiya na instituta po tsarevitsata* [*Corn Institute Notifications*] 4: 179-183. (In Bulgarian).

Zamfirov Ts. 1962c. Srokat na seitbata na pshenitsata i echemika kato sredstvo za borba s zhitnite muhi [The time of sowing of the wheat and barley as a means of fighting against wheat flies]. *Rastitelna zashtita* [*Plant protection*] 10 (4): 29-38. (In Bulgarian).

Zamfirov Ts. 1963. Kam vaprosa za sroka na seitbata i naprashvaneto na poseva s heksahloran v borbata s zhitnite muhi pri pshenitsata i ovesa [To the question of time sowing and seed treatment with hexachloran in the fight against wheat flies in wheat and oats]. *Izvestiya na instituta po tsarevitsata* [*Corn Institute Notifications*] 6: 119-125. (In Bulgarian).

Zhelev I. 1948. Plodovi hobotnitsi *Rhynchites* sp., i krushovo komarche, *Contarinia pirivora* [Fruit weevils, *Rhynchites* sp., and Pear mosquito, *Contarinia pirivora*]. *Gradinarstvo* [*Gardening*] 29 (8): 238-242. (In Bulgarian).

Ziegler J. 2003. Ordnung Diptera, Zweiflügler (Fliegen und Mücken) [Order Diptera, two-winged flies (flies and mosquitoes)]. In: Dathe H. (Ed.). *Lehrbuch der Speziellen Zoologie,* Band I: Wirbellose Tiere, 5. Teil: Insecta [*Textbook of Special Zoology,* Volume I: Invertebrates, Part 5: Insecta]. Heidelberg, Spektrum. pp. 756-860.

Zilahi G. 1934. Beiträge zur Fliegenfauna Bulgariens. I. Chironomidae [Contributions to the Diptera fauna of Bulgaria. I. Chironomidae]. *Mitteilungen der Bulgarischen Entomologischen Gesellschaft in Sofia* [*Messages of the Bulgarian Entomological Society in Sofia*] 8: 152-158.

Zlatanov S. 1971. *Insektenschädlinge der Eiche in Bulgarien [Insect pests of the oak in Bulgaria*]. Sofia, Verlag der Bulgarischen Akademie der Wissenschaften [Publishing house of the Bulgarian Academy of Sciences]. 250 p. (In Bulgarian with German summary).

In: Biodiversity of the Bulgarian-Romanian Section
Editor: Shurulinkov, Hubenov et al.
ISBN: 978-1-53615-663-8

Chapter 6

THE BUTTERFLIES (LEPIDOPTERA: PAPILIONOIDEA) OF THE DANUBIAN PLAIN AND ADJACENT TERRITORIES OF NORTH BULGARIA

Zdravko Kolev*
National Museum of Natural History, Bulgarian Academy of Sciences, Sofia, Bulgaria

ABSTRACT

This review is a much-needed update of the known distribution of butterflies (Lepidoptera: Papilionoidea) in the least-known part of Bulgaria: the Danubian Plain and much of the Fore-Balkan hilly country north of the Stara Planina mountain chain (in total, over a third of the country's territory). I combine previously published records with results from my own extensive field research in the area, my revision of public collections, and other sources. A brief history of lepidopterological research in the area is provided. A revised systematic checklist and distribution maps of all species are provided, with notes on the more notable species. In total, 153 species are recorded (70% of Bulgarian butterflies) of which two (*Euphydryas maturna* and *Lopinga achine*) are not found elsewhere in Bulgaria. Ten of the recorded species are legally protected by EU and/or national legislation. The results underscore the importance of the area for nature conservation and invite further intensive butterfly research to fill in the remaining, unfortunately still extensive, blank spots.

Keywords: butterflies, Lepidoptera, Papilionoidea, Danubian Plain, Fore-Balkan, checklist, distribution, conservation

* Corresponding Author's E-mail: zkolev@nmnhs.com.

Introduction

The butterflies are doubtless the most studied and thus best-known group of insects in the world. Bulgaria is no exception, with research beginning in the latter half of 19-th century, and especially after the onset of the 20-th century. Already several of the earlier explorers of Bulgarian butterflies published data from a few easily accessible localities and their wider surroundings on or near the Danube, e.g., Rousse (Kowatschew 1898), Lom (Rebel 1903) and Svishtov (Nedelkow 1909), as well as the wider region under consideration here, e.g., Razgrad (Markowitsch 1904, 1909), Varna (Lederer 1863, Rebel 1903, Buresch 1926, 1930, Buresch & Karnojizky 1942), Lovetch (Iwanov 1926), Veliko Tarnovo (Bachmetjew 1902), etc. The major two-volume compendium of Bulgarian Lepidoptera (Buresch & Tuleschkow 1929, 1930), also added numerous unpublished records from this region. The southern part of Dobroudzha, which was incorporated in Romania between 1913 and 1940, attracted the attention of researchers mostly concentrating on the coastal area near Balchik (Caradja 1930, 1931, 1932, 1934, Zukowsky 1937, Popescu-Gorj 1964).

Between 1944 and 1990, under a Socialist dictatorship, butterfly research stagnated throughout the country, but even more so in the region under consideration. Only sporadic records were published from North Bulgaria during that time (e.g., Karnojizky 1954, Tuleschkow 1955, Slivov 1968, Slivov & Loukov 1976, Krzywicki 1981, Ganev & Beshkov 1987, Ganev 1988).

After 1990, interest in the butterfly fauna of lowland North Bulgaria recovered somewhat, with regional assessments (Beshkov 1998, 2000) as well as sporadic data (Abadjiev 1992, 1993, 1995, 2001, Essayan & Jugan 1993) having been published in the following decade or so. Nevertheless, owing to the preceding decades of neglect this territory remained the least-recorded region in Bulgaria together with the Upper Thracian Lowland (Abadjiev 2001).

In the nearly two decades following the latest national compendium of butterfly data (Abadjiev 2001), a wealth of new data has accumulated. Only a small part thereof has been published in traditional publications, mostly dealing with single rare species (e.g., Kolev 2002b, 2003, Abadjiev & Beshkov 2003, Beshkov 2011, Rákosy & Weidlich 2017, Kolev & Tsvetanov 2018) or with certrain areas of particular interest for butterfly conservation (Abadjiev & Beshkov 2007). The accumulated research data necessitate an updated overview of the butterfly fauna of the lowland parts of North Bulgaria. This is the purpose of the present publication which provides a checklist and distribution maps for all species.

The maps show previously published records as well as: 1. recent unpublished data from my own surveys in the region; 2. unpublished material, determined by myself, in the collections of the Institute of Biodiversity and Ecosystem Research (IBER) – Sofia and the National Museum of Natural History – Sofia (NMNHS); data from coll. IBER have been partly plagued by labelling errors and thus rendered questionable (Kolev 2002a) but I have decided to include them here in order to draw attention to localities in need of confirmation (which is mentioned under the respective species); 3. unpublished records provided to me by several competent collectors and recorders, notably Markus Lindberg (Turku, Finland), Yordan Kutsarov (Tutrakan, Bulgaria) and Dimitre Staykov (Vienna, Austria); 4. for the first time in Bulgaria, moderated data uploaded to citizen-science platforms, notably Observation.org and the Facebook group 'Butterflies and moths of Bulgaria', the latter founded and moderated by myself. To all contributors and correspondents, I herewith extend my heartfelt thanks. Regrettably, owing to severe constrains of space, it has been possible to list and reference localities only for the most local species.

Geographical Scope

The geographical region covered mostly coincides with the Danubian Plain, but as the latter's southern border is geomorphologically ill-defined, I have taken the liberty to include portions of the so-called Fore-Balkan (chains of hilly country between the Danubian Plain and the Stara Planina mountain chain). The main rationale to add these adjacent territories here is to highlight their, in part, better-known butterfly faunas which include species that may be found closer to the Danube in the future. As delimited, the area is 38744 km^2, over a third (34.9%) of the territory of Bulgaria. The altitude is overwhelmingly below 300 m, only in places rising above 500 m (maximum 863 m a.s.l., near Belogradchik).

Results and Discussion

General Notes on the Butterfly Fauna

The systematic checklist follows the order, taxonomy and nomenclature of the latest European checklist (Wiemers et al. 2018), with the following exceptions: 1. based on my own research, *Pieris balcana* is considered a junior subjective synonym of *Pieris napi*; 2. the genus *Rubrapterus*, as opposed to *Pseudophilotes*, is retained for the species *R. bavius* (Kolev 2017); 3. *Polyommatus eroides* is provisionally retained as a species-level taxon instead of a synonym of *P. eros*.

A total of 153 species are recorded from the area under consideration, representing 70% of the national butterfly fauna. Three previously reported species - *Pyrgus alveus*, *Pieris balcana*, and *Hipparchia semele* - are excluded: *P. balcana* for reasons explained above; the old records of the montane *P. alveus* almost certainly pertain to *P. armoricanus*; *H. semele* is taken off the list because in my critical examination of male and female genitalia of old material in NMNHS published as "*semele*" from North Bulgaria I have only found *H. volgensis*.

Two butterfly species can be considered emblematic for this region, as they have not been recorded elsewhere in Bulgaria: *Euphydryas maturna* and *Lopinga achine*. Two other species, *Colias hyale* and *Rubrapterus bavius*, have at least half of their known Bulgarian populations here. A total of 10 of the recorded species are legally protected by EU (EU 1992) and/or national legislation (BDA 2002), and several more possibly merit conservation concern regionally, as detailed under the respective species.

This review highlights the considerable diversity of the butterfly fauna of lowland North Bulgaria, and it is hoped that it will serve to further accelerate butterfly research and recording across this region.

Checklist of Butterfly Species

The number preceding each species corresponds to the number of its distribution map.

Superfamily Papilionoidea Latreille, [1802]

Family Papilionidae Latreille, [1802]

1. *Iphiclides podalirius* (Linnaeus, 1758). Widespread.
2. *Papilio machaon* Linnaeus, 1758. Widespread.
3. *Parnassius mnemosyne* (Linnaeus, 1758). Local. Listed on Council Directive 92/43/EEC (Annex IV) and Biological Diversity Act (Annex 3).
4. *Zerynthia cerisy* (Godart, [1824]) ssp. *ferdinandi* Stichel, 1907. Local. The subspecies is endemic to the Balkan Peninsula; the Danube is the northernmost boundary of its range.
5. *Zerynthia polyxena* ([Denis & Schiffermüller], 1775). Widespread. Listed on Council Directive 92/43/EEC (Annex IV) and Biological Diversity Act (Annex 3).

Family Hesperiidae Latreille, 1809

6. *Heteropterus morpheus* (Pallas, 1771). Very local: Lovetch (Drenowski 1909a), Svishtov (Nedelkow 1909), Belogradchik (coll. IBER – Sofia, unpublished). A hygrophilous specialist; possibly of conservation concern.
7. *Ochlodes sylvanus* (Esper, 1777). Widespread.
8. *Hesperia comma* (Linnaeus, 1758). Local.
9. *Thymelicus acteon* (Rottemburg, 1775). Local.
10. *Thymelicus sylvestris* (Poda, 1761). Widespread.
11. *Thymelicus lineola* (Ochsenheimer, 1808). Widespread.
12. *Spialia orbifer* (Hübner, [1823]). Widespread.

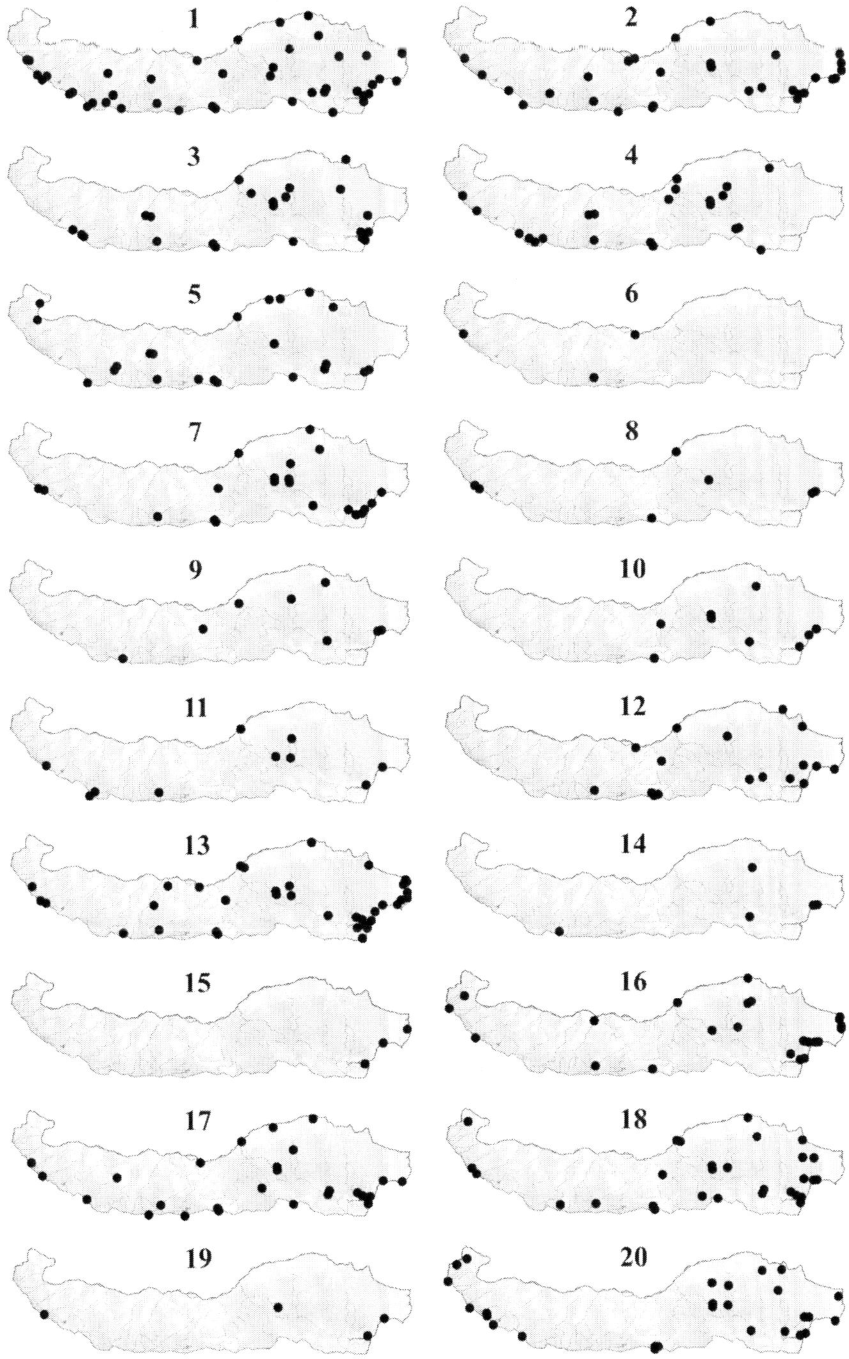

Figure 1. Distribution of butterflies of Danube Plain –plate 1.

13. *Carcharodus alceae* (Esper, 1780). Widespread.
14. *Carcharodus lavatherae* (Esper, 1783). Local.
15. *Carcharodus orientalis* Reverdin, 1913. Only reported from the sea coast (Abadjiev 2001), although a wider range is probable.

16. *Carcharodus floccifera* (Zeller, 1847). Widespread.
17. *Erynnis tages* (Linnaeus, 1758). Widespread.
18. *Pyrgus malvae* (Linnaeus, 1758). Widespread.
19. *Pyrgus carthami* (Hübner, [1813]). Very local: Razgrad (Bachmetjew 1902); Varna (Rebel 1903); Balchik (Caradja 1930); Mitrovtsi (own data).
20. *Pyrgus sidae* (Esper, 1784). Widespread.
21. *Pyrgus serratulae* (Rambur, 1839). Very local: Svishtov (Buresch & Tuleschkow 1930), Euxinograd (Buresch 1930), Balchik (Caradja 1931). Needs confirmation.
22. *Pyrgus armoricanus* (Oberthür, 1910). Widespread.
23. *Pyrgus cinarae* (Rambur, 1839). In N Bulgaria, reported only from Balchik (Caradja 1931), but needs confirmation.

Family Pieridae Duponchel, [1835]

24. *Leptidea duponcheli* (Staudinger, 1871). Very local: Razgrad (Markowitsch 1909), Pleven (Buresch 1912, Buresch & Tuleschkow 1929), Balchik (Popescu-Gorj 1964), Dobrudzha (material in coll. IBER, Z. Kolev det.), and the western Fore-Balkan (M. Lindberg, pers. comm.).
25. *Leptidea sinapis* (Linnaeus, 1758). Widespread.
26. *Gonepteryx rhamni* (Linnaeus, 1758). Widespread.
27. *Colias hyale* (Linnaeus, 1758). Widespread (own data, coll. NMNHS), but its distribution has been previously misunderstood due to confusion with *C. alfacariensis*, especially as the two species often occur together. Until now, the only certain published locality from the entire country has been Razgrad (Abadjiev 1992).
28. *Colias alfacariensis* Ribbe, 1905. Widespread.
29. *Colias erate* (Esper, [1805]). Widespread albeit sporadic.
30. *Colias crocea* (Geoffroy, 1785). Widespread.
31. *Aporia crataegi* (Linnaeus, 1758. Widespread.
32. *Pontia edusa* (Fabricius, 1777). Widespread.
33. *Pieris krueperi* Staudinger, 1860. Recorded from Preobrazhensky Manastir only (Buresch & Tuleschkow 1929).
34. *Pieris brassicae* (Linnaeus, 1758). Widespread.
35. *Pieris rapae* (Linnaeus, 1758). Widespread.
36. *Pieris mannii* (Mayer, 1851). Extremely local: Balchik (Zukowsky 1937), Lovetch, Preobrazhensky Manastir, Mitrovtsi (Abadjiev 2001).
37. *Pieris napi* (Linnaeus, 1758). Widespread.
38. *Euchloe ausonia* (Hübner, [1804]). Widespread albeit local. The Danube is the northern range limit in the Balkan Peninsula.
39. *Anthocharis cardamines* (Linnaeus, 1758). Widespread.

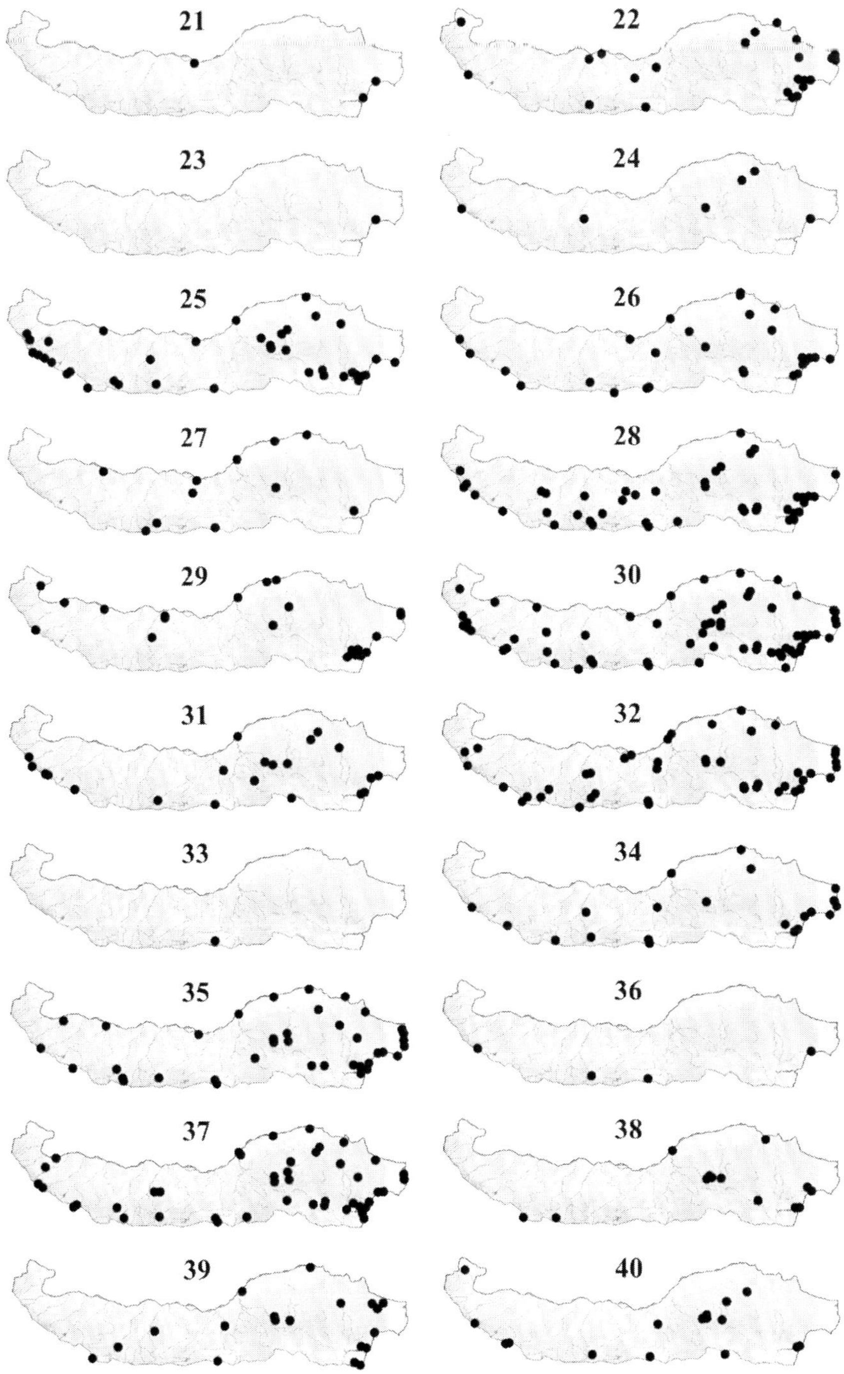

Figure 2. Distribution of butterflies of Danube Plain –plate 2.

Family Riodinidae Grote, 1895

40. *Hamearis lucina* (Linnaeus, 1758). Widespread.

Family Lycaenidae [Leach], [1815]

41. *Lycaena alciphron* (Rottemburg, 1775). Widespread.
42. *Lycaena thersamon* (Esper, 1784). Widespread.
43. *Lycaena dispar* ([Haworth], 1802). Widespread. Listed on Council Directive 92/43/EEC (Annexes II and IV) and Biological Diversity Act (Annexes 2 and 3).
44. *Lycaena candens* (Herrich-Schäffer, [1844]). Extremely local: Varna (Rebel 1903), Voden near Razgrad (Markowitsch 1904). Needs confirmation.
45. *Lycaena phlaeas* (Linnaeus, [1760]). Widespread.
46. *Lycaena tityrus* (Poda, 1761). Widespread.
47. *Thecla betulae* (Linnaeus, 1758). Extremely local: Krasimir (Buresch 1940), 'Palamara' game preserve [near Venets], Kubrat (coll. IBER), Bozhi Most southwest of Chiren (own data). However, this species is difficult to detect as an adult and may turn out to be widespread.
48. *Favonius quercus* (Linnaeus, 1758). Widespread.
49. *Callophrys rubi* (Linnaeus, 1758). Widespread.
50. *Satyrium pruni* (Linnaeus, 1758). Extremely local: Veliko Tarnovo (Bachmetjew 1902), 'Demir Baba Teke' tomb [near Sveshtari, Isperih municipality] (own data), Onogur, Malak Kanagyol gorge (Y. Kutsarov, pers. comm.). However, this species is difficult to detect as an adult and may turn out to be widespread.
51. *Satyrium ilicis* (Esper, 1779). Widespread.
52. *Satyrium w-album* (Knoch, 1782). Local.
53. *Satyrium spini* ([Denis & Schiffermüller], 1775). Widespread.
54. *Satyrium acaciae* (Fabricius, 1787). Widespread.
55. *Leptotes pirithous* (Linnaeus, 1767). Sporadic; a highly mobile migrant.
56. *Lampides boeticus* (Linnaeus, 1767). Sporadic; a highly mobile migrant.
57. *Cacyreus marshalli* Butler, 1898. Vinarovo: the only record from North Bulgaria (Kolev & Tsvetanov 2018). A native to South Africa, invasive in Europe.
58. *Celastrina argiolus* (Linnaeus, 1758). Widespread.
59. *Phengaris alcon* ([Denis & Schiffermüller], 1775). Extremely local: 'Demir Baba Teke' tomb [near Sveshtari, Isperih municipality] (Buresch & Tuleschkow 1930), 'Palamara' game preserve (coll. IBER: Kolev 2002b). Needs confirmation.
60. *Phengaris arion* (Linnaeus, 1758). Widespread albeit local (Kolev 2002b). Listed on Council Directive 92/43/EEC (Annex IV) and Biological Diversity Act (Annex 3).
61. *Rubrapterus bavius* (Eversmann, 1832). Extremely local. Despite intensive search across North Bulgaria (Beshkov 2011, Kolev 2017), only discovered in one locality near Balik (Rákosy & Weidlich 2017). The only other known Bulgarian locality is in Mt. Sakar (Kolev 2017). Listed on Council Directive 92/43/EEC (Annexes II and IV).
62. *Pseudophilotes vicrama* (Moore, 1865). Widespread.

63. *Scolitantides orion* (Pallas, 1771). Local.
64. *Iolana iolas* (Ochsenheimer, 1816). Local; confined to the sea coast between Varna and Cape Kaliakra (Caradja 1929, Buresch 1930, Popescu-Gorj 1964, Observation.org 2019, Butterflies and moths of Bulgaria 2019, own data).

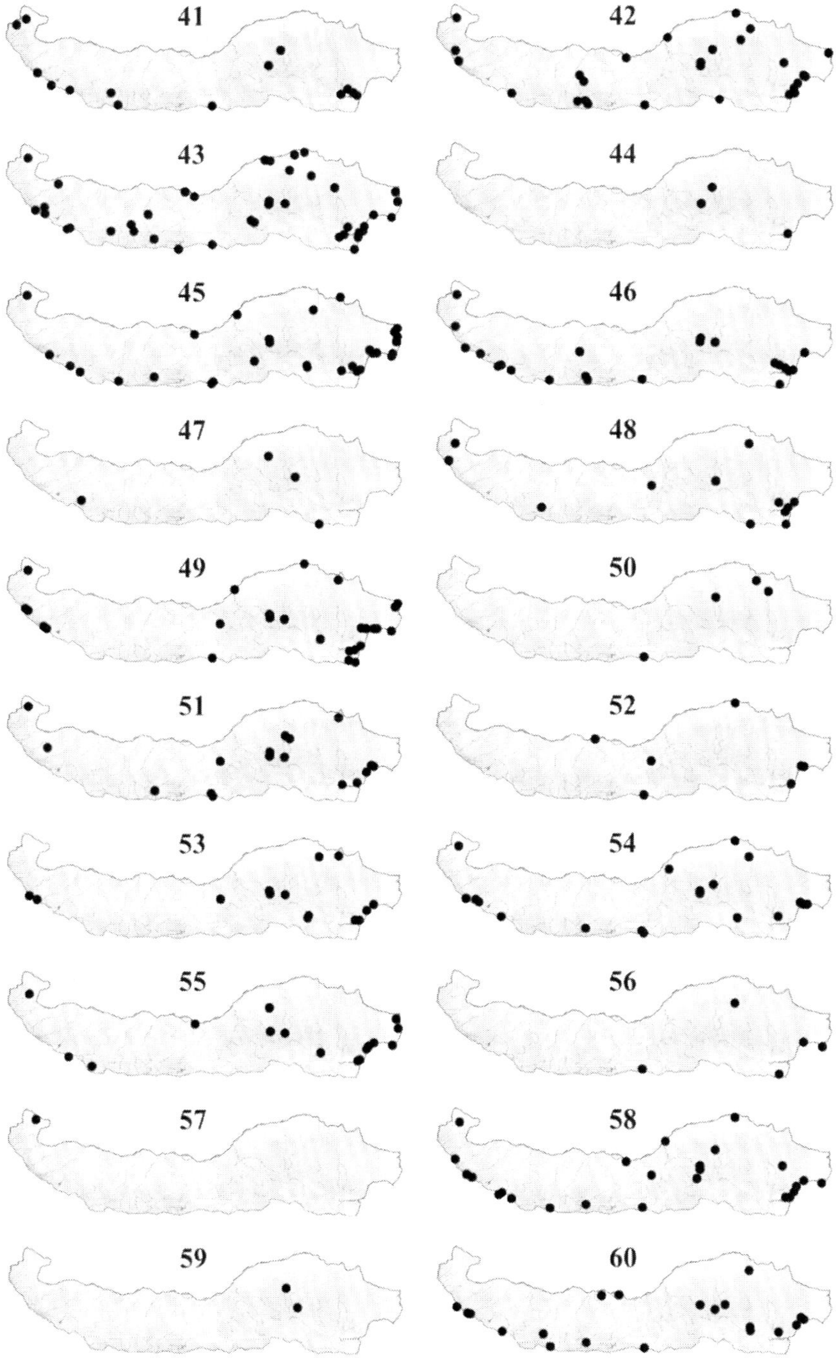

Figure 3. Distribution of butterflies of Danube Plain –plate 3.

65. *Glaucopsyche alexis* (Poda, 1761). Widespread.
66. *Cupido argiades* (Pallas, 1771). Widespread.
67. *Cupido decoloratus* (Staudinger, 1886). Local but likely to be widespread. Endemic to East Europe.
68. *Cupido alcetas* (Hoffmansegg, 1804). Local but likely to be widespread.
69. *Cupido osiris* (Meigen, 1829). Extremely local: Razgrad (Markowitsch 1909), Balchik (Popescu-Gorj 1964), Alfatar (coll. IBER), vicinity of Nevsha, the western Fore-Balkan (own data).
70. *Cupido minimus* (Fuessly, 1775). Widespread albeit local.
71. *Plebejus argus* (Linnaeus, 1758). Widespread.
72. *Plebejus argyrognomon* (Bergsträsser, 1779). Widespread.
73. *Eumedonia eumedon* (Esper, 1780). Extremely local: Kamenna Riksa (M. Lindberg, pers. comm.). An upland/montane species.
74. *Kretania sephirus* (Frivaldszky, 1835). Extremely local: Lom (Rebel 1903), Balchik, Fish-Fish Dere (Caradja 1931), Cape Kaliakra (Observation.org 2019), Ivanovo (Y. Kutsarov, pers. comm.).
75. *Cyaniris semiargus* (Rottemburg, 1775). Local, possibly widespread.
76. *Aricia anteros* (Freyer, 1838). Widespread albeit local. The Danube is the northern range limit in Europe.
77. *Aricia agestis* ([Denis & Schiffermüller], 1775). Widespread.
78. *Lysandra bellargus* (Rottemburg, 1775). Widespread.
79. *Lysandra coridon* (Poda, 1761). Widespread.
80. *Polyommatus escheri* (Hübner, [1823]) ssp. *dalmaticus* (Speyer, 1882). Newly identified for North Bulgaria. Extremely local: Shumen (material in coll. NMNHS, Z. Kolev det.), vicinity of Nevsha (own data). The subspecies is endemic to the Balkan Peninsula.
81. *Polyommatus thersites* (Cantener, 1835). Widespread albeit local.
82. *Polyommatus daphnis* ([Denis & Schiffermüller], 1775). Widespread albeit local.
83. *Polyommatus amandus* (Schneider, 1792). Local, possibly widespread.
84. *Polyommatus icarus* (Rottemburg, 1775). Widespread.
85. *Polyommatus eroides* (Frivaldszky, 1835). Extremely local: Razgrad, Gara Razgrad (Markowitsch 1909), 'Palamara', 'Karakuz' [near Alfatar] and 'Zlidol' [near Irnik] game preserves, Boynitsa (coll. IBER). All of these records need confirmation. Certain recent records in the western Fore-Balkan (M. Lindberg, pers. comm.). Listed on Council Directive 92/43/EEC (Annexes II and IV) and Biological Diversity Act (Annexes 2 and 3).
86. *Polyommatus admetus* (Esper, 1783). Extremely local: Lovetch (Iwanov 1926), Preobrazhensky Manastir (Buresch & Tuleschkow 1930), Dulovo (coll. IBER, Z. Kolev det.; needs confirmation), vicinity of Nevsha (own data).

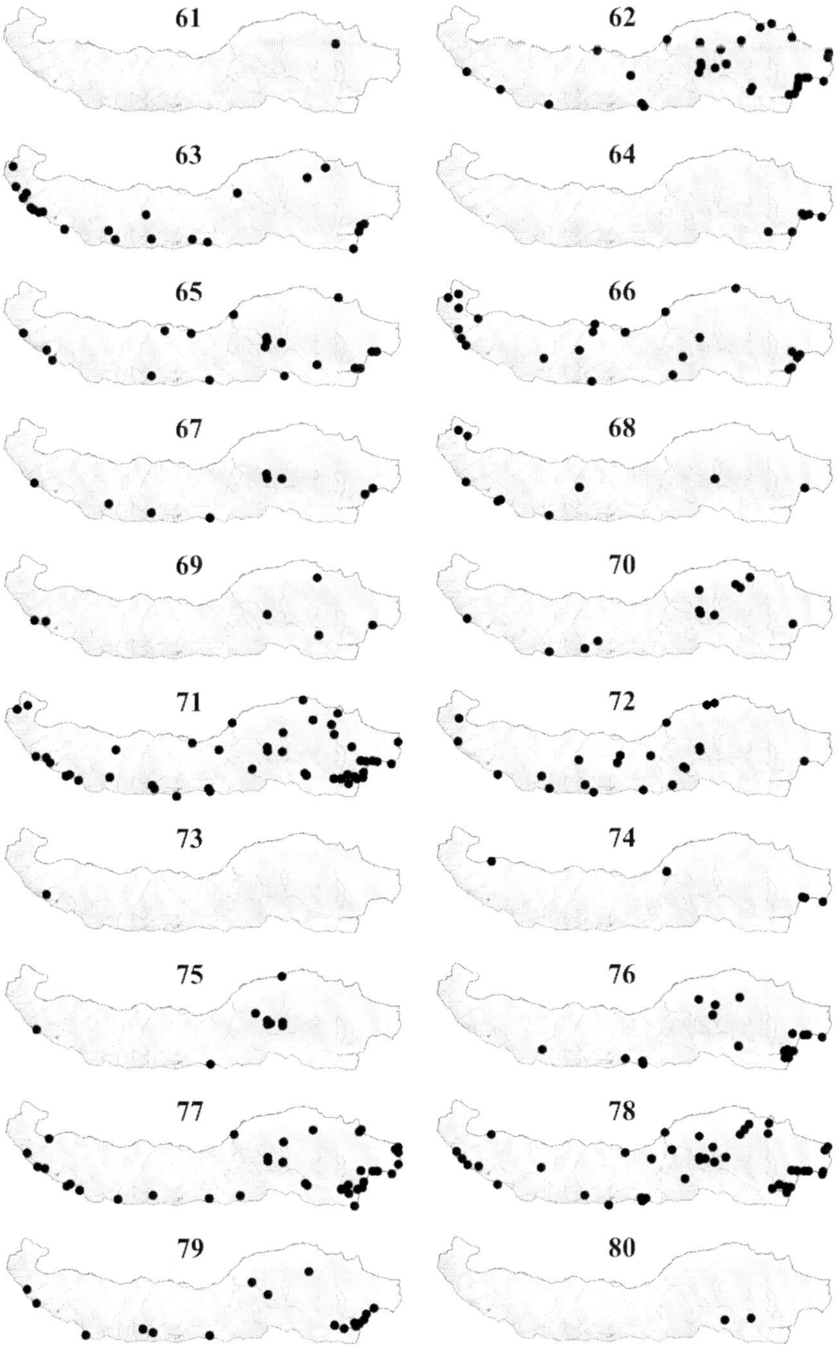

Figure 4. Distribution of butterflies of Danube Plain –plate 4.

Family Nymphalidae Swainson, 1827

87. *Neptis sappho* (Pallas, 1771). Widespread.
88. *Neptis rivularis* (Scopoli, 1763). Magurata cave (own data). A montane species; newly recorded from this region.

89. *Limenitis reducta* Staudinger, 1901. Widespread albeit local.
90. *Limenitis camilla* (Linnaeus, 1764). A montane species newly identified from this region: Dulovo, Isperih (coll. IBER). Confirmation needed.
91. *Issoria lathonia* (Linnaeus, 1758). Widespread.
92. *Brenthis hecate* ([Denis & Schiffermüller], 1775). Widespread albeit local.
93. *Brenthis ino* (Rottemburg, 1775). Extremely local in the western Fore-Balkan: Dolni Lom, Granitovo (Observation.org 2019).
94. *Brenthis daphne* (Bergsträsser, 1780). Widespread.
95. *Argynnis paphia* (Linnaeus, 1758). Widespread.
96. *Argynnis pandora* ([Denis & Schiffermüller], 1775). Widespread.
97. *Speyeria aglaja* (Linnaeus, 1758). Widespread.
98. *Fabriciana niobe* (Linnaeus, 1758). Widespread.
99. *Fabriciana adippe* ([Denis & Schiffermüller], 1775). Widespread.
100. *Boloria euphrosyne* (Linnaeus, 1758. Widespread albeit local.
101. *Boloria dia* (Linnaeus, 1767). Widespread.
102. *Apatura iris* (Linnaeus, 1758). A montane species, extremely local in this region: Krasimir (Buresch 1940), Oslen Krivodol (Slivov 1968), Prevala (own data).
103. *Apatura metis* Freyer, 1829. Widespread along the Danube and deep inland along its tributaries (Abadjiev 1995, 2001; coll. IBER; own data; M. Lindberg, pers. comm.; Y. Kutsarov, pers. comm.). The overwhelming part of the range of this priority species in the EU is centred on the lower Danube. Listed on Council Directive 92/43/EEC (Annex IV) and Biological Diversity Act (Annex 3).
104. *Apatura ilia* ([Denis & Schiffermüller], 1775). Very local, the following records need confirmation due to old age or issues with the source collection: Rousse (Kowatschew 1898), Varna, Pleven (Buresch & Tuleschkow 1929), Vidin, Alfatar (coll. IBER, Z. Kolev det.). The only recent certain locality in this region is Rabisha (own data).
105. *Araschnia levana* (Linnaeus, 1758). Widespread.
106. *Vanessa cardui* (Linnaeus, 1758). Widespread.
107. *Vanessa atalanta* (Linnaeus, 1758). Widespread.
108. *Aglais io* (Linnaeus, 1758). Widespread.
109. *Aglais urticae* (Linnaeus, 1758). Widespread.
110. *Polygonia egea* (Cramer, [1775]). Extremely local: Reselets (Abadjiev 1995; own data), Dulovo (coll. IBER, needs confirmation).
111. *Polygonia c-album* (Linnaeus, 1758). Widespread.
112. *Nymphalis polychloros* (Linnaeus, 1758). Widespread.
113. *Nymphalis xanthomelas* (Esper, [1781]). Extremely local and sporadic, without recent records in this region following the apparent extinction of this species from the entire country in the 1970-1980's (Abadjiev 2001, own data) and its recent reappearance in the beginning of this century (own data).

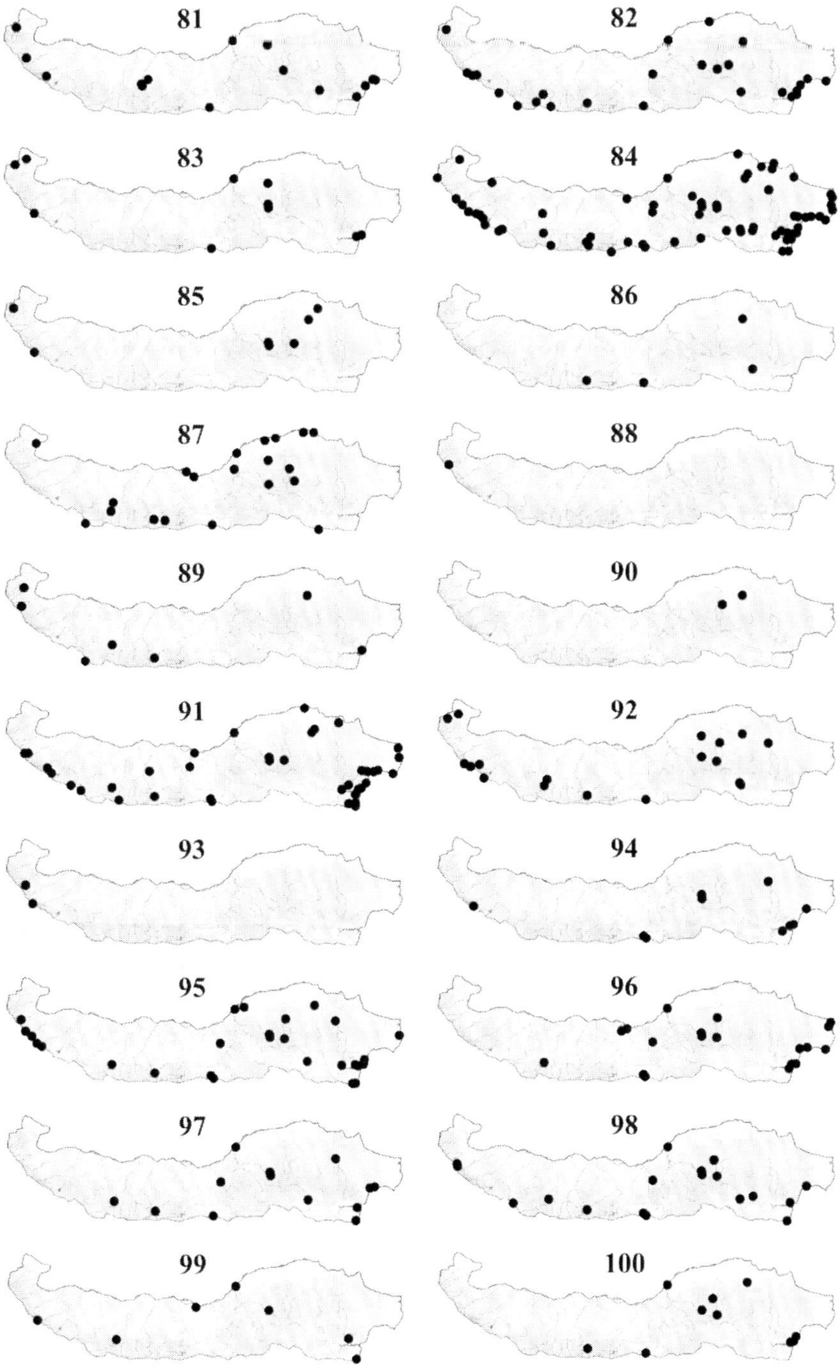

Figure 5. Distribution of butterflies of Danube Plain –plate 5.

114. *Nymphalis antiopa* (Linnaeus, 1758). Widespread albeit sporadic.

115. *Euphydryas aurinia* (Rottemburg, 1775). Local but apparently widespread (Beshkov 2011, Observation.org 2019, Butterflies and moths of Bulgaria 2019).

Listed on Council Directive 92/43/EEC (Annex II) and Biological Diversity Act (Annex 2).

116. *Euphydryas maturna* (Linnaeus, 1758). An emblematic species for this region which is very local in sheltered woodland margins and gorges in the eastern parts of the region (Markowitsch 1909, Abadjiev 1995, Abadjiev & Beshkov 2003; coll. IBER; own data). Listed on Council Directive 92/43/EEC (Annexes II and IV) and Biological Diversity Act (Annexes 2 and 3). The Bulgarian populations represent a large part of the range of the endemic ssp. *opulenta* Rákosy & Varga, 2012.

117. *Melitaea trivia* ([Denis & Schiffermüller], 1775). Widespread.

118. *Melitaea didyma* (Esper, 1778). Widespread.

119. *Melitaea arduinna* (Esper, 1783). 'Demir Baba Teke' tomb near Sboryanovo [Sveshtari] (Buresch & Tuleschkow 1929), needs confirmation; relatively widespread and often very abundant in the western Fore-Balkan and adjacent lowland areas nearer the Danube (Slivov & Loukov 1976, Beshkov 2011, Observation.org 2019; M. Lindberg, pers. comm.; own data).

120. *Melitaea phoebe* ([Denis & Schiffermüller], 1775). Widespread.

121. *Melitaea cinxia* (Linnaeus, 1758). Widespread.

122. *Melitaea diamina* (Lang, 1789). Extremely local: Voden near Razgrad (Buresch & Tuleschkow 1929), 'Karakuz' game preserve (coll. IBER, Z. Kolev det.). Needs confirmation; possibly of conservation concern.

123. *Melitaea britomartis* Assmann, 1847. Extremely local, though possibly overlooked due to confusion with *M. aurelia*. Known from the lakeside wetlands near Varna (Lederer 1963) and the vicinity of Nevsha (own data).

124. *Melitaea athalia* (Rottemburg, 1775. Widespread.

125. *Melitaea aurelia* Nickerl, 1850. Widespread albeit local.

126. *Libythea celtis* (Laicharting, 1782). Local.

127. *Coenonympha pamphilus* (Linnaeus, 1758). Widespread.

128. *Coenonympha glycerion* (Borkhausen, 1788). Widespread.

129. *Coenonympha leander* (Esper, 1784). Widespread albeit local.

130. *Coenonympha arcania* (Linnaeus, [1760]). Widespread.

131. *Kirinia roxelana* (Cramer, 1777). Widespread albeit local.

132. *Kirinia climene* (Esper, 1783). Extremely local: Gorna Kremena (Essayan & Jugan 1993); Balchik (Kolev 2003).

133. *Lopinga achine* (Scopoli, 1763). Extremely local: Samuil (Markowitsch 1909, Buresch & Tuleschkow 1929), Razgrad (Abadjiev 2001), 'Karakuz' game reserve (coll. IBER), Trapotansko locality ca. 12 km north of Vratsa (Abadjiev 1995). All these records need confirmation. Listed on Council Directive 92/43/EEC (Annex IV), Biological Diversity Act (Annex 3) and the Red Data Book of Bulgaria as 'Extinct' (Abadjiev 2011).

134. *Pararge aegeria* (Linnaeus, 1758). Widespread.

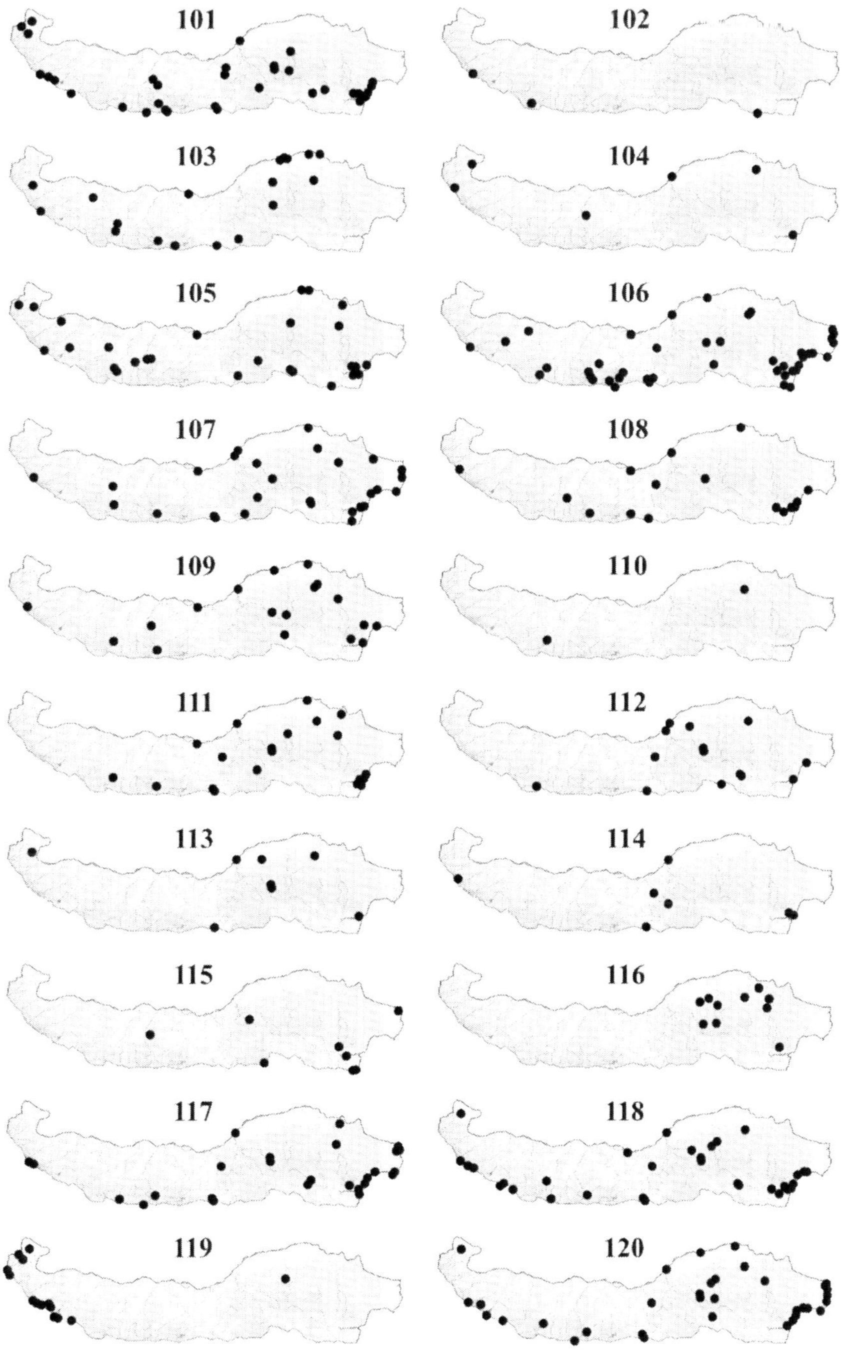

Figure 6. Distribution of butterflies of Danube Plain –plate 6.

135. *Lasiommata maera* (Linnaeus, 1758). Widespread.

136. *Lasiommata petropolitana* (Fabricius, 1787). Extremely local: 'Demir Baba Teke' tomb [near Sveshtari] (Buresch & Tuleschkow 1929). A montane species; needs confirmation.

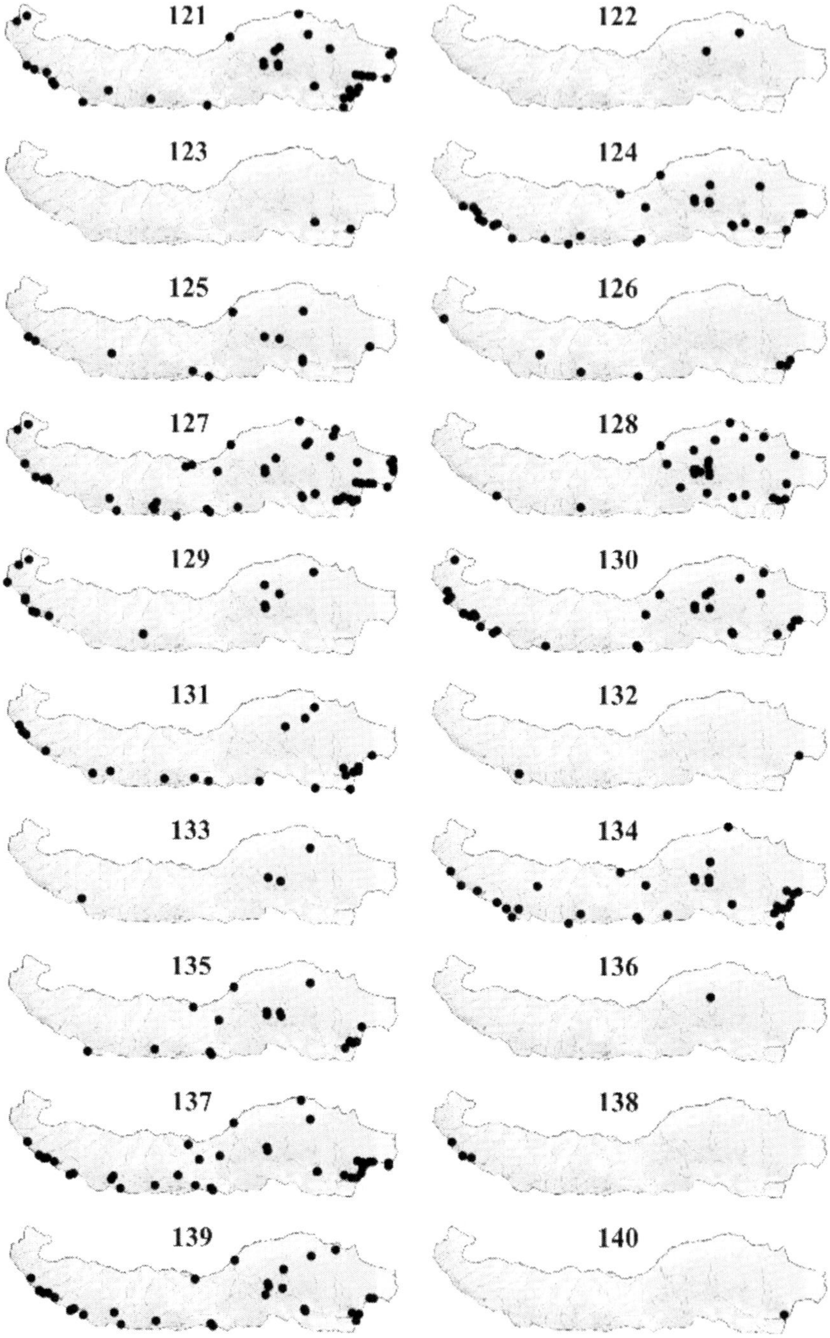

Figure 7. Distribution of butterflies of Danube Plain –plate 7.

137. *Lasiommata megera* (Linnaeus, 1767). Widespread.

138. *Melanargia larissa* (Geyer, [1828]). Extremely local in the western Fore-Balkan (Ganev & Beshkov 1987; M. Lindberg, pers. comm.). The Danube is the northernmost range limit of this species in Europe.

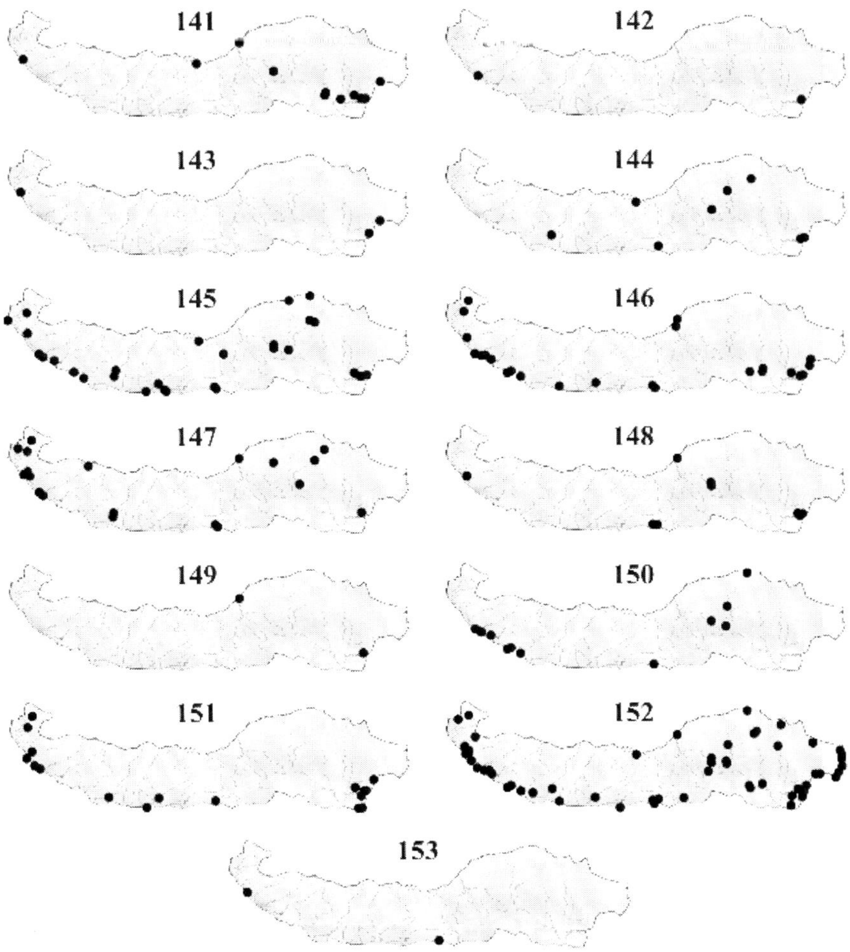

Figure 8. Distribution of butterflies of Danube Plain –plate 8.

139. *Melanargia galathea* (Linnaeus, 1758). Widespread.
140. *Hipparchia fatua* Freyer, 1843. Extremely local: Varna (Ganev 1988). The only record from the entire North Bulgaria; needs confirmation.
141. *Hipparchia statilinus* (Hufnagel, 1766). Widespread in the eastern part of the region; in the west, found only near Belogradchik (own data).
142. *Hipparchia syriaca* (Staudinger, 1871). Extremely local: Varna (Ganev 1988), between Prevala and Gorna Luka (Observation.org 2019).
143. *Hipparchia fagi* (Scopoli, 1763). Extremely local: Aladzha Manastir (Ganev 1988), Balchik (Caradja 1930, Popescu-Gorj 1964 – needs confirmation), Magurata cave (own data).
144. *Hipparchia volgensis* (Mazokhin-Porshnyakov, 1952). Local. My examination of genitalia of old material from North Bulgaria previously reported as *Hipparchia semele* (Linnaeus, 1758) show it to belong to *H. volgensis*.
145. *Minois dryas* (Scopoli, 1763). Widespread.

146. *Brintesia circe* (Fabricius, 1775). Widespread.
147. *Arethusana arethusa* ([Denis & Schiffermüller], 1775). Widespread.
148. *Chazara briseis* (Linnaeus, 1764). Local.
149. *Hyponephele lycaon* (Kühn, 1774). Extremely local: Obraztsov Chiflik near Rousse (Kowatschew 1898), Varna (Rebel 1903). A montane species, these records need confirmation.
150. *Aphantopus hyperantus* (Linnaeus, 1758). Local.
151. *Pyronia tithonus* (Linnaeus, 1771). Widespread.
152. *Maniola jurtina* (Linnaeus, 1758). Widespread.
153. *Erebia medusa* ([Denis & Schiffermüller], 1775). Extremely local: Veliko Tarnovo (Krzywicki 1981), Belogradchik (Observation.org 2019). An upland/montane species.

References

Abadjiev S. 1992. *Butterflies of Bulgaria, Part 1, Papilionidae & Pieridae*. Veren Publishers, Sofia, 91 pp.

Abadjiev S. 1993. *Butterflies of Bulgaria, Part 2, Nymphalidae: Libytheinae & Satyrinae*. Veren Publishers, Sofia, 127 pp.

Abadjiev S. 1995. *Butterflies of Bulgaria,* Part 3, *Nymphalidae: Apaturinae & Nymphalinae*. Publisher S. Abadjiev, Sofia, 159 pp.

Abadjiev S. 2001. *An Atlas of the Distribution of the Butterflies in Bulgaria*. Pensoft Publishers, Sofia – Moscow, 335 pp.

Abadjiev S. 2011. Woodland brown Lopinga achine (Scopoli, 1763). In: *Red Data Book of the Republic of Bulgaria*. Digital edition, http://e-ecodb.bas.bg/rdb/en/vol2/Loachine.html (accessed 18.02.2019).

Abadjiev S. & Beshkov S. 2003. More about Euphydryas maturna in Bulgaria (Lepidoptera: Nymphalidae, Nymphalinae). *Phegea* 31 (1), 33—37.

Abadjiev S. & Beshkov S. (2007). *Prime Butterfly Areas in Bulgaria*. Pensoft Series Faunistica 69, Pensoft Publisher, Sofia – Moscow, 222 pp. + CD.

Bachmetjew P. 1902. The Butterflies of Bulgaria. *Works of the Russian Entomological Society* 35: 356—466 [In Russian].

BDA. 2002. Biological Diversity Act, *State Gazette*, 77: 99, August 2002.

Beshkov S. 1998. Order Lepidoptera. In: Michev T. (Ed.), *Biodiversity of the Srebarna Biosphere Reserve. Checklist and Bibliography*. Ministry of Environment and Waters, Sofia, 81—83.

Beshkov S. 2000. Lepidoptera species (Macrolepidoptera) recorded and reported from the Western Stara Planina Mountain, NW Bulgaria. *Entomologica Romanica* 5: 11—43.

Beshkov S. 2011. *Butterflies and moths in Bulgaria included in Natura 2000*. A guide for identification in the field. Library 'Vitosha', Sofia, 151 pp. [In Bulgarian].

Beshkov S. & Abadjiev S. 2000. The butterfly and moth fauna of the areas around the lakes of Durankulak, Ezerets, Shabla and Shablenska Tuzla in NE Bulgaria (Insecta: Lepidoptera). Atalanta 31 (3/4): 543—573.

Buresch I. 1926. Contribution to the butterfly fauna of the castle park Euxinograd near the city Varna on the Black Sea. *Communications of the Bulgarian Entomological Society* 3:159—192 [In Bulgarian].

Buresch I. 1930. Second Contribution to the butterfly fauna of the castle park Euxinograd near the city Varna on the Black Sea. *Communications of the Bulgarian Entomological Society* 5: 207—252 [In Bulgarian].

Buresch I. 1940. Official part: Presentations and communications. *Communications of the Bulgarian Entomological Society* 11: 243—249 [In Bulgarian].

Buresch I. & Karnojizky N. 1942. Biological observations on the butterflies and moths of the Varna area. *Communications of the Bulgarian Entomological Society* 12: 127—162 [In Bulgarian].

Buresch I. & Tuleschkow K. 1929. The horizontal distribution of butterflies and moths in Bulgaria, I. *Communications of the Royal Natural Science Institutes in Sofia* 2: 145—250 [In Bulgarian].

Buresch I. & Tuleschkow K. 1930. The horizontal distribution of butterflies and moths in Bulgaria, II. *Communications of the Royal Natural Science Institutes in Sofia* 3: 145—248 [In Bulgarian].

Butterflies and moths of Bulgaria. 2019. https://www.facebook.com/groups/122766351094491/?ref=group_header, accessed on 28.02.2019.

Caradja A. 1930. Contributions to the Lepidopteran fauna of the southern Dobrogea, in particular the so-called "Silver Coast". *Bulletin of the Society of Sciences*, *Romanian Academy* 13 (3): 1—21, pls. I—VI [In German].

Caradja A. 1931. Contributions to the Lepidopteran fauna of Greater Romania for the year 1930. *Romanian Academy*, *Memoirs of the Scientific Section* 7 (8): 293—344 [In German].

Caradja A. 1932. Contributions to the Lepidopteran fauna of Great Romania for the year 1931. I. Southern Dobrogea. *Bulletin of the Society of Sciences*, *Romanian Academy*, 15(1-2): 35—46 [In German].

Caradja A. 1934. New contribution to the knowledge of the Lepidoptera fauna of Romania. *German Entomological Journal Iris*, 48: 185—191 [In German].

Essayan R. & Jugan D. 1993. The genus Kirinia Moore, 1893 in Europe (Satyridae). *Linneana belgica*, *14*, 135—142 [In French].

EU 1992. Council Directive 92/43/EEC of 21 May 1992 on the conservation of natural habitats and of wild fauna and flora. *OJ L, 206, 22.7.1992, pp. 7—50.*

Ganev J. 1988. The distribution of species of the Genus Hipparchia Fabricius, 1807 (Lepidoptera: Satyridae) in Bulgaria. Articulata 3 (4): 151—153.

Ganev J. & Beshkov S. 1987. Records of Macrolepidoptera from Bulgaria (Lepidoptera). Phegea, 15 (2): 113—117.

Iwanov K. 1926. Contribution to the butterfly and moth fauna of the area around Lovech and the Troyan section of the Balkan Mts. (Bulgaria). *Communications of the Bulgarian Entomological Society* 3: 211—224 [In Bulgarian].

Karnojizky N. 1954. Biological, ecological and systematic studies on new and rare Lepidoptera of the Black Sea Coast of Bulgaria. *Communications of the Zoological Institute in Sofia* 3: 161—200 [In Bulgarian].

Kolev Z. 2002a. Critical notes on some recent butterfly records from Bulgaria and their source collection. Phegea 30 (3): 95—101.

Kolev Z. 2002b. The species of Maculinea van Eecke, 1915 in Bulgaria: distribution, state of knowledge and conservation status (Lycaenidae). *Nota lepidopterologica* 25 (2/3): 177—190.

Kolev Z. 2003. On the distribution, ecology and conservation status of two little-known Bulgarian butterflies: Brenthis ino (Rottemburg, 1775) and Kirinia climene (Esper, [1783]) (Lepidoptera: Nymphalidae). *Linneana belgica* 19 (4): 165—172.

Kolev Z. 2017. Rubrapterus bavius (Eversmann, 1832), a butterfly genus and species new to Bulgaria (Lepidoptera: Lycaenidae). *ZooNotes* 114: 1—4.

Kolev Z. & Tsvetanov T. 2018. Clarifications and new data on the distribution of Cacyreus marshalli Butler, 1898 in Bulgaria (Lepidoptera: Lycaenidae). *ZooNotes* 122: 1—4.

Kowatschew V. T. 1898. On the fauna of the environs of Ruse. *Yearbook of the Bulgarian Naturalists' Association* 2: 20—28 [In Bulgarian].

Krzywicki M. 1981. Comments on the butterfly fauna of Bulgaria. *Nota lepidopterologica* 4 (1-2): 29—46 [In German].

Lederer J. 1863. Directory of Lepidoptera collected by Mr. Johann and Mrs. Ludmilla Haberhauer in 1861 and 1862 near Varna in Bulgaria and Sliven in Rumelia. *Viennese Entomological Monthly* 7 (1): 17—27, 7 (2): 40—47, pl. 1.

Markowitsch A. 1904. Materials from the insect fauna in the vicinity of Razgrad. *Works of the Bulgarian Naturalists' Association* 2: 220—252 [In Bulgarian].

Markowitsch A. 1909. The Lepidoptera fauna in the vicinity of Razgrad. *Compendium of Folklore, Science and Literature* 25: 1—38 [In Bulgarian].

Nedelkov N. 1909. Our entomofauna. *Archives of the Ministry of National Enlightenment*, 1 (3): 83—135 [In Bulgarian].

Observation.org. 2019. https://observation.org/, accessed on 28.02.2019.

Popescu-Gorj A. 1964. Catalog of the Lepidoptera Collection "Prof. A. Ostrogovich" at the Museum of Natural History "Grigore Antipa", Bucharest. Bucharest, 293 pp., 18 pls.

Rákosy L. & Weidlich M. 2017. Rubrapterus bavius from north-eastern Bulgaria and new data on its conservation status in Romania. *Entomologica Romanica* 21: 15—22.

Rebel H. 1903. Studies on the lepidopteran fauna of the Balkan countries. Part I: Bulgaria and Eastern Rumelia. *Annals of the Natural History Museum in Vienna*: 18 (2—3): 123—347, pl. III [In German].

Slivov A. V. 1968. Butterflies and moths (Lepidoptera) of the Iskar gorge. I. Macrolepidoptera. *Communications of the Zoological Institute in Sofia* 28: 141—186 [In Bulgarian].

Slivov A. V. & Loukov H. 1976. Contribution to the study of butterflies and moths in Bulgaria. *Terrestrial Fauna of Bulgaria: Materials* 234—242 [In Bulgarian].

Tuleschkow K. 1930. First contribution to the lepidopteran fauna of the city of [Veliko] Tarnovo and its surroundings. *Communications of the Bulgarian Entomological Society* 5: 125—162 [In Bulgarian].

Wiemers M., Balletto E., Dincă V., Fric Z. F., Lamas G., Lukhtanov V., Munguira M. L., van Swaay C. A. M., Vila R., Vliegenthart A., Wahlberg N. & Verovnik R. 2018. An updated checklist of the European Butterflies (Lepidoptera, Papilionoidea). *ZooKeys* 811: 9—45.

Zukowsky B. 1937. Trip to the Black Sea and Herkulesbad (Lep[idoptera].). *Entomological review* 54 (45): 549—553, (46): 557—559, (47): 565—568, (48): 573—576.

In: Biodiversity of the Bulgarian-Romanian Section
Editor: Shurulinkov, Hubenov et al.
ISBN: 978-1-53615-663-8

Chapter 7

ICHTHYOFAUNA OF THE BULGARIAN STRETCH OF THE DANUBE RIVER AND LOWER COURSES OF ITS TRIBUTARIES

Tihomir Stefanov *
National Museum of Natural History,
Bulgarian Academy of Sciences, Sofia, Bulgaria

ABSTRACT

The ichthyofauna of Bulgarian stretch of Danube River and lower courses of its tributaries is reviewed based on the published and original data. A total of 83 fish species belonging to 21 families have been reported. Of these, 69 species are native for Danube River basin and 12 are alien and introduced. The status of two species is unclear. Eight species are considered endemic in the Danube River basin. Full faunistic list with notes on distribution, origin and conservation status of the fish species is given.

Keywords: freshwater fish, Danube River, distribution, origin, conservation status

INTRODUCTION

The Bulgarian stretch of the Danube River is part of the Lower Danube located between river kilometres 845 and 375. The ichthyofauna of the Lower Danube including the Danube Delta and its floodplains has been characterized by high species diversity

*Corresponding Author's Email: tishos@gmail.com.

(Busnita 1961, Bănărescu 1964, Otel 2007, Năstase et al. 2017) and lower level of endemism.

First data about the fish fauna of the Bulgarian section of the Danube were published in the summery works of Kovatcheff (1923), Morov (1931) and Drensky (1948, 1951), where up to 54 different species and subspecies are published. Later, Kolarov (1963) studied the biology of shad along the Bulgarian shore of Danube River. The first comprehensive survey of the Danube ichthyofauna in Bulgaria has been made by Marinov (1966, 1978). Based on own data and literature review he reported the occurrence of 61 species. Oundjian (1964) and Stefanov & Holčik (2007) published information about the lampreys in Danube River. Data on the distribution of different gobiid species in Danube River is published by Gheorghiev (1966). Manolov & Sivkov (1977) and Marinov (1979) found the invasive *Pseudorasbora parva* for the first time in Danube River and some adjacent marshlands and fish farms. Information about the downstream migration of the ichthyoplankton along the Bulgarian shore of the river was published by Vassilev (1994). Vassilev (2003) surveyed the spawning sites of Beluga sturgeon (*Huso huso*) in the Bulgarian-Romanian Danube River stretch. Later, Vassilev & Pehlivanov (2003) published information about the structural changes of sturgeon catches in Bulgaran Danube section. Velkov et al. (2004) found the species *Umbra krameri* in the Danube riverside lake Srebarna, which had been treated as extinct in Bulgaria. Jurajda et al. (2006) found the invasive species *Perccottus glenii* for the first time in the Bulgarian stretch of Danube River. Information about the occurrence of *Salmo labrax* in the surveyed area is given by Vassilev & Trichkova (2007) and Holčik & Stefanov (2008). Tsekov et al. (2008) and Tsekov & Tsekov (2013) gave information about the natural hybridization of different sturgeon species in Bulgarian stretch of Danube River. Compiled information about the ichthyofauna of Bulgarian stretch of Danube River and some of the adjacent marshes was provided by Kutsarov (2010). Zarev et al. (2013) made a review of the distribution of gobiid fish in the Bulgarian Danube River tributaries. Recently, Honţ et al. (2018) studied the long distance migration of Beluga and Stellate sturgeons in Lower Danube River.

The ichthyofauna of adjacent riverside lakes has been studied by Drensky (1923), Bulgurkov (1958), Karapetkova (1998), Pehlivanov (2000) and Pehlivanov et al. (2011).

Summarized data on the fish species composition of the Danube River were published also by Karapetkova & Zivkov (1995), Karapetkova et al. (1998), Vassilev & Pehlivanov (2005) and Stefanov (2007). According to Karapetkova et al. (1998) 67 out of 83 fish species inhabiting the entire Danube River are found in its Bulgarian section and the adjacent marshlands, which is ca. 47.2% of all freshwater ichthyofauna of Bulgaria.

There are 13 tributaries of the Danube River in Bulgaria. In the present study we give information about the fish fauna of the lowermost courses on 11 of them. They all take their sources from the northern slope of Stara Planina Mountains with the only exception of the Iskar River, which originates from Rila Mountains.

The first data about the ichthyofauna of the lower courses of these rivers come from the works of Kovatcheff (1909, 1923), Drensky (1928, 1932, 1948, 1951) and Chichkoff (1929, 1937, 1939). Later, information about the fish species diversity in the same river courses could be found in Marinov (1967, 1987), Karapetkova (1972, 1979), Sivkov (1979, 1985, 1987a, 1987b, 1988a, 1988b, 1989a, 1989b, 1991) and Karapetkova & Undžian (1988).

MATERIAL AND METHODS

The present study gives information about the ichthyofauna of the whole Bulgarian stretch of Danube River located between rkm 845 and 375. The survey covers also the lowermost courses of eleven of the rivers in Bulgaria, which are right tributaries of the Danube. From west to east they are as follows: Voynishka, Vidbol, Archar, Lom, Tsibritsa, Ogosta, Iskar, Vit, Osam, Yantra and Russenski Lom (Figure 1).

The study was conducted in the period 2005 – 2017. The material in Danube River was collected mainly by bottom floating drift nets as well as with different beach seine nets with mesh size between 4-7 mm. In the Danube River tributaries the main method was electrofishing using portable backpack electrofishing gear by HANS-GRASSL, Germany (output 200-500 V), equipped with circle anode of stainless steel (netting of 4 mm mesh size).

A total of 89 published sources were used in the present study, as well as an original data and revision of the ichthyological collections of the National Museum of Natural History in Sofia and the Naturhistorische Museum in Vienna.

The classification of families followed that of Nelson et al. (2016) till the taxonomy is based mainly on the review of Kottelat & Freyhof (2007).

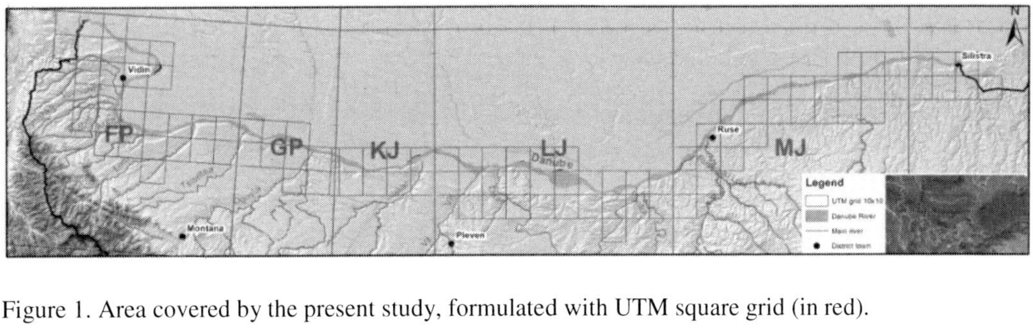

Figure 1. Area covered by the present study, formulated with UTM square grid (in red).

RESULTS AND DISCUSSION

A total of 83 species belonging to 21 families have been reported for the Bulgarian stretch of Danube River and the lower courses of its tributaries so far. The family Cyprinidae is represented by 35 fish species, which is 42.2% of all the fish species, found in the region, followed by family Percidae with 8 species (9.6%) and Cobitidae and Gobiidae with 7 species (8.4%).

The species are as follows:

Northern Lampreys Family (Petromyzontidae)

Ukrainian Brook Lamprey – *Eudontomyzon mariae* (Berg, 1931)

A total of four species of lampreys are reported to occur in Bulgaria according to the ichthyological literature. These are the European brook lamprey – *Lampetra planeri* (Bloch, 1784), the European river lamprey – *L. fluviatilis* (Linnaeus, 1758), the Carpathian lamprey – *Eudontomyzon danfordi* Regan, 1911, and the Ukrainian brook lamprey – *E. mariae* (Berg, 1931) (Kovatcheff 1923, Morov 1931, Drensky 1935, 1948, 1951, Kolarov 1960, Oundjian 1964, Marinov 1978, Vassilev & Pehlivanov 2005). After the revision made by Stefanov & Holčik (2007) only the species *E. mariae* is considered to inhabit the Bulgarian stretch of Danube River. In the past the species has been distributed in Danube River and some of its tributaries such as Lom, Iskar, Vit, Osam, Yantra and Russenski Lom (Karapetkova 1994, Stefanov &Holčik 2007). Nowadays, it is found only in Danube River near Lom, Nikopol, Belene, Russe, Tutrakan and Silistra (Stefanov & Holčik 2007, Kutsarov 2010, our data).

Status and distribution in the study area: Rare. Only in the Danube River.

Origin: Native to the Danube River region.

Sturgeons Family (Acipenseridae)

Russian Sturgeon – *Acipenser gueldenstaedtii* (Brandt & Ratzeburg, 1833)

Kovatcheff (1923) found the species in the Danube River and its tributary Yantra and reported that the numbers of this species hugely decreased, which has been confirmed by Drensky (1951). According to Vassilev & Pehlivanov (2003) the Russian sturgeon annual catch decreased from 25.4 tons in 1948 down to 0.6 tons in 2001. The species was published for the Bulgarian stretch of the Danube River by Marinov (1978). Nowadays, there are no published records of Russian sturgeon for the last 30 years. One record of a hybrid specimen defined as *A. gueldenstaedtii* X *A. ruthenus*, caught in October 2005 in the Danube River near the village of Gorni Vadin at rkm 654 has been published by Tsekov et al. (2008). Later the same fish was determined as *A. nudiventris* X *A. gueldenstaedtii* (Tsekov & Tsekov, 2013). Another specimen with morphological signs of *A. gueldenstaedtii* and *A. ruthenus* has been caught in November 2010 in the Danube River near the town of Oryahovo at 678 rkm (Tsekov & Tsekov, 2013).

Status and distribution in the study area: Very rare, probably on the edge of extinction. Only in the Danube River.

Origin: Ponto-Caspian relic. Native to the Danube River region.

Ship Sturgeon – *Acipenser nudiventris* (Lovetsky, 1828)

In the past, it was reported as a rare species in the Bulgarian stretch of the Danube River (Kovatcheff 1923, Drensky 1951). Marinov (1966) reported for the catch of one specimen at rkm 776 near Simeonovo. Later the species was listed for the Danube River by Marinov (1978). There is no evidence for the presence of this species last 50 years and nowadays it is treated as extinct. Only one hybrid specimen defined as *A. nudiventris* X *A. gueldenstaedtii* has been caught in October 2005 near the village of Gorni Vadin at rkm 678 (Tsekov & Tsekov 2013).

Status and distribution in the study area: Extinct.

Origin: Ponto-Caspian relic. Native to the Danube River region.

Sterlet – *Acipenser ruthenus* (Linnaeus, 1758)

The species was reported for the Bulgarian stretch of the Danube River by Kovatcheff (1923), Drensky (1948, 1951) and Marinov (1966, 1978). It has been found also in the lower courses of some tributaries like Iskar (Chichkoff 1939), Ogosta and Yantra (Karapetkova 1972, 1994). According to Vassilev & Pehlivanov (2003) the Sterlet annual catch decreased from 21.2 tons in 1960/1964 down to 1.2 tons in 2001. Nowadays this is the most widespread sturgeon species in Lower Danube and the only one which does not migrate to the Black Sea.

Status and distribution in the study area: Fairly common. Only in the Danube River.

Origin: Ponto-Caspian relic. Native to the Danube River region.

Stellate Sturgeon – *Acipenser stellatus* (Pallas, 1771)

According to Kovatcheff (1923) the species occurred in the Danube River and the lower courses of Vit, Iskar and Yantra. It has been published for the Bulgarian stretch of the river also by Drensky (1948, 1951) and Marinov (1978). According to Vassilev & Pehlivanov (2003) the Stellate sturgeon annual catch decreased from 16.9 tons in 1947 down to 0.7 tons in 2001. Last years, the species was found in Danube River near Archar, Belene and Vetren (our data). Recent survey shows that Stellate sturgeon still moves upstream to Iron Gate Dam migrating more than 860 km from the Black Sea due to its homing instinct (HONŢ et al. 2018).

Status and distribution in the study area: Rare. Only in the Danube River.

Origin: Ponto-Caspian relic. Native to the Danube River region.

Atlantic Sturgeon – *Acipenser sturio* (Linnaeus, 1758)

It is reported for Bulgarian stretch of Danube River by Drensky (1948), but according to Marinov (1978) its presence is doubtful. Kutsarov (2010) reported for the catch of one specimen in the Romanian side of rkm 400, but according to him this data was later treated as unreliable. There is one adult specimen of Atlantic sturgeon in the Ichthyological collections of the National Museum of Natural History in Sofia (NMNHS, no number) labeled "from Danube River in Bulgaria".

Status and distribution in the study area: Extinct.

Origin: Ponto-Caspian relic. Native to the Danube River region.

Beluga – *Huso huso* (Linnaeus, 1758)

The species was found in Danube River by Kovatcheff (1923), Drensky (1948, 1951) and Marinov (1978). Drensky (1951) published that its catch last 25 years highly decreased due to intensive fishing. According to Vassilev & Pehlivanov (2003) the Beluga annual catch increased from 1.4 tons in 1945 up to 9.9 tons in 2002, with a peak of 31.2 tons in 1998. In the period 1995-2002 it formed 79.1% of the sturgeon catch in the Bulgarian stretch of Danube River (Vassilev & Pehlivanov 2003). Vassilev (2003) found thirteen spawning sites of Beluga sturgeon located in the Lower Danube River between rkm 755 (Petrich Island) and rkm 840 (Vrav Village) in 2002. A hybrid *H. huso* X *A. ruthenus* was caught near Oryahovo at rkm 678 in 2006 (Tsekov et al. 2008). Last years, the species was found near Vidin, Vardim, Vetren and Silistra (our data). Recent survey shows that the Beluga sturgeon still moves upstream to Iron Gate Dam migrating more than 860 km from the Black Sea trying to pass over the dam due to its homing instinct (Honţ et al. 2018).

Status and distribution in the study area: Rare. Only in the Danube River.

Origin: Ponto-Caspian relic. Native to the Danube River region

Paddlefish Family (Polyodontidae)

American Paddlefish – *Polyodon spathula* (Walbaum, 1792)

The species is cultivated in fishponds in Romania and Bulgaria from where it escaped in Danube River. In 2000 one specimen was caught near Pozharevo at rkm 423 (Kutsarov 2010).

Status and distribution in the study area: Unknown. Only in the Danube River.

Origin: Introduced.

Freshwater Eels Family (Anguillidae)

European Eel – *Anguilla anguilla* (Linnaeus, 1758)

During the 17th century the traveler and scientist Count Marsigli reported for the presence of eels in the upper and middle sections of the Danube River (in Berg 1961). According to Kovachev (1923) the species occurred in the Bulgarian stretch of Danube River and the adjacent marshlands but it was not numerous. Drensky (1948, 1951) reported that the presence of eels in the Lower Danube is due to stocking in the Upper Danube. The question about the origin of eels in the Danube River is still open. The Scientific, Technical and Economic Committee for Fisheries (STEFC) of EU suggested the Black Sea region should be excluded from the list of natural areas, where the European eel is distributed, because the eels caught in the Black Sea region (e. g. Danube River) represent escaped specimens from fish farming in some Danubian countries. According to Kutsarov (2010) there are reliable data that the species was abundant in the past in the Eastern part of the Bulgarian stretch and Kalimok-Brushlen Marsh system which is an evidence for the native origin of eels in Danube River. Apostolou et al. (2013) stated that the Black Sea is the end of natural distribution of European eel and the species was never abundant in the region but it presence should be a result of natural distribution. Anyway, probably all of the recently caught eels in Danube River are stocked or escaped from aquaculture. In 2017 a total of 25 000 eels have been stocked near Lom (S. Studenkov, pers. comm.).

Status and distribution in the study area: Unknown, probably the native population is extinct nowadays and only stocked fish could be found.

Origin: Native to the Black Sea region.

Herrings Family (Clupeidae)

Black Sea Shad – *Alosa immaculata* (Bennet, 1835)

The species migrate to the Bulgarian stretch of Danube River for spawning and it is reported (as *A. pontica*, *A./Caspialosa kessleri pontica*) by Kovatcheff (1923), Drensky (1948, 1951), Kolarov (1963), Stojanov et al. (1963) and Marinov (1966, 1978). Karapetkova (1972, 1994) found the species in the lowermost part of Yantra River. Recently, it annually enters in the whole Bulgarian part of the Lower Danube for spawning.

Status and distribution in the study area: Seasonally common. Only in the Danube River.

Origin: Ponto-Caspian relic. Native to the Danube River region.

Small Black Sea Shad – *Alosa tanaica* (Grimm, 1901)

This species was reported for the Bulgarian stretch of the Danube River (as *A. nordmanni* and *A. caspia nordmanni*) by Kovatcheff (1923), Drensky (1948, 1951),

Stojanov et al. (1963) and Marinov (1978). There are no recent data to confirm the presence of Small Black Sea shad in Bulgarian part of the Danube.

Status and distribution in the study area: Unknown. Only in the Danube River.

Origin: Ponto-Caspian relic. Native to the Danube River region.

Carps Family (Cyprinidae)

Bitterling – *Rhodeus amarus* (Bloch, 1782)

The species was found in the Danube River tributaries and adjacent marshlands (Kovatcheff 1923, Drensky 1951, Bulgurkov 1958, Marinov 1966, 1978, Karapetkova 1972, 1994, Karapetkova & Undžian 1988, Pehlivanov et al. 2011, Polačik et al. 2008). It was published also for the main course of the Danube River (Marinov 1978, Vassilev 1994, Polačik et al., 2008). Nowadays, it is widespread and numerous in the shoreline zone of the Danube River, lower courses of its tributaries and some of the adjacent wetlands.

Status and distribution in the study area: Common. In the Danube River, tributaries and adjacent wetlands.

Origin: Native to the Danube River region.

Danube Gudgeon – *Gobio obtusirostris* (Valenciennes, 1842)

Probably all published gudgeons (as *G. gobio*) for the Bulgarian part of the Danube River and its tributaries (Chichkoff 1939, Drensky 1951, Paspalev & Peshev 1955, Marinov 1966, 1978, Karapetkova 1972, 1994, Karapetkova & Undžian 1988, Vassilev 1994) refer to the species *R. vladykovi* for Danube River and *G. obtusirostris* – for the tributaries. Recently, we found the species in the lower courses of the rivers Voynishka, Vidbol, Archar, Lom, Tsibritsa, Iskar, Vit, Yantra and Russenski Lom.

Status and distribution in the study area: Common. Only in the Danube River tributaries.

Origin: Native to the Danube River region.

Pseudorasbora – *Pseudorasbora parva* (Temminck & Schlegel, 1846)

The species was found for the first time in Bulgaria in 1975 in "Mechka" Fishponds near the town of Russe (Marinov 1979) and later in 1976 it was caught in Malak Preslavets Marsh as well as in Danube River near Krivina Village, Rousse district (Manolov & Sivkov 1977). Recently, the species was found also in Danube River (Polačik et al. 2008) and in some of its tributaries in the country – Archar, Lom, Tsibritsa, Ogosta and Skat (Trichkova et al. 2009). It is reported also from some of the riverside wetland along Danube River – Orsoya, Kalimok and Malak Preslavets marshes as well as from Srebarna Lake (Pehlivanov et al. 2011)

Status and distribution in the study area: Common. In the Danube River, tributaries and adjacent wetlands.

Origin: Alien.

Danube Delta Gudgeon – *Romanogobio antipai* (Bănărescu, 1953)

Danube delta gudgeon occurred in the Lower Danube from the confluence of Arges River at rkm 432 (near Tutrakan) downstream to the delta. It has been considered extinct since no recent findings were known. For long time it was treated as a subspecies of *R. kesslerii*. According to us, the report of *R. kesslerii* in Danube River by Marinov (1966) (see next species) is a misidentification of *R. antipai* due to the small eye diameter of the caught specimens – 19.1 and 24.3% of the head length and the presence of 8 branched rays in the dorsal fin. The hidden way of living in the deepest parts of Lower Danube could be a reason for the lack of recent catches of this species.

Status and distribution in the study area: Unknown, probably extinct. Only in the Danube River.

Origin: Native to the Danube River region.

Sand Gudgeon – *Romanogobio kesslerii* (Dybowski, 1862)

Marinov (1966) reported the catch of two specimens of this species in the Bulgarian stretch of Danube River – one near Ostrov at rkm 661 and one near Krivina at rkm 536. According to us this is a misidentification of *R. antipai,* because *R. kesslerii* in Bulgaria is known only from the middle courses of Danube River tributaries. The specimen stored in Russe Regional Museum and labeled as *R. kesslerii* is actually *R. vladykovi* (Kutsarov 2010). In the study area the species is known only from the lower courses of Lom and Tsibritsa Rivers (our data).

Status and distribution in the study area: Very rare. Only in the tributaries of the Danube River.

Origin: Native to the Danube River region.

Stone Gudgeon – *Romanogobio uranoscopus* (Agassiz, 1828)

In the study area the species is known only from Lom River near Drenovets village (our data).

Status and distribution in the study area: Very rare. Only in the tributaries of the Danube River.

Origin: Native to the Danube River region.

Danube Whitefin Gudgeon – *Romanogobio vladykovi* (Fang, 1943)

The species (reported as *Gobio albipinnatus*) was found in Bulgaria for the first time in Ogosta River near Lehchevo village (Marinov 1967). Later, it was found in the lower course of Yantra River (Karapetkova 1972) as well as in Danube River (Marinov 1978,

Sivkov 1989). Polačik et al. (2008) found it in various locations between Vrav (rkm 840) and Vetren (rkm 395). Nowadays, it is widespread in the whole Bulgarian stretch of Danube River and the lowest courses of its tributaries.

Status and distribution in the study area: Common. In the Danube River and lower courses of its tributaries.

Origin: Native to the Danube River region.

Large Spot Barbel – *Barbus balcanicus* (Kotlík, Tsigenopoulos, Ráb & Berrebi, 2002)

The species is recently described as a genetically different barbel from the Western and Southern parts of Balkan Penninsula. According to Kottlik et al. (2002) in Bulgaria it is known from the rivers Archar and Beli Vit. It is not clear wether the distribution of this species overlaps with the one of *B. petenyi* or they are separated. Most probably the barbels from the lower courses of western Bulgarian Danube tributaries belong to *B. balcanicus* and the ones from the rest of the rivers to *B. petenyi*.

Status and distribution in the study area: Unknown, probably locally common. Only in the tributaries.

Origin: Native to the Danube River region.

Common Barbel – *Barbus barbus* (Linnaeus, 1758)

The species was published for the Danube River and its tributaries (Kovatcheff 1923, Chichkoff 1939, Drensky 1948, 1951, Paspalev & Peshev 1955). Later it has been found in the Danube River (Marinov 1966, 1978, Vassilev 1994) as well as in the lower courses of most of the tributaries – Voynishka, Vidbol, Archar, Lom, Tsibritsa, Ogosta, Iskar, Vit, Osam, Yantra and Russenski Lom (Karapetkova 1972, 1994, Karapetkova & Undžian 1988). Polačik et al. (2008) found it in various locations between Vrav (rkm 840) and Marten (rkm 482). Nowadays, it is widespread in the whole Bulgarian section of the Danube River and lower courses of its tributaries.

Status and distribution in the study area: Common. In the Danube River and its tributaries.

Origin: Native to the Danube River region.

Romanian Barbel – *Barbus petenyi* (Heckel, 1852)

The species was published for the lower courses of the Bulgarian Danube River tributaries by Karapetkova (1994). In the study area we found barbels in the lowermost parts of the rivers Voynishka, Vidbol, Archar, Lom, Tsibritsa, Ogosta, Iskar, Vit and Russenski Lom. It is not clear whether they belong to this species or to *B. balcanicus*, or both.

Status and distribution in the study area: Common. Only in the tributaries.

Origin: Native to the Danube River region.

Crucian Carp – *Carassius carassius* (Linnaeus, 1758)

In the past, the species has been widely spread in the Danube River and adjacent marshes (Kovatcheff 1923, Drensky 1948, 1951) as well as in the lower course of Iskar River (Chichkoff 1939, Paspalev & Peshev 1955). Later, it was found in the old arms near the mouth of Yantra River (Karapetkova 1972). According to Karapetkova (1994) it occurs in the lower courses of some Danube tributaries – Iskar, Vit, Osam and Yantra. Vassilev (1994) recorded it in Danube River near Baykal. Crucian carp has been found also in Srebarna Lake and Kalimok Marsh (Pehlivanov 2000, Pehlivanov et. al. 2000), the mouth of Timok River (Kutsarov 2010) as well as in Malak Preslavets Marsh (our data).

Status and distribution in the study area: Very rare. Only in some marshes adjacent to the Danube River.

Origin: Native to the Danube River region.

Prussian Carp – *Carrasius gibelio* (Bloch, 1782)

It was found in the Danube River (Drensky 1948, Marinov 1966, 1978, Polačik et al. 2008, Kutsarov 2010) and adjacent marshlands (Kutsarov 2010, Pehlivanov et al. 2011) as well as in the lower courses of its tributaries (Paspalev & Peshev 1955, Karapetkova 1972, 1994, Karapetkova & Undžian 1988). Nowadays, it is widespread and numerous everywhere in the study area.

Status and distribution in the study area: Common. In the Danube River, adjacent wetlands and tributaries.

Origin: Introduced to the Danube River region.

Common Carp – *Cyprinus carpio* (Linnaeus, 1758)

It used to be a common species in the Danube River and adjacent wetlands in the past (Kovatcheff 1923, Morov 1931, Drensky 1948, 1951). Chichkoff (1939) published the species for the lower course of Iskar River. After the drainage of the riverside marshlands in the first half of 20th century it numbers decreased in great extent. According to Marinov (1978) in the period 1925-1958 Common carp formed 40.8% of the fish catch in Bulgarian section of Danube River and decreased down to 4.2% in the period 1970-1974. Gradually, the wild form got extinct and replaced by the stocked culture forms of the species. Nowadays, the aquaculture carps inhabit the whole part of Lower Danube and its tributaries, as well as in many adjacent wetlands.

Status and distribution in the study area: Wild form is considered extinct.

Origin: Native to the Danube River region.

Common Bream – *Abramis brama* (Linnaeus, 1758)

The species was found in the Danube River and adjacent marshlands (Kovatcheff 1923, Drensky 1948, 1951, Bulgurkov 1958, Marinov 1966, 1978, Vassilev 1994, Polačik et al. 2008, Pehlivanov et al. 2011) as well as in the lower courses of the rivers Vidbol, Archar,

Lom, Tsibritsa, Ogosta, Iskar, Vit, Osam, Yantra and Russenski Lom (Chichkoff 1939, Paspalev & Peshev 1955; Karapetkova 1972, 1994, Karapetkova & Undžian 1988). Nowadays, it is widespread and numerous in Danube River, lower courses of its tributaries and some adjacent wetlands.

Status and distribution in the study area: Very common. Mainly in the Danube River, but also in most of the tributaries and some adjacent marshlands.

Origin: Native to the Danube River region.

Danube Spirlin – *Alburnoides sp. 2* (sensu Stierandová et al. 2016)

It is not reported for the Lower Danube. In the study area the species is distributed only in the lower courses of the following Danube tributaries: Voynishka, Vidbol, Archar, Lom, Tsibritsa, Ogosta and Vit (our data).

Status and distribution in the study area: Locally common. Only in the tributaries.

Origin: Native to the Danube River region.

Common Bleak – *Alburnus alburnus* (Linnaeus, 1758)

The species was found in the Danube River (Kovatcheff 1923, Drensky 1948, 1951, Marinov 1966, 1978, Vassilev 1994, Polačik et al. 2008) and adjacent marshlands (Pehlivanov et al. 2011) as well as in the lower courses of most of its tributaries (Kovatcheff 1923, Chichkoff 1939, Drensky 1951, Paspalev & Peshev 1955, Karapetkova 1972, 1994, Karapetkova & Undžian 1988). Nowadays, it is widespread and numerous in the study area.

Status and distribution in the study area: Very common in the Danube River, adjacent wetlands and tributaries.

Origin: Native to the Danube River region.

Danube Shemaya – *Alburnus danubicus* (Antipa, 1909)

The species was reported (as *Chalcalburnus chalcoides danubicus*) for the Bulgarian stretch of the Danube River (Kovatcheff 1923, Drensky 1943, 1951, Marinov 1978), Srebarna Lake (Bulgurkov 1958) and the tributaries Iskar, Vit, Osam and Yantra (Drensky 1921, 1951, Paspalev & Peshev 1955). Only Drensky (1943, 1951) provided morphologic data and show that shemayas with more than 17 branched rays in the anal fin (typical for *A. danubicus*) existed in the Bulgarian part of Lower Danube. According to Freyhof & Kottelat (2007) this species is considered extinct.

Status and distribution in the study area: Unclear, probably extinct.

Origin: Native to the Danube River region.

Pontian Shemaya – *Alburnus sarmaticus* (Kottelat & Freyhof, 2007)

Karapetkova (1972) reported the presence of this species (as *Chalcalburnus chalcoides danubicus*) in the lowermost part of Yantra River. According to her data, the collected three specimens possessed between 15-16 branched rays in the anal fin, which is typical for *A. sarmaticus*. Later, Karapetkova (1994) published it also for Iskar River. Recently, we found Pontian shemaya in Danube River near Tutrakan and Vetren.

Status and distribution in the study area: Very rare. Only in the Danube River.

Origin: Native to the Danube River region.

Asp – *Aspius aspius* (Linnaeus, 1758)

In the past, the species has been found in the Danube River and adjacent wetlands, as well as in the lower courses of its tributaries – Iskar, Vit and Osam (Kovatcheff 1923, Chichkoff 1939, Drensky 1930, 1948, 1951). Later, it was reported for Danube River (Marinov 1966, 1978, Vassilev 1994), Srebarna Lake (Bulgurkov 1958) and the lower stretches of Lom, Ogosta, Iskar, Vit, Osam and Yantra (Paspalev & Peshev 1955, Karapetkova, 1972, 1994). Polačik et al. (2008) found it in various locations between Vrav (rkm 840) and Dolno Ryahovo (rkm 418). Pehlivanov et al. (2011) reported it for Orsoya fishponds and Srebarna Lake. Nowadays it is widespread in the whole Bulgarian section of the Danube River and lower courses of its tributaries.

Status and distribution in the study area: Common. In the Danube River and its tributaries.

Origin: Native to the Danube River region.

Blue Bream – *Ballerus ballerus* (Linnaeus, 1758)

According to Kovatcheff (1923) the species occurred in the Danube River and its adjacent marshlands and rarely penetrated into Yantra River. Later, it has been reported for the Danube River also by Drensky (1948, 1951), Marinov (1966, 1978) and Vassilev (1994) and for the lower courses of some of the tributaries – Iskar, Vit, Osam and Yantra (Paspalev & Peshev 1955, Karapetkova 1972, 1994). Recently, it has been found only by Polačik et al. (2008) near Archar. According to Kutsarov (2010) there are no recent data for the presence of this species in the Bulgarian stretch of Danube River. It was not found by us during our survey.

Status and distribution in the study area: Very rare. Only in the Danube River.

Origin: Native to the Danube River region.

Whiteye Bream – *Ballerus sapa* (Pallas, 1814)

The species was found in the Danube River (Kovatcheff 1923, Drensky 1948, 1951, Marinov 1966, 1978, Vassilev 1994, Pehlivanov et al. 2011, Polačik et al. 2008) and Srebarna Lake (Bulgurkov 1958). It was reported also for the lower courses of the rivers Iskar, Vit, Osam and Yantra (Chichkoff 1939, Paspalev & Peshev 1955, Karapetkova 1972, 1994). Nowadays, it is widespread and numerous in the Danube River.

Status and distribution in the study area: Common in the Danube River.

Origin: Native to the Danube River region.

Silver Bream – *Blicca bjoerkna* (Linnaeus, 1758)

The species was found in the Danube River and adjacent marshlands (Kovatcheff 1923, Drensky 1948, 1951, Bulgurkov 1958, Marinov 1966, 1978, Vassilev 1994, Pehlivanov et al. 2011, Polačik et al. 2008) as well as in the lower courses of the rivers Vidbol, Ogosta, Iskar, Vit, Osam, Yantra and Russenski Lom (Drensky 1951, Karapetkova 1972, 1994). Nowadays, it is widespread and numerous in the Danube River, but it also occur in the lowermost sections of its tributaries and some adjacent wetlands.

Status and distribution in the study area: Very common. In the Danube River, adjacent marshlands and tributaries.

Origin: Native to the Danube River region.

Common Nase – *Chondrostoma nasus* (Linnaeus, 1758)

The species was published for the Danube River and its tributaries (Kovatcheff 1923, Chichkoff 1939, Drensky 1951, Paspalev & Peshev 1955). Later, it has been found in the Danube River (Marinov 1966, 1978, Vassilev 1994) as well as the middle and lower courses of most of the tributaries – Vidbol, Archar, Lom, Tsibritsa, Ogosta, Iskar, Vit, Osam, Yantra and Russenski Lom (Karapetkova 1972, 1994, Karapetkova & Undžian 1988). Polačik et al. (2008) found it in various locations between Vrav (rkm 840) and Silistra (rkm 376). Nowadays, it is widespread in the whole Bulgarian section of the Danube River and its tributaries.

Status and distribution in the study area: Very common. In the Danube River and its tributaries.

Origin: Native to the Danube River region.

Silver Carp – *Hypophthalmichthys molitrix* (Valenciennes, 1844)

The species was reported for the Danube River by Marinov (1978) and Karapetkova (1994). According to the authors some specimens originating from riverside fishponds in Bulgaria and Romania got into the river. It is still not clear whether the species could reproduce naturally in Danube. In 2006 a big school of 0+ fish with size of 3-4 cm was found near Radetzki Island at rkm 436 (Kutsarov 2010), which might be an offspring of natural spawners. In April 2006 the Danube was flooded, reaching between 850 and 900 cm water level maximum at Vidin. In June 2006, the water level was still high ranging from 600 to 700 cm (Polačik et al. 2008).This high water level of the river for a comparatively long period of time with temperatures ranged between 15-25°C (according to the data of the Executive Agency for Exploration and Maintenance of the Danube River – Rousse) could be suitable conditions for natural reproduction of the species. In 2009 another 0+ fish were found in the channel connecting Srebarna Lake with Danube River (Kutsarov 2010). Silver carp was found in Srebarna Lake also by Pehlivanov et al. (2011). Recently, we found the species in Danube River near Vidin, Belene, Vardim, Rousse and

Vetren. Nowadays it occurs in the whole Bulgarian stretch of the river where it is regularly stocked.

Origin and distribution in the study area: Fairly common. In the Danube River and some of the adjacent wetlands.

Origin: Alien.

Bighead Carp – *Hypophthalmichthys nobilis* (Richardson, 1845)

The species was reported for the Danube River by Marinov (1978), Karapetkova (1994) and Kutsarov (2010). There are no reliable data about natural reproduction of this species in the Lower Danube. Nowadays it occurs in the whole Bulgarian stretch of the river where it is regularly stocked together with hybrids with Silver carp.

Status and distribution in the study area: Fairly common in the Danube River.

Origin: Alien.

Sun Bleak – *Leucaspius delineatus* (Heckel, 1843)

The species has been found in the Danube River for the first time by Vassilev (1994) and later recorded also by Polačik et al. (2008). According to Kutsarov (2010) it is a common and numerous species in the Danube River and the lower course of its tributary Vit. It has been found also in some Danube wetlands – Orsoya, Belene, Kalimok, Malak Preslavets and Srebarna (Kutsarov 2010, Pehlivanov et al. 2011, our data).

Status and distribution in the study area: Locally common. In the Danube River and most of the adjacent wetlands.

Origin: Native to the Danube River region.

Ide – *Leuciscus idus* (Linnaeus, 1758)

The species was found in the Danube River (Kovatcheff 1923, Drensky 1948, 1951, Marinov 1966, 1978, Vassilev 1994, Polačik et al. 2008), Srebarna Lake (Bulgurkov 1958) and Kalimok Marsh (Kutsarov 2010). It was reported also for the lower courses of the tributaries Vidbol, Archar, Lom, Tsibritsa, Ogosta, Iskar, Vit, Osam and Yantra (Chichkoff 1939, Paspalev & Peshev 1955, Karapetkova 1972, 1994). Nowadays, it is widespread in the Bulgarian stretch of the Danube River and the lowermost courses of its tributaries, as well as in some of the riverside wetlands.

Status and distribution in the study area: Common. In the Danube River and its tributaries and some adjacent wetlands.

Origin: Native to the Danube River region.

Ziege – *Pelecus cultratus* (Linnaeus, 1758)

It has been found in the Danube River and adjacent marshlands, as well as in some of the tributaries like Iskar River (upstream to Dolni Lukovit) and Yantra River (upstream to Polsko Kosovo) (Kovatcheff 1923, Morov 1931, Chichkoff 1939, Drensky 1948, 1951).

Later, it has been reported for the whole Bulgarian stretch (Marinov 1966, 1978, Sivkov & Manolov 1978, Vassilev 1994), as well as near the mouths of Iskar and Yantra Rivers (Paspalev & Peshev 1955, Karapetkova 1972, 1994). Nowadays, it occurs in the whole Bulgarian stretch of the Danube River.

Status and distribution in the study area: Fairly common. Nowadays, only in the Danube River.

Origin: Native to the Danube River region.

Common Roach – *Rutilus rutilus* (Linnaeus, 1758)

The species was found in the Danube River and adjacent marshlands (Kovatcheff 1923, Drensky 1948, 1951, Bulgurkov 1958, Marinov 1966, 1978, Pehlivanov et al. 2011) as well as in the lower courses of the rivers Vidbol, Archar, Ogosta, Iskar, Vit, Osam, Yantra and Russenski Lom (Drensky 1948, 1951, Paspalev & Peshev 1955, Karapetkova 1972, 1994, Karapetkova & Undžian 1988). Nowadays, it is widespread and numerous in the study area.

Status and distribution in the study area: Common. In the Danube River and its adjacent marshlands and tributaries.

Origin: Native to the Danube River region.

Rudd – *Scardinius erythrophthalmus* (Linnaeus, 1758)

The species was found in the Danube River and adjacent marshlands (Kovatcheff 1923, Drensky 1948, 1951, Bulgurkov 1958, Marinov 1978, Vassilev 1994, Pehlivanov et al. 2011). It is published also for the lower courses of the rivers Vidbol, Lom, Ogosta, Iskar, Vit, Osam, Yantra and Russenski Lom (Chichkoff 1939, Paspalev & Peshev 1955, Karapetkova 1972, 1994, Karapetkova & Undžian 1988). We found the species in Danube River near Vetren (395 rkm) as well as in Orsoya and Malak Preslavets wetlands.

Status and distribution in the study area: Fairly common. In the Danube River and the adjacent wetlands.

Origin: Native to the Danube River region.

Common Chub – *Squalius cephalus* (Linnaeus, 1758)

The species was published for the Danube River (Kovatcheff 1923, Drensky 1948, 1951, Marinov 1966, 1978, Vassilev 1994). It has been found also in the lower courses of most of its tributaries (Chichkoff 1939, Drensky 1948, 1951, Paspalev & Peshev 1955, Karapetkova 1972, 1994, Karapetkova & Undžian 1988). Polačik et al. (2008) found it in the Danube River at Vidin (796 rkm) and Rousse (493 rkm). Nowadays, it is widespread and numerous in the tributaries. It occurs also in the Danube River, mainly in the western part of the Bulgarian section, but single specimens could be found also in the rest of the river.

Status and distribution in the study area: Very common in tributaries and rare in the Danube River.

Origin: Native to the Danube River region.

Common Vimba – *Vimba vimba* (Linnaeus, 1758)

The species was published for the Danube River and adjacent marshlands (Kovatcheff 1923, Drensky 1948, 1951, Marinov 1978, Vassilev 1994, Polačik et al. 2008) as well as in the lower courses of the rivers Ogosta, Iskar, Vit, Osam and Yantra (Chichkoff 1939, Drensky 1951, Paspalev & Peshev 1955, Karapetkova 1972, 1994). Recently, we found it in the Danube River near Novo selo, Koshava, Vidin, Vardim, Rousse and Vetren, as well as in the lower courses of Tsibritsa, Lom, Ogosta, Iskar, Vit and Yantra. Nowadays, it is widespread and numerous in the study area.

Status and distribution in the study area: Very common. In the Danube River and its tributaries.

Origin: Native to the Danube River region.

Grass Carp – *Ctenopharyngodon idella* (Valenciennes, 1844)

The species was reported for the Danube River by Marinov (1978) and Karapetkova (1994). According to the authors some specimens originating from riverside fishponds in Bulgaria and Romania got into the river. It is still not clear whether the species could reproduce naturally in the Danube. Kutsarov (2010) found a school of 0+ fish which entered into Srebarna Lake in 2009. We found drifted larvae of this species near the islands of Vardim (rkm 541) and Lyulyak (rkm 510) in 2013. It was also reported for Srebarna Lake by Pehlivanov et al. (2011). Nowadays the species occurs in the whole Bulgarian stretch of the river where it is regularly stocked.

Status and distribution in the study area: Fairly common in Danube River and some of the adjacent wetlands.

Origin: Alien.

Black Carp – *Mylopharyngodon piceus* (Richardson, 1846)

According to Karapetkova (1994) the species occurred in the Bulgarian stretch of the Danube River. Kutsarov (2010) reported for escaped fish from aquaculture.

Status and distribution in the study area: Unclear, most probably not existing nowadays.

Origin: Alien.

Tench – *Tinca tinca* (Linnaeus, 1758)

The species was reported for the Danube River and riverside marshlands (Kovatcheff 1923, Drensky 1948, 1951, Bulgurkov 1958). Later it has been found in the Danube River (Marinov 1978), some of the adjacent wetlands (Kutsarov 2010, Pehlivanov et al. 2011) and the lower courses of the tributaries Ogosta, Iskar, Osam and Yantra (Chichkoff 1939, Paspalev & Peshev 1955, Karapetkova 1972, 1994).

Status and distribution in the study area: Rare. Nowadays only in the Danube River and some adjacent wetlands.

Origin: Native.

Loaches Family (Cobitidae)

Balkan Spined Loach – *Cobitis elongata* (Heckel & Kner, 1858)

The species has not been found in the Bulgarian stretch of the Danube River. In the study area it is known only from the lower courses of Lom River (near the villages Vasilovtsi, Kriva Bara, Topolovets and Dabova mahala) and Vit River, near Kreta (our data).

Status and distribution in the study area: Rare. Only in the tributaries.

Origin: Native to the Danube River region.

Danubian Spined Loach – *Cobitis elongatoides* (Băcescu & Maier, 1969)

In the study area the species was published (as *C. taenia*) for the Danube River (Drensky 1928, 1948, 1951, Danailov et al. 1998, Ivanova et al. 2003) and the lower courses of its tributarics (Kovatcheff 1923, Drensky 1928, Karapetkova 1994, Karapetkova &Undžian 1988). Pehlivanov et al. (2011) found it in some of the riverside wetlands like Orsoya, Kalimok, Malak Preslavets and Srebarna. Recently, we have caught it in Danube River near Novo selo, Koshava, Vidin, Simeonovo, Oryahovo, Tutrakan, Pozharevo and Dunavets, as well as in the lower courses of the rivers Vidbol, Archar, Lom, Tsibritsa, Ogosta, Vit and Yantra.

Status and distribution in the study area: Fairly common. In the Danube River, its tributaries and some of the adjacent wetlands.

Origin: Native to the Danube River region.

Struma Spined Loach – *Cobitis strumicae* (Karaman, 1955)

This species is often mixed with *C. elongatoides* in the old literature. It is published (as *C. albicoloris* or *C. peshevi*) for the rivers Voynishka, Lom and Tsibritsa (Karapetkova 1994, Ivanova & Dobrovolov 2002, Danailov et al. 1998). Choleva et al. (2008) found it in the rivers Tsibritsa and Archar. Recently, we have caught it in the lower courses of the rivers Lom, Tsibritsa and Vit.

Status and distribution in the study area: Rare. Only in the tributaries.

Origin: Unclear, most probably introduced to the Danube River region.

Don Spined Loach – *Cobitis tanaitica* (Băcescu & Mayer, 1969)

The species was discovered in the Danube River near Koshava (811 rkm) by Choleva et al. (2008) using molecular methods. The authors found also the gynogenetic hybrid *Cobitis elongatoides* X *Cobitis tanaitica* at the same locality.

Status and distribution in the study area: Very rare. Only in the Danube River.

Origin: Native to the Danube River region.

Weather Loach – *Misgurnus fossilis* (Linnaeus, 1758)

It was reported for the Danube River and its adjacent marshlands, as well as for the lower courses of most of its tributaries – Voynishka, Iskar, Vit, Ogosta, Osam, Yantra and Russenski Lom (Kovatcheff 1923, Drensky 1928, 1948, 1951, Marinov 1966, 1978, Karapetkova 1972, 1994). It is found recently in the shoreline zone of the Danube River (Polačik et al. 2008) and some of the adjacent wetlands – Orsoya, Belene, Kalimok, Malak Preslavets Marshes and Srebarna Lake (Pehlivanov et al., 2011).

Status and distribution in the study area: Rare. In the Danube River and the adjacent wetlands.

Origin: Native to the Danube River region.

Balkan Golden Loach – *Sabanejewia balcanica* (Karaman, 1922)

In the study area the species is known only from the lower courses of the rivers Voynishka, Vidbol, Archar, Lom, Tsibritsa and Yantra (our data).

Status and distribution in the study area: Rare but locally numerous. Only in the tributaries.

Origin: Native to the Danube River region.

Bulgarian Golden Loach – *Sabanejewia bulgarica* (Drensky, 1928)

The species was described from the Danube River near Vidin (Drensky, 1928) and later was found in the whole Bulgarian section of the river from where it penetrated into the lower courses of its bigger tributaries – Ogosta, Iskar, Vit, Osam and Yantra (Drensky 1948, 1951). It was published for Danube River also by Marinov (1966, 1978), Sivkov (1991) and Vassilev (1994) as well as for the lower section of Yantra River (Karapetkova 1972). The species is recently distributed in the whole stretch of the Danube River in Bulgaria but its occurrence in the lower parts of the tributaries needs to be confirmed.

Status and distribution in the study area: Fairly common in the Danube River.

Origin: Native to the Danube River region.

Stone Loaches Family (Nemacheilidae)

Stone Loach – *Barbatula barbatula* (Linnaeus, 1758)

In the study area the species is known only from the lower courses of the rivers Voynishka, Vidbol, Archar, Lom and Tsibritsa (our data).

Status and distribution in the study area: Rare. Only in the tributaries of the Danube River.

Origin: Native to the Danube River region.

North American Catfishes Family (Ictaluridae)

Black Bullhead – *Ameiurus melas* (Rafinesque, 1820)

In Bulgaria the species was recorded for the first time in the autumn of 2013 in Srebarna Lake, where two specimens have been caught. They had a total length of 13.5 and 14.0 cm, respectively. In 2014 six young-of-the-year specimens with total length of 2.1-2.2 cm were caught by electrofishing in the discharge canal of the Malak Preslavets Marsh (rkm 414). The presence of *A. melas* at both sites was confirmed in 2015 and 2016 and nowadays the species has established self-sustaining and abundant populations in some constant oxbow lakes and wetlands along the Bulgarian sector of the Danube River (Pehlivanov ct al. 2016).

Status and distribution in the study area: Rare. In the Danube River and adjacent wetlands.

Origin: Alien.

Brown Bullhead – *Ameiurus nebulosus* (Leseur, 1819)

The species has been found in the Danube River near Vidin and Tutrakan (Kutsarov 2010). We have also found it near Pozharevo village (rkm 424) and Vidin (rkm 793) in 2013 and 2017.

Status and distribution in the study area: Rare. Probably only in the Danube River.

Origin: Alien.

Catfishes Family (Siluridae)

European Catfish – *Silurus glanis* (Linnaeus, 1758)

The species was found in the Danube River and adjacent marshlands (Kovatcheff 1923, Drensky 1948, 1951, Bulgurkov 1958, Marinov 1966, 1978, Pehlivanov et al. 2011) as well as in the lower courses of the rivers Archar, Ogosta, Iskar, Vit, Osam, Yantra and Russenski

Lom (Chichkoff 1939, Drensky 1948, 1951, Paspalev & Peshev 1955, Karapetkova 1972, 1994, Karapetkova & Undžian 1988). Nowadays, it is widespread and numerous all over the study area.

Status and distribution in the study area: Very common. In the Danube River and its adjacent marshlands and tributaries.

Origin: Native to the Danube River region.

Pikes Family (Esocidae)

European Pike – Esox lucius (Linnaeus, 1758)

The species was found in the Danube River and adjacent marshlands (Kovatcheff 1923, Drensky 1948, 1951, Bulgurkov 1958, Marinov 1966, 1978, Polačik et al. 2008, Pehlivanov et al. 2011), as well as in the lower courses of the rivers Tsibritsa, Ogosta, Iskar, Vit, Osam and Yantra (Drensky 1951, Karapetkova 1972, 1994). Nowadays, it is widespread and numerous in the study area.

Status and distribution in the study area: Very common. In the Danube River and its adjacent marshlands and tributaries.

Origin: Native to the Danube River region.

Mudminnows Family (Umbridae)

European Mudminnow – *Umbra krameri* (Walbaum, 1792)

According to Kovatcheff (1923) it was widely distributed in most of the Danube riverside lakes. After that in a period of over 80 years the species was not found in Bulgaria till it was rediscovered in Srebarna Lake in 2004 (Velkov et al. 2004).

Status and distribution in the study area: Very rare. Found only in Srebarna Lake.

Origin: Native to the Danube River region.

Trouts Family (Salmonidae)

Black Sea Trout – *Salmo labrax* (Pallas, 1814)

Drensky (1951) reported the catch of *Hucho hucho* (Linnaeus, 1758) in the Bulgarian stretch of the Danube River. Later Marinov (1978) proposed that these specimens should belong to the species *Salmo labrax*, which was known to occur near Silistra. The analysis of salmonid fish samples from the Bulgarian stretch of the river made by Holčik & Stefanov (2008) resulted in the conclusion that this river stretch is inhabited only by the Black Sea trout and not by the Huchen. The species is also found near Rousse by Vassilev & Trichkova (2007).

Status and distribution in the study area: Rare. Only in the Danube River.

Origin: Ponto-caspian relic. Native to the Danube River region.

European Brown Trout – *Salmo cf. trutta* (Linnaeus, 1758)

Solitary specimens are rarely washed from the upper courses of the tributaries downstream to the Danube River. It is unknown if they are part of the native populations or introduced. It is also not clear if they can adapt and live in the Danube River or migrate back upstream.

Distribution in the study area: Very rare.

Origin: Unclear.

Cods Family (Gadidae)

Burbot – *Lota lota* (Linnaeus, 1758)

It used to be a numerous species in the Lower Danube in the past, occurring also quite upstream in some of its tributaries – in Iskar River up to Mezdra, Roman and even Sofia and in Yantra River up to Dolna Studena as well as in its inflow Rositsa (Kovatcheff 1923, Drensky 1930, 1948, 1951, Chichkoff 1939). Later it was reported for the Danube River, Iskar River upstream to Cherven bryag and the mouth of Yantra River (Paspalev & Peshev 1955, Karapetkova 1972, 1994, Marinov 1978). Polačik et al. (2008) found the species in the Danube River near Novo selo (rkm 833), Yasen (rkm 825), Koshava (rkm 811), Vidin (rkm 791), Simeonovo (rkm 776) and Tutrakan (rkm 433). Nowadays it is distributed only in the Danube River where its populations are growing in numbers.

Status and distribution in the study area: Fairly common. Only in the Danube River.

Origin: Native to the Danube River region.

Livebearers Family (Poecilidae)

Eastern Mosquitofish – *Gambusia holbrooki* (Girard, 1859)

Polačik et al. (2008) reported the first record of the species from the Lower Danube, since it was not reported from the Romanian section (Navodaru & Nastase 2006) neither from the Danube Delta and adjacent lakes (Navodaru et al. 2002; Otel 2007). They found the species near Belene (rkm 573), Batin (rkm 526), Rousse (rkm 495) and Ryahovo (rkm 466). According to Kutsarov (2010) the species was introduced to Antimovo Reservoir in the 1990-ies through water pump station using water from the Danube River. Eastern mosquitofish occurs also in the channel system of Belene Marsh and Srebarna Lake (Kutsarov 2010).

Distribution in the study area: Rare. In the Danube River and some adjacent wetlands.

Origin: Alien.

Sticklebacks Family (Gasterosteidae)

Threespine Stickleback – *Gasterosteus aculeatus* (Linnaeus, 1758)

It was reported for the Danube River marshes only by Kovatcheff (1923). Polačik et al. (2008) provided the first record of the species in the main Bulgarian Danube channel. They found the species near Koshava (rkm 811) and Mechka (rkm 516). The Threespine stickleback is reported for the Lower Danube also by Kutsarov (2010).

Status and distribution in the study area: Rare. Only in the Danube River.

Origin: Native to the Danube River region.

Ninespines Stickleback – *Pungitius platygaster* (Kessler, 1859)

It was reported as numerous species in the Lower Danube marshes (Kovatcheff 1923). Later it was reported for the Danube River (Marinov 1978) and its tributary Vidbol River, near Dunavtsi (Karapetkova 1994). Recently, it has been found in some riverside wetlands – Kalimok, Malak Preslavets and Srebarna (Pehlivanov 2000, Kutsarov 2010, Pehlivanov et al. 2011). We found the species in the main channel near Nova Cherna, the outflow channel of Malak Preslavets Marsh (rkm 414), as well as in the Danube River near Tutrakan (rkm 432).

Status and distribution in the study area: Rare. In the Danube River and adjacent wetlands.

Origin: Native to the Danube River region.

Pipefishes Family (Syngnathidae)

Shore Pipefish – *Syngnathus abaster* (Risso, 1827)

The species has been found in the Bulgarian stretch of the Danube River for the first time in 1982 near Silistra (Karapetkova 1994). Later, Karapetkova et al. (1998) reported it as rare in the Danube. Polačik et al. (2008) found it in various locations between Vrav (840 rkm) and Silistra (376 rkm). Nowadays, the species is widespread and numerous in the whole Bulgarian section of the river, especially in the shoreline zone.

Status and distribution in the study area: Common in the Danube River.

Origin: Native to the Black Sea from where it recently penetrated in the Lower Danube.

Sunfishes Family (Centrarchidae)

Pumpkinseed – *Lepomis gibbosus* (Linnaeus, 1758)

The species was found for the first time in the Svishtov Marsh (Drensky 1923). Later, it was reported also for Belene Marsh and Srebarna Lake (Drensky 1948, 1951, Bulgurkov 1958). It was published for the Danube River (Drensky 1948, Marinov 1966, 1978, Polačik et al. 2008), adjacent marshlands (Pehlivanov et al. 2011) and the lower courses of its tributaries (Paspalev & Peshev 1955, Karapetkova 1972, 1994). Nowadays, it is common in Danube River, adjacent marshlands and most of the tributaries.

Status and distribution in the study area: Common. In the Danube River and its adjacent marshlands and tributaries.

Origin: Alien.

Perches Family (Percidae)

Danube Ruffe – *Gymnocephalus baloni* (Holčik & Hensel, 1974)

The species has been found for the first time in the Bulgarian stretch of the Danube River near Archar, Lom, Svishtov and Tutrakan as well as in the mouth of Yantra River near Krivina village (Sivkov, 1985). Later it has been reported for the whole section of Danube River and lower courses of the tributaries Ogosta, Iskar, Vit, Osam, Yantra and Russenski Lom (Karapetkova 1994). Polačik et al. (2008) found it in Danube River near Vrav (rkm 840), Novo selo (rkm 833), Vidin (rkm 791), Stanevo (rkm 724), Marten (rkm 482) and Sandrovo (rkm 477). Nowadays, it occurs in the whole Bulgarian section.

Status and distribution in the study area: Fairly common. Only in the main course of the Danube River.

Origin: Native to the Danube River region.

Ruffe – *Gymnocephalus cernua* (Linnaeus, 1758)

The species is well known from the Bulgarian stretch of Danube River and its adjacent wetlands (Kovatcheff 1923, Drensky 1948, 1951, Marinov 1966, 1978, Sivkov 1988a, Pehlivanov 2011). It is found also in the lower courses of the rivers Vidbol, Lom, Ogosta, Iskar, Vit, Osam, Yantra and Russenski Lom (Chichkoff 1939, Paspalev & Peshev 1955, Karapetkova 1994). According to Karapetkova (1972) in Yantra River it reached upstream to Polski Trambesh. From the 1980-ies the Ruffe is introduced in many reservoirs in different parts of the country and recently it is established in many wetlands near the Danube River in North Bulgaria.

Status and distribution in the study area: Rare in the Danube River. Numerous in the reservoirs in North Bulgaria where it was introduced.

Origin: Native only to the Danube River and its tributaries. Introduced in some reservoirs in the study area.

Striped Ruffe – *Gymnocephalus schraetser* (Linnaeus, 1758)

The species was reported from the whole Bulgarian section of Danube River and Iskar River upstream to Dolni Lukovit (Kovatcheff 1923, Chichkoff 1939, Drensky 1948, 1951). Later it was confirmed for the Danube River and reported also for the lower courses of its tributaries – Vidbol, Lom, Ogosta, Iskar, Vit, Osam and Yantra (Chichkoff 1939, Paspalev & Peshev 1955, Marinov 1966, Karapetkova 1972, Sivkov 1986, Karapetkova 1994). According to Karapetkova (1972) the species is distributed in Yantra River upstream to Polsko Kosovo. Polačik et al. (2008) found it in Danube River between Vrav (rkm 836) and Vetren (rkm 395). Nowadays, the species occurs in the whole Bulgarian section of the Danube River.

Status and distribution in the study area: Common. Only in the main course of the Danube River.

Origin: Native to the Danube River region.

European Perch – *Perca fluviatilis* (Linnaeus, 1758)

The species was found in the Danube River and adjacent marshlands (Kovatcheff 1923, Drensky 1948, 1951, Bulgurkov 1958, Marinov 1966, 1978, Vassilev 1994, Polačik et al. 2008, Pehlivanov et al. 2011) as well as in the lower courses of the rivers Vidbol, Lom, Tsibritsa, Ogosta, Iskar, Vit, Osam, Yantra and Russenski Lom (Karapetkova 1972, 1994, Karapetkova & Undžian 1988). Nowadays, it is widespread and numerous all over the study area.

Status and distribution in the study area: Very Common. In the Danube River and its adjacent marshlands and tributaries.

Origin: Native to the Danube River region.

Common Pikeperch – *Sander lucioperca* (Linnaeus, 1758)

The species was reported for the Danube River and adjacent marshlands (Kovatcheff 1923, Drensky 1948, 1951, Marinov 1966, 1978, Vassilev 1994, Polačik et al. 2008, Pehlivanov et al. 2011), as well as in the lower courses of the tributaries Tsibritsa, Ogosta, Iskar, Vit, Osam, Yantra and Russenski Lom (Chichkoff 1939, Paspalev & Peshev 1955, Karapetkova 1972, 1994, Karapetkova & Undžian 1988). Nowadays, it is widespread and numerous in the Danube River and its tributaries.

Status and distribution in the study area: Common. In the Danube River and its tributaries.

Origin: Native to the Danube River region.

Volga Pikeperch – *Sander volgensis* (Gmelin, 1789)

In the past, the species was found in the Danube River from Vidin downstream to Silistra (Kovatcheff 1923, Drensky 1948, 1951, Marinov 1978, Sivkov 1987a, Vassilev 1994) and rarely in the lower course of Yantra River (Karapetkova 1972). Nowadays, it is a rare species, mostly found in the Eastern part of the Bulgarian stretch of the Danube River (Kutsarov 2010).

Satus and distribution in the study area: Rare. Only in the main course of the Danube River.

Origin: Native to the Danube River region.

Streber – *Zingel streber* (Siebold, 1863)

The species was reported for the Danube River and some of its biggest tributaries – Iskar, Vit, Osam and Yantra (Kovatcheff 1923, Chichkoff 1939, Drensky 1948, 1951, Karapetkova 1972, 1994, Marinov 1966, 1978, Sivkov 1987b, Vassilev, 1994). It used to migrate much upstream in the tributaries in the past – it has been found in Iskar River near Mezdra, Roman (Chichkoff 1939) and Karlukovo (Paspalev & Peshev 1955). Nowadays, it is distributed only in the main course of the Danube River in the whole Bulgarian stretch.

Status and distribution in the study area: Fairly common. Only in the Danube River.

Origin: Native to the Danube River region.

Zingel – *Zingel zingel* (Linné, 1766)

The species has been reported for the Danube River and some of its tributaries like Iskar, Vit, Osam and Yantra (Kovatcheff 1923, Chichkoff 1939, Drensky 1948, 1951, Paspalev & Peshev 1955, Karapetkova 1972, 1994, Marinov 1966, 1978). Nowadays, it occurs only in the main course of the Danube River in the whole Bulgarian stretch.

Status and distribution in the study area: Fairly common. Only in the Danube River.

Origin: Native to the Danube River region.

Freshwater Sleepers Family (Odontobutidae)

Amur Sleeper – *Perccottus glenii* (Dybowski, 1877)

In Bulgaria the species has been found for the first time in Danube River in the section between Vrav and Lom (rkm 840-744) in 2005 (Jurajda et al. 2006). Later, it was found also in the Danube River near the village of Dunavets (rkm 423) as well as in Kalimok-Brushlen Protected area and the channels around Srebarna Lake (Kutsarov 2010). Pehlivanov et al. (2011) reported it for Malak Preslavets Marsh and Srebarna Lake. Recently it has been found in the Danube River near Vidin (rkm 795) and adjacent channel systems and fishponds (Trichkova & Stefanov 2017) as well as in Danube River near Vardim (our data).

Status and distribution in the study area: Locally common. In the Danube River and adjacent wetlands.

Origin: Alien.

Gobies Family (Gobiidae)

Brauner's Tadpole Goby – *Benthophiloides brauneri* (Beling & Iljin, 1927)

The species was found in 2011 in the Bulgarian stretch of the Danube River at rkm 422, East of the town of Tutrakan, near Dunavets Village (Stefanov & Kutsarov 2018). Later, in 2018 it was found also in the Danube River on rkm 395 near the village of Vetren (V. Koev, pers. comm.)

Status and distribution in the study area: Very rare. Only in the Danube River.

Origin: Ponto-caspian relic. Native to the Lower Danube River.

Black Sea Tadpole Goby – *Benthophilus nudus* (Berg, 1898)

It was reported (as *B. stellatus*) as a rare species in the Danube River, near Svishtov and Nikopol (Drensky 1924, 1948, 1951). Later it was found also near Vidin, Simeonovo, Baikal, Svishtov, Rousse and Silistra (Gheorghiev 1966, Marinov 1966, Michajlowa 1967, Vassilev 1994). Polačik et al. (2008) discovered it in various localities in Danube River between Vidin (rkm 796) and Vetren (rkm 395). We found the species also near Vardim (rkm 546) and Rousse (rkm 496).

Status and distribution in the study area: Unknown, probably fairly common. Only in the Danube River.

Origin: Ponto-caspian relic. Native to the Lower Danube River.

Pontian Monkey Goby – *Neogobius fluviatilis* (Pallas, 1814)

Kovatcheff (1923) found the species in the adjacent marshlands along Danube River. According to Drensky (1948, 1951) the species is distributed in the Danube River and the lower courses of the rivers Ogosta, Iskar, Vit, Osam and Yantra. Later, the species was found in Danube River (Gheorghiev 1966, Marinov 1966, 1978, Vassilev 1994, Polačik et al. 2008, Kutsarov 2010, Szalóky et al. 2015), some of the adjacent wetlands (Bulgurkov 1958, Pehlivanov 2011) and its tributaries (Paspalev & Peshev 1955, Karapetkova 1972, 1994, Zarev et al. 2013). Nowadays, it is widespread and numerous species in the whole Bulgarian stretch of the Danube River, as well as in some of the adjacent wetlands and tributaries.

Status and distribution in the study area: Very common. In the Danube River, adjacent wetlands and tributaries.

Origin: Ponto-caspian relic. Native to the Lower Danube River.

Racer Goby – *Neogobius gymnotrachelus* (Kessler, 1857)

The species was found in the Bulgarian stretch of the Danube River (Drensky 1948, Gheorghiev 1966, Marinov 1978, Vassilev 1994, Polačik et al. 2008, Kutsarov 2010, Szalóky et al. 2015). It is reported also for some of the tributaries – Archar, Vit, Yantra and Russenski Lom (Karapetkova 1994, Karapetkova & Undžian 1988, Zarev et al. 2013). We found it also in Danube River and the tributaries Ogosta, Vit and Yantra.

Status and distribution in the study area: Common. In the Danube River and some of the tributaries.

Origin: Ponto-caspian relic. Native to the Lower Danube River.

Pontian Bighead Goby – *Neogobius kessleri* (Günther, 1861)

The species was found in the Danube River (Drensky 1948, 1951) and the adjacent marshlands along it (Kovatcheff 1923). Later, it was reported for the Danube River (Gheorghiev 1966, Marinov 1966, 1978, Vassilev 1994; Polačik et al. 2008, Kutsarov 2010) and the lower courses of some of the tributaries like Ogosta, Vit, Osam and Yantra (Gheorghiev 1966, Karapetkova 1972, 1994, Zarev et al. 2013). Nowadays, the species is widespread in the whole Bulgarian stretch of the Danube River and the lower courses of some of the tributaries.

Status and distribution in the study area: Fairly common. In the Danube River and its tributaries.

Origin: Ponto-caspian relic. Native to the Lower Danube River.

Round Goby – *Neogobius melanostomus* (Pallas, 1814)

The species was reported for the first time for the Bulgarian stretch of the Danube River by Drensky (1948, 1951). He found it also in the lower course of Vit River. Gheorghiev (1966) published it only for the Bulgarian Black Sea coast and Marinov (1978) doubted that the Round goby is distributed in the Danube River. The species invaded Lower Danube in the 1990-ies and recently has been found in the Danube River (Vassilev 1994, Polačik et al. 2008, Kutsarov 2010, Szalóky et al. 2015), some of the adjacent wetlands (Pehlivanov 2011) and the tributaries Archar, Ogosta, Iskar, Vit and Yantra (Zarev et al. 2013). Nowadays, it is widespread and one of the most numerous fish species in the whole Bulgarian stretch of the Danube River, as well as in some of the adjacent wetlands and tributaries.

Status and distribution in the study area: Very common. In the Danube River, adjacent wetlands and tributaries.

Origin: Ponto-caspian relic. Recently invaded the Lower Danube River.

Western Tubenose Goby – *Proterorhinus semilunaris* (Heckel, 1837)

The species was found in the Danube River and its adjacent marshlands (Kovatcheff 1923, Drensky 1948, 1951, Gheorghiev 1966, Marinov 1966, 1978). Recently, it was found also in the Danube River (Vassilev 1994, Polačik et al. 2008, Kutsarov 2010), some of the adjacent wetlands (Pehlivanov 2011) and tributaries (Karapetkova 1994, Zarev et al. 2013). Nowadays, the species is widespread and numerous in the whole Bulgarian stretch of the Danube River, as well as in some of the adjacent wetlands and tributaries.

Status and distribution in the study area: Very common. In the Danube River, adjacent wetlands and tributaries.

Origin: Ponto-caspian relic. Native to the Lower Danube River.

The obtained results show that a total of 83 species belonging to 21 families have been found in the Bulgarian stretch of Danube River, adjacent wetlands and the lowermost courses of its tributaries.

From the reported species, 69 are considered native to the region. This total includes the Ponto-Caspian gobies from genus *Neogobius*, some of which recently invaded the Danube River, as well as the Shore pipefish (*S. abaster*). The status of two species (*A. anguilla* and *S. cf. trutta*) is unclear (Table 1). Eight species are considered endemic in the Danube River basin: *G. obtusirostris*, *R. antipai*, *R. uranoscopus*, *Alburnoides* sp. 2, *A. danubicus*, *C. elongata*, *S. bulgarica* and *G. schraetser*.

A total of 12 species found in the region are not typical for the Danube River ichthyofauna (Table 1). Most of them were hatched and reared in fish ponds and introduced intentionally into the area with different purposes including consumption, sport fishing, control of phytoplankton and aquatic vegetation.

Several species have been erroneously reported in the ichthyological literature for the Bulgarian stretch of the Danube River. From the fourth reported species of lampreys (*L. planeri*, *L. fluviatilis*, *E. danfordi* and *E. mariae*) Stefanov & Holčik (2007) proved that only the species *E. mariae* inhabits the Bulgarian part of the river. The Huchen *Hucho hucho* (Linnaeus, 1758) was reported by Drensky (1948, 1951), but later Marinov (1978) doubted that this species occurs in the Lower Danube. Holčik & Stefanov (2008) analyzed salmonid fish samples from the Bulgarian stretch of the river and stated that they all belong to the species *S. labrax*. Kovatcheff (1923) included the species *Neogobius constructor* (Nordmann, 1840) (as *Gobius cephalarges*) in the Bulgarian ichthyofauna and stated that it is distributed in the Danube River. Later, Drensky (1951) and Marinov (1966, 1978) published it again (as *G. cephalarges constructor*) for the Danube River and its tributaries Vit and Osam. Marinov (1966) even reported the catch of two specimens near the mouth of Topolovets River at rkm 784 but did not provide any data about their morphology. The presence of this species was not proved in any of the following surveys and its presence in Lower Danube is considered unlikely now. Drensky (1951) reported the presence of the species *Rutilus meidingeri* (Heckel, 1851) (as *R. frisii meidingeri*) in the Danube River near

Svishtov, which was later repeated by Marinov (1978). There are no preserved specimens or any other type of evidence to prove this find; as a result the species is excluded from the Bulgarian ichthyofauna due to the long distance of its habitats in the Upper Danube.

Drensky (1948) published the species *Rutilus heckelii* (Nordmann, 1840) (as *R. rutilus heckeli*) for the western part of the Bulgarian stretch of the Danube River. In his later work (Drensky 1951) he again published this species, but this time for the Danube River near Silistra and Srebarna Lake, located in the easternmost part of the Bulgarian Danube. This species was listed also by Marinov (1978). Although, there are no preserved specimens or any other type of evidence, the presence of this species in Bulgarian waters is not impossible since it occurs in the area of the Danube Delta. Same is valid also for the species *Alburnus sava*, *Leuciscus leuciscus* (Linnaeus, 1758) and *Rutilus virgo* (Heckel, 1852), which occur in the Middle Danube. Although they are still not recorded in the Bulgarian stretch of the Danube River, their presence could be expected if some specimens were washed downstream by high waters. The Siberian sturgeon (*Acipenser baerii* Brandt, 1869) is also a potential species for the Lower Danube. It was stocked in the Middle and Upper Danube and later was found in the Slovak-Hungarian section of the river in the period 2005-2012 (Masár et al. 2006, Farský et al. 2013). Another potential species for the Lower Danube is the introduced *Ictalurus punctatus* (Rafinesque, 1818), which is widely stocked in Europe.

Some of the fish species are not permanent inhabitants of the Danube and occur in the study area only temporary. These are the anadromous shads (*A. immaculata* and *A. tanaica*) and all species of sturgeons (except *A. ruthenus*). They live in the Black Sea and migrate upstream for spawning in the Danube River. The only catadromous fish in the area is *A. anguilla*, but its status in Lower Danube is unclear. It is still not proven that the adult specimens caught in the area, originated from the Atlantic Ocean.

There are some species of high conservation importance among the indigenous fish in the Bulgarian stretch of the Danube River (Table 1). The species *R. antipai* is included in the category "Extinct" of the IUCN Red List of Threatened Species. Another 6 species are included in the List under the Critically Endangered category. These are all species of sturgeons (except *A. ruthenus*) and *A. anguilla*. The species *A. sarmaticus* is listed in the category "Endangered" till the species *A. ruthenus*, *A. immaculata*, *C. carpio* and *U. krameri* are included in the category "Vulnerable." A total of 43 species are included in the different categories of the Red Data Book of Bulgaria. Twenty nine species are included in the Biodiversity Protection Law (2002). A total of 38 species are included in Annexes II and III of the Bern Convention and 31 species in Annexes II, IV and V of the Council Directive 92/43 (Table 1).

Table 1. List of fish species recorded in the study area and their conservation status

Family/species	Origin	Conservation status				
		IUCN	RDB	BPL	BERN	Dir 92/43
Petromyzontidae						
Eudontomyzon mariae	Native	LC	CR	2	III	II
Acipenseridae						
Acipenser gueldenstaedtii	P-C relic	CR	CR	4		V
Acipenser nudiventris	P-C relic	CR	EX	3		V
Acipenser ruthenus	P-C relic	VU	EN	4	III	V
Acipenser stellatus	P-C relic	CR	CR	4	III	V
Acipenser sturio	Native	CR	EX	3	II	II, IV
Huso huso	P-C relic	CR	CR	4	II, III	V
Polyodontidae						
Polyodon spathula	Alien					
Anguillidae						
Anguilla anguilla	Unclear	CR	EN			
Clupeidae						
Alosa immaculata	P-C relic	VU	VU	2, 4	III	II, V
Alosa tanaica	P-C relic	LC	VU	2, 4		II, V
Cyprinidae						
Rhodeus amarus	Native	LC		2	III	II
Gobio obtusirostris	Native	LC				
Pseudorasbora parva	Alien					
Romanogobio antipai	Native	EX				
Romanogobio kesslerii	Native	LC	EN	2	III	II
Romanogobio uranoscopus	Native	LC	EN	2	III	II
Romanogobio vladykovi	Native	LC	VU	2	III	II
Barbus balcanicus	Native	LC				V
Barbus barbus	Native	LC	VU			V
Barbus petenyi	Native	LC	DD	2, 4		V
Carassius carassius	Native	LC	EN			

Table 1. (Continued)

Family/species	Origin	Conservation status				
		IUCN	RDB	BPL	BERN	Dir 92/43
Carassius gibelio	Alien					
Cyprinus carpio	Native	VU	CR			
Abramis brama	Native	LC				
Alburnoides sp. 2	Native	LC	DD		III	
Alburnus alburnus	Native	LC				
Alburnus danubicus	Native	NE	EN	2	III	II
Alburnus sarmaticus	Native	EN	EN	2	III	II
Aspius aspius	Native	LC	VU	2, 4	III	II, V
Ballerus ballerus	Native	LC	DD		III	
Ballerus sapa	Native	LC	DD		III	
Blicca bjoerkna	Native	LC				
Chondrostoma nasus	Native	LC			III	
Hypophthalmichthys molitrix	Alien					
Hypophthalmichthys nobilis	Alien					
Leucaspius delineatus	Native	LC	VU		III	
Leuciscus idus	Native	LC				
Pelecus cultratus	Native	LC	VU	2, 4	III	II, V
Rutilus rutilus	Native	LC				
Scardinius erythrophthalmus	Native	LC				
Squalius cephalus	Native	LC				
Vimba vimba	Native	LC	DD		III	
Ctenopharyngodon idella	Alien					
Mylopharyngodon piceus	Alien					
Tinca tinca	Native	IUCN	RDB	BPL	BERN	Dir 92/43
Cobitidae						
Cobitis elongata	Native	LC	DD	2	III	II
Cobitis elongatoides	Native	LC		2	III	II
Cobitis strumicae	Native	LC		2		
Cobitis tanaica	Native	LC			III	II
Misgurnus fossilis	Native	LC	EN	2	III	II
Sabanejewia balcanica	Native	LC	VU	2	III	II
Sabanejewia bulgarica	Native	LC	VU	2	III	II

Family/species	Origin	Conservation status				
		IUCN	RDB	BPL	BERN	Dir 92/43
Nemacheilidae						
Barbatula barbatula	Native	LC	VU			
Ictaluridae						
Ameiurus melas	Alien					
Ameiurus nebulosus	Alien					
Siluridae						
Silurus glanis	Native	LC			III	
Esocidae						
Esox lucius	Native	LC				
Umbridae						
Umbra krameri	Native	VU	CR	2	II	II
Salmonidae						
Salmo labrax	Native	LC	CR			
Salmo cf. trutta	Unclear	LC				
Gadidae						
Lota lota	Native	LC	VU			
Poecilidae						
Gambusia holbrooki	Alien					
Gasterosteidae						
Gasterosteus aculeatus	Native	LC	VU			
Pungitius platygaster	Native	LC	CR		III	
Syngnathidae						
Syngnathus abaster	Native	LC			III	
Centrarchidae						
Lepomis gibbosus	Alien					
Percidae						
Gymnocephalus baloni	Native	LC	VU	2, 3	III	II, IV
Gymnocephalus cernua	Native	LC				
Gymnocephalus schraetser	Native	LC	VU	2, 4	III	IV, V
Perca fluviatilis	Native	LC				
Sander lucioperca	Native	LC				
Sander volgensis	Native	IUCN	RDB	BPL	BERN	Dir 92/43
Zingel streber	Native	LC	EN	2	III	II
Zingel zingel	Native	LC	EN	2, 4	III	II, V
Odontobutidae						
Perccottus glenii	Alien					
Gobiidae						
Benthophiloides brauneri	P-C relic	DD	EX			

Table 1. (Continued)

Family/species	Origin	Conservation status				
		IUCN	RDB	BPL	BERN	Dir 92/43
Benthophilus nudus	P-C relic	LC	VU			
Neogobius fluviatilis	P-C relic	LC			III	
Neogobius gymnotrachelus	P-C relic	LC				
Neogobius kessleri	P-C relic	LC			III	
Neogobius melanostomus	P-C relic	LC				
Proterorhinus semilunaris	P-C relic	LC			III	

Legend: P-C relic – Ponto-Caspian relic species; IUCN – IUCN Red List of Threatened Species (European Red List): EX – extinct, CR – critically endengared, VU – vulnerable, LC – least concern, NE – not evaluated; RDB – Red Data Book of Bulgaria: EX – extinct, CR – critically endengared, VU – vulnerable, DD – data deficient; BPL – Biodiversity Protection Law: Annexes 2, 3 & 4; BERN – Bern Convention: Annexes II & III; Dir 92/43 – European Council Directive 92/43: Annexes II, IV & V.

References

Bănărescu P. 1964. *Fauna R. P. R. Pisces – Osteichthyes*. Ed. Academiei R. P. R., Bucuresti, 13, 959 p.

Berg L. S. 1961. *Selected works, Ichthyology*. IV, Leningrad, 746 pp. (In Russian).

Bulgurkov K., 1958. Hydrological peculiarities of the reserve of Lake Srebarna and composition of its fish fauna. *Bulletin de l'Institut zoologique de l'Académie des sciences de Bulgarie* [*Bulletin of the Institute of Zoology, Bulgarian Academy of Sciences*] 7: 251–268. (In Bulgarian).

Busnita T. 1961. Die Wandlungen der Fischfauna der unteren Donau während der letzten hundert Jahre [Changes in the fish fauna of the Lower Danube during the last hundred years]. *Verhandlungen Internationale Vereinigung für Theoretische und Angewandte Limnologie* [*International Association for Theoretical and Applied Limnology*] 14: 114-126.

Chichkoff G. 1929. Sur un nouveau poisson du genre *Gobio* Cuvier, *Gobio similis* n. sp. [On one new species from genus *Gobio* Cuvier, *Gobio similis* n. sp.]. *Annuaire de l'Universite de Sofia, Faculte physico-mathematique* [*Annual of Sofia University, Faculty of Physics and Mathematics*] 25 (3): 158-174. (In Bulgarian).

Chichkoff G. 1937. [Fishes of Vit River]. *Ribarski pregled [Fisheries review]* 7 (5): 66-68. (In Bulgarian).

Chichkoff G. 1939. [On the fisheries of Iskar River]. *Ribarski pregled [Fisheries review]* 9 (8): 4-7. (In Bulgarian).

Choleva L., Apostolou A., Ráb P. & Janko K. 2008. Making it on their own: sperm-dependent hybrid fishes (*Cobitis*) switch the sexual hosts and expand beyond the ranges of their original sperm donors. *Philosophical Transactions of the Royal Society* 363: 2911-2919.

Danailov M., Terzieva P. & Dobrovolov I. 1998. Electrophoretical survey of *Cobitis taenia, Cobitis peshevi* and their hybrid from some Bulgarian rivers. *Acta zoologica bulgarica* 50 (2-3): 127-132.

Drensky P. 1921. [The fish fauna of Iskar River and the fisheries in it]. *Estestvoznanie i geografija [Natural history and geography]* 6 (2/3): 49-58. (In Bulgarian).

Drensky P. 1923. [*Eupomotis gibbosus* L. one new fish in the Lake Svishtovsko]. *Svedeniya po zemledelieto [Reports on agriculture]* 11: 6-7. (In Bulgarian).

Drensky P. 1924. *Benthophilus macrocephalus* - new fish species for the Danube River. *Travaux de la Societe Bulgare des sciences naturelles, Sofia [Proceedings of the Bulgarian Society for Natural History, Sofia]* 11: 1-4. (In Bulgarian).

Drensky P. 1928. Die Fische der Familie Cobitidae in Bulgarien [Fishes from the family Cobitidae in Bulgaria]. *Mitteilungen aus den Königlichen naturwissenschaftlichen Instituten in Sofia-Bulgarien, Bulletin des Institutions royales d'histoire naturelle à Sofia-Bulgarie [Bulletin of the Royal Institute of Natural Sciences in Sofia-Bulgaria]* 1: 156-181. (In Bulgarian).

Drensky P. 1930. Zur Kenntnis des Süsswasserfischfauna Bulgariens [On the knowledge of the freshwater fish fauna of Bulgaria]. *Zoologische Jahrbücher, Abteilung für Systematik, Ökologie und Geographie der Tiere [Zoological Yearbooks, Department of Systematics, Ecology and Geography of Animals]* 59: 63-680.

Drensky P. 1932. [Fishes and fisheries of Osam River]. *Lovech i Lovchansko [Lovech and its districts]* 4: 171-200. (In Bulgarian).

Drensky P. 1943. *Chalcalburnus chalcoides* Güld. in Bulgarien. *Annulaire de l'Universite de Sofia, Faculte de Sciences [Annual of Sofia University, Faculty of Sciences]* 39 (3): 343-360. (In Bulgarian).

Drensky P. 1948. Synopsis and distribution of fishes in Bulgaria. *Annulaire de l'Université de Sofia, Faculté de Sciences [Annual of Sofia University, Faculty of Sciences]* 44 (3): 11-71. (In Bulgarian).

Drensky P. 1951. [*Fishes of Bulgaria*]. Fauna of Bulgaria. II. Sofia, BAS, 270 p. (In Bulgarian).

Farský M., Hajdú J., Pekárik L. & Kautman J. 2013. On the occurrence of the Siberian sturgeon (*Acipenser baerii* Brandt, 1869) in Slovak-Hungarian section of the Danube. *Pisces Hungarici* 7: 139-140.

Gheorghiev J. 1966. Composition d'especé et caracteristique des Gobiides (Pisces) en Bulgarie [Species composition and characteristics of the Gobiidae (Pisces) in Bulgaria]. *Proceedings of the Research Institute of Fisheries and Oceanography* – Varna 7: 159-228. (In Bulgarian).

Holčik J. & Stefanov T. 2008. Taxonomic status of salmonids in the Bulgarian stretch of the Danube River and their bionomic strategy. *Journal of Applied Ichthyology* 24 (5): 605-609.

Honţ Ş., Paraschiv M., Iani M. I., Stefanov T., Lendhardt M. & Oprea L. 2018. Long Distance Migration of Beluga (*H. huso*) and Stellate Sturgeon (*Acipenser stellatus*) in Lower Danube River in Relation with Iron Gate II Dam. *Bulletin UASVM, Animal Science and Biotechnologies* 75 (1): 28-39.

Ivanova P. & Dobrovolov I. 1999. Morphological and electrophoretical comparision of some representatives of genus *Cobitis* Linne (Pisces). *Comptes rendus de l'Académie bulgare des Sciences* [*Proceedings of the Bulgarian Academy of Sciences*] 52 (11-12): 79-82.

Ivanova P., Erk'akan F., Ozeren S. C. & Dobrovolov I. 2006. Biochemical-genetic Comparison of *Cobitis simplicispina* from Turkey, and *C. taenia*, *C. albicoloris* and *C. elongata* from Bulgarian Freshwaters. *Folia biologica* 51 (Supplement): 79-84.

Jurajda P., Vassilev M., Polacik M. & Trichkova T. 2006. A first record of *Perccottus glenii* (Perciformes: Odobtobutidae) in the Danube River in Bulgaria. *Acta zoologica bulgarica* 58 (2): 279-282.

Karapetkova M. 1972. L'ichthyofaune de la rivière Yantra [The Ichthyofauna of Yantra River]. *Bulletin de l'Institut de Zoologie et Museé, Academie Bulgare des Sciences* [*Bulletin of the Institute of zoology with museum, Bulgarian Academy of Sciences*] 36: 149-182. (In Bulgarian).

Karapetkova M. 1979. Ubersicht der Ichthyofauna der bulgarischen Donauzuflusse [Overview of the ichthyofauna of the Bulgarian Danube River tributaries]. In: *Limnologische Berichteder* 21. Tagung der IAD. Novi Sad, Yugoslavia, pp. 378–384.

Karapetkova M. 1994. Vertebrate Animals. In: Roussev B. (Ed.). *Limnologie der Bulgarischen Donauzuflüsse* [*Limnology of the Bulgarian Danube River tributaries*]. MEW, BAS, Sofia, 175-186. (In Bulgarian).

Karapetkova M. 1998. Class Pisces (Fishes). In: Michev T., Georgiev B., Petrova A. & Stoyneva M. (eds). *Biodiversity of the Srebarna Biosphere Reserve. Checklist and bibliography.* Context & Pensoft, Sofia: 84–85.

Karapetkova M. & Undžian E. 1988. The ichthyofauna of the river valley of the Roussenski Lom River. *Hydrobiology* 32: 44-50. (In Bulgarian).

Karapetkova M. & Živkov M. 1995. [*The fishes of Bulgaria*]. Gealibris, Sofia, 246 p. (In Bulgarian).

Karapetkova M., Zivkov M. & Alexandrova-Kolemanova K. 1998. The freshwater fishes of Bulgaria. – In: C. Meine (ed.). *Bulgaria's Biological Diversity: Conservation Status and Needs Assessment*. Vol. 1 and 2. Pensoft Publishers, 375-393.

Kolarov P. 1960. [One rare fish finding of the Danube River]. *Priroda* [*Nature*] 3: 70-71. (In Bulgarian).

Kolarov P. 1963. Biological notes on Pontic shad caught along Bulgarian Danube River shore in 1962. *Ribno Stopanstvo* [*Fish farming*] 8 (In Bulgarian).

Kottelat M. & Freyhof J. 2007. Handbook of European Freshwater Fishes. Kottelat, Cornol, Switzerland and Freyhof, Berlin, Germany, 646 pp.

Kottlík P., Tsigenopoulos C., Ráb P. & Berrebi P. 2002. Two new *Barbus* species from the Danube River basin, with redescription of *B. petenyi* (Teleostei: Cyprinidae). *Folia zoologica* 51 (3): 227-240.

Kovatcheff V. 1909. Contribution a la ichthyofaune bulgare [Contribution to the Bulgarian ichthyofauna]. *Annuaire du gymnase de garcons de Rousse* [*Annual of the Men's High School in Rousse*]: 3-7. (In Bulgarian).

Kovatcheff V. 1923. Faune des poissons d'eau douce en Bulgarie [Freshwater fish fauna of Bulgaria]. *Archives du Ministère de l'Agriculture et des Domaines du Royaume de Bulgarie* [*Archives of the Ministry of Agriculture and Domains of the Kingdom of Bulgaria*] 3: 1-164. (In Bulgarian).

Kutsarov Y. 2010. Changes in the species composition and number in the ichthyology of the Danube River. In: Atanasov V. (ed.). *The river and the time*. Tutrakan, 2010: 259-286. (In Bulgarian).

Ludwig A., Lippold S., Debus L. & Reinartz R. 2009. First evidence of hybridization between endangered sterlets (*Acipenser ruthenus*) and exotic Siberian sturgeons (*Acipenser baerii*) in Danube River. *Biological Invasions* 11 (3): 753-760.

Manolov Ž. & Sivkov J. 1977. Morphometrische charakteristik eines fur die Bulgarische ichthyofauna neuen fisches Pseudorasbora parva (Schlegel) [Morfometric characteristics of one new fish for the Bulgarian ichthyofauna Pseudorasbora parva (Schlegel)]. *Proceedings of the National Museum, Varna* 13 (28): 219-222. (In Bulgarian).

Marinov B. 1966. Über die fischfauna des Bulgarischen Donausectors [Ichthyofauna of the Bulgarian sector of the Danube River]. *Bulletin de l'Institut de Zoologie et Museé, Academie Bulgare des Sciences* [*Bulletin of the Institute of zoology with museum, Bulgarian Academy of Sciences*] 20: 139-155. (In Bulgarian).

Marinov B. 1967. *Gobio albipinnatus* Lukasch, 1933, a fish unknown to Bulgaria. *Annuaire de l'Université de Sofia "Kliment Ohridski," Faculté de Biologie* [*Annual of Sofia University "Kliment Ohridski," Faculty of Biology*] 59 (1): 39-41. (In Bulgarian).

Marinov B. 1978. Die Fischfauna im bulgarischen Donauabschnitt und ihre Bedeutung für die Wirtschaft [The fish fauna in the Bulgarian section of Danube River and its importance to the economy]. In: Roussev B. & Naidenow W. (Eds.) Limnologie des bulgarischen Donauabschnitts. Bulgarische Akademie der Wissenschaften [Lymnology of the Bulgarian Danube stretch. Bulgarian Academy of Sciences], Sofia, 201-228. (In Bulgarian).

Marinov B. 1979. *Pseudorasbora parva* (Schlegel, 1942) (Pisces, Cyprinidae) – a new representative of the ichthyofauna of Bulgaria. *Hydrobiology,* Sofia 8: 75-78. (In Bulgarian).

Marinov B. 1987. Growth of the asp (*Aspius aspius* L.) from the Bulgarian sector of the Danube. In: Botev B. (ed.). *Modern Achievements in the Bulgarian Zoology. Materials of the Fourth Jubilee Scientific Conference on Zoology*. Publishing House of the Bulgarian Academy of Sciences, Sofia: 111-116. (In Bulgarian).

Masár J., Turanský R., Krupka I. & Kautman J. 2006. The first record of the Siberian sturgeon (*Acipenser baerii*) in Slovak-Hungarian stretch of the Danube river. *Acta rerum naturalium Musei Nationalis Slovaci Bratislava* 52: 50-55.

Michajlowa L. 1967. Seltene Fischarten aus der Süsswasserfauna Bulgariens [Rare fish species from the freshwater fauna of Bulgaria]. *Zeitschrift für Fischerei und deren Hilfswissenschaften* [*Journal of Fisheries and its Subsidiary Sciences*] 15 (1/2): 153-160.

Morov T. 1931. [*The freshwater fishes of Bulgaria*]. Sofia, 93 p. (In Bulgarian).

Navodaru I. & Nastase A. 2006. Qualitative inventory of fish fauna from Danube River around Cama Dinu islets. *Scientific Annals of the Danube Delta National Institute for Research and Development – Tulcea, Romania* 12: 97-102.

Navodaru I., Buijse A.D. & Staras M. 2002. Effects of hydrology and water quality on the fish community in Danube Delta Lakes. *International Review of Hydrobiology* 87 (2-3): 329-348.

Năstase A., Oţel V. & Năvodaru I. 2017. Ecological Status of Fish Fauna in Arms of the Danube Delta (Danube Delta Biosphere Reserve, Romania) at the Beginning of the Third Millennium. *Acta zoologica bulgarica* 69 (3): 349-360.

Nelson J., Grande T. & Wilson M. 2016. *Fishes of the World.* Fifth Edition. John Wiley & Sons, 707 pp.

Oţel V. 2007 *Atlasul pestilor din rezervatia biosferei Delta Dunarii* [*Atlas of fishes of the Danube Delta Biosphere reserve*]. Institutul National de Cercetare-Dezvoltare Delta Dunarii, Tulcea [National Institute for Research and Development of the Danube Delta, Tulcea] 481 p. (In Romanian).

Oundjian E. 1964. [About a find of lamprey in Danube River]. *Bulletin de Museé National a Rousse* [*Bulletin of the National Museum in Rousse*] 1: 227-232. (In Bulgarian)

Paspalev G. & Peshev T. 1955. Contribution to the study of the ichthyofauna in the Iskar River. *Annuaire de L'Universite de Sofia, Faculte de Biologie, Geologie et Geographie* [*Annual of Sofia University, Faculty of Biology, Geology and Geography*] 48 (1): 1-39. (In Bulgarian).

Pehlivanov L. 2000. Ichthyofauna in the Srebarna Lake, the Danube Basin: state and significance of the management and conservation strategies of this wetland. *International Association for Danube Research* 33: 317-322.

Pehlivanov L., Stefanov T., Mihov S., Biserkov V., Vassilev M., Apostolou A. & Velkov B. 2011. Recent ichthyofauna in the wetlands along the Bulgarian section of the Danube. *Scientific Annals of the Danube Delta Institute, Tulcea – Romania* 17: 83-88.

Pehlivanov L., Stefanov T., Todorov M., Kutsarov Y. & Trichkova T. 2016. First records of the Black bullhead Ameiurus melas (Rafinesque, 1820) along the Bulgarian section of the Danube River. The 41st International Association for Danube Research (IAD) Conference, Sibiu, Romania, 2016 – Tributaries as key elements in sustainable management of the Danube River basin. Book of abstracts, p. 34.

Polačik M., Trichkova T., Janáč M., Vassilev M. & Jurajda P. 2008. The Ichthyofauna of the Shoreline Zone in the Longitudinal Profile of the Danube River, Bulgaria. *Acta zoologica bulgarica* 60 (1): 77-88.

Sivkov J. 1979. Morphological characteristics of the common perch *Perca fluviatilis* L., from the Danube and the Shabla Lake. *Bulletin de Musée National de Varna* [*Bulletin of the National Museum in Varna*] 15 (30): 149-157. (In Bulgarian).

Sivkov J. 1985. Morphological characteristics of *Gymnocephalus baloni* Holcik et Hensel, 1974 (Pisces, Percidae), a new species to the Bulgarian ichthyofauna. *Acta zoologica bulgarica* 27: 10-20. (In Russian).

Sivkov Y. 1986. Biometrical study on the *Gymnocephalus schraetser* (L.) (Pisces, Percidae) from the Bulgarian section of the Danube. *Bulletin du Musee National de Varna* [*Bulletin of the National Museum in Varna*] 22 (37): 161-168. (In Bulgarian).

Sivkov J. 1987a. Morphometric characteristics of *Stizostedion volgense* (Gmelin, 1788) (Pisces, Percidae) from the Bulgarian section of the Danube. *Bulletin du Musee National de Varna* [*Bulletin of the National Museum in Varna*] 23 (38): 204-211. (In Bulgarian).

Sivkov J. 1987b. Morphometric characteristic of Zingel streber Siebold, 1863 (Pisces, Percidae) from the Bulgarian section of the Danube. *Bulletin du Musee National de Varna* [*Bulletin of the National Museum in Varna*] 23 (38): 198-204. (In Bulgarian).

Sivkov J. 1988a. Morphological characteristics of the ruffe *Gymnocephalus cernuus* (Linnaeus, 1758) (Pisces, Percidae) from the Bulgarian section of the Danube River. *Bulletin de Musée National de Varna* [*Bulletin of the National Museum in Varna*] 24 (39): 180-185. (In Bulgarian).

Sivkov J. 1988b. Morphological characteristic of the pikeperch *Stizostedion lucioperca* (L.) from the Danube River and the Shabla Lake. *Bulletin de Musée National de Varna* [*Bulletin of the National Museum in Varna*] 24 (39): 174-179. (In Bulgarian).

Sivkov J. 1989a. Morphological characterization of the gudgeon (*Gobio albipinnatus* Lukasch, 1933) (Pisces, Cyprinidae) from the Bulgarian section of the Danube River. *Acta zoologica bulgarica* 38: 11-15. (In Bulgarian).

Sivkov J. 1989b. Morphological characterization of the *Cobitis elongata* Heckel et Kner, 1858 (Cobitidae) from Bulgaria. *Acta zoologica bulgarica* 38: 16-23. (In Bulgarian).

Sivkov J. 1991. Morphological characteristic of the Danubian loach *Sabanejewia bulgarica* (Drensky, 1928) (Pisces, Cobitidae). *Acta zoologica bulgarica* 42: 34-43. (In Bulgarian).

Sivkov Y. & Manolov Z. 1978. Morphologische Charakteristik des Donaufisches *Pelecus cultratus* (Linne) [Morphological characteristic of the Danube fish *Pelecus cultratus* (Linne)]. *Bulletin du Musée National de Varna* [*Bulletin of the National Museum in Varna*] 14 (29): 224-229. (In Bulgarian).

Stefanov T. 2007. Species diversity and distribution of fishes in Bulgaria. In: Popov A. & Fett V. (Eds.) *Ecology and Biogeography of Bulgaria*. Springer: 109-139.

Stefanov T. & Holčik J. 2007. The lampreys of Bulgaria. *Folia Zoologica* 56 (2): 213-224.

Stierandová S., Vukic J., Vasil'eva E. D., Zogaris S., Shumka S., Halacka K., Vetešník L., Švátora M., Nowak M., Stefanov T., Košco J. & Mendel J. 2016. A multilocus assessment of nuclear and mitochondrial sequence data elucidates phylogenetic relationships among European spirlins (Alburnoides, Cyprinidae). *Molecular Phylogenetics and Evolution* 94: 479-491.

Stojanov S., Gheorghiev J., Ivanov L., Hristov D., Kolarov P., Aleksandrova K., & Karapetkova M. 1963. [*Fishes of the Black Sea*]. Varna, 246 p. (In Bulgarian).

Szalóky Z., Bammer V., György A. I., Pehlivanov L., Schabuss M., Zornig H., Weiperth A., & Erős T. 2015. Offshore distribution of invasive gobies (Pisces: Gobiidae) along the longitudinal profile of the Danube River. *Fundamental and Applied Limnology* 187 (2): 127-133.

Tsekov A., Ivanova P., Angelov M., Atanasova S. & Bloesch J. 2013. Natural Sturgeon Hybrids along Bulgarian Black Sea Coast and in Danube River. *Acta zoologica bulgarica* 60 (3): 311-316.

Tsekov N. & Tsekov A. 2013. Natural Hybrid Posterities – Problem for the Existence of the Acipenseridae Family Species in Nature. *Ecologia Balkanica* 5 (2): 77-80.

Trichkova T., Stefanov T., Vassilev M. & Zivkov M. 2009. Fish species diversity in the rivers of the North-West Bulgaria. *Transylvanian Review of Systematical and Ecological Research – Wetlands Biodiversity* 8: 161-168.

Trichkova T. & Stefanov T. 2017. *Peccottus glenii* Dybowski, 1877. In: Trichkova T., Vladimirov V., Tomov R. & Todorov M. (eds.). *Guide to invasive alien species of European Union concern*. IBER-BAS, ESENIAS, Sofia, 98-101. (In Bulgarian).

Vassilev M. 1994. On the downstream migration of ichthyoplankton along the Bulgarian shore of the Danube River. *Vie Milieu* 44 (3/4): 273-280.

Vassilev M. 2003. Spawning sites of beluga sturgeon (*Huso huso* L.) located along the Bulgarian-Romanian Danube River stretch. *Acta zoologica bulgarica* 55 (2): 91-94.

Vassilev M. & Pehlivanov L. 2003. Structural changes of sturgeon catches in the Bulgarian Danube section. *Acta zoologica bulgarica* 55 (3): 97-102.

Vassilev M. & Pehlivanov L. 2005. Checklist of Bulgarian freshwater fishes. *Acta zoologica bulgarica* 57 (2): 161-190.

Vassilev M., & Trichkova T. 2007. *Salmo labrax* Pallas, 1811 (Pisces: Salmonidae) – a new record for the Bulgarian sector of the Danube River. *Acta zoologica bulgarica* 59 (1): 97-99.

Vassilev M., Apostolou A., Velkov B., Dobrev D. & Zarev V. 2012. *Atlas of the Gobies (Gobiidae) in Bulgaria*. Institute of Biodiversity and Ecosystem Research, Bulgarian Academy of Sciences, Sofia, 112 pp.

Velkov B., Pehlivanov L. & Vassilev M. 2004. *Umbra krameri* (Pisces, Umbridae): a reinstated species for the Bulgarian ichthyofauna. *Acta zoologica bulgarica* 56 (2): 233-235.

Zarev V., Apostolou A., Velkov B. & Vassilev M. 2013. Review of the Distribution of the Family Gobiidae (Pisces) in the Bulgarian Danube Tributaries. *Ecologia Balkanica* 5 (2): 81-89.

Živkov M., Prodanov K., Trichkova T., Raikova-Petrova G. & Ivanova P. 2005. Fishes in Bulgaria – research, priorities, conservation and sustainable use. In: Petrova A. (Ed.), *Current state of Bulgarian biodiversity – problems and perspectives*. Drakon Publishers, Sofia, 247-282. (In Bulgarian).

In: Biodiversity of the Bulgarian-Romanian Section
Editor: Shurulinkov, Hubenov et al.
ISBN: 978-1-53615-663-8

Chapter 8

Diversity and Distribution of Amphibians and Reptiles in the Bulgarian Part of the Lower Danube

Georgi Popgeorgiev[1,*], Borislav Naumov[2], Yurii V. Kornilev[1,3], Vladislav Vergilov[1], Miroslav Slavchev[1,2], Simeon Lukanov[2], Angel Dyugmegdzhiev[2], Andrey Stoyanov[1], Dobrin Dobrev[2] and Nikolay Tzankov[1]

[1]National Museum of Natural History,
Bulgarian Academy of Sciences, Sofia, Bulgaria
[2]Institute of Biodiversity and Ecosystem Research,
Bulgarian Academy of Sciences, Sofia, Bulgaria
[3]Department of Integrative Zoology, Faculty of Life Sciences,
University of Vienna, Vienna, Austria

— *To Andro, Dobri, and Niki, true friends*

Abstract

Here we present the first attempt to combine the collected data, map localities, and analyze the distribution of the herpetofauna in the area stretching ca. 10 km south of the Bulgarian Danube River. We identified 687 records from publications that we attributed to 62 cells of the 10 × 10 km MGRS grid. We included previously unpublished data collected

[*] Corresponding Author's Email: georgi.popgeorgiev@gmail.com.

through various visits throughout the past ca. 13 years, comprising 1918 records with real coordinates, which fall in 1269 cells of the 1 × 1 MGRS grid. We have confirmed 34 native species – 15 amphibians (4 salamanders and 11 frogs) and 19 reptiles (1 turtle, 2 tortoises, 8 lizards and 8 snakes), and one invasive turtle. Although finding other species is unlikely, additional field studies could confirm the presence of *Pseudopus apodus* (for now, this species cannot be accepted as part of the regional fauna). Further detailed studies are needed, especially in areas that were likely not sampled, and in the detection of cryptic species such as *Pelophylax lessonae*. We also caution that future changes to the river's hydrodynamics must really consider the highly likely multitude of negative effects to the biodiversity and specifically to the herpetofauna.

Keywords: Amphibia, Reptilia, herpetofauna, species composition, river, conservation

Introduction

The Danube River is the second longest European river, crossing into ten countries. The section of the Danube between the Iron Gate dam in Serbia and the mouth is usually called the Lower Danube and for the most part falls within the Romanian and Bulgarian territory. Compared to most major rivers in Europe, the Lower Danube is less influenced by human activity, although major changes have been made in the 20th century (e.g., draining of wetlands for agriculture or against mosquitoes) (Kopralev 2002). Multiple permanent and temporary water bodies (e.g., tributary rivers, lakes, ponds, spills) along with the Danube form a complex of wetlands that is essential to preserve biodiversity (and in particular amphibians and aquatic reptiles) on a pan-European scale.

Several publications have summarized the distribution of species along the Romanian Danube (such as Fuhn 1960, Covaciu-Marcov et al. 2006, Cogălniceanu et al. 2013a, b). For Bulgaria, relevant information has been published in major works such as Buresch & Zonkov (1941, 1942), Undjian (2000), Stoyneva & Michev (2007), Naumov et al. (2011), Stojanov et al. (2011), Kornilev et al. (2017), but for the most part the data have not been aggregated and mapped. However, the river has not yet been sufficiently studied in herpetological terms.

Here we present the first attempt to combine the collected data, map localities, and analyze the distribution of the herpetofauna in the Bulgarian Danube River. Further, we note major conservation challenges for the herpetofauna.

Material and Methods

Study Area

We chose to define the study area as the territory falling within 109 squares of the 10 × 10 km MGRS grid (Military Grid Reference System, UTM zone 35N, datum WGS

1984; see Figure 1). To be included in the study, each of these squares is either intersected by the Danube or is within 10 km of its southern bank. The overall length of the Danube within Bulgaria is 472 km.

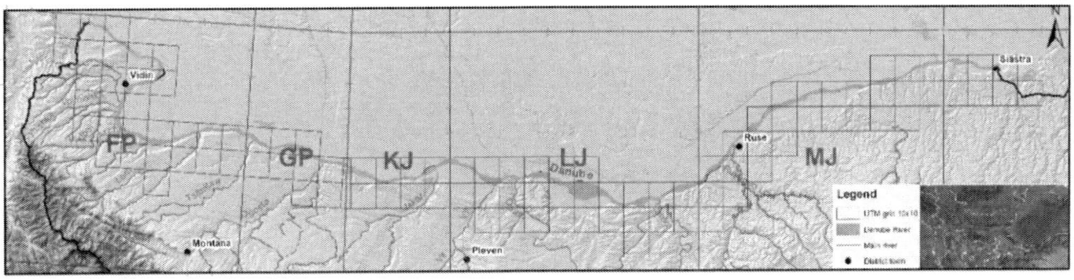

Figure 1. Overview map of the study area, the Bulgarian part of the Lower Danube, with some of the main rivers and cities, and the 10 × 10 km MGRS grid.

Fourty-four percent of the Bulgarian Danube bank is high and steep, at times vertical (Kopralev 2002), in contrast to the low Romanian bank, where hydrological impacts of high water levels are more pronounced. The elevation of the study site roughly lays between 6–297 m a.s.l., with an average of 90 m.

The climate is moderately continental (Köppen-Geiger class Dfa), characterized by a warm summer (mean temperatures in July: 22–24°C) and cold winter (mean temperatures in January: 0.0 – -1.5°C), a spring-summer maximum and a winter minimum of the rainfall, and a yearly relatively stable snow cover depending on the elevation (Kopralev 2002).

Within the study area exist multiple potentially suitable habitat types for the herpetofauna. Although the main channel of the Danube itself is not optimal for most species (except for the *Natrix* sp.), the river shapes and impacts various aquatic habitats of great importance to the herpetofauna: dead arms and oxbows, spills, wetlands complexes, marshes, and lakes (such as Persin, Srebarna, Garvan), tributary rivers. Additionally, the Danube lowlands had one of the most developed systems for irrigation of agricultural lands in the country by 1989 (Kopralev 2002); currently, the dams and artificial channels are at different states of maintenance, from completely abandoned to functional, but many still provide suitable habitats for aquatic species. Further information on the habitats, especially the terrestrial ones, is provided in the phytocoenological study by Tzonev et al. (2019).

About 20 major rivers are right tributaries to the Danube, mostly flowing from Stara Planina Mountains and the Predbalkan (Figure 1). In the west, they are mostly relatively short and with limited flow, whose tributaries form in the lowlands, have inconsistent inflow and dry out during the summer. In the central part the rivers are generally large. In the east, the largest river is Roussenski Lom. Lakes and wetlands have formed near the rivers, in oxbows in the middle and lower courses. Gradually, during the first half of the 20th century, in relation to corrections to the river beds and drying of wetlands as part of sanitary-hygienic considerations they were drained. Due to its unique hydrology and hydrobiology, the Srebarna Lake has been preserved (Kopralev 2002). Another important hydrological component are marshes, formed by the spring high-waters or succession of lakes or dams.

Fourty-nine SCIs (Site of Community Importance, defined by the Habitats Directive, 92/43/EEC), part of the European Natura 2000 network, fall completely or partially within the study site. Some of these are designated for the protection of certain herpetofauna species (in particular *Triturus dobrogicus*, *Bombina bombina*, *B. variegata*, *Elaphe sauromates*, *Emys orbicularis*, *Testudo graeca*, *T. hermanni*).

Data Acquisition

We combined previously unpublished observations and data obtained through an exhaustive literature review.

Unpublished data were primarily collected personally by the authors or provided by colleagues in 2005–2018, using handheld GPS devices that provided precise coordinates (±10 m). Data were collected mostly during mapping for research projects (e.g., area or species surveys, environmental assessments), using various methods (e.g., visual and auditory observations supplemented or not with trapping), by multiple people, throughout the active season (March–October), and with varying sampling effort. In this sense, we assume that data on the presence of individual species were collected haphazardly. The same applies to the cited literature data as they come from multiple publications with different aims during an extensive time interval.

Still, during targeted surveys, we predominantly used the track method for collecting observational data. We attempted to traverse varying habitats when possible, in order to get a representative sample from a range of suitable environments and to maximize the potential number of species. We employed mostly diurnal visual searches while walking along standing and running water bodies, or through flood plains and (semi-)natural fields, sparse deciduous forests, and other habitats we presume harbor herpetofauna. We focused on known favorable micro-habitats and potential refuges, which we specifically sampled (e.g., through turning over large rocks and fallen logs, searching cavities in the bank).

When possible, we supplemented those with nocturnal visual searches along favorable habitats to allow detection of nocturnally active species. At times we set up funnel traps (dimensions: L50/W25/H25 cm; mesh size: 0.5 cm) to detect aquatic species in bodies of standing water and channels with slow moving water.

Mapping

Mapping and analyses were based on the 10 km grid selected for the study area (Figure 1). We used the codes of the Military Grid Reference System (MGRS; spatially identical with UTM) for map visualization and listing the localities. The unpublished localities were almost exclusively collected using hand-held GPS devices (e.g., by using the mobile application SmartBirds Pro; Popgeorgiev et al. 2015), while published localities (except those that were previously published by us) were generally provided at the MGRS level or we attributed them to a corresponding MGRS square based on the given geographic description (usually a named human settlement).

With regards to the literature data, any original description (or geographical coordinates) for an occurrence were taken as a separate locality, except when two or more of the literary sources provide the same description for an observation. In a few cases, a locality for a species was provided duplicating one in a previous work but without it being cited.

Analyses

The data were compiled into a GIS database. Analyses and maps were done using software ArcGIS 10.4.1 (ESRI, Redlands, CA, USA) and Excel 2013 (Microsoft, Redmond, WA, USA).

Taxonomic Notes

Some recent changes in taxonomy have still not been completely resolved phylogeographically for Bulgaria, and for the study area. Due to the unclear boundary between *Bufotes viridis* and *B. variabilis* (see Stöck et al. 2006; Dubois & Bour 2010), and the unclear systematic status of *Mediodactylus kotschyi* (see Kotsakiozi et al. 2018), we present them respectively as *B. viridis* complex and *M. kotschyi* complex. We use *Darevskia praticola* sensu lato, given the unknown taxonomic status of the populations in Bulgaria (Saberi-Pirooz et al. 2018).

In northern Bulgaria only *Hyla orientalis* is present (split up from *H. arborea*; Stöck et al. 2008, 2010). We attribute all former *Anguis fragilis* in the Danube lowlands to the taxon *A. colchica* (see Jablonski et al. 2016).

Results and Discussion

The native herpetofauna in the Bulgarian part of the Lower Danube comprises a total of 34 species – 15 amphibians (4 salamanders and 11 frogs) and 19 reptiles (1 turtle, 2 tortoises, 8 lizards and 8 snakes; Figure 2). Overall, we have identified 687 records from publications that we attributed to 62 cells of the 10 × 10 km MGRS grid. The previously unpublished data that we have collected consists of 1918 records with real coordinates, which fall in 1269 cells of the 1 × 1 MGRS grid. The number of "published data" could be a bit misleading, since some of these data come from relatively recent publications on the national distribution of species authored by us as well (Naumov et al. 2011: *Natrix tessellata*; Kornilev et al. 2017: *Emys orbicularis*).

Species Composition of the Herpetofauna

Here we present in the form of an annotated checklist the known species occurrence based on published and unpublished data (Figure 2).

Class: Amphibia
Order: Caudata
Family: Salamandridae

***Salamandra salamandra* (Linnaeus, 1758):** Undjian (2000) & this study.

***Lissotriton vulgaris* (Linnaeus, 1758):** Genov (1969), Beshkov (1998), Undjian (2000), Stoyneva & Michev (2007), Biserkov & Naumov (2012), Wielstra et al. (2018) & this study.

***Triturus dobrogicus* (Kiritzescu, 1903):** Kovatscheff (1903), Kowatscheff (1905), Kovachev (1912), Schumann (1930), Buresch & Zonkow (1941), Genov (1969), Wallis & Arntzen (1989), Beshkov (1998), Undjian (2000), Arntzen et al. (2007), Stoyneva & Michev (2007), Biserkov & Naumov (2012), Naumov & Biserkov (2013), Wielstra et al. (2013a, b, 2016), Arntzen et al. (2014) & this study.

***Triturus ivanbureschi* Arntzen et Wielstra, 2013:** Kovachev (1912).

Order: Anura
Family: Bombinatoridae

***Bombina bombina* (Linnaeus, 1761):** Kovachev (1894, 1905a, b, 1912), Kovatscheff (1903), Buresch & Zonkow (1942), Paspaleva-Antonova (1961), Genov (1969), Beškov (1972), Tomov (1991), Beshkov (1998), Undjian (2000), Stoyneva & Michev (2007), Biserkov & Naumov (2012), Mollov et al. (2013) & this study.

***Bombina variegata* (Linnaeus, 1758):** Buresch & Zonkow (1942), Genov (1969), Beškov (1972) & Beshkov (1998).

Family: Pelobatidae

***Pelobates fuscus* (Laurenti, 1768):** Kovachev (1905a, b, 1912), Buresch & Zonkow (1942), Paspaleva-Antonova (1961), Beškov (1972), Beshkov (1998), Undjian (2000), Stoyneva & Michev (2007), Biserkov & Naumov (2012) & this study.

Pelobates syriacus **Boettger, 1889:** Buresch & Zonkow (1942), Beškov (1972), Beshkov (1998), Undjian (2000), Stoyneva & Michev (2007), Biserkov & Naumov (2012) & this study.

Family: Bufonidae

Bufo bufo **(Linnaeus, 1758):** Kovachev (1912), Buresch & Zonkow (1942), Genov (1969), Beškov (1972), Beshkov (1998), Undjian (2000), Stoyneva & Michev (2007), Biserkov & Naumov (2012), Mollov et al. (2013) & this study.

Bufotes viridis **(Laurenti, 1768) complex:** Kovatscheff (1903), Kovachev (1912), Buresch & Zonkow (1942), Paspaleva-Antonova (1961), Boschkow (1965), Genov (1969), Beškov (1972), Tomov (1990), Beshkov (1998), Undjian (2000), Stoyneva & Michev (2007), Biserkov & Naumov (2012), Mollov et al. (2013) & this study.

Family: Hylidae

Hyla orientalis **Bedriaga, 1890:** Kovachev (1894), Kovatscheff (1903), Buresch & Zonkow (1942), Paspaleva-Antonova (1961), Boschkow (1965), Genov (1969), Beškov (1972), Chochev (1980), Beshkov (1998), Undjian (2000), Stoyneva & Michev (2007), Biserkov & Naumov (2012), Mollov et al. (2013), Dufresnes et al. (2015), Gvoždík et al. (2015) & this study.

Family: Ranidae

Rana dalmatina **Fitzinger, 1839:** Kovatscheff (1903), Kovachev (1905a, b, 1912), Buresch & Zonkow (1942), Genov (1969), Beškov (1972), Beshkov (1998), Undjian (2000), Stoyneva & Michev (2007), Biserkov & Naumov (2012) & this study.

Pelophylax lessonae **(Camerano, 1882):** Tzankov & Popgeorgiev (2014) & Lukanov et al. (2017).

Pelophylax ridibundus **(Pallas, 1771):** Kovachev (1894), Buresch & Zonkow (1942), Paspaleva-Antonova (1961), Paspalev et al. (1962), Beschkov (1965), Boschkow (1965), Božkov (1967), Genov (1969), Beškov (1972), Simeonov & Boev (1988), Tomov (1989), Beshkov (1998), Undjian (2000), Kirin (2003), Stoyneva & Michev (2007), Biserkov & Naumov (2012), Lukanov et al. (2014), Metcheva et al. (2016) & this study.

Pelophylax **kl. *esculentus*** **(Linnaeus, 1758):** Kovachev (1910, 1912), Beschkov (1965), Boschkow (1965), Genov (1969), Beškov (1972), Popov (1975), Bachvarov (1988), Beshkov (1998), Undjian (2000), Stoyneva & Michev (2007), Biserkov & Naumov (2012) & this study.

Class: Reptilia

Order: Testudines

Family: Emydidae

***Emys orbicularis* (Linnaeus, 1758):** Kovachev (1894, 1905a, 1912), Werner (1898), Bolkay (1924), Călinescu (1931), Buresch & Zonkow (1933), Paspaleva-Antonova (1961), Undjian (2000), Stoyneva & Michev (2007), Biserkov & Naumov (2012), Mollov et al. (2013), Kornilev et al. (2017) & this study.

Family: Testudinidae

***Testudo graeca* Linnaeus, 1758:** Kovachev (1907, 1912), Buresch & Zonkow (1933), Beshkov (1998), Undjian (2000), Biserkov & Naumov (2012) & this study.

***Testudo hermanni* Gmelin, 1789:** Kovachev (1894, 1912), Biserkov & Naumov (2012) & this study.

Order: Squamata

Suborder: Sauria

Family: Anguidae

***Anguis colchica* (Nordmann, 1840):** Kowatscheff (1905), Kovachev (1912), Biserkov & Naumov (2012), Jablonski et al. (2016) & this study.

Family: Gekkonidae

***Mediodactylus kotschyi* (Strauch, 1887) complex:** Kowatscheff (1905) & Petkov (2002).

Family: Lacertidae

***Darevskia praticola* s.l. (Eversmann, 1834):** Kowatscheff (1905), Kovatcheff (1906), Kovachev (1912), Biserkov (1987), Beshkov (1998), Undjian (2000) & this study.

***Lacerta trilineata* Bedriaga, 1886:** Kowatscheff (1905), Kovatcheff (1906), Kovachev (1912), Biserkov (1987), Beshkov (1998), Undjian (2000) & this study.

***Lacerta viridis* (Laurenti, 1768):** Kovachev (1894), Bolkay (1924), Genov (1969), Simeonov & Boev (1988), Beshkov (1998), Schlüter (2005), Böhme et al. (2007), Biserkov & Naumov (2012), Mollov et al. (2013), Marzahn et al. (2016) & this study.

***Podarcis muralis* (Laurenti, 1768):** Kovachev (1894, 1905a, 1912), Buresch & Zonkow (1933), Beshkov (1998), Undjian (2000), Kirin (2002), Biserkov & Naumov (2012), Mollov et al. (2013) & this study.

***Podarcis tauricus* (Pallas, 1814):** Werner (1898), Kovatscheff (1903), Kovachev (1912), Bolkay (1924), Buresch & Zonkow (1933), Genov (1969), Beshkov (1998), Biserkov & Naumov (2012), Mollov et al. (2013) & this study.

Family: Scincidae

***Ablepharus kitaibelii* (Bibron et Bory St-Vincent, 1833):** Kovatscheff (1903), Kovachev (1905a, 1912), Buresch & Zonkow (1933), Beshkov (1998), Undjian (2000), Mollov et al. (2013), Vergilov & Tzankov (2014), Vergilov et al. (2017, 2018) & this study.

Order: Squamata

Suborder: Serpentes

Family: Boidae

***Eryx jaculus* (Linnaeus, 1758):** Beschkov (1961) & Beschkov (2015a).

Family: Colubridae

***Coronella austriaca* Laurenti, 1768:** Kovatscheff (1903), Kovachev (1905a, 1912), Kowatscheff (1905), Beshkov (1998), Undjian (2000) & Mollov et al. (2013).

***Dolichophis caspius* (Gmelin, 1789):** Kovatscheff (1903), Kovachev (1905a, 1912), Kowatscheff (1905), Buresch & Zonkow (1934), Genov (1969), Beshkov & Dushkov (1981), Beshkov (1998), Undjian (2000), Biserkov & Naumov (2012), Mollov et al. (2013) & this study.

***Elaphe sauromates* (Pallas, 1811):** Kovatscheff (1903), Kowatscheff (1905), Kovachev (1912), Buresch & Zonkow (1934), Beshkov (1998), Beschkov (2015b), Naumov (2005), Stoyneva & Michev (2007) & this study.

***Zamenis longissimus* (Laurenti, 1768):** Kovachev (1905a, 1912), Paspaleva-Antonova (1961), Genov (1969), Beshkov (1998), Stoyneva & Michev (2007), Biserkov & Naumov (2012), Metcheva et al. (2016) & this study.

Family: Natricidae

***Natrix natrix* (Linnaeus, 1758):** Kovachev (1894, 1912), Werner (1898), Kovatscheff (1903), Kowatscheff (1905), Bolkay (1924), Hecht (1930), Buresch & Zonkow (1934), Tuleshkov (1954), Michev (1958), Paspaleva-Antonova (1961), Genov (1969), Beshkov & Dushkov (1981), Beshkov (1986, 1998), Biserkov (1996), Stoyneva & Michev (2007), Biserkov & Naumov (2012), Mollov et al. (2013), Kindler et al. (2017) & this study.

***Natrix tessellata* (Laurenti, 1768):** Kovatscheff (1903), Kovachev (1912), Hecht (1930), Paspaleva-Antonova (1961), Genov (1969), Biserkov (1996), Beshkov (1998), Undjian (2000), Stoyneva & Michev (2007), Naumov et al. (2011), Biserkov & Naumov (2012), Andonov et al. (2017) & this study.

Family: Viperidae

***Vipera ammodytes* (Linnaeus, 1758):** Werner (1898), Kovatscheff (1903), Kovachev (1905a, 1912), Bolkay (1920, 1924), Beshkov & Dushkov (1981), Biserkov et al. (1985), Beshkov (1998), Undjian (2000) & this study.

In addition, unconfirmed data exists on the presence of two native species, *Pseudopus apodus* (Pallas, 1775) and *Vipera berus* (Linnaeus, 1758). We chose to exclude them for now from the Lower Danube herpetofauna.

1. Only a single individual of *P. apodus* was observed near Tsarevets Village (N43.59649° E25.32055°; MGRS: LJ62) on 21.05.2012 (S. Boneva, personal observation). The characteristics of the contemporary range of *P. apodus* are such that its presence at this site is unlikely (see the distribution map in Stojanov et al. 2011: 321); the closest known locations to Tsarevets are 50–60 km to the east (in the basin of Roussenski Lom River: near Pepelina Village and between Nisovo and Svalenik Villages) and are accepted as relict (see Beškov et al. 1967 and Undjian 2000). It is possible that the observed individual has been transported to the site by accident (i.e., human activity) or has been erroneously identified (it was not captured, but only seen). Nevertheless, the possibility that in this location a relict population exists (such as these in the Roussenski Lom basin) cannot be excluded.
2. Kovachev (1894) writes that two specimens of *V. berus* have been collected near Rousse (MGRS: MJ15), and Bolkay (1924) mentions one specimen, collected near Svishtov (MGRS: LJ63) in 1890. In their study Buresch & Zonkow (1932) express doubt in the validity of these reports. Furthermore, the fact that the species has not been registered for the Lower Danube (incl. Romanian and Serbian banks, see Cogălniceanu et al. 2013b and Tomović et al. 2014, respectively) for the last ca. 125 years should be enough to discard the possibility of *V. berus* being present in the study area.

Furthermore, we do not include in the native fauna the allochthonous species, *Trachemys scripta* (Thunberg in Schoepff, 1792). Individuals belonging to this invasive species have been observed in several sites near the Danube River (Tzankov et al. 2015, this study). It is yet unknown whether they breed successfully there, as they do in some southern parts of Bulgaria (Tzankov et al. 2017).

In view of the above it could be surmised that the native herpetofauna in the Bulgarian part of the Lower Danube includes a total of 34 species (15 amphibian and 19 reptile species). Thirty-three species are known from the Romanian (see maps in Cogălniceanu et al. 2013a, b), and 29 from the Serbian part (see maps in Vukov et al. 2013 and Tomović et al. 2014). As could be expected, most species are common for all three parts, but there are also some differences – e.g., the species *Triturus cristatus* (Laurenti, 1768) and *Lacerta agilis* Linnaeus, 1758 are absent from the Bulgarian part, but present in both the Romanian and Serbian parts; *T. ivanbureschi* and *M. kotschiy* complex are only found along the Bulgarian bank. When the characteristics of the ranges of these taxa are considered on a continental scale, the highlighted differences appear logical – i.e., it is more likely that the absence of these species is real and not a result of insufficient studies.

Distribution

Overall, we have compiled data (published and unpublished) for 91 cells in the 10 km grid, which represents 83% of the cells of the study area. Expectedly, species are unequally distributed (see Figure 2 for species specific maps). If we provisionally divide the Bulgarian Lower Danube into three sections of approximately the same length (western: from the Timok River at the border to the Ogosta River; middle: between Ogosta and Yantra River; eastern: east of Yantra; see Figure 1), the species would fall in several groups based on their distribution within these sections.

The ubiquitous species, found in all sections, consist of 25 species (74% of all detected native species). Most of these species are widely distributed in Bulgaria, or their distribution is related to the Danube to a greater (*T. dobrogicus* and *P.* kl. *esculentus*) or lesser extent (*P. fuscus* and *B. bombina*). Much more intriguing is the distribution of two other species. The new localities for *Bombina variegata* are not in the immediate vicinity of the Danube, but are closest to the Stara Planina Mountain and to the Serbian populations. In contrast, published data are for Vidin, Svishtov, and the Srebarna, which could be considered as isolated localities, situated far from the main species range (see map in Stojanov et al. 2011:163). For the Romanian counterpart of the study site there is no published record (see map in Cogălniceanu et al. 2013a: 47). For *E. sauromates* the Bulgarian localities in the Lower Danube represent the north-western edge of the range. The most north-western locality is noted as "Lom" without further detail (Kovachev 1912); there are few unpublished data from two localities, only the one between the city of Svishtov and the village of Tsarevets being new.

Six species were detected in two sections: four species in the central and eastern, and two species in the western and eastern. The first combination seems logical for *T. graeca* and *L. trilineata* (given their known ranges). Furthermore, here we provide previously unpublished data for their most north-western localities in Bulgaria – respectively north of the city of Miziya (N43.73395° E23.88523°) and between Ostrov and Gorni Vadin Villages (N43.67556° E24.21625°). The circumstances related to the other two species, *P. syriacus* and *P. lessonae*, are different. The first species is known from the Serbian section of the Lower Danube, and further west to Belgrade (see Ljubisavljević & Dzukić 2015), from the Romanian coast (including the western part); especially given its nocturnal activity and explosive short-term breeding, the limited records in Bulgaria likely reflect insufficient sampling and not true absence. The lack of data for *P. lessonae* is also logical. The species was proven for Bulgaria relatively recently (Tzankov & Popgeorgiev 2014). It is very difficult for correctly identify without applying specialized methods (e.g., precise morphometry, genotyping, or bioacoustics) and the species is likely present at multiple other localities along the Bulgarian Danube (also suggested in Lukanov et al. 2017).

The second combination (species found in the W and E sections only) is more difficult to interpret. We suppose that further targeted sampling will lead to finding *T. hermanni* in the central section as well. The previously unpublished localities of *S. salamandra* seem logical, occurring in the region closest to the Stara Planina Mountain and to the Serbian populations and are not in the immediate proximity to the Danube. The only known locality from the eastern section (Undjian 2000) is questionable and needs to be confirmed: it was not observed personally by E. Undjian, but an employee of the forestry administration reported one individual from the yard of a house in Marten Village in 1985.

Three species can be considered as sporadically distributed, being detected in a single section. *Triturus ivanbureschi* is widely spread in Bulgaria except the north-western part of the country and the Danube coast (see Stojanov et al. 2011 and Wielstra et al. 2014), where we stipulate the species is likely outcompeted by the congeners *T. dobrogicus/T. cristatus*. The locality next to Balgarka Village (NJ27; "Eni-kyoyu" in Kovachev 1912), is the northern-most and the closest known to the Danube. *Mediodactylus kotschyi* complex is found predominantly in SW and SE Bulgaria, but along the Black Sea Coast reaches far to the north (see Stojanov et al. 2011 and Koynova et al. 2017). The observations at the city of Rousse (MJ15; originally reported by Kowatscheff 1905 and confirmed by Petkov 2002) is the only one known from the Danube. *Eryx jaculus* has been recorded sporadically in southern Bulgaria, but in the north is known only from two close localities (LJ33 and LJ52, respectively Vabel Village and Tatari Village, given by Beschkov 2015a and Beschkov 1961). The only known locality of *E. jaculus* from the Romanian Danube is situated close to the Bulgarian ones: 15–20 km NW of Vabel (see Cogălniceanu et al. 2013b). These three localities together form the north-westernmost edge of the species range.

If we consider the number of cells with known localities as a percent of the total number of cells in the study area, we could use this to obtain an indirect indicator for the species abundance. Thus, *P. ridibundus* and *L. viridis* stand out as the most common species (found in > 50% of the cells), while very rare (in < 5%) are six species, including *T. ivanbureschi* and *M. kotschyi* complex, found in only one cell each (Table 1). The high abundance of *P. ridibundus* and *L. viridis* is to be expected: they are wide spread in Bulgaria, and in most places they are the species with the highest numbers amongst amphibians and reptiles, respectively). The two "rarest" species are at the edges of their range, while in other parts of the country they are frequently observed.

The cells with the most species are either well-studied sites of known biological importance and diversity (Srebarna Lake, 30 species, NJ08), or cover major cities whose vicinities have been visited more (Rousse, 24 species, MJ15; Svishtov, 18 species, LJ63; Figure 3). Further, the cells without species are likely ones that have not been visited at all.

Given this unequal and biased sampling, the distribution maps will likely change in the future.

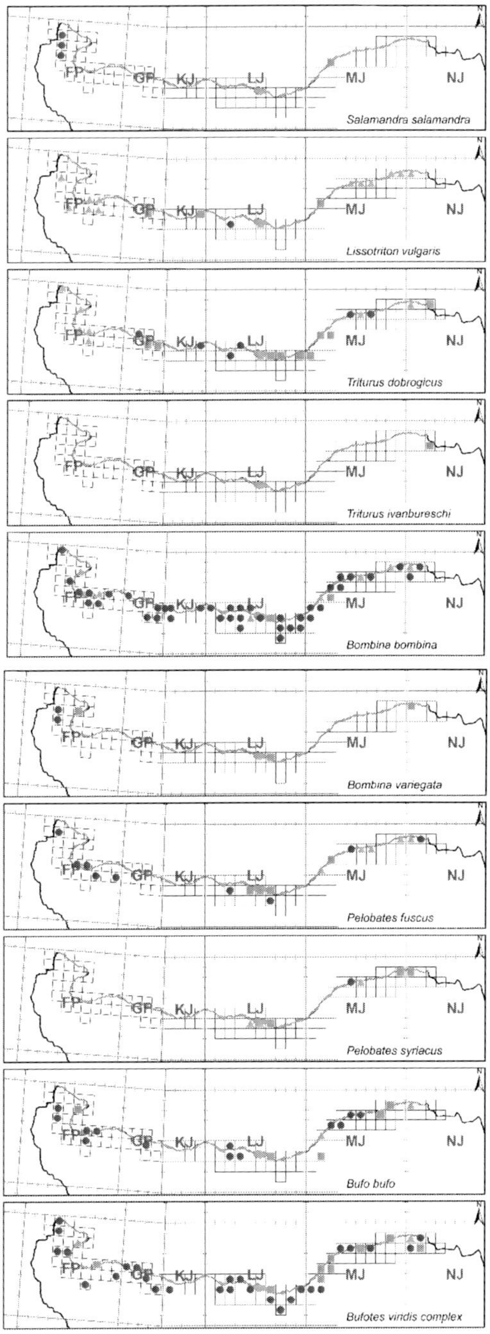

Figure 2. (Continued).

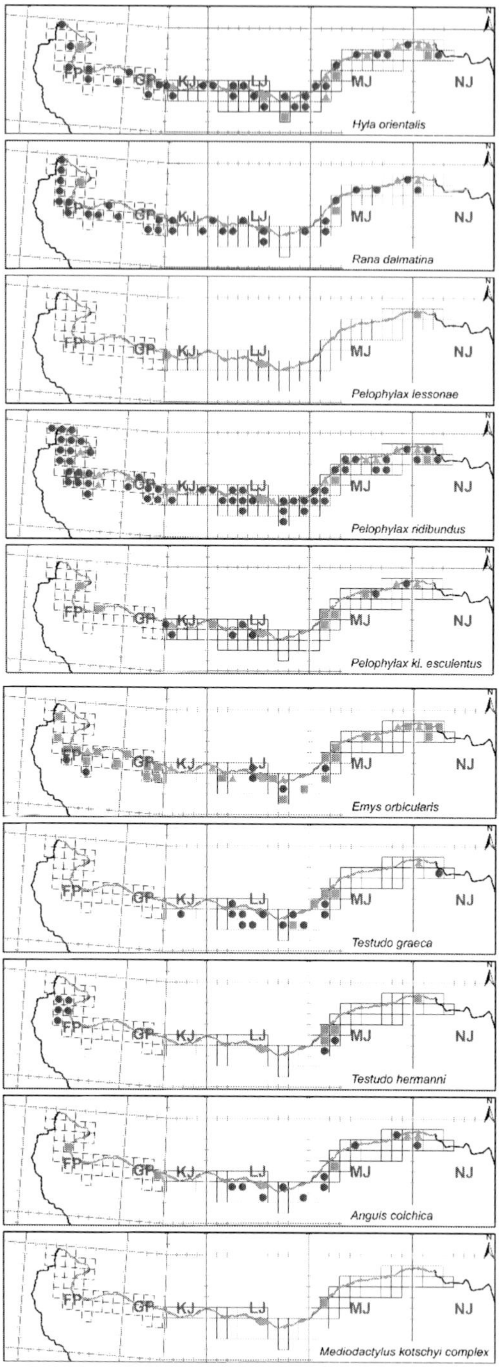

Figure 2. (Continued).

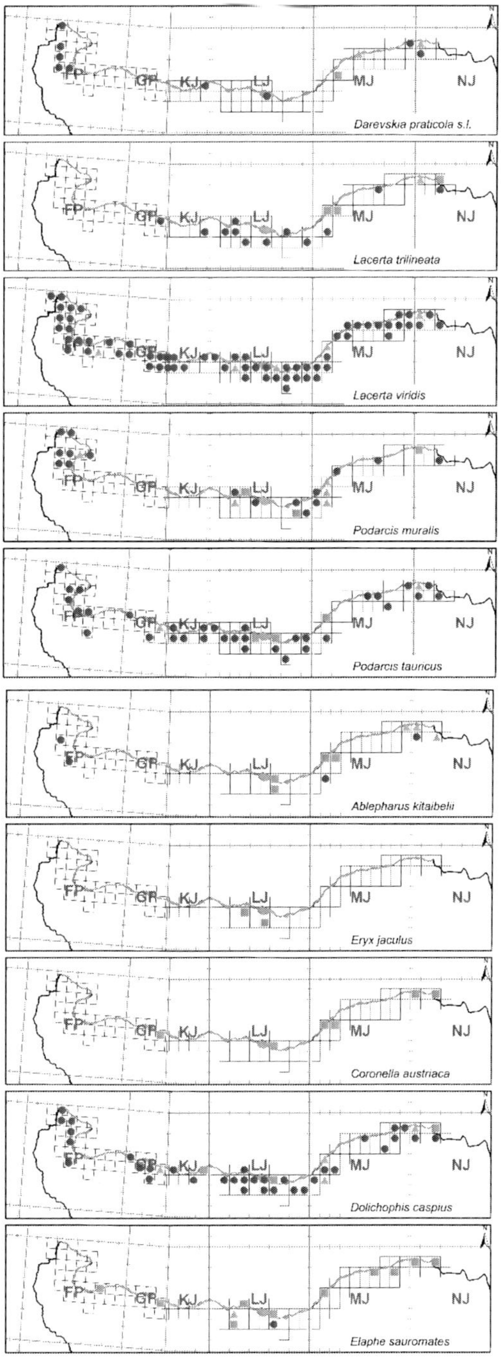

Figure 2. (Continued).

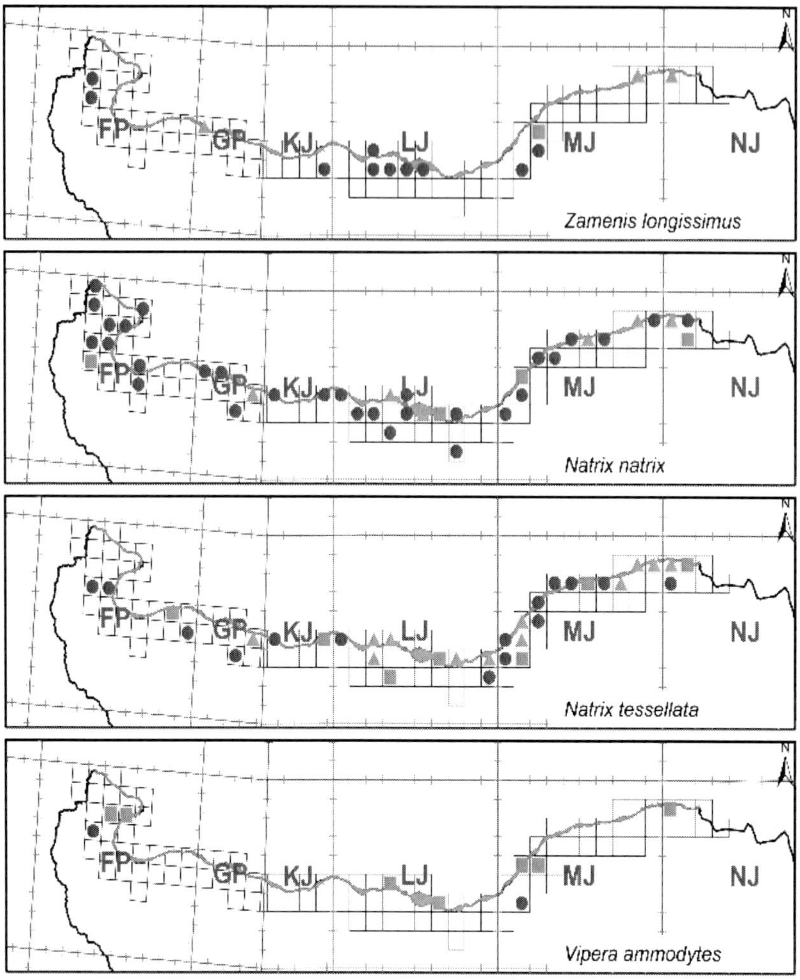

Figure 2. Distribution of amphibians and reptiles (by species) in the study area. The symbols on the maps are as follows: • – new, previously unpublished data for the respective square, ▲ – confirmed, previously published data supplemented by our observations, and ■ – published, but unconfirmed by our new observations.

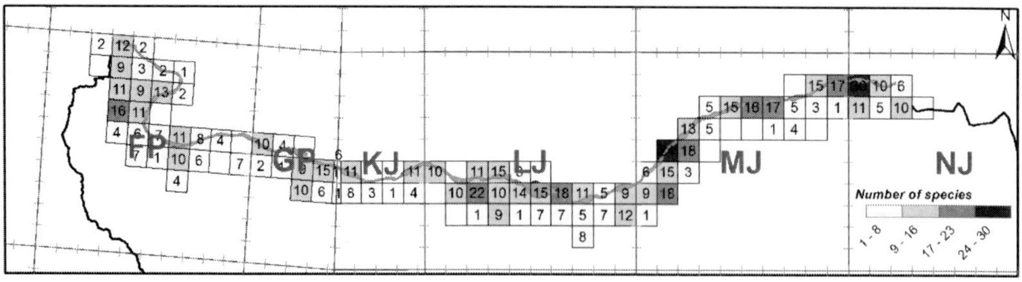

Figure 3. Number of species detected per MGRS cell in the study area.

Table 1. Number of UTM squares (N – new; C – confirmed; U – unconfirmed; T – total) of occurrence for each detected herpetofaunal species. The total number of squares within the study area is 109. The percent occurrence within the squares (%) equals T/109

Species	N	C	U	T	%
Salamandra salamandra	3		1	4	3.7
Lissotriton vulgaris	1	13	3	17	15.6
Triturus dobrogicus	6	8	9	23	21.1
Triturus ivanbureschi			1	1	0.9
Bombina bombina	41	7	2	50	45.9
Bombina variegata	2		3	5	4.6
Pelobates fuscus	9	5	4	18	16.5
Pelobates syriacus	1	3	3	7	6.4
Bufo bufo	13	2	5	20	18.3
Bufotes viridis complex	27	2	9	38	34.9
Hyla orientalis	30	8	5	43	39.4
Rana dalmatina	31	2	2	35	32.1
Pelophylax lessonae			2	2	1.8
Pelophylax ridibundus	55	13	2	70	64.2
Pelophylax kl. esculentus	7	3	7	17	15.6
Emys orbicularis	6	9	25	40	36.7
Testudo graeca	13	1	4	18	16.5
Testudo hermanni	7		4	11	10.1
Anguis colchica	10	2	3	15	13.8
Mediodactylus kotschyi complex			1	1	0.9
Darevskia praticola s.l.	9	2	1	12	11.0
Lacerta trilineata	12	1	3	16	14.7
Lacerta viridis	65	5		70	64.2
Podarcis muralis	16	5	3	24	22.0
Podarcis tauricus	33	2	3	38	34.9
Ablepharus kitaibelii	4	3	4	11	10.1
Eryx jaculus			2	2	1.8
Coronella austriaca			6	6	5.5
Dolichophis caspius	34	4	3	41	37.6
Elaphe sauromates	1	1	10	12	11.0
Zamenis longissimus	10	3	1	14	12.8
Natrix natrix	30	6	4	40	36.7
Natrix tessellata	15	12	7	34	31.2
Vipera ammodytes	2		7	9	8.3

Conservation Threats and Recommendations

The herpetofauna along the Lower Danube faces multiple threats, commonly observed through Bulgaria such as direct killing (e.g., intentional killing of snakes and freshwater turtles by fishermen), road mortality, loss of habitat due to changes in land use (conversion to agricultural lands), runoff pollution by agriculture, etc. Some of those threats have been described in further details in Popgeorgiev (2008), Popgeorgiev & Kornilev (2009), Stojanov et al. (2011), Popgeorgiev et al. (2014a, b), Tzankov & Popgeorgiev (2014), and Naumov et al. (2016). Here, we focus on ongoing and potential risks more specific to the Danube (which still might apply to other riverine systems). One must keep in mind that often actions that impact the hydrology of a river (especially a major river such as the Danube) could often have complex and cumulative effects and should be considered in such a framework.

However, for ease of discussion, we roughly divide them into several types.

Loss and Habitat Deterioration of Temporary and Permanent Water Bodies

Worldwide, among the key threats for the herpetofauna is habitat loss, especially the disappearance of aquatic habitats, particularly important for the amphibians. In the study area, natural wetlands can be either "Natural or semi-natural mesotrophic to eutrophic lakes and marshes with macrophytic vegetation (04C1)" or "Shallow drying-up water bodies with floating vegetation (05C1)". Both are categorized as Endangered (EN) in the Red Data Book of Bulgaria, being imperiled by draining, use for watering, water table control, embankment creation, gravel and sand extraction, pollution with waste waters or household waste, fertilizer and pesticide runoff, destruction of riverside forests, drying up as a result of climate aridization or accumulation of dead plant material (Tzonev et al. 2015, Tzonev & Valchev 2015).

The area of the Danube wetlands from the beginning of the 20th c. to 1943 has decreased almost double (from ca. 69 km^2 to ca. 37 km^2), predominantly due to embankment construction (Batakliev 1943). Batakliev lists 15 larger wetlands, from which only 2 remain today in their natural state (Garvan and Srebarna), another 2 have been converted into fishponds that no longer function (Orsoya and Kalimok), and the rest do not exist anymore. Undoubtedly, this progressive reduction in the area of marshes has had a strong negative impacts on the amphibian populations and aquatic reptiles.

Changes to the River Bed and the Hydrodynamics

Another major set of threats for the herpetofauna in the Lower Danube area can be unified under man-made changes to the (semi-)natural conditions of the river, especially as a result of large scale projects for improvement of the conditions for shipping, in flood control, or in urban development (e.g., dredging; embankment, sluice, or dam creation; straightening of the river course). Based on the examples from multiple large European rivers, whose bed and water course have been altered for the purposes of navigation, the effects on the primary water habitats and the near-river ecosystems are in all cases highly negative.

The adverse effects of "river improvements" include: shortening of the river course; increase in the velocity in the main channel combined with concentration of erosive forces, deepening of the river bed, and removal downstream of the lighter components of the riverbed substrate to leave large, rocky components; decreased water temperature; decrease of the water table level and subsequent drying up of soils and wetlands, loss of floodplains and sections of slow moving water, and reduced connectivity between the river and the floodplains; loss of flood retention capacity and increased intensity of flooding, especially further downstream.

Particularly for the Danube, major changes such as the construction on Iron Gates I in the early 1970s led to changes in the species composition and likely contributed to the local disappearance of four species (*P. syriacus*, *T. cristatus*, *L. vulgaris*, and *Zootoca vivipara*) in Romania (Stănescu et al. 2015). The multiple changes in the hydrology of the Danube in Austria upstream from Vienna have likely led to the disappearance of *E. orbicularis*, with the only breeding population remaining in the relatively natural river section between Vienna and Bratislava, the Donau-Auen National Park (M. Schindler, pers. comm.). The deepening of the main channel in relation to the floodplain areas and the inflows into side channels has considerably affected the timing and volume of the amount of water entering the side arms and floodplain (Schiemer 2006), which are the main habitats for the turtle.

Specifically for amphibians, in floodplain environments the availability and accessibility of suitable breeding places are controlled by the water level. Decreases in water level result in fragmentation of the breeding population across multiple breeding sites and are often unpredictable and vary in location from one breeding season to another.

An important statement was presented by Cogălniceanu & Miaud (2002) – if the present trend of damming, draining, and embankment creation continue, the range of water level fluctuations will increase in the remaining naturally flooded areas of the Lower Danube floodplain. Amphibians inhabiting areas subjected to flooding and with irregular water levels further experience physiological stress, impacting their biological functions. For example, the Danube crested newt (*T. dobrogicus*) has a shorter life cycle in flooding territories with irregular water levels in the lower portions of the Danube (Cogălniceanu & Miaud 2002). Similar results have been demonstrated for *L. vulgaris* and *B. bombina* (Cogălniceanu & Miaud 2003; 2004). During this period amphibians feed less frequently, with a subsequent decrease in body condition (Cogălniceanu 1997) (e.g., green water frogs, *Pelophylax spp.*) and reproductive output (Cogălniceanu 1999) (e.g., the smooth newt, *L. vulgaris*).

Flood events also allow predators of amphibians, such as fish, to colonize breeding ponds normally not connected with the river. The highest amphibian species richness in

Beschkov V. 2015a. Sand boa *Eryx jaculus* (Linnaeus, 1758). In: Golemanski V. et al. (Eds.): *Red Data Book of the Republic of Bulgaria*. Volume 2. Animals. Sofia: BAS & MoEW, p. 204.

Beschkov V. 2015b. Blotched snake *Elaphe sauromates* (Pallas, 1814). In: Golemanski V. et al. (Eds.): *Red Data Book of the Republic of Bulgaria*. Volume 2. Animals. Sofia: BAS & MoEW, p. 206.

Beshkov V. & Dushkov D. 1981. Materials on the batrachophagy and herpetophagy of snakes in Bulgaria. *Ecology* 9: 43–50. (In Bulgarian; English summary).

Beshkov V. 1986. Striped and non-striped type of colouring of the ringed snake *Natrix natrix* (L.) (Colubridae, Serpentes) in Bulgaria. *Acta Zoologica Bulgarica* 31: 32–36. (In Bulgarian; English summary).

Beshkov V. 1998. Classes Amphibia and Reptilia (Amphibians and Reptiles). In: Michev T. et al. (Eds.): *Biodiversity of the Srebarna Biosphere Reserve*. Checklist and Bibliography. Sofia: Context & Pensoft, pp. 85–86.

Beškov V. 1972. Zwischenartenkontakte und Zusammenleben der Frösche in Bulgarien [Inter-species contacts and coexistence of frogs in Bulgaria]. *Bulletin de l'Institut de Zoologie et Musée, Sofia [Bulletin of the Institute of Zoology and Museum, Sofia*] 34: 85–95. (In Bulgarian; German summary).

Beškov V., Undshian E., Simeonov S. & Daraktschiev, A. 1967. Neue Angaben über die Verbreitung einiger Amphibien und Reptilien in Bulgarien [New information on the distribution of some amphibians and reptiles in Bulgaria]. *Bulletin de l'Institut de Zoologie et Musée, Sofia* [*Bulletin of the Institute of Zoology and Museum, Sofia*] 25: 5–10. (In Bulgarian; German summary).

Biserkov V. & Naumov B. 2012. Changes after 1948 in the habitats of amphibians and reptiles in the area of the Srebarna lake biosphere reserve. In: Uzunov Y., Georgiev B, Varadinova E., Ivanova N, Pehlivanov L. & Vasilev V. (Eds.) *Ecosystems of the Biosphere Reserve Srebarna Lake*. Sofia: Professor Marin Drinov Academic Publishing House, pp. 163–183.

Biserkov V. 1987. The role of reptiles as additional and reservoir hosts in the helminth circulation in Bulgaria. In: Botev B. (Ed.) *Contemporary Achievements of Bulgarian Zoology*. Sofia: BAS, pp. 81–84. (in Bulgarian).

Biserkov V. 1996. New records of platyhelminth parasites from snakes in Bulgaria. *Comptes Rendus de l'Académie Bulgare des Sciences* [*Accounts of the Bulgarian Academy of Sciences*] 49 (1): 73–75.

Biserkov V., Georgiev B. & Genov T. 1985. *Paracapillaria viperae* sp. n. (Nematoda, Capillariidae) - a parasite of *Vipera ammodytes* in Bulgaria. *Comptes Rendus de l'Académie Bulgare des Sciences* [*Accounts of the Bulgarian Academy of Sciences*] 38 (2): 251–254.

Böhme M., Fritz U., Kotenko T., Džukić G., Ljubisavljević K., Tzankov N. & Berendonk T. U. 2007. Phylogeography and cryptic variation within the *Lacerta viridis* complex (Lacertidae, Reptilia). *Zoologica Scripta* 36: 119–131.

Bolkay S. 1920. O razvojnom redu Vipera Gedulyi By. - *Vipera ammodytes* L - *Vipera meridionalis* Blgr [On the phylogenetical series Vipera Gedulyi By. - *Vipera ammodytes* L - *Vipera meridionalis* Blgr]. *Glasnik Zemaljskog Muzeja u Bosni i Hercegovini, Sarajevo* [*Courier of the National Museum in Bosnia and Herzegovina, Sarajevo*] 32: 1–12. (in Serbo-Croatian; English summary).

Bolkay S. 1924. A list of the amphibians and reptiles in the Bosnian-Hercegovian land-museum, with morphological, biological and zoogeographical notes. *Spomenik Srpske Kralyevske Akademiye* [*Memoirs of the Serbian Royal Academy*] 61 (1): 1–37. (In Serbo-Croatian; English summary).

Boschkow D. 1965. Über die helminthenfauna von *Rana ridibunda* Pall. aus dem Srebărna-sumpf (bezirk Silistra) [On the helminth fauna of *Rana ridibunda* Pall. from the Srebarna swamp (district Silistra). *Bulletin de l'Institut de Zoologie et Musée, Sofia* [*Bulletin of the Institute of Zoology and Museum, Sofia*] 19: 151–165. (In Bulgarian; German summary).

Božkov D. 1967. Experimentelle untersuchungen über das schicksal einiger helminthen beim kannibalismus der *Rana ridibunda* Pall. [Experimental studies on the destiny of some helminths after cannibalism by *Rana ridibunda* Pall.]. *Bulletin de l'Institut de Zoologie et Musée, Sofia* [*Bulletin of the Institute of Zoology and Museum, Sofia*] 25: 79–85. (In Russian; German summary).

Buresch I. & Zonkow J. 1932. Die Verbreitung der Giftschlangen (Viperidae) in Bulgarien und auf der Balkanhalbinsel [The distribution of poisonous snakes (Viperidae) in Bulgaria and on the Balkan Peninsula]. *Travaux de la Société Bulgare des Sciences Naturelles* [*Works of the Bulgarian Society of Natural Sciences*] 15/16: 189–206. (In Bulgarian; German summary).

Buresch I. & Zonkow J. 1933. Untersuchungen über die Verbreitung der Reptilien und Amphibien in Bulgarien und auf der Balkanhalbinsel. I. Schildkrötten (Testudinata) und Eidechsen (Sauria) [Research on the distribution of reptiles and amphibians in Bulgaria and the Balkan Peninsula. I. Turtles (Testudinata) and lizards (Sauria)]. *Mitteilungen aus den Königlichen Naturwissenschaftlichen Instituten in Sofia* [*Communications from the Royal Institute of Natural Sciences in Sofia*] 6: 150–207. (In Bulgarian; German summary).

Buresch I. & Zonkow J. 1934. Untersuchungen über die Verbreitung der Reptilien und Amphibien in Bulgarien und auf der Balkanhalbinsel. II Teil: Schlangen (Serpentes) [Research on the distribution of reptiles and amphibians in Bulgaria and the Balkan Peninsula. II part: Snakes (Serpentes)]. *Mitteilungen aus den Königlichen Naturwissenschaftlichen Instituten in Sofia* [*Communications from the Royal Institute of Natural Sciences in Sofia*] 7: 106–188. (In Bulgarian; German summary).

Buresch I. & Zonkow J. 1941. Untersuchungen über die Verbreitung der Reptilien und Amphibien in Bulgarien und auf der Balkanhalbinsel. III. Teil: Schwanzlurche (Amphibia, Caudata) [Research on the distribution of reptiles and amphibians in Bulgaria and the Balkan Peninsula. III part: Tailed amphibians (Amphibia, Caudata]. *Mitteilungen aus den Königlichen Naturwissenschaftlichen Instituten in Sofia* [*Communications from the Royal Institute of Natural Sciences in Sofia*] 14: 171–273. (In Bulgarian; German summary).

Buresch I. & Zonkow J. 1942. Untersuchungen über die Verbreitung der Reptilien und Amphibien in Bulgarien und auf der Balkanhalbinsel. IV. Teil: Froschlurche (Amphibia, Salientia) [Research on the distribution of reptiles and amphibians in Bulgaria and the Balkan Peninsula. IV part: Frogs (Amphibia, Salientia)]. *Mitteilungen aus den Königlichen Naturwissenschaftlichen Instituten in Sofia* [*Communications from the Royal Institute of Natural Sciences in Sofia*] 15: 68–165. (In Bulgarian; German summary).

Călinescu R. 1931. Contribuţiuni sistematice şi zoogeografice la studiul amphibiilor şi reptilelor din România [Systematic and zoogeographic contributions to the study of amphibians and reptiles in Romania]. *Academia Română Memoriile Secţiunii Ştiinţifice* [*Romanian Academy Scientific Section Memos*] Seria III, 7: 119–291.

Chochev B. 1980. *Hyla arborea* /Linné/ a new host for *Polystoma integerrimum* in Bulgaria. *Université de Plovdiv "Paisii Hilendarski" Travaux Scientifiques* [*Plovdiv University "Paisii Hilendarski" Scientific Works*] 18 (4): 191–194. (In Bulgarian; English summary).

Cogălniceanu D. & Miaud C. 2002. Age, survival and growth in *Triturus dobrogicus* (Amphibia, Urodela) from the lower Danube floodplain. *International Association for Danube Research* 34: 777–783.

Cogălniceanu D. & Miaud C. 2003. Population, age structure and growth of four amphibian species inhabiting a large river floodplain. *Canadian Journal of Zoology* 81: 1096–1106.

Cogălniceanu D. & Miaud C. 2004. Variations in life history of a *Bombina bombina* population from the lower Danube floodplain. *Amphibia-Reptilia* 25: 115–118.

Cogălniceanu D. 1997. Fitness of green frog populations (*Rana esculenta* complex) from the lower Danube floodplain. In: 32 Konferenz der IAD, Wien. Wissenschaftliche Referate [32nd Conference of IAD, Vienna. Scientific papers]: 485–489.

Cogălniceanu D. 1999. Egg deposition strategies of the Smooth newt (*Triturus vulgaris*) in an unpredictable environment. *Herpetological Journal* 9: 119–123.

Cogălniceanu D., Székely P., Samoilă C., Iosif R., Tudor M., Plăiaşu R., Stănescu F. & Rozylowicz L. 2013a. Diversity and distribution of amphibians in Romania. *ZooKeys* 296: 35–57.

Cogălniceanu D., Rozylowicz L., Székely P., Samoilă C., Stănescu F., Tudor M., Székely D. & Iosif R. 2013b. Diversity and distribution of reptiles in Romania. *ZooKeys* 341: 49–76.

Covaciu-Marcov S. D., Ghira I., Cicort-Lucaciu A. S., Sas I., Strugariu A. & Bogdan H. 2006. Contributions to knowledge regarding the geographical distribution of the herpetofauna of Dobrudja, Romania. *North-Western Journal of Zoology* 2 (2): 88–125.

Dubois A. & Bour R. 2010. The nomenclatural status of the nomina of amphibians and reptiles created by Garsault (1764), with a parsimonious solution to an old nomenclatural problem regarding the genus Bufo (Amphibia, Anura), comments on the taxonomy of this genus, and comments on some nomina created by Laurenti (1768). *Zootaxa* 2447: 1–52.

Dufresnes C., Brelsford A., Crnobrnja-Isailović J., Tzankov N., Lymberakis P. & Perrin N. 2015. Timeframe of speciation inferred from secondary contact zones in the European tree frog radiation (*Hyla arborea* group). *BMC Evolutionary Biology* 15: 155. DOI 10.1186/s12862-015-0385-2.

Fuhn I. 1960. *Fauna Republicii Populare Române* [*Fauna of the People's Republic of Romania*]. Vol. 14. Fascicle 1. Amphibia. Bucharest: Editura Academiei Republicii Populare România. 288 p.

Genov T. 1969. Natural infestation of the vertebrates from the reservation "Srebarna" with plerocercoids of *Spirometra erinacei-europaei* (Rudolphi, 1819) - Diphyllobothriidae (Cestoidea). *Bulletin of the Central Helminthological Laboratory* 13: 197–209. (In Bulgarian; English summary).

Gvoždík V., Canestrelli D., García-París M., Moravec J., Nascetti G., Recuero E., Teixeira J. & Kotlík P. 2015. Speciation history and widespread introgression in the European short-call tree frogs (*Hyla arborea* sensu lato, *H. intermedia* and *H. sarda*). *Molecular Phylogenetics and Evolution* 83: 143–155.

Hecht G. 1930. Systematik, Ausbreitungsgeschichte und Oekologie der europäischen Arten der Gattung Tropidonotus (Kuhl) H. Boie [Systematics, distrobution history and ecology of the European species of the genus Tropidonotus (Kuhl) H. Boie]. *Mitteilungen aus dem Zoologischen Museum in Berlin* [*Communications of the Zoological Museum in Berlin*] 16: 244–397.

Jablonski D., Jandzik D., Mikulíček P., Džukić G., Ljubisavljević K., Tzankov N., Jelić D., Thanou E., Moravec J. & Gvoždík V. 2016. Contrasting evolutionary histories of the legless lizards slow worms (*Anguis*) shaped by the topography of the Balkan Peninsula. *BMC Evolutionary Biology* 16: 99. DOI 10.1186/s12862-016-0669-1.

Joly P. & Morand A. 1994. Theoretical habitat templets, species traits, and species richness: amphibians in the Upper Rhône River and its floodplain. *Freshwater Biology* 31 (3): 455–468.

Kindler C., Chèvre M., Ursenbacher S., Böhme W., Hille A., Jablonski D., Vamberger M. & Fritz U. 2017. Hybridization patterns in two contact zones of grass snakes reveal a new Central European snake species. *Scientific Reports* 7: 7378. DOI: 10.1038/s41598-017-07847-9.

Kirin D. 2002. New data on the Helminth fauna of *Lacerta viridis* Laurenti, 1768, and *Podarcis muralis* (Laurenti, 1768) in Bulgaria. *Acta Zoologica Bulgarica* 54 (1), 43–48.

Kirin D. 2003. Biological diversity and ecological measurements of the helminth communities of *Rana ridibunda* from district of town of Vidin. *Experimental Pathology and Parasitology* 6 (11): 37–43.

Kopralev I. (Ed.) 2002. *Geography of Bulgaria. Physical and socio-economic geography.* Sofia: ForKom. 760 p. (In Bulgarian).

Kornilev Y. V., Popgeorgiev G., Naumov B., Stoyanov A. & Tzankov N. 2017. Updated distribution and ecological requirements of the native freshwater turtles in Bulgaria. *Acta Zoologica Bulgarica* Suppl. 10: 65–76.

Kotsakiozi P., Jablonski D., Ilgaz Ç., Kumlutaş Y., Avcı A., Meiri S., Itescu Y., Kukushkin O., Gvoždík V., Scillitani G., Roussos S. A., Jandzik D., Kasapidis P., Lymberakis P. & Poulakakis N. 2018. Multilocus phylogeny and coalescent species delimitation in Kotschy's gecko, *Mediodactylus kotschyi*: Hidden diversity and cryptic species. *Molecular Phylogenetics and Evolution* 125: 177–187.

Kovachev V. 1894. Materials for studying the Bulgarian fauna. *Periodichesko Spisanie na Balgarskoto Knizhovno Druzhestvo* [*Periodical Journal of the Bulgarian Literary Society*] 47 (9): 742–749. (In Bulgarian).

Kovachev V. 1905a. A contribution to the study of amphibians and reptiles in Bulgaria. *Sbornik za Narodni Umotvoreniya, Nauka i Knizhnina* [*Collection of Folk Mind-Creations, Science, and Literacy*] 21: 1–13. (In Bulgarian).

Kovachev V. 1905b. *A brief identification guide of the freshwater fishes and amphibians of Bulgaria*. Ruse: Publ. House "Bratya Zlatevi". 50 p. (In Bulgarian).

Kovachev V. 1907. Several species and varieties new for the Bulgarian fauna. *Periodichesko Spisanie na Balgarskoto Knizhovno Druzhestvo* [*Periodical Journal of the Bulgarian Literary Society*] 68: 217–218. (In Bulgarian).

Kovachev V. 1910. *Rana esculenta* typica in Bulgaria. *Estestvoznanie* [*Natural science*] 1 (9): 588. (In Bulgarian).

Kovachev V. 1912. *The herpetofauna of Bulgaria*. Plovdiv: "H. G. Danov" Publ.. 90 p. (In Bulgarian).

Kovatcheff V. 1906. Quelques espèces nouvelles pour la faune bulgare [Some new species for the Bulgarian fauna]. *Trudove na Balgarskoto Prirodoizpitatelno Druzhestvo* [*Works of the Bulgarian Naturalist Society*] 3: 127–128. (In Bulgarian; French summary).

Kovatscheff V. 1903. Beiträge zur Kenntnis der Reptilien- und Amphibienfauna Bulgariens [Contributions to the knowledge of the reptile and amphibian fauna of Bulgaria]. *Verhandlungen der kaiserlich-königlichen zoologisch-botanischen*

Gesellschaft in Wien [*Proceedings of the Imperial-royal Zoological-botanical Society in Vienna*] 53: 171–173.

Kowatscheff V. 1905. Beiträge zur Kenntnis der Reptilien- und Amphibienfauna Bulgariens [Contributions to the knowledge of the reptile and amphibian fauna of Bulgaria]. *Verhandlungen der kaiserlich-königlichen zoologisch-botanischen Gesellschaft in Wien* [*Proceedings of the Imperial-royal Zoological-botanical Society in Vienna*] 55: 31–32.

Koynova T., Tzankov N., Popgeorgiev G., Naumov B. & Natchev N. 2017. A new distribution record of the Kotschy's Gecko (*Mediodactylus kotschyi*) from inland north-eastern Bulgaria. *Herpetology Notes* 10: 1–2.

Ljubisavljević K. & Dzukić G. 2015. *Pelobates syriacus*. In: Kalezić M., Tomović L. & Dzukić G. (Eds.): *Red Book of Fauna of Serbia I - Amphibians*. Belgrade: Planeta Print, pp. 172–178.

Lukanov S., Tzankov N. & Naumov B. 2017. First documented records of *Pelophylax lessonae* (Camerano, 1882) (Amphibia: Ranidae) from Bulgaria. *Acta Zoologica Bulgarica* 69 (4): 483–488.

Lukanov S., Tzankov N. & Simeonovska-Nikolova D. 2014. Effects of environmental factors on mating call characteristics of the Marsh frog *Pelophylax ridibundus* (Pallas, 1771) (Amphibia: Ranidae) in Bulgaria. *Acta Zoologica Bulgarica* 66 (2): 209–216.

Marzahn E., Mayer W., Joger U., Ilgaz C., Kindler C., Kumlutas Y., Nistri A., Schneeweiss N., Vamberger M., Zagar A. & Fritz U. 2016. Phylogeography of the *Lacerta viridis* complex: mitochondrial and nuclear markers provide taxonomic insights. *Journal of Zoological Systematics and Evolutionary Research* 54 (2): 85–105.

Metcheva R., Beltcheva M., Naumov B., Yankov Y., Michev T., Profirov L., Mitov P., Kenderov L., Georgieva E., Petrov P. & Goranov S. 2016. Faunistic study of the Tsibar Danube Island. Forum Geografic. *Studii şi Cercetări de Geografie şi Protecţia Mediului* [*Studies and Research of Geography and Environmental Protection*] 15 (2): 152–161.

Michev T. 1958. The Grass snake (*Natrix natrix*). *Priroda i Znanie* [*Nature and knowledge*] 9 (2): 17–18. (In Bulgarian).

Mollov I., Kirov K., Petrova S., Georgiev D. & Velcheva I. 2013. Assessing the influence of the automobile traffic on the amphibians and reptiles in the buffer zone of biosphere reserve "Srebarna" (NE Bulgaria). *Ecologia Balkanica* 5 (2): 31–39.

Morand A. & Joly P. 1995. Habitat variability and space utilization by amphibian communities of the French Upper-Rhone floodplain. *Hydrobiologia* 300/301: 249–257.

Naumov B. & Biserkov V. 2013. On the distribution and subspecies affiliation of *Triturus dobrogicus* (Amphibia: Salamandridae) in Bulgaria. *Acta Zoologica Bulgarica* 65 (3): 307–313.

Naumov B. 2005. New records of some herpetofauna species in Bulgaria. *Acta Zoologica Bulgarica* 57 (3): 391–396.

Naumov B., Tzankov N., Donchev K., Petrov B., Stojanov A., Popgeorgiev G., Mollov I. & Beshkov V. 2016. The herpetofauna (Amphibia and Reptilia) of Vrachanska Planina Mountains - species composition, distribution and conservation. *ZooNotes* Suppl. 3: 231–257.

Naumov B., Tzankov N., Popgeorgiev G., Stojanov A. & Kornilev Y. 2011. The Dice snake (*Natrix tessellata*) in Bulgaria: distribution and morphology. *Mertensiella* 18: 288–297.

Paspalev G., Dokov V., Tschacarov E. & Boschkov D. 1962. Eine bisher unbekannte, durch Rickettsienähnlichen Mikroorganismus hervorgerufene Erkrankung bei Fröschen in Bulgarien [A previously unknown Rickettsia-like micro-organism found in frogs in Bulgaria]. *Bulletin de l'Institut de Zoologie et Musée, Sofia* [*Bulletin of the Institute of Zoology and Museum, Sofia*] 12: 5–24. (In Bulgarian; German summary).

Paspaleva-Antonova M. 1961. Beitrag zur Ornithofauna des Reservats "Srebarna" im Bezirk Silistra [Contributions to the ornithofauna of the "Srebarna" Reserve, district Silistra]. *Bulletin de l'Institut de Zoologie et Musée, Sofia* [*Bulletin of the Institute of Zoology and Museum, Sofia*] 10: 139–163. (In Bulgarian; German summary).

Petkov V. 2002. Second record of the Balkan gecko (*Cyrtodactylus kotschyi danilewskii* Strauch 1887) in the region of Russe. *Historia Naturalis Bulgarica* 15: 156–157. (In Bulgarian).

Popgeorgiev G. & Kornilev Y. V. 2009. Effects of a high intensity fire on the abundance and diversity of reptiles in the Eastern Rhodopes Mountains, southeastern Bulgaria. *Ecologia Balkanica* 1: 41–50.

Popgeorgiev G. 2008. The effects of a large-scale fire on the demographic structure of a population of Hermann's (*Testudo hermanni boettgeri* Mojsisovics, 1889) and Spur-thighed (*Testudo graeca ibera* Pallas, 1814) tortoises in Eastern Rhodopes Mountains, Bulgaria. *Historia Naturalis Bulgarica* 19: 115–127.

Popgeorgiev G., Spasov S. & Kornilev Y. V. 2015. SmartBirds – Information system with biological information, BSPB. Available at http://www.smartbirds.org.

Popgeorgiev G., Tzankov N., Kornilev Y. V., Naumov B. & Stoyanov A. 2014a. Amphibians and reptiles in Ponor Special Protection Area (Natura 2000), western Bulgaria: Species diversity, distribution and conservation. *Acta Zoologica Bulgarica* Suppl. 5: 85–96.

Popgeorgiev G., Tzankov N., Kornilev Y. V., Plachyiski D., Naumov B. & Stoyanov A. 2014b. Changes in agri-environmental practices pose a threat to the herpetofauna: A case study from Besaparski Ridove Special Protection Area (Natura 2000), southern Bulgaria. *Acta Zoologica Bulgarica* Suppl. 5: 157–169.

Popov P. 1975. Investigation on the absolutely length of the chromosomes and the genome of *Rana esculenta* L. *Université de Plovdiv "Paisii Hilendarski" Travaux Scientifiques*

[*Plovdiv University "Paisii Hilendarski" Scientific Works*] 13 (4): 19–23. (In Bulgarian; English summary).

Saberi-Pirooz R., Ahmadzadeh F., Ataei S., Taati M., Turk Qashqaei A. & Carretero M. 2018. A phylogenetic assessment of the meadow lizard *Darevskia praticola* (Eversmann, 1834) from Iran. *Zootaxa* 4441 (1): 046–058.

Schiemer F. 2006. Ecological status and problems of the Danube and its fish fauna. In: Proceedings 36th International Conference of IAD. ISBN 13: 978-3-9500723-2-7. Vienna: Austrian Committee DanubeResearch/IAD, pp. 6–11.

Schlüter U. 2005. Die Smaragdeidechsen der Dobrudscha [The green lizard in Dobrudzha]. *Die Eidechse* [*The Lizard*] 16 (2): 46–61.

Schumann A. 1930. Die Schausammlungen des Königlichen Naturhistorischen Museums in Sofia [The collections of the Royal Natural History Museum in Sofia]. *Mitteilungen aus den Königlungen Naturwissenschaftlichen Instituten in Sofia* [*Communications of the royal societies of science institutes in Sofia*] 3: 1–60.

Simeonov S. & Boev Z. 1988. A study on the nutritive spectrum of the eagle owl (*Bubo bubo* L.) in Bulgaria. *Ecology* 21: 47–56. (In Bulgarian; English summary).

Stănescu F., Buhaciuc E., Székely P., Székely D., Rozylowicz L. & Cogălniceanu D. 2015. The impact of dam construction on amphibians and reptiles. Study case – Iron Gates I. *Analele Ştiinţifice ale Universităţii "Alexandru Ioan Cuza" din Iaşi, s. Biologie animală* [*Scientific Annals of "Alexandru Ioan Cuza" University of Iasi, Animal Biology*] 61: 19–24.

Stöck M., Dubey S., Klütsch C., Litvinchuk S., Scheidt U. & Perrin N. 2008. Mitochondrial and nuclear phylogeny of circum-Mediterranean tree frogs from the *Hyla arborea* group. *Molecular Phylogenetics and Evolution* 49: 1019–1024.

Stöck M., Dufresnes S., Litvinchuk S., Lymberakis P., Biollay S., Berroneau M., Borzée A., Ghali K., Ogielska M. & Perrin N. 2012. Cryptic diversity among Western Palearctic tree frogs: postglacial range expansion, range limits, and secondary contacts of three European tree frog lineages (*Hyla arborea* group). *Molecular Phylogenetics and Evolution* 65: 1–9.

Stöck M., Moritz C., Hickerson M., Frynta D., Dujsebayeva T., Eremchenko V., Macey J., Papenfuss T. & Wake D. 2006. Evolution of mitochondrial relationships and biogeography of Palearctic green toads (*Bufo viridis* subgroup) with insights in their genomic plasticity. *Molecular Phylogenetics and Evolution* 41: 663–689.

Stojanov A., Tzankov N. & Naumov B. 2011. *Die Amphibien und Reptilien Bulgariens* [*The Amphibians and Reptiles of Bulgaria*]. Frankfurt am Main: Chimaira. 588 p.

Stoyneva M. & Michev T. 2007. Database of Bulgarian non-lotic wetlands and their biodiversity. In: Michev T. & Stoyneva M. (Eds.): *Inventory of Bulgarian wetlands and their biodiversity. Part 1: Non-lotic Wetlands*. Sofia: Publ. House Svetlostrouy. 364 p. + CD supplement.

Tomov V. 1989. On the food of *Rana ridibunda* Pall. in the region of Lom. *Université de Plovdiv "Paisii Hilendarski" Travaux Scientifiques* [*Plovdiv University "Paisii Hilendarski" Scientific Works*] 27 (5): 143–151. (In Bulgarian; English summary).

Tomov V. 1990. On the food of *Bufo viridis* Laur. *Université de Plovdiv "Paisii Hilendarski" Travaux Scientifiques* [*Plovdiv University "Paisii Hilendarski" Scientific Works*] 28 (6): 121–129. (In Bulgarian; English summary).

Tomov V. 1991. On the food of Fire-bellied toad (*Bombina bombina* L.) in NW Bulgaria. *Université de Plovdiv "Paisii Hilendarski" Travaux Scientifiques* [*Plovdiv University "Paisii Hilendarski" Scientific Works*] 29 (6): 49–54. (In Bulgarian; English summary).

Tomović L., Ajtić R., Ljubisavljević K., Urošević A., Jović D., Krizmanić I., Labus N., Đorđević S., Kalezić M., Vukov T. & Džukić G. 2014. Reptiles in Serbia - distribution and diversity patterns. *Bulletin of the Natural History Museum* 7: 129–158.

Tuleshkov K. 1954. The snakes and the national economy. *Priroda i Znanie* [*Nature and Knowledge*] 7 (10): 3–6. (In Bulgarian).

Tzankov N. & Popgeorgiev G. 2014. Conservation and declines of Amphibians in Bulgaria. In: Heatwole H. & Wilkinson J. (Eds.): *Amphibian Biology, Volume 11 Part 4: Status of Conservation and Decline of Amphibians: Eastern Hemisphere: Southern Europe & Turkey*. Exeter: Pelagic Publishing, pp. 131–139.

Tzankov N., Popgeorgiev G., Kornilev Y., Natchev N., Stoyanov A., Naumov B. & Ivanchev I. 2015. First survey on the invasive Pond slider (*Trachemys scripta*) in Bulgaria: historic development and current situation. *Hyla* 2015 (1): 18–27.

Tzankov N., Stoyanov A., Dobrev D., Lukanov S., Kornilev Y. V., Andonov K., Vergilov V. & Popgeorgiev G. 2017. *Trachemys scripta* (Thunberg in Schoepff, 1792). In: Trichkova T., Vladimirov V., Tomov R. & Todorov M. (Eds.) *Guide to Invasive Alien Species of European Union Concern*. Sofia: IBER-BAS & ESENIAS, pp. 110–113.

Tzonev R. & Valchev V. 2015. 05C1 Shallow, drying-up water bodies with floating vegetation. In: Biserkov V., Gussev Ch., Popov V., Hibaum G., Roussakova V., Pandurski I., Uzunov J., Dimitrov M., Tzonev R. & Tsoneva S. (Eds.): *Red Data Book of the Republic of Bulgaria, vol. 3. Natural Habitats*. Sofia: MOEW-BAS, pp. 87–88.

Tzonev R., Gussev Ch., Georgiev V. & Tsoneva S. 2019. Important Plant Areas along the Bulgarian stretch of the Danube River – contemporary knowledge and gaps in their determination. In: Shurulinkov P. (ed.) *Biodiversity of the Bulgarian-Romanian section of the Lower Danube*. Chapter 1. Nova Sceince Publishers (manuscript, in print).

Tzonev R., Valchev V. & Georgiev V. 2015. 04C1 Natural or semi-natural mesotrophic to eutrophic lakes and marshes with macrophytic vegetation. In: Biserkov V., Gussev Ch., Popov V., Hibaum G., Roussakova V., Pandurski I., Uzunov J., Dimitrov M., Tzonev R. & Tsoneva S. (Eds.) *Red Data Book of the Republic of Bulgaria, vol. 3. Natural Habitats*. Sofia: MOEW-BAS, pp. 84–87.

Undjian E. 2000. *Die Lurche und Kriechtiere des Lomtales und des Naturparkes "Russenski Lom" Bezirke Russe und Rasgrad Nordostbulgarien* [*The amphibians and reptiles of the Lom valley and the Nature Park "Russenski Lom" Districts Russe and Razgrad, Northeast Bulgaria*]. Ruse: Priroden Park "Rusenski Lom". 88 p. (In Bulgarian; German summary).

Vergilov V. & Tzankov N. 2014. Influence of environmental factors on the morphology of the Snake-eyed Skink *Ablepharus kitaibelii* (Reptilia: Scincidae) in Eastern Balkans. *Acta Zoologica Bulgarica* 66 (2): 217–224.

Vergilov V., Tzankov N. & Zlatkov B. 2018. Age structure and growth in Bulgarian populations of *Ablepharus kitaibelii* (Bibron & Bory de Saint-Vincent, 1833) (Squamata: Sauria: Scincidae). *Herpetozoa* 30 (3/4): 179–185.

Vergilov V., Zlatkov B. & Tzankov N. 2017. Hemipenial differentiation in the closely related congeners *Ablepharus kitaibelii* (Bibron & Bory de Saint-Vincent, 1833), and *Ablepharus budaki* Göçmen, Kumlutaş & Tosunoğlu, 1996 (Squamata: Sauria: Scincidae). *Herpetozoa* 30 (1/2): 39–48.

Vukov T., Kalezić M., Tomović L., Krizmanić I., Jović D., Labus N. & Džukić G. 2013. Amphibians in Serbia - distribution and diversity patterns. *Bulletin of the Natural History Museum* 6: 90–112.

Wallis G. & Arntzen J. 1989. Mithochondrial-DNA variation in the crested newt superspecies: limited cytoplasmic gene flow among species. *Evolution* 43 (1): 88–104.

Werner F. 1898. Prilozi poznavanju faune reptilija i batrahija Balkanskog poluostrva [Contributions to the knowledge of fauna of reptiles and amphibians of the Balkan peninsula]. *Glasnik Zemaljskog Muzeja u Bosni I Hercegovini, Sarajevo* [*Courier of the National Museum in Bosnia and Herzegovina, Sarajevo*] 10 (1): 131–156.

Wielstra B., Baird A. & Arntzen J. 2013a. A multimarker phylogeography of crested newts (*Triturus cristatus* superspecies) reveals cryptic species. *Molecular Phylogenetics and Evolution* 67: 167–175.

Wielstra B., Crnobrnja-Isailovic J., Skidmore A., Sotiropoulos K., Toxopeus A., Tzankov N., Vukov T. & Arntzen J. 2013b. Tracing glacial refugia of *Triturus* newts based on mitochondrial DNA phylogeography and species distribution modeling. *Frontiers in Zoology* 10: 13. DOI: 10.1186/1742-9994-10-13.

Wielstra B., Canestrelli D., Cvijanović M., Denoël M., Fijarczyk A., Jablonski D., Liana M., Naumov B., Olgun K., Pabijan M., Pezzarossa A., Popgeorgiev G., Salvi D., Si Y., Sillero N., Sotiropoulos K., Zielinski P. & Babik W. 2018. The distributions of the six species constituting the smooth newt species complex (*Lissotriton vulgaris* sensu lato and *L. montandoni*) - an addition to the New Atlas of Amphibians and Reptiles of Europe. *Amphibia-Reptilia* 39 (2): 252–259.

Wielstra B., Sillero N., Vörös J. & Arntzen J. 2014. The distribution of the crested and marbled newt species (Amphibia: Salamandridae: *Triturus*) - an addition to the New Atlas of Amphibians and Reptiles of Europe. *Amphibia-Reptilia* 35: 376–381.

Wielstra B., Vörös J. & Arntzen J. 2016. Is the Danube crested newt *Triturus dobrogicus* polytypic? A review and new nuclear DNA data. *Amphibia-Reptilia* 37 (2): 167–177.

In: Biodiversity of the Bulgarian-Romanian Section
Editor: Shurulinkov, Hubenov et al.
ISBN: 978-1-53615-663-8

Chapter 9

Heron and Cormorant Colonies along the Bulgarian-Romanian Section of the Danube River: Status and Trends, 2010-2014

Peter Shurulinkov[1,*], Girgina Daskalova[1], Svilen Cheshmedzhiev[2], Krassimir Kirov[4], Veselin Koev[6], Hristo Dinkov[3], Ivan Hristov[5], Ivailo Nikolov[3], Stoyan Mihov[5] and Yordan Kutsarov[2]
[1]National Museum of Natural History, Bulgarian Academy of Sciences, Sofia, Bulgaria
[2]Bulgarian Society for the Protection of Birds, Sofia, Bulgaria
[3]Balkani Wildlife Society, Sofia, Bulgaria
[4]Rousse Regional Museum, Rousse, Bulgaria
[5]World Wildlife Fund- Danube-Carpathian programme, Sofia, Bulgaria
[6]Directorate of Persina" Nature Park, Belene, Bulgaria

Abstract

Distribution and numbers of colonially nesting herons, cormorants, Eurasian Spoonbills and Glossy Ibises were studied along the whole Bulgarian –Romanian section of the Danube River in the period 2010-2014. A total of 15 colonies, 12 of them mixed, were located and their species composition and trends in numbers were followed during the study. The total numbers by species differed considerably between years as follows: Great Cormorant (*Phalacrocorax carbo*) - 1573 -2299 pairs, Pigmy Cormorant

[*]Correspodnding Author's Email: p.shurulinkov@gmail.com.

(*Microcarbo pygmaeus*) – 283 - 877 pairs, Grey Heron (*Ardea cinerea*) – 237-399 pairs, Purple Heron (*Ardea purpurea*) – 4 - 18 pairs, Great White Heron (*Ardea alba*) – 6 - 12 pairs, Squacco Heron (*Ardeolla ralloides*) – 60- 181 pairs, Night Heron (*Nycticorax nycticorax*) – 234 - 484 pairs, Eurasian Spoonbill (*Platalea leucorodia*) – 89 -147 pairs, Glossy Ibis (*Plegadis falcinellus*) – 10 - 186 pairs. The most numerous colonies were at Ibisha Island, small islet off Golyama Burzina Island (on Romanian territory) and in Srebarna Lake. Trends of decrease in numbers for the period 2006 -2014 were detected for Pigmy Cormorants and Grey Herons. Increasing numbers were recorded for Night Herons, Eurasian Spoonbills and Glossy Ibises. The remaining species showed more or less stable population numbers.

Keywords: population trends, heronries, distribution, numbers, fluctuations

INTRODUCTION

Heron and cormorant colonies along the Lower Danube River include part of the most important and priority bird species of conservation point of view. Most of the colonies along the Bulgarian-Romanian section of the Danube River are situated on islands covered by willow and poplar forests. The birds of the colonies nest on the trees. Lower number of colonies were known from the riverside marshes, lakes and fishponds. From the past there are data on the species composition and numbers of breeding pairs from particular colonies and only in concrete years (Paspaleva & Mitschev 1968). The only colony which has been monitored lees or more constantly is in Srebarna Lake (Michev 1968, Kambourova 2005, Kambourova 2012).

Full count of all colonies of herons and cormorants of Bulgarian-Romanian section of the Danube River was made in 2006 (Shurulinkov et al. 2007). In the present study we add new data from four full yearly counts performed in the period 2010-2014. These materials were gathered with the aim to obtain knowledge about the trends of the species composition and numbers for the whole period of nine years, 2006-2014. For more general, long -term trends we used other published studies mostly after 1990. The catastrophic negative trends in the numbers of these colonial birds and the reasons for them before 1990 were already presented and discussed (Botev & Peshev Eds. 1985, Simeonov et al. 1990).

MATERIAL AND METHODS

All known colonies of herons, cormorants, ibises and spoonbills along the Bulgarian-Romanian section of Danube were included in the counts during eight expeditions in the period 2010-2014. Island colonies were studied by boat or on foot. Colonies in the wetlands on the Bulgarian side of the river were studied by car. In 2010 one full count was conducted in mid-June. During the period 2011-2013 two expeditions were organized annually - first in the second half of May and second in mid - June. The second count had the purpose to lighten whether the breeding of the birds has been successful and is there any considerable changes in the numbers of the birds and the location of the colonies. The counts included all Bulgarian and Romanian Danube islands along the studied section as well as all wetlands on the Bulgarian side of the river. The wetlands on the Romanian river bank were not covered during this study.

In 2014 nine of the colonies were covered by one census made in June. These data were used for determining more precisely the trends for particular colonies. They were not used in the analysis for the total numbers of the studied bird species or for the comparison of the importance of the different colonies for every species.

In most of the cases the island colonies were counted from a motor boat. Entering the island and inside the colonies was implemented only in three colonies with poor or lacking visibility from the river flow – the colonies at Kutovo Island, Ibisha Island, and Vardim island. That was made only in the first (May) expeditions, in each year of the study.

Results and Discussion

In total we detected 15 colonies of herons, cormorants, ibises and spoonbills along the studied Danube River section. The list of the colonies, their location and species composition are presented on Table 1.

Great Cormorant (*Phalacrocorax carbo*)

Distribution and numbers of the species are presented on Table 2. Great Cormorant showed significant fluctuations of its numbers but generally with a trend of slight increase (Figure 1). It had stable or increasing numbers in all colonies except on the islands Ibisha, Byalata Prust and Veresht. The colony at Byalata Prust Island (close to the Srebarna Lake) was exterminated due to disturbance as a result of logging activity during the breeding period of the birds. Logging was conducted in a neighboring poplar plantation. Since 2006 part of the birds from that major colony moved to nest on the closely situated (5 km.) Romanian island Veresht but the others left the area.

The colony on Mishka Island also decreased after 2010 when an intensive clear cutting was implemented in the poplar plantations resulting to removal of 80% of the island's forest coverage. Since 2013 part of the Great Cormorants from Mishka Island moved to a secure small island upstream the river, at 800 m. from their previous breeding location. Thus that colony was divided to two colonies.

Table 1. Heron and cormorant colonies along the Bulgarian-Romanian section of Danube River, 2010-2014

Locality	Coordinates	UTM-grid (10X10 km) square	Full species composition of the nesting birds for the study period
Kutovo island	N 44°00'51.93" E 22°59'31.14"	FP57	Pigmy Cormorant (*Microcarbo pygmaeus*), Great Cormorant (*Phalacrocorax carbo*), Grey Heron (*Ardea cinerea*), Little Egret (*Egretta garzetta*), Night Heron (*Nycticorax nycticorax*), Squacco Heron (*Ardeola ralloides*), Eurasian Spoonbill (*Platalea leucorodia*)
Malak Bliznak Island	43°51'51.76" N 22°50'23.87" E	FP65	Pigmy Cormorant (*Microcarbo pygmaeus*), Great Cormorant (*Phalacrocorax carbo*), Grey Heron (*Ardea cinerea*), Little Egret (*Egretta garzetta*), Night Heron (*Nycticorax nycticorax*), Eurasian Spoonbill (*Platalea leucorodia*)
Ibisha Island (of Dolni Tsibar village)	43°49'34.07" N 23°31'11.47" E	GP05	Great Cormorant (*Phalacrocorax carbo*), Grey Heron (*Ardea cinerea*), Little Egret (*Egretta garzetta*), Night Heron (*Nycticorax nycticorax*), Squacco Heron (*Ardeola ralloides*), Eurasian Spoonbill (*Platalea leucorodia*)
Ostrov marsh	43°40'7.72"С 24° 5'44.10"И	KJ63	Grey Heron (*Ardea cinerea*)
Kalnovats Island (Romania)	43°42'12.35" N 24°46'57.91" E	LJ24	Great Cormorant (*Phalacrocorax carbo*), Grey Heron (*Ardea cinerea*), Little Egret (*Egretta garzetta*), Night Heron (*Nycticorax nycticorax*), Squacco Heron (*Ardeola ralloides*)
Small island of Golyama Burzina Island (Romania)	43°41'51.81" N 25°09'08.63" E	LJ53	Pigmy Cormorant (*Microcarbo pygmaeus*), Great Cormorant (*Phalacrocorax carbo*), Grey Heron (*Ardea cinerea*), Little Egret (*Egretta garzetta*), Night Heron (*Nycticorax nycticorax*), Squacco Heron (*Ardeola ralloides*), Eurasian Spoonbill (*Platalea leucorodia*), Glossy Ibis (*Plegadis falcinellus*)
Vardim Island	43°37'33.85" N 25°29'06.12" E	LJ73	Pigmy Cormorant (*Microcarbo pygmaeus*), Great Cormorant (*Phalacrocorax carbo*), Grey Heron (*Ardea cinerea*), Little Egret (*Egretta garzetta*), Night Heron (*Nycticorax nycticorax*), Eurasian Spoonbill (*Platalea leucorodia*)
Paserika Island (of Mechka	43°42'28.52" N 25°46'05.99" E	MJ04	Pigmy Cormorant (*Microcarbo pygmaeus*), Great Cormorant (*Phalacrocorax carbo*), Grey Heron (*Ardea cinerea*), Little Egret

Locality	Coordinates	UTM-grid (10X10 km) square	Full species composition of the nesting birds for the study period
fishponds) (Romania)			(*Egretta garzetta*), Night Heron (*Nycticorax nycticorax*), Squacco Heron (*Ardeola ralloides*), Eurasian Spoonbill (*Platalea leucorodia*)
Small island upwards the river flow from Mishka island	44°00'38.58" N 26°16'43.22" E	MJ47	Great Cormorant (*Phalacrocorax carbo*)
Mishka Island	44°00'47.52" N 26°17'23.57" E	MJ47	Pigmy Cormorant (*Microcarbo pygmaeus*), Great Cormorant (*Phalacrocorax carbo*), Grey Heron (*Ardea cinerea*), Little Egret (*Egretta garzetta*), Night Heron (*Nycticorax nycticorax*), Eurasian Spoonbill (*Platalea leucorodia*)
Malak Kosuj Island	44°04'36.58" N 26°42'47.90" E	MJ78	Great Cormorant (*Phalacrocorax carbo*), Grey Heron (*Ardea cinerea*), Little Egret (*Egretta garzetta*), Night Heron (*Nycticorax nycticorax*), Squacco Heron (*Ardeola ralloides*)
Veresht (Veresti) Island (Romania)	44° 8'16.65"N 26°59'43.90"E	MJ98	Great Cormorant (*Phalacrocorax carbo*)
Srebarna Lake	44°06'55.47" N 27°03'53.67" E	NJ08	Pigmy Cormorant (*Microcarbo pygmaeus*), Great Cormorant (*Phalacrocorax carbo*), Grey Heron (*Ardea cinerea*), Little Egret (*Egretta garzetta*), Night Heron (*Nycticorax nycticorax*), Squacco Heron (*Ardeola ralloides*), Eurasian Spoonbill (*Platalea leucorodia*), Glossy Ibis (*Plegadis falcinellus*), Purple Heron (*Ardea purpurea*), Great White Egret (*Ardea alba*)
Kalimok fishponds- west	44° 1'2.65"N 26°26'3.32"E	MJ57	Pigmy Cormorant (*Microcarbo pygmaeus*), Little Egret (*Egretta garzetta*), Night Heron (*Nycticorax nycticorax*), Squacco Heron (*Ardeola ralloides*), Purple Heron (*Ardea purpurea*), Great White Egret (*Ardea alba*)
Kalimok fishponds –east	44° 1'53.72"N 26°30'11.09"E	MJ67	Pigmy Cormorant (*Microcarbo pygmaeus*), Little Egret (*Egretta garzetta*), Night Heron (*Nycticorax nycticorax*), Squacco Heron (*Ardeola ralloides*), Purple Heron (*Ardea purpurea*), Glossy Ibis (*Plegadis falcinellus*)

The Great Cormorants from the colony at Vardim Island also didn't nest in 2012 because of disturbance due to logging and fishing by local people using boats in the vicinity. Since 2014 that colony was abandoned.

Most significant increase of the Great Cormorant numbers was recorded in some of the colonies of the western part of the studied river section –at Malak Bliznak, Kutovo, and Kalnovats as well as the eastern island of Malak Kosuj. Since 2018 the Great Cormorants started to nest also in Kalimok –fishponds - 40-50 pairs.

Table 2. Breeding distribution and numbers (br. pairs) of Great Cormorants (*Phalacrocorax carbo*), 2006-2014. (Data for 2006 cited from Shurulinkov et al. 2007)

Locality	2006	2010	2011	2012	2013	2014	Trend, 2010-2013 (2014)
Kutovo island	70	67	128	100	120	127	Increase
Malak Bliznak Island	28	33	84	61	77	85	Increase
Ibisha Island	475	735	900	700	425	295	Decrease
Kalnovats Island	280	137	200	350	405		Increase
Small island close to Golyama Burzina Island	300	237	175	375	275	310	Increase
Vardim Island	17	30	10	0	31	**0**	Decrease
Paserika Island (Mechka)	140	56	126	220	105		Increase
Mishka Island	200	88	97	260	80	148	Stable
Small island upwards the river flow from Mishka Island	0	0	0	0	59		New locality
Malak Kosuj Island	60	40	60	100	90	90	Increase
Srebarna Lake	0	0	0	3	0	3	New locality
Byalata prust Island	550	0	0	0	0	0	Extinct
Veresht Island	0	150	100	130	90		Decrease
Total	2120	1573	1880	2299	1757		Possible increase

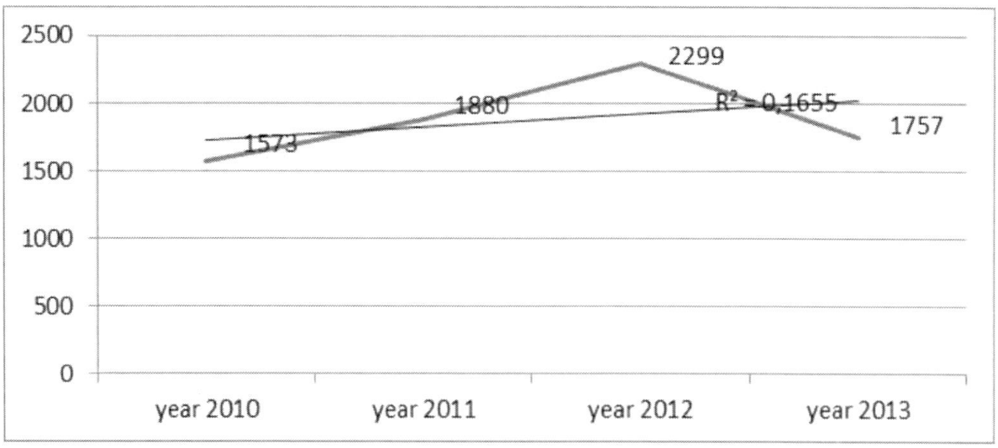

Figure 1. Dynamics of the population numbers (in breeding pairs) of Great Cormorant (*Phalacrocorax carbo*) along the Bulgarian-Romanian section of Danube River, 2010-2013.

Figure 2. Great Cormorants (*Phalacrocorax carbo*) colony on Paserika Island, Romania, June, 2013. Photo by Peter Shurulinkov.

The total Great Cormorant population of the studied river section varied between 1573 and 2299 pairs at 13 localities (Figure 1), fluctuating without clear relation to the river water levels. About 61% of the population is concentrated in three colonies - at Ibisha Island, the small island of Golyama Burzina Island (part of Belene island complex), and Kalnovats Island (Figure 3).

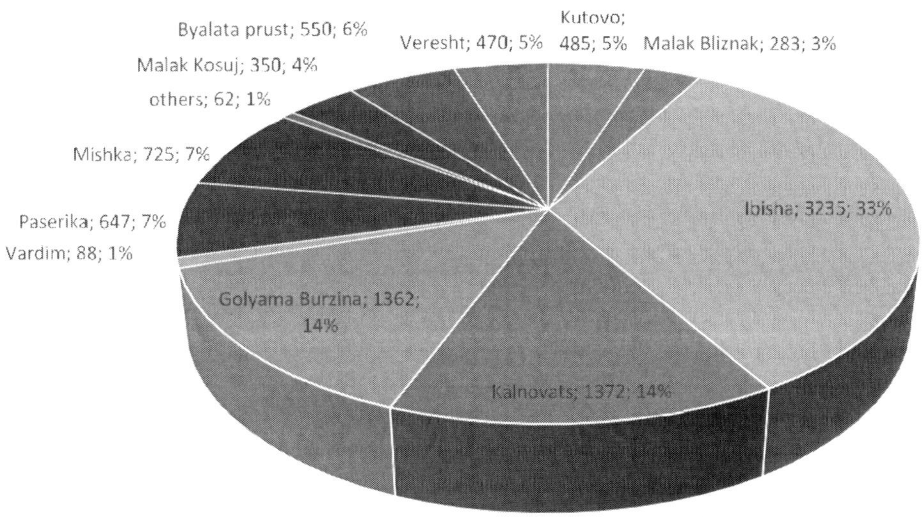

Figure 3. Importance of different colonies for the Great Cormorant (*Phalacrocorax carbo*) population, 2006-2013 (pulled number of pairs for a colony for all study years as a share from all recorded pairs).

In the second half of 20-th century Great Cormorant was much more localized and less abundant species along the Danube River. Until 1985 it was known as a breeding bird from three Danube colonies (on the islands Kutovo, Persina, and Vardim) where only 339 nests were counted (Botev & Peshev Eds. 1985). Afterwards a period of steady increase of the Great Cormorant population followed (1985-2007) leading to formation of at least three new colonies – at Ibisha Island, Paserika Island (close to Mechka fishponds) and Byalata prust Island in "Srebarna" important bird area (Kostadinova & Gramatikov Eds. 2007).

Pigmy Cormorant (*Microcarbo pygmaeus*)

The species was registered in nine colonies but regular breeding was recorded only at three of them – on a small islet close to Golyama Burzina Island, on Paserika Island and in Srebarna Lake (Table 3). The total population varied between 283 and 877 pairs with a decreasing trend (Figure 4). Population of the Pigmy Cormorant decreased significantly on Paserika Island (at Mechka fishponds) and on Mishka Island. In the first case that was caused by considerable loss of feeding habitat for the species in Mechka fishponds which were abandoned, dried and ploughed. In Mishka Island the decrease was probably because of logging activities in the vicinity and movement of part of the birds to nest in the newly flooded wetlands at Kalimok fishponds. Since 2010 breeding on Mishka Island has irregular character.

Table 3. Breeding distribution and numbers (br.pairs) of Pigmy Cormorant (*Microcarbo pygmaeus*), 2006-2014. (Data for 2006 cited from Shurulinkov et al. 2007)

Locality	2006	2010	2011	2012	2013	2014	Trend, 2010-2013 (2014)
Kutovo Island	0	0	5	0	0	0	Irregular breeding
Malak Bliznak Island	0	6	31	0	0	0	Irregular breeding
Ibisha Island	57	0	0	0	0	0	Irregular breeding
Small island close to Goly ama Burzina Island	75	70	70	75	65	80	possible increase
Paserika Island	47	17	27	4	11		Decrease
Mishka Island	35	2	0	6	0	0	Decrease. Irregular breeding
Srebarna Lake	215	350	150	215	190	220	Decrease
Kalimok fishponds–west	0	32	0	9	70	90	Increase
Kalimok fishponds –east	0	400	0	0	0	0	Irregular breeding
Total	429	877	283	309	338		Decrease

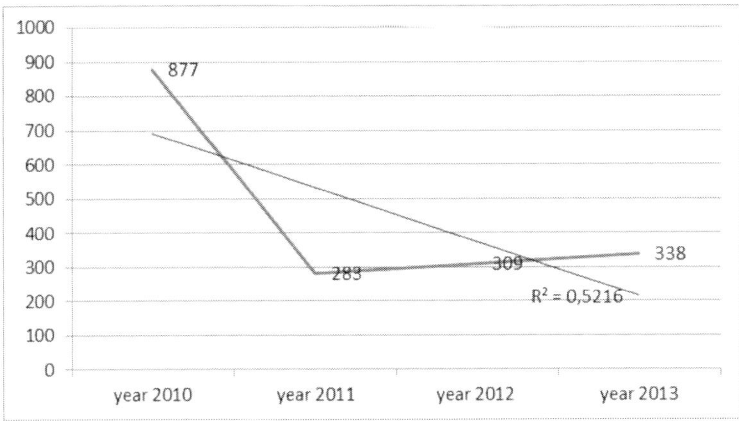

Figure 4. Dynamics of the population numbers (in breeding pairs) of Pigmy Cormorant (*Microcarbo pygmaeus*) along the Bulgarian-Romanian section of Danube River, 2010-2013.

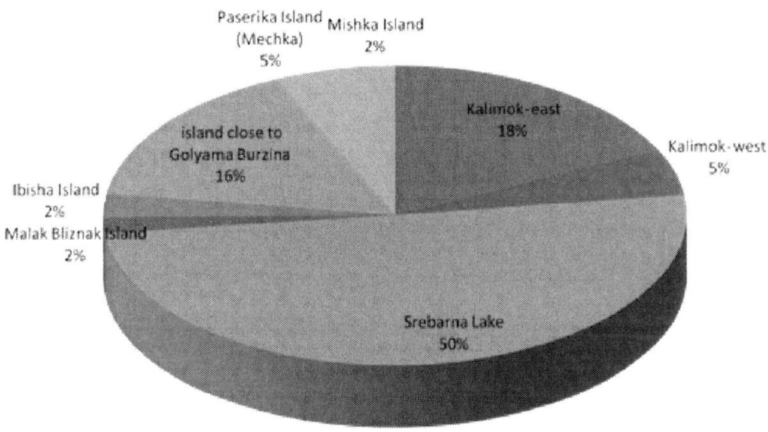

Figure 5. Importance of different colonies for the Pigmy Cormorant (*Microcarbo pygmaeus*) population, 2006-2013 (pulled number of pairs for a colony for all study years as a share from all recorded pairs).

Figure 6. Pigmy Cormorant (*Microcarbo pygmaeus*). Photo by Peter Shurulinkov.

The significance of particular colonies for the Pigmy Cormorant population as a whole is presented on Figure 5. The highest part of the Pigmy Cormorant population nests in Srebarna Lake (50%), Kalimok fishponds, and the islet close to Golyama Burzina Island.

In the near past (1985-2006) the Pigmy Cormorant enlarged its distribution and its breeding numbers increased substantially. Until 1985 it was a nesting bird in low numbers only at two colonies along the studied Danube River section –at Srebarna Lake - 10 pairs in 1980 and Persina Island - 40 pairs in 1979 (Botev & Peshev Eds. 1985, Michev et al. 2007). Until 2005 it started to nest at four new colonies (on the islands Malak Bliznak, Ibisha, Kalnovats, and Mishka) and the total population including the birds breeding at Srebarna Lake and the non-named island at Golyama Burzina reached 220 pairs (Nikolov et al. 2011).

Grey Heron (*Ardea cinerea*)

The species was found to nest in 12 out of all 15 colonies (Table 4). The total numbers varied between 237 and 399 breeding pairs. Overall numbers showed decreasing trend (Figure 7).

Table 4. Breeding distribution and numbers (br.pairs) of Grey Heron (*Ardea cinerea*), 2006-2014. (Data for 2006 cited from Shurulinkov et al. 2007)

Locality	2006	2010	2011	2012	2013	2014	Trend
Kutovo island	5	27	16	6	11	14	Decrease
Malak Bliznak Island	41	20	18	19	26	23	Increase
Ibisha Island	65	39	115	49	37	25	Decrease
Kalnovats Island	33	25	15	10	15		Decrease
Small island close to Golyama Burzina Island	10	36	25	30	33	48	Increase
Vardim Island	21	13	20	8	12	0	Possible Decrease
Paserika Island (Mechka)	17	20	33	38	30		Increase
Mishka Island	15	10	37	22	15	31	Possible increase
Ostrov marsh	0	52	70	20	23	14	Decrease
Malak Kosuj Island	17	16	25	10	17	14	Possible decrease
Srebarna Lake	35	50	25	25	17	60	Stable
Kalimok fishponds-west	0	0	0	0	1	3	New locality
Total	259	308	399	237	237		Decrease

Colonies of Grey Herons showing most clear decrease were those on the islands Ibisha and Kalnovats and in Ostrov marsh. The last colony is newly formed, it was not present during the census made in 2006. At Vardim Island the species didn't nest since 2014. At Malak Bliznak Island, non-named islet at Golyama Burzina Island (Belene Island complex) and Paserika Island Grey Heron's numbers were increasing. The remaining colonies were generally stable, despite the high annual fluctuations.

The importance of the different colonies for the species is compared on Figure 8. The most important colonies for the Grey Heron are situated on the Ibisha Island, Srebarna Lake, and Ostrov marsh, although no even one colony had real dominance, with more than 30% of all breeding pairs.

In 1982 the breeding population of the Grey Heron in Bulgarian part of the Lower Danube region was counted at 250 pairs (Michev & Petrov 1984). Afterwards it remained more or less stable until 2006 (259 pairs) despite the population on national level was considered as growing (Kovachev et al. 2007a).

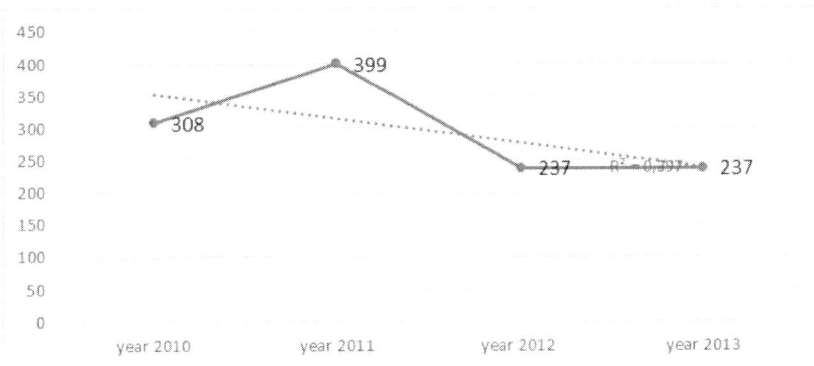

Figure 7. Dynamics of the population numbers (in br. pairs) of Grey Heron (*Ardea cinerea*) along the Bulgarian-Romanian section of Danube River, 2010 - 2013.

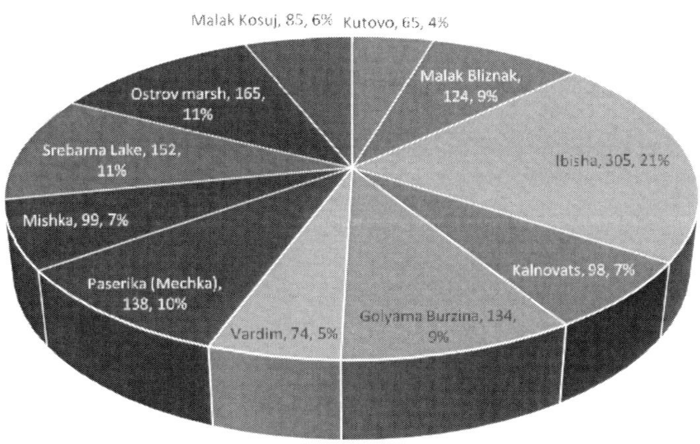

Figure 8. Importance of different colonies for the Grey Heron population, 2006-2013 (pulled number of pairs for a colony for all study years as a share from all recorded pairs).

Figure 9. Grey Heron (*Ardea cinerea*). Photo by Peter Shurulinkov.

Purple Heron (*Ardea purpurea*)

The species was found to nest in three colonies –in the Srebarna Lake, Kalimok fishponds –west, and Kalimok fishponds – east. At Srebarna it breeds annually, with numbers varying between 3 and 9 pairs. In 2006 there were 8 pairs (Shurulinkov et al. 2007). At the Kalimok fishponds Purple Heron occurs in both known colonies – in the western parts of the wetland - 0-4 pairs, in the eastern parts - 1-5 pairs. In 2006 there were no heron colony in these fishponds (Shurulinkov et al. 2007) but still some solitary pairs of Purple Herons were present. Solitary pairs of that species breed sparsely also in some other wetlands along Danube as the marshes on Persina Island and Hadzhidimitrovo fishponds.

Great White Egret (*Ardea alba*)

Breeding of few pairs of that species was recorded in the Srebarna Lake. Between 2 and 6 pairs nested in a mixed colony among the reed and bushes. A solitary pair was observed entering the colony in the western part of Kalimok fishponds, in 2013. More recently breeding of 3-5 pairs of Great White Egrets was recorded also in the colony on Ibisha Island (Mecheva et al. 2017).

Little Egret (*Egretta garzetta*)

The species was found to nest in 12 out of all 15 colonies (Table 5). The total numbers for the period 2010-2013 varied between 189 and 421 breeding pairs. Numbers were fluctuating to a great extent showing a trend of slight increase (Figure 11). Two colonies showed decreasing numbers – at Kutovo and Malak Kosuj Islands. In all the remaining colonies the numbers of the nesting Little Egrets were stable or showed positive short-term trend. Little Egrets remained stable in numbers in the biggest colony at Srebarna Lake.

About 35% of the breeding population of the Little Egret is concentrated in Srebarna Lake (Figure 12). Other two most important colonies are on the islet close to Golyama Burzina Island (15%) and on Ibisha Island (12%).

In the period 1985-2005 the Little Egret was stable or decreasing in most of the Danube colonies. The Little Egrets nesting in the colony at Persina Island decreased dramatically between 60-ties and 80-ties of 20-th century (Simeonov et al. 1990) but afterwards it survived on a small island close to Golyama Burzina Island and remained more or less stable until 2005. The main colony in Srebarna decreased substantially during the 90-ties of 20-th century (own observations) and its recovery started after 2000-2001.

Figure 10. Little Egret (*Egretta garzetta*) fledglings. Photo by Peter Shurulinkov.

Table 5. Breeding distribution and numbers (br.pairs) of Little Egret (*Egretta garzetta*), 2006-2014. (Data for 2006 cited from Shurulinkov et al. 2007)

Locality	2006	2010	2011	2012	2013	2014	Trend 2010-2013(2014)
Kutovo island	0	30	3	3	6	5	Decrease
Malak Bliznak Island	11	26	16	20	30	37	Increase
Ibisha Island	120	14	15	18	18	20	Increase
Kalnovats Island	15	5	30	5	13		Stable
Small island close to Golyama Burzina Island	65	35	35	45	52	60	Increase
Vardim Island	25	1	0	0	4	0	Single pairs, irregular breeding
Paserika Island (Mechka)	33	20	26	18	27		Slight increase
Mishka Island	35	8	8	6	11	12	Increase
Malak Kosuj Island	23	28	10	8	12	7	Decrease
Srebarna Lake	100	110	75	60	205	50	Stable
Kalimok fishponds-west	0	15	0	6	43	30	Increase
Kalimok fishponds-east	0	40	0	0	0	0	Irregular breeding
Total	427	332	218	189	421		Possible increase

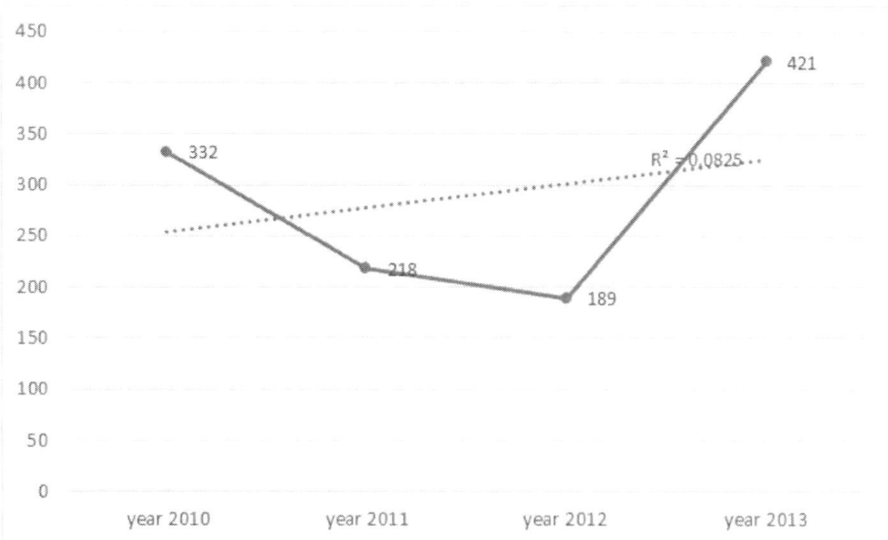

Figure 11. Dynamics of the population numbers (in br.pairs) of Little Egret (*Egretta garzetta*) along the Bulgarian-Romanian section of Danube River, 2010 -2013.

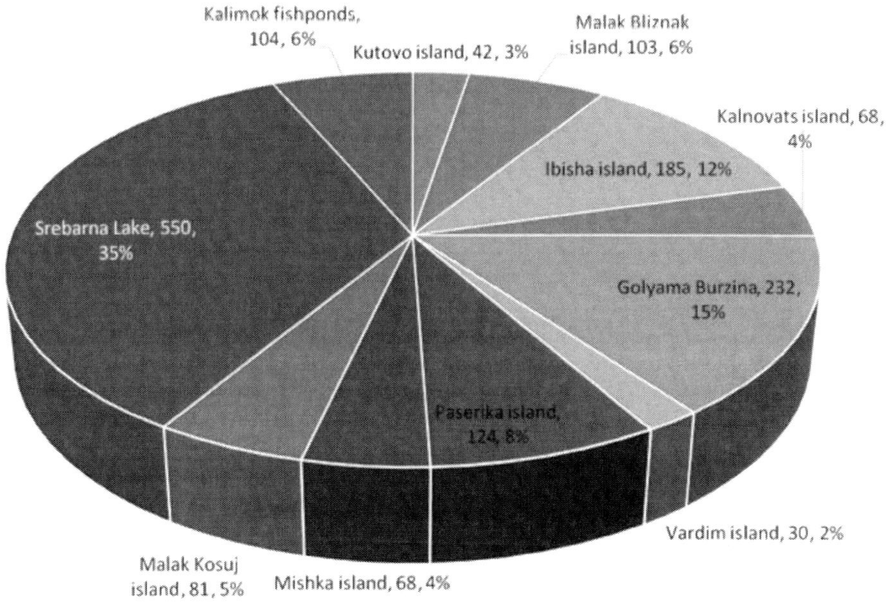

Figure 12. Importance of different colonies for the Little Egret (*Egretta garzetta*) population, 2006-2013 (pulled number of pairs for a colony for all study years as a share from all recorded pairs).

Night Heron (*Nycticorax nycticorax*)

The species is widely distributed, found in 12 colonies (Table 6). The total population varied between 234 and 484 pairs. Taking in account also the results of the study conducted in 2006 the trend of the numbers showed increase, although passing through a period with very low numbers in 2011 and 2012 (Figure 13). The colonies showing most clear increase are situated on Ibisha Island and at Kalimok fishponds. At Ibisha Island the species population is recovering after a dramatic drop from 130 pairs in 2006 (Shurulinkov et al. 2007) to 18 pairs in 2010. It was caused probably by logging and disturbance of the breeding birds during these activities and by local fishermen. Considerable decrease of the breeding Night Herons was registered on Kutovo Island, on the islet close to Golyama Burzina Island and on Malak Kosuj Island. At Vardim Island the last breeding pair of the species was seen in 2010 and afterwards it became extinct.

The importance of the different colonies for the Night Heron population is compared on Figure 14. The results are very similar to these for Little Egret - the major colony holding more than 1/3 of the population is that in the Srebarna Lake.

For the period 1985-2005 Night Heron population in the studied Danube River section decreased. The estimations for the numbers in some colonies from the period 1990-2000 were much higher compared to the present counts – at Ibisha island – 250 pairs (Antonov 1997), at Malak Kosuj Island - 50-150 pairs (Kutsarov et al. 1997) and complex Kalimok – 200-250 pairs (Kostadinova & Gramatikov Eds. 2007).

Table 6. Breeding distribution and numbers (br.pairs) of Night Heron (*Nycticorax nycticorax*), 2006-2014. (Data for 2006 cited from Shurulinkov et al. 2007)

Locality	2006	2010	2011	2012	2013	2014	Trend, 2010-2013(2014)
Kutovo island	0	17	4	1	3	4	Decrease
Malak Bliznak Island	9	28	13	9	22	26	Stable
Ibisha Island	130	18	36	43	28	50	Increase
Kalnovats Island	15	19	25	15	23		Stable
Small island close to Golyama Burzina Island	27	70	50	40	18	27	Decrease
Vardim Island	43	1	0	0	0	0	extinct
Paserika Island (Mechka)	23	21	37	28	28		possible increase
Mishka Island	50	5	25	13	14	15	Stable
Malak Kosuj Island	13	26	14	10	13	9	Decrease
Srebarna Lake	230	80	40	60	250	60	Possible increase
Kalimok fishponds-west	0	15	0	15	85	20	Increase
Kalimok fishponds-east	0	10	0	0	0	0	Irregular breeding
total	540	310	244	234	484		Increase

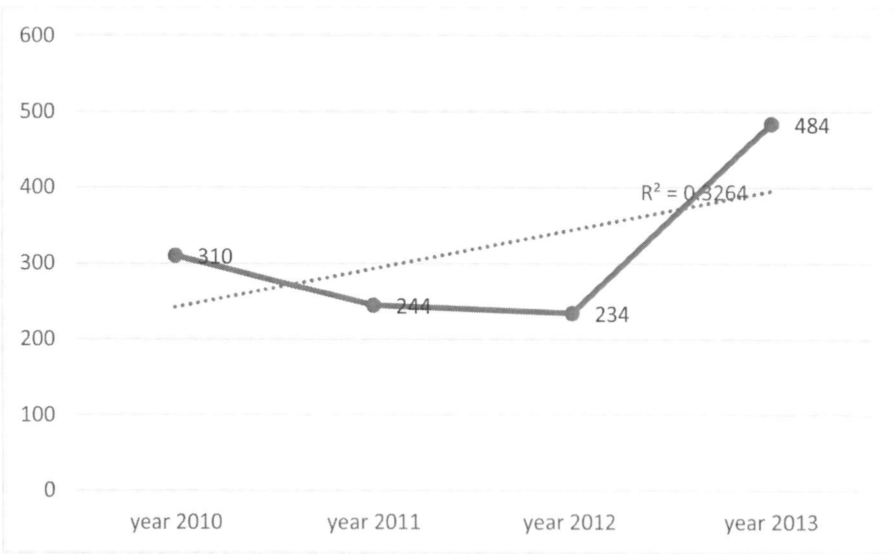

Figure 13. Dynamics of the population numbers of Night Heron (*Nycticorax nycticorax*) along the Bulgarian-Romanian section of Danube River, 2010-2013.

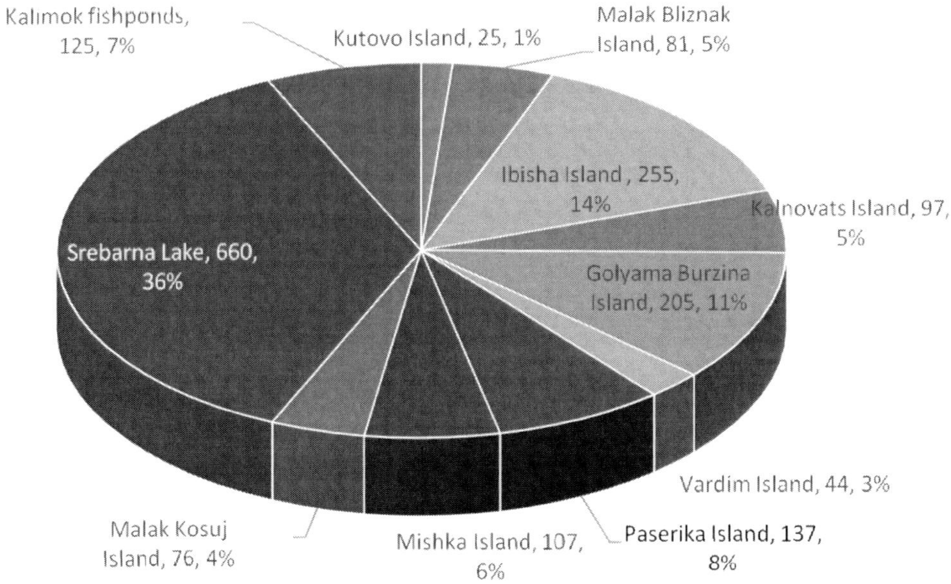

Figure 14. Importance of different colonies for Night Heron (*Nycticorax nycticorax*) population, 2006-2013 (pulled number of pairs for a colony for all study years as a share from all recorded pairs).

Squacco Heron (*Ardeola ralloides*)

Table 7. Breeding distribution and numbers (br.pairs) of Squacco Heron (*Ardeola ralloides*), 2006-2014. (Data for 2006 cited from Shurulinkov et al. 2007)

Locality	2006	2010	2011	2012	2013	2014	Trend, 2010-2013(2014)
Kutovo island	0	4	0	0	0	0	Irregular breeding
Ibisha Island	3	0	4	2	2	9	Increase
Kalnovats Island	2	0	3	0	2		Irregular breeding
Small island close to Golyama Burzina Island	5	15	29	30	35	19	Possible increase
Paserika Island (Mechka)	9	8	9	5	22		Increase
Mishka Island	7	0	0	0	0	0	Irregular breeding
Malak Kosuj Island	0	1	0	0	0	0	Irregular breeding
Srebarna Lake	45	70	15	30	100	30	Stable (with high fluctuations)
Kalimok fishponds-west	0	10	0	3	20	8	Possible increase
Kalimok fishponds-east	0	20	0	0	0	0	Irregular breeding
Total	71	128	60	70	181		Possible increase

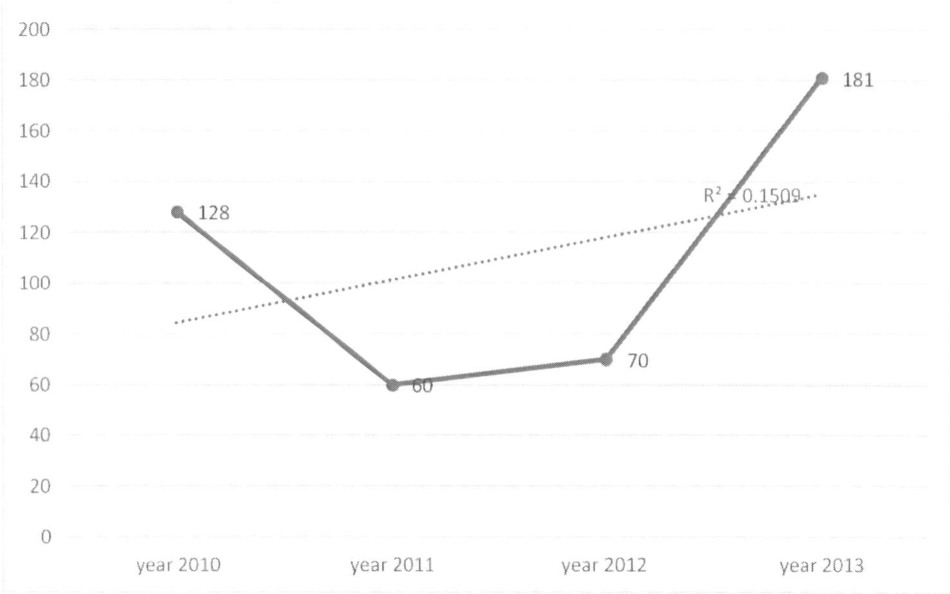

Figure 15. Dynamics of the population numbers of Squacco Heron (*Ardeola ralloides*) along the Bulgarian-Romanian section of Danube River, 2010-2013.

It is not a numerous breeder. It was recorded to nest in 10 of the studied colonies. It nests regularly only in five of these colonies (Table 7). The numbers of the breeding pairs (2010-2013) varied between 60 and 181 pairs. After long period of steady decrease of its breeding numbers in Bulgaria and along the Lower Danube River (Michev et al. 2007, Shurulinkov et al. 2007, Michev 2015) Squacco Heron showed stability and even increase of its population after 2010 (Figure 15). Especially the numbers of the species are increasing at Ibisha Island and Paserika Island. The biggest colony is situated in Srebarna Lake, with 51% of all recorded breeding pairs (Figure 16). It showed high fluctuations in numbers, without clear trend.

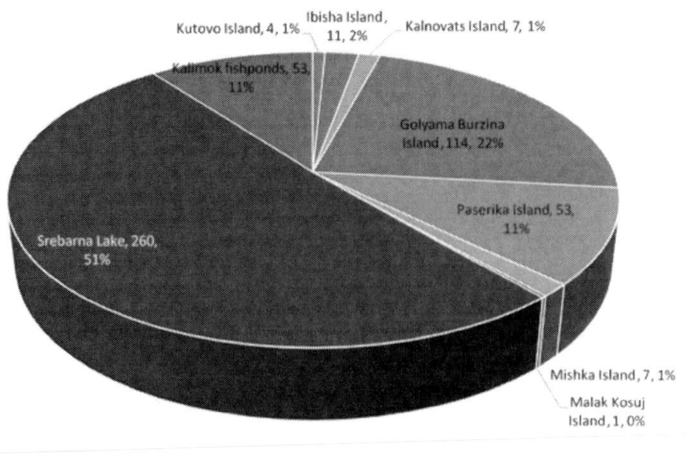

Figure 16. Importance of different colonies for Squacco Heron (*Ardeola ralloides*) population, 2006-2013 (pulled number of pairs for a colony for all study years as a share from all recorded pairs).

Eurasian Spoonbill (*Platalea leucorodia*)

Table 8. Breeding distribution and numbers (br.pairs) of Eurasian Spoonbill (*Platalea leucorodia*), 2006-2014. (Data for 2006 cited from Shurulinkov et al. 2007)

Locality	2006	2010	2011	2012	2013	2014	Trend, 2010-2013 (2014)
Kutovo island	0	3	0	6	3	0	Irregular breeding
Malak Bliznak island	9	17	22	13	18	22	Possible increase
Ibisha Island	6	14	6	13	9	5	Decrease
Kalnovats Island	0	0	0	0	3		Irregular breeding
Small island close to Golyama Burzina Island	15	15	23	23	24	25	Increase
Vardim Island	3	0	0	0	3	0	Irregular breeding
Paserika Island (Mechka)	11	10	24	15	11		Stable
Mishka Island	7	3	4	2	4	0	Decrease
Srebarna Lake	17	25	30	75	30	25	Stable
Total	68	89	109	147	105		Increase

Figure 17. Squacco Heron (*Ardeola ralloides*). Photo by Ivailo Zafirov.

The species nests in nine colonies, of which regularly in six (Table 8). In the remaining three colonies (Kutovo, Kalnovats, and Vardim) it nests only is some years. The numbers of the breeding pairs (2010-2013) varied between 89 and 147. The general trend for the period 2010-2013 showed increase of the Eurasian Spoonbill numbers (Figure 18) which can be confirmed also by a comparison to the total numbers counted in 2006 –only 68 pairs.

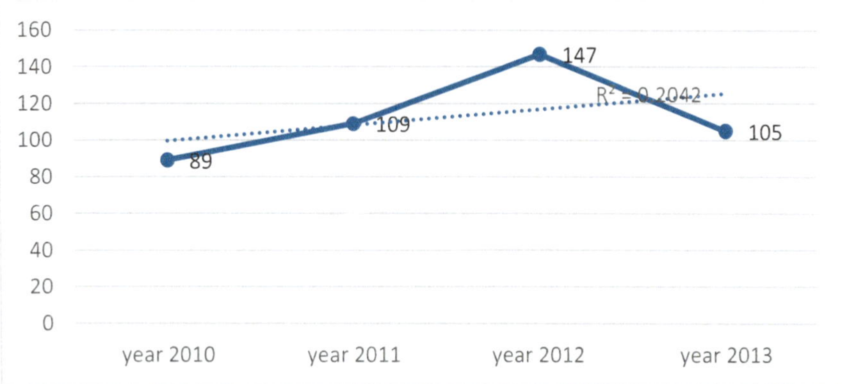

Figure 18. Dynamics of the population numbers of Eurasian Spoonbill (*Platalea leucorodia*) along the Bulgarian-Romanian section of Danube River, 2010-2013.

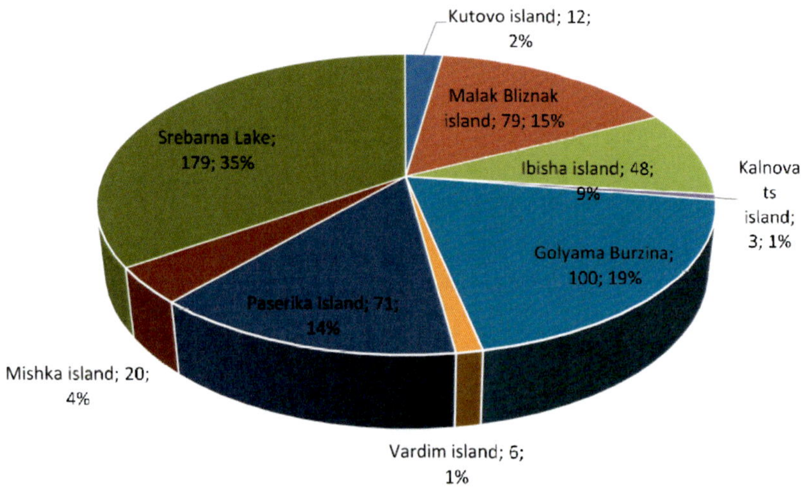

Figure 19. Importance of different colonies for Eurasian Spoonbill (*Platalea leucorodia*) population, 2006-2013 (pulled number of pairs for a colony for all study years as a share from all recorded pairs).

Figure 20. Eurasian Spoonbill (*Platalea leucorodia*). Photo by Peter Shurulinkov.

The most important colonies for the Eurasian Spoonbill are situated at Srebarna Lake (35% of all recorded pairs), on a small island close to Golyama Burzina Island (19%), on Malak Bliznak Island (15%) and Paserika Island (14%) (Figure 19).

By 1980 the whole Bulgarian population of the Eurasian Spoonbill was estimated at 50-70 pairs (Botev & Peshev 1985). Afterwards in the period 1985-2005 it showed stable numbers (Kovachev et al. 2007b) which can be concluded also for the studied Danube River section.

Glossy Ibis (*Plegadis falcinellus*)

It is a regular breeding bird only at one locality along the studied river section - at Srebarna Lake. The numbers of the breeding pairs of Glossy Ibises there varied highly between 10 and 175 pairs for the period 2010 -2014. The highest number of 175 pairs was recorded in 2013.

Other two new localities were found in 2013 which indicates possible start of a territorial expansion of the species. In 2013 on the small island at Golyama Burzina Island there were 4 breeding pairs of Glossy Ibises. At Kalimok fishponds (western part) seven pairs nested in 2013 and six pairs in 2014.

The total short-term trend of the breeding numbers of Glossy Ibises showed increase (Figure 21), including in the major locality –Srebarna Lake. This is confirmed also by a comparison to the numbers of Glossy Ibises recorded in the lake during the period 2001-2003 - 10 - 28 pairs (Kambourova 2005).

That revival of the population comes after a quite long period of steady decrease of the breeding numbers of the species along the Danube shore of Bulgaria and elsewhere in the country –from 100-700 pairs in 1985 to 52-57 pairs in 2006 (Botev & Peshev 1985, Shurulinkov et al. 2007).

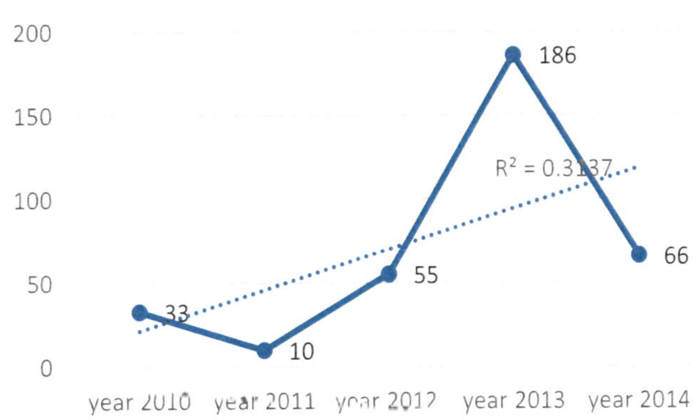

Figure 21. Dynamics of the population numbers of Glossy Ibis (*Plegadis falcinellus*) along the Bulgarian-Romanian section of Danube River, 2010-2014.

Figure 22. Glossy Ibis (*Plegadis falcinellus*). Photo by Ivailo Zafirov.

Conclusion

Heron and cormorant colonies along the Bulgarian –Romanian section of Danube consist of ten species, of which eight breed in at least three localities. In Table 9 are presented average numbers of most common eight species and their short-term and long-term trends. Long-term trends were obtained on the basis of the existing published sources and own observations of the authors.

Both cormorant species had long period of population increase between 1985 and 2005. For Great Cormorant, the most numerous breeding species in the studied colonies, that period continued during the recent years although the growth of the population is already moderate. Among the recorded negative trends is the decrease of Pigmy Cormorant population – it was recorded since 2006. Despite the significant fluctuations connected mainly with the Danube water levels this negative trend was confirmed. Another species which population showed negative trend during the recent years is the Grey Heron. These negative trends can be explained with such factors as disturbance by people, drought of the fishponds at Mechka and Orsoya villages, rapid changes of the water level during the nesting period and mostly for the Pigmy Cormorant –shooting of individuals at some fishponds from both sides of the river.

Positive short-term trends prevail for the rest of the studied species. It is important to note that during the period of the present study species as Night Heron, Little Egret, Squacco Heron, and Glossy Ibis increased in numbers or were stable after a long lasting population decrease. The possible reasons for these positive trends are the successful restoration of wetlands on Persina Island and Kalimok fishponds and existence of years with very high water levels during the study period. These factors lead to improvement and enlargement of the shallow water habitats used for collecting food by the birds and increase in the abundance of food sources (amphibians, fish, and water insects).

Table 9. Aveage total numbers of the colonial herons and cormorants along Bulgarian-Romanian section of Danube and population long-term and short-term trends

Species	number of localities	average number of br.pairs, 2006-2014	actual short term trend, 2010-2014	long term trend, 1985-2005
Great Cormorant, *Phalacrocorax carbo*	13	1926	Possible increase	Increase
Pigmy Cormorant, *Microcarbo pygmaeus*	9	447	Decrease	Increase
Grey Heron, *Ardea cinerea*	12	288	Decrease	Stable
Little Egret, *Egretta garzetta*	12	317	Possible increase	Decrease
Night Heron, *Nycticorax nycticorax*	12	362	Increase	Decrease
Squacco Heron, *Ardeola ralloides*	10	102	Possible increase	Decrease
Eurasian Spoonbill, *Platalea leucorodia*	9	104	Increase	Stable
Glossy Ibis, *Plegadis falcinellus*	3	65	Increase	Decrease

Among the most important threats to the colonially breeding herons and cormorants along the Lower Danube River are water drainage of wetlands in the floodplains, logging of the forests on the islands, disturbance by local fishermen and tourists, dredging the river bottom, fires in the reed massifs, rapid fluctuations of the water levels of the Danube, water pollution leading to eutrophication in some marshlands, and possibly – limited food resources, especially low abundance of fish.

ACKNOWLEDGMENTS

The authors are grateful to the World Wildlife Fund (WWF) – Danube - Carpathian program and the Directorates of "Persina" and "Roussenski Lom" Nature Parks who kindly supplied funding and equipment for the terrain expeditions along the Danube River. We thank personally also to Georgi Georgiev ("Roussenski Lom" Nature Park) and Stela Apostolova (WWF) for their assistance in part of the field expeditions.

References

Antonov A. 1997. Ibisha Island. In: Kostadinova I. (Ed.) *Important bird areas in Bulgaria*. BSPB Conservation series, Book 1, Sofia, p.176.

Botev B. & Peshev T. (Eds.) 1985. *Red Data Book of PR Bulgaria*. vol. 2 *Animals*. BAS. Sofia.

Kambourova N. 2005. The recent status of breeding bird communities of Srebarna Biosphere reserve, North-east Bulgaria. *Acrocephalus* 26 (125): 81-97.

Kambourova N. 2012. Ornithofauna of Biosphere Reserve "Srebarna" Lake. In: Uzunov Y., Georgiev B., Varadinoiva E., Kambourova N., Pehlivanov L. & Vasilev V. (Eds.) *Ecosystems of Biosphere Reserve Srebarna Lake*. pp. 129-147.

Kostadinova I. & Gramatikov M. (Eds.) 2007. *Important bird areas and Natura 2000 in Bulgaria*. BSPB. Conservation series, Book 11, Sofia.

Kovachev A., Michev T. & Petrov T. 2007a. Grey Heron (*Ardea cinerea*). In: Iankov P. (Ed.) *Atlas of the breeding birds in Bulgaria*. BSPB, Conservations Series, Book 10, pp.78-79.

Kovachev A., Michev T., Boev Z., Todorov E. & Alexandrov T. 2007b. Spoonbill (*Platalea leucorodia*). In: Iankov P. (Ed.) *Atlas of the breeding birds in Bulgaria*. BSPB, Conservations Series, Book 10, pp.88-89.

Kutsarov I., Zehtindjiev P., Iankov P., Marinov M. & Kurtev M. 1997. Complex Kalimok. In: Kostadinova I. (Ed.) *Important bird areas in Bulgaria*. BSPB Conservation series, Book 1, Sofia, 101-103 p.

Mecheva R., Beltcheva M., Naumov B., Yankov Y., Michev T., Mitov P. & Kenderov L. 2017. Actual data on the terrestrial fauna of the Danube Island Tsibar (Ibisha). *Comptes rendus de l'Academie bulgare des Sciences [Reports of the Bulgarian Academy of Sciences*]. 70(1): 53-60.

Michev T. 1968. New data on the avifauna of the Srebarna nature reserve, Southern Dobroudzha. *Izvestia na Zoologicheskia Institut s muzej [Reports of the Institute of Zoology with Museum*] 27: 13–21. (in Bulgarian, English summary).

Michev T. 2015. Squacco Heron (*Ardeola ralloides*). In: Golemansky V. et al. (Eds.) *Red Data Book of Republic of Bulgaria*. vol.2, *Animals*, BAS, MOEW, Sofia, p.161.

Michev T. & Petrov T. 1984. Breeding distribution of the Grey Heron (*Ardea cinerea*) in Bulgaria. *Bull. Museums of Southern Bulgaria*. 10: 17-26. (in Bulgarian, English summary).

Michev T., Kovachev A., Todorov E. & Alexandrov T. 2007. Squacco Heron (*Ardeola ralloides*) In: Iankov P. (Ed.) *Atlas of the breeding birds in Bulgaria*. BSPB, Conservation Series, Book 10, pp.70-71.

Michev T., Petkov N., Nikolov C., Kutsarov Y., Demerdzhiev D. & Mitev D. 2007. Pigmy Cormorant (*Phalacrocorax pygmaeus*). In: Iankov P. (Ed.) *Atlas of the breeding birds in Bulgaria*. BSPB, Conservation Series, Book 10, pp.58-59.

Nikolov H., Marin S., Kambourova N. & Demerdzhiev D. 2011. Pigmy Cormorant (*Phalacrocorax pygmaeus*). In: Golemansky V. et al. (Eds.) *Red Data Book of Republic of Bulgaria*. vol.2, *Animals*, BAS, MOEW, Sofia, p.158.

Paspaleva M. & Mitschev T. 1968. Ornithological investigations along Bulgarian Danube section. pp.445 - 454. In: *Limnological Reports of X-th Aniversary Meeting of Danube Research*. Bulgaria, Sofia, 18-20.X.1966.

Shurulinkov P., Nikolov I., Demerdjiev D., Bedev K., Dinkov H., Daskalova G., Stoychev S., Hristov I. & Ralev A. 2007. Die aktuelle Brutverbreitung und die Bestande koloniebrutender Reiher und Kormorane in Bulgarien [Actual breeding distribution and numbers of colonial herons and cormorants in Bulgaria]. *Ornithilogische Mitteilungen* [Ornithological observations] 59 (11): 370-378. (in German).

Simeonov S., Michev T. & Nankinov D. 1990. *Fauna of Bulgaria*. vol. 20. Aves. BAS, Sofia. (in Bulgarian,English summary).

In: Biodiversity of the Bulgarian-Romanian Section
Editor: Shurulinkov, Hubenov et al.
ISBN: 978-1-53615-663-8

Chapter 10

RECENT DATA ON THE DISTRIBUTION AND NUMBERS OF THE WATER BIRDS IN THE WETLANDS ALONG THE BULGARIAN SECTION OF THE DANUBE RIVER

Peter Shurulinkov*[1,*], *Svilen Cheshmedzhiev*[2], *Girgina Daskalova*[1,2], *Hristo Dinkov*[3], *Krasimir Kirov*[4], *Ivan Hristov*[5], *Yordan Kutsarov*[2], *Veselin Koev*[6] and *Stoyan Mihov*[5]

[1]National Museum of Natural History, Bulgarian Academy of Sciences, Sofia, Bulgaria
[2]Bulgarian Society for the Protection of Birds Bulgaria. Sofia, Bulgaria
[3]Balkani Wildlife Society, Sofia, Bulgaria
[4]Rousse Regional Museum, Rousse, Bulgaria
[5]World Wildlife Fund- Danube-Carpathian programme, Sofia, Bulgaria
[6]Directorate of "Persina" Nature Park, Belene, Bulgaria

ABSTRACT

The distribution and numbers of 34 species of breeding water birds were studied along the Bulgarian section of the Danube River between 2006 and 2013. Seven synchronized counts of breeding water birds were conducted in all of the wetlands in that particular stretch. The most numerous species were found to be the Great Crested Grebe (*Podiceps cristatus*), 34-78 pairs; Little Bittern (*Ixobruchus minutus*), 40-93 pairs; Mallard (*Anas platyrhynchos*), 84-148 pairs; Gadwall (*Anas strepera*), 30-51 pairs, Garganey (*Anas querquedula*), 38- 81 pairs; Common Pochard (*Aythya ferina*), 72 141 pairs; Ferruginous Duck (*Aythya nyroca*), 88-157 pairs; Coot (*Fulica atra*), 290-552 pairs; and Whiskered

* Corresponding Author's Email: p.shurulinkov@gmail.com.

Tern (*Chlidonias hybrida*), 339 - 1637 pairs. Observations made during the breeding period and findings of proved breeding cases of rare or irregularly breeding species such as the White Pelican (*Pelecanus onocrotalus*), Ruddy Shelduck (*Tadorna ferruginea*), Teal (*Anas crecca*), Tufted Duck (*Aythya fuligula*), Red-crested Pochard (*Netta rufina*), Pied Avocet (*Recurvirostra avosetta*) and White-winged Tern (*Chlidonias leucopterus*) are also presented in this study. A long-term (last 30 years) trend indicating an increase in water bird numbers can be confirmed for the Mute Swan, Great Crested Grebe, Ferruginous Duck, Gadwall, Common Pochard and Whiskered Tern; in contrast, a decrease has been observed in the numbers of Greylag Goose, Garganey and Black Tern (*Chlidonias niger*).

Keywords: waterfowl, distribution, wetlands, trends, numbers, waders, terns

INTRODUCTION

Breeding water birds have been studied in particular wetlands along the Bulgarian section of the Danube River (Reiser 1894, Striburni 1930, Stanev 1956, Geissler 1962, Donchev 1967). The only work covering the specific avifauna of the wetlands along this section of the river was written in 1961 (Paspaleva – Antonova 1961). Since that time, further ornithological surveys were conducted on wetlands but they were not coordinated in a timely manner and were often too local in their territorial coverage. Nearly permanent ornithological monitoring and intensive research was organized for a long duration at Srebarna Lake (Michev 1968, Michev 1981, Hibaum et al. 2001, Kambourova 2005, Kambourova 2012).

To a lesser extent and only recently, studies on breeding waterfowl and other water birds were undertaken in several other wetlands declared as important bird areas, including the Kalimok fishponds (Vassilev 2003), Mechka fishponds (Vassilev 1999, Vassilev 2003), and marshes on Persina Island (Ivanov 1985, Shurulinkov et al. 2005). Much of the data on the breeding birds in the major wetlands along the Danube were summarized and published in two books on the important bird areas of Bulgaria (Kostadinova 1997, Kostadinova & Gramatikov 2007). All of these investigations, however, were not synchronized and systematic. As a result, there is no exact information on the size of the breeding populations of water birds along the Bulgarian section of the Danube; only rough estimations based on expert opinion. The only exceptions to this are the Ferruginous Duck (*Aythya nyroca*) and Common Pochard (*Aythya ferina*), for which specific full counts were conducted along the Danube and in other parts of the country (Petkov 1997, Petkov 2000, Petkov 2015, Petkov 2015b, Cheshmedzhiev & Petkov 2014). Total estimations for the national breeding populations of water birds in Bulgaria were published in several general sources (Nankinov et al. 2004, Iankov ed. 2007, Golemansky et al. 2015).

There is no concrete data that shows past trends in the numbers of water birds in marshes located along the studied river section (except for Ferruginous Duck and Common Pochard). The aim of this study was to gather actual data on the distribution and numbers of nesting populations of all water birds along the Bulgarian section of the Danube River and to trace the trends and fluctuations in their numbers. We take into account that the study period of eight years is not sufficient to detect sustainable trends. However, utilizing the gathered data we aimed to establish if there was a considerable decrease or increase in particular populations as compared to the findings in previously published investigations. From 2000 to 2013, several important changes impacted the studied wetlands; a project for large-scale restoration of two of the wetlands was successfully completed between 2002 and 2008 and water connections between the marshes of Persina Island and the Kalimok fishponds and the Danube's main course were created. This allowed large populations of fish and other water animals to enter and reproduce in these marshes, thus improving the food supply for water birds. Additionally, a significant improvement in the water quality of the marshes was detected. In contrast to the positive trends in some wetlands, others such as the Mechka and Orsoya fishponds have suffered from drainage after inappropriate management, leading to a substantial decrease in their water bird populations.

Material and Methods

Seven expeditions to count breeding water birds were conducted along the entire section of the Bulgarian Danube (471 km). The counts were made in June 2006, June 2010, May and June 2011, May and June 2012, and June 2013. In 2011 and 2012, two separate complementary counts were made and the total number of breeding pairs was calculated from the combined counts. Additionally, several observations from 2014 to 2018 were added to the overall statistics of breeding pairs, despite the lack of full coverage of the wetlands during that time period. All known marshes, swamps, lakes and fishponds were included in the study. A total of fourteen wetlands covering a territory of 5579 hectares were studied during the five-year span of the counts (Figure 1). These are all freshwater shallow wetlands partially covered (10-90%) by water vegetation. The Orsoya, Mechka, Kalimok and Hadzidimitrovo fishponds were established in the areas that were former marshes; currently, only the Hadzidimitrovo fishponds are still functioning, while the others have been abandoned.

We conducted total counts utilizing all accessible parts of the shoreline of the wetlands. In the marshes of Persina Island, boats (canoes) were used to penetrate some areas that were not visible from the shore. Due to dense vegetation, there was no visibility into the central parts of several of the wetlands. The Little Grebe, Red-necked Grebe, Moorhen and Coot were only registered acoustically; some bird species were also missed due to the lack of visibility. Thus we do not pretend that we have obtained absolute numbers of the studied bird populations but we have made the maximum effort to reach values closer to them. The studies during every field expedition were completed using the same methodology and routes and the recorded numbers can be used for comparisons in time.

Several small and temporal wetlands were included in the seven overall counts: flooded areas in Baley village; the floodplain at the mouth of the Timok River; Vidin marsh, Brashlen marsh; flooded areas to the west of the town of Belene and between Zagrazhden and Dabovan villages, and in Slivo Pole village; and four small dam lakes in Zafirovo, Antimovo, Chernogor and Nova Popina villages in Silistra District (not included on the map of Figure 1).

All wetlands included in the study area are in very close vicinity to the Danube River except for the Hadzhidimitrovo fishponds in Svishtov municipality, which are situated 11 km from the river. Their avifauna is in close connection with the other Danube wetlands and thus they were included in the findings.

Water birds breeding out of the wetlands and those breeding on the small islands located in the main course of the Danube River were not included in the study. It should be noted that some of the breeding numbers reported in the study are in fact lower that the actual numbers; this is due to the low detectability of several species living in dense reed massifs and the limited amount of data gathered in particular wetlands due to short fieldwork periods. As a result, the numbers of species such as the Little Grebe, Little Bittern, Moorhen and Coot are underestimated. Water Rail (*Rallus aquaticus*) and Crakes (genus *Porzana*) were not included in the study because we had no possibility and time to apply the appropriate specific methods for their recording including night time transects.

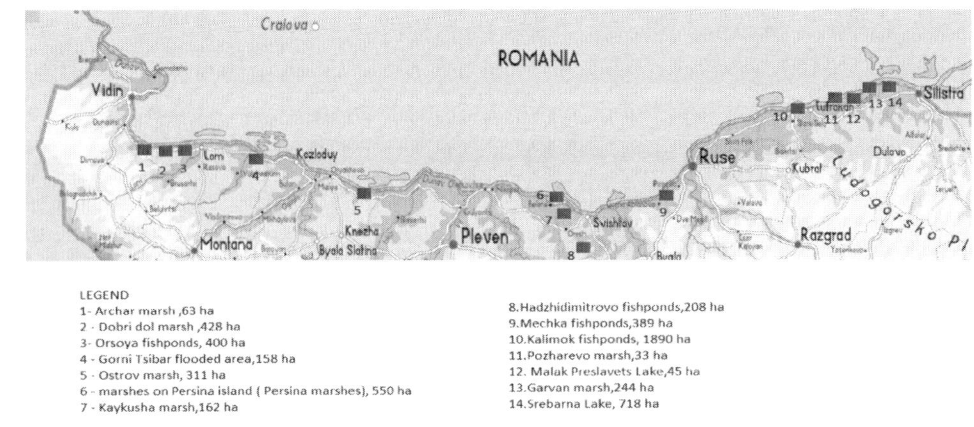

Figure 1. Map of the Bulgarian section of the Danube River and Northern Bulgaria including the studied wetlands.

In the graphs for each bird species, we note the numbers of recorded pairs, as well as all three codes for breeding evidence: confirmed, probable and possible breeding (according to Yeatman 1976).

We used the codes of the Military Grid Reference System (MGRS - spatially identical with UTM) for map visualization and listing the localities of some of the species of high conservation importance.

RESULTS

A total of 34 species of water birds were registered as breeding pairs in the study area.

Great Crested Grebe (*Podiceps cristatus*)

It is a widely distributed and common breeding species. It was recorded to nest at nine of the studied wetlands. The total numbers varied between 46 and 78 breeding pairs (Figure 3). The main breeding site was Srebarna Lake (33% of all recorded pairs, n = 350), followed by Hadzhidimitrovo fishponds (25%) and the marshes of Persina Island (15%). Another major breeding site, Mechka fishponds, lost its importance for the species after the ponds were turned into arable land in 2012.

Figure 2. Great Crested Grebe (*Podiceps cristatus*) with a juvenile. Photo by P. Shurulinkov.

The total breeding numbers of the Great Crested Grebe for the period 2006-2013 showed stability or slight increase. A stable increase of the numbers of the species was recorded at Persina marshes, Malak Preslavets Lake and a moderate increase at Srebarna Lake. These positive trends compensate for the rapid decline of the species numbers at Mechka fishponds and the extinction of the small subpopulation of the species at the Orsoya fishponds after 2011.

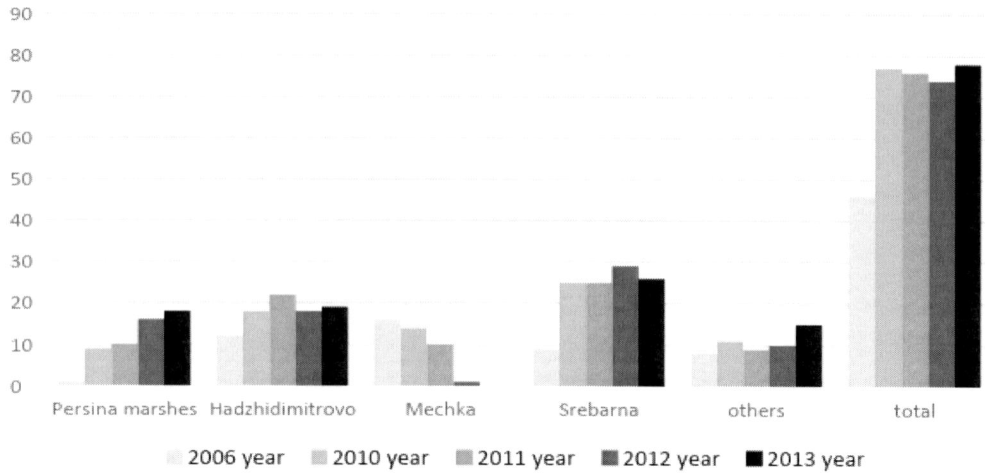

Figure 3. Distribution and numbers of Great Crested Grebe (*Podiceps cristatus*) along the Bulgarian section of the Danube River by years. Numbers on ordinate are in breeding pairs.

Although our data showed stable numbers or only slight increase in numbers of the Great Crested Grebe this trend can be supported by some references published in the near past. The breeding numbers of the Great Crested Grebe at Srebarna Lake in the period 2001-2003 varied between 12 and 14 pairs (Kambourova 2005), which, compared to 29 pairs recorded in the present study for 2012, illustrates an increase of the population of that species there. At Hadzhidimitrovo fishponds, 6 pairs were breeding in year 2000 (Shurulinkov et al. 2005) which increased to 12-22 pairs during the period of the present study.

Red-Necked Grebe (*Podiceps grisegena*)

Scarce breeding species. It was recorded at seven of the studied wetlands. The numbers of its total breeding population varied by years between 7 and 72 pairs (Figure 5). The main breeding localities are the marshes on Persina Island (68% of all observed pairs, n = 164) and Kalimok fishponds (16%). Concentration of the Red-necked Grebes in these wetlands clearly showed that the species prefers comparatively shallow marshes covered with dense water vegetation and in advanced stage of succession.

The total trend of the numbers of the species for the period 2006-2013 was positive despite the fact that in 2011 the level of the Danube was extremely low and most of the traditional breeding localities of the species became dry and the species didn't nest there. The increase was formed mainly because of the improvement of the quality of the habitats in the Persina marshes and Kalimok fishponds after the creation of new water connections of these marshlands with the Danube River in 2008.

Figure 4. Red-necked Grebe (*Podiceps grsegena*). Photo by Boris Belchev.

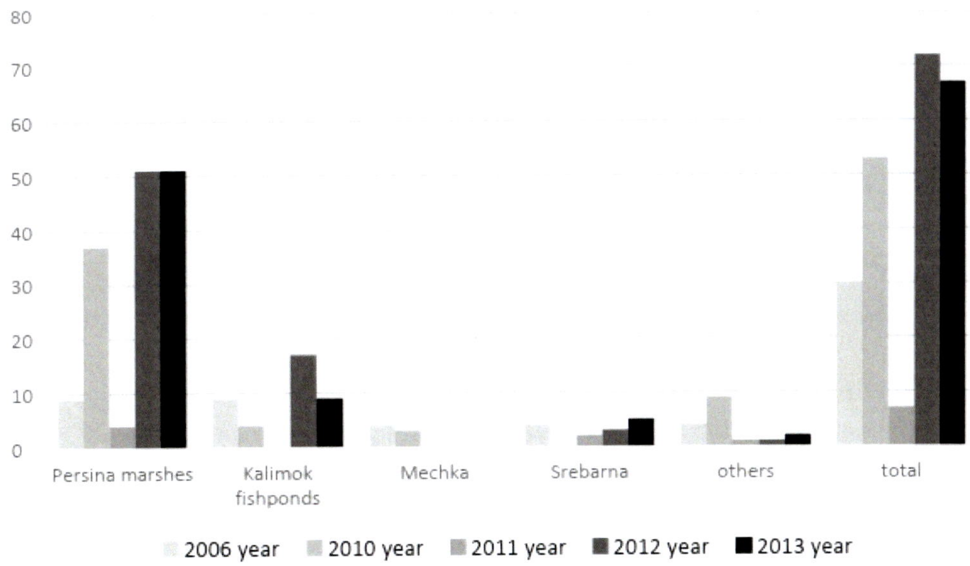

Figure 5. Distribution and numbers of Red-necked Grebe (*Podiceps grisegena*) along the Bulgarian section of the Danube River by years. Numbers on ordinate are in breeding pairs.

The positive trend in numbers of Red-necked Grebe is visible for the last 30 years. During the 1980s, stable and permanent nesting localities of the Red-necked Grebe were found at Srebarna Lake, Garvan marsh and the Persina marshes, but there were no data on the presence of the species at Kalimok, Mechka, Orsoya and Hadzhidimitrovo fishponds (Simeonov et al. 1990). Thus, the spread of the species towards new localities and its reoccupation of abandoned localities is only documented after 1990. In the Hadzhidimitrovo fishponds, the species was found to nest in 2004 (Shurulinkov et al. 2005) and solitary pairs nested there during the period of the present study. During the 1980s, approximately 20-25 pairs were found in the Persina marshes (Ivanov 1985b); in 1998-2001 the numbers there were even lower: 7-18 pairs (Shurulinkov et al. 2005). The numbers of the species in these marshes rose to 51 pairs in 2012 and 2013 (present work). Between 2001 and 2006, one to five pairs of Red-necked Grebes nested in Srebarna Lake (Kambourova 2005, Kambourova 2012), mirroring the results obtained in the present study. At Kalimok fishponds, Red-necked Grebe was almost extinct as a breeder in the first years of the 21th century because of the overgrowth of the water basins with water vegetation and lack of spring floods (own data). The recovery of the species in this location began after 2005/2006 (own data).

The entire Bulgarian population of the Red-necked Grebe is concentrated along the Danube River. According to the newest national estimates, its numbers are between 20 and 50 breeding pairs (Ivanov et al. 2015) or 55-80 pairs (Birdlife International 2015). Taking into account the data from the present study, these numbers should be corrected to 20-80 pairs, which reflect the high possible fluctuations, especially the decrease in number during years with low water levels of the Danube River, like in 2011.

Black-Necked Grebe (*Podiceps nigricollis*)

It is not breeding regularly and is in highly variable numbers at most of its localities. Breeding was recorded at eight wetlands. The only locality with permanent breeding of the species are the marshes on Persina Island but the numbers there also are highly variable– 1- 26 pairs were found to nest there in different years. Others important localities for breeding are Srebarna Lake – 0-8 pairs (2006 – 8 pairs), Kalimok fishponds – 0-6 pairs (2013 - 6 pairs), and Garvan marsh (2010 - 4 pairs). The species was mostly found nesting in the colonies of Whiskered Terns (*Chlidonias hybrida*) among the surface water vegetation or in a close vicinity to them. In some cases, the Black-necked Grebes formed sparse monospecific colonies. The total numbers along the studied section of the Danube River varied between 4 and 38 breeding pairs. There is no trend in the numbers for the period of the study but the numbers showed high fluctuations.

Black-necked Grebe has been a numerous breeding bird along the Danube in the past, especially in the marshes on Persina Island where 34 pairs nested in 1969 and 31 pairs in 1971, although there were no nesting pairs recorded in some of the study's years (Ivanov 1985a). No clear trend can be traced for the species because of the highly fluctuating character of its numbers. In Srebarna Lake, 1-5 pairs of Black-necked Grebes nested in the period 2001-2006 (Kambourova 2005, Kambourova 2012).

Currently, the Bulgarian population of the Black necked Grebe is concentrated along the Danube River; existing cases of breeding at other places been sporadic (Shurulinkov et al. 2006, Ivanov 2015).

Figure 6. Nest with eggs of Black-necked Grebe (*Podiceps nigricollis*), Peschina marsh, Persina Island, June 2010. Photo by P. Shurulinkov.

Little Grebe (*Tachybaptus ruficollis*)

The species is distributed in almost all suitable habitats but it is not numerous. It was registered in wetlands covered with dense reed massifs and found in 12 of the studied wetlands. The numbers varied between 24 and 50 breeding pairs (Figure 7), but, taking into account its habitat and low detectability, we consider these numbers to be much lower than the real ones. The numbers showed positive trend during the period of the study. The most important places for the breeding of the Little Grebe are Srebarna Lake (32% of all pairs, n = 180), marshes on Persina Island (24%), and the fishponds at Hadzhidimitrovo (17%), Mechka (7%) and Orsoya (6%).

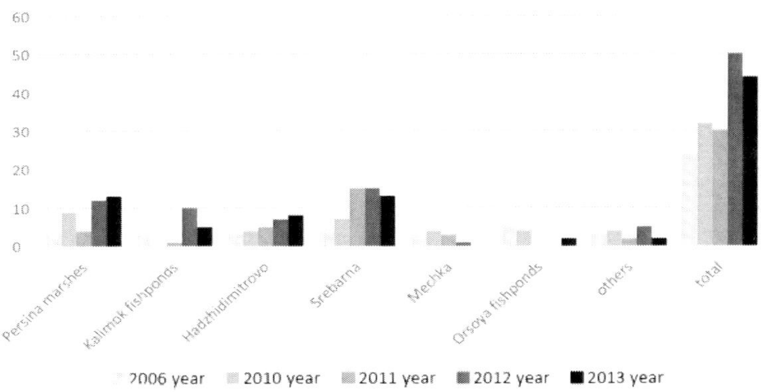

Figure 7. Distribution and numbers of Little Grebe (*Tachybaptus ruficollis*) along the Bulgarian section of the Danube River by years. Numbers are in breeding pairs.

Dalmatian Pelican (*Pelecanus crispus*)

The species nests only in Srebarna Lake and Persina marshes. In Srebarna the numbers of the breeding pairs for the period of the study was as follows: 2006 - 25 pairs, 2010 - 70 pairs, 2011 - 30 pairs, 2012 - 40 pairs, and 2013 - 65 pairs. In 2011 the colony was divided into 3-4 parts and remained in low numbers. There is no clear trend of the total numbers of the birds for the period 2006-2013; compared to the period 2001-2003 when the numbers varied between 80 and 128 pairs (Kambourova 2005), a decrease in numbers is obvious. In the Persina marshes, the species was attracted to breed on artificial platforms since 2016. The numbers of the nesting pairs there by years were as follows: 2016 – 7 pairs, 2017 - 4 pairs, 2018 – 9 pairs. In the past, Dalmatian Pelicans bred in the Persina marshes, until the 1960s (Paspaleva-Antonova 1961).

White Pelican (*Pelecanus onocrotalus*)

It is an irregular breeder found only at Srebarna Lake and, since 2016, at the Persina marshes. Breeding of two pairs was recorded in Srebarna Lake only in 2012 and in the colony of the Dalmatian Pelicans. In the Persina marshes, 2-4 pairs started to nest after 2016 on the artificial platforms, together with Dalmatian Pelicans.

Little Bittern (*Ixobrychus minitus*)

Numerous and widely distributed species. It nests even in very small wetlands including channels partially covered with water vegetation and thus its numbers reported here are most probably underestimated. Little Bittern was found to nest in 13 out of 14 main wetlands included in the present study. The numbers varied between 40 and 93 breeding pairs with a positive trend. The most important wetlands for the species are the marshes at Persina island (20% of all pairs, n = 353), fishponds Hadzhidimitrovo (17%), Kaykusha marsh (16%) and Srebarna Lake (14%).

Bittern (*Botaurus stellaris*)

Very rare breeding species (Figure 8). Despite the perfect habitats for the species available in the study area it was absent at most of them. Bittern was registered in seven of the studied wetlands: Orsoya fishponds, Ostrov marsh, in the marshes on Persina Island, Hadzidimitrovo fishponds, Kaykusha marsh, Kalimok fishponds and Srebarna Lake. Booming males were registered in only two localities annually: Hadzhidimitrovo and

Kalimok fishponds (Table 3). After 2014, one to two booming males were also registered annually in Kaykusha marsh. The total numbers of the booming males was between 4 and 12. The numbers showed an increasing trend which was not broken despite the very dry spring season in 2011.

In the past, the species was found in the Mechka fishponds during the breeding period (Vassilev 2003), but it was not present there during our study. In Srebarna Lake, for the period 2001-2003 the species was not included in the breeding avifauna (Kambourova 2005). In contrast, for almost the same period (2001-2002) one breeding "pair" of the species was reported for Srebarna (Vassilev 2003) which was confirmed also during the present study. The species was recorded as a breeding species in the second half of 20th century in Persina marshes, Hadzidimitrovo and Kalimok fishponds (Boev 1985, Boev & Petkov 2015).

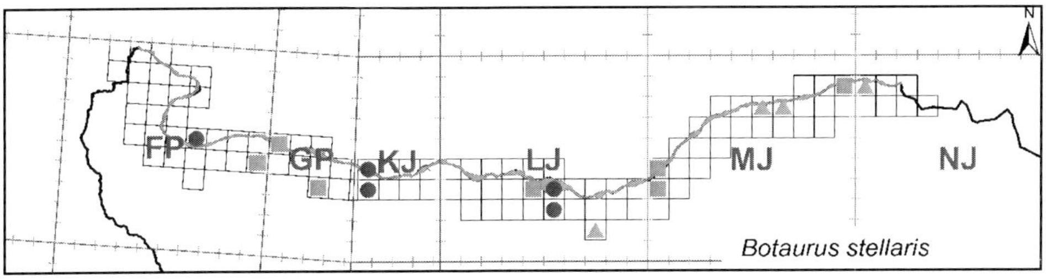

Figure 8. Breeding distribution of Bittern (*Botaurus stellaris*) along Bulgarian section of the Danube River (circles – new localities, triangles – confirmed published localities, squares – unconfirmed published localities).

Purple Heron (*Ardea purpurea*)

Widely distributed but not numerous breeding species. It nests solitarily or in small sparse colonies among the dense reed massifs, sometimes together with other species of herons, as in Srebarna Lake. It was recorded to nest in 13 of the studied main wetlands but found annually in two of them – Srebarna Lake and Hadzhidimitrovo fishponds (Figure 9). After 2010, breeding of the species was permanent also in the marshes of Persina Island as a result of their recovery after creation of a new connection with the Danube. The total numbers of the Purple Herons in the Danube wetlands varied between 12 and 42 pairs in different years. The numbers show high fluctuations without a clear trend. The considerable decrease and even extinction of the species in the wetlands along the western part of the study area after 2011 should be noted, especially in the Orsoya fisponds, Dobri Dol marsh, and Gorni Tsibar. At the opposite in the central and eastern parts of the study area the numbers of the breeding Purple Herons are increasing. The key wetlands for that species holding about 80% of the population are Srebarna Lake (26% of all pairs, n = 116),

Kalimok fishponds (21%), the marshes on Persina island (17%) and Hadzhidimitrovo fishponds (16%).

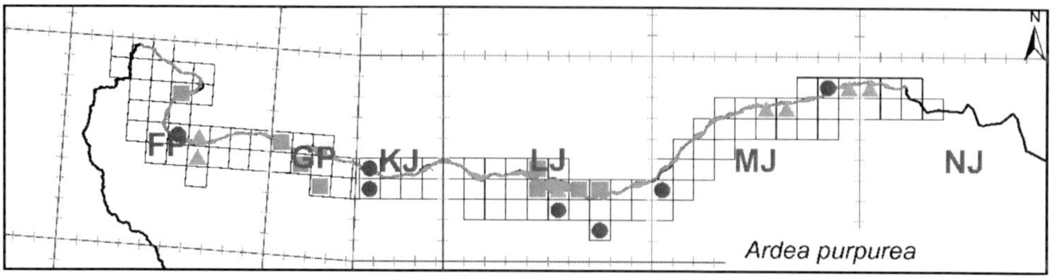

Figure 9. Breeding distribution of Purple Heron (*Ardea purpurea*) along Bulgarian section of the Danube River (circles – new localities, triangles – confirmed published localities, squares – unconfirmed published localities).

Mute Swan (*Cygnus olor*)

It is a rare breeding species in eight wetlands, mostly along the eastern part of the study area (Figure 10). Nesting of the species was recorded at Dobri Dol marsh (confirmed breeding of one pair in 2006, 2010 and 2013), Persina marshes (1-3 pairs annually), Kalimok fishponds (2010 - 2 pairs, 2013 – one pair), Malak Preslavets Lake (2011 - 2 pairs, 2012 and 2013 – one pair, 2016 - 2 pairs), Garvan marsh (one pair in 2006, 2012, 2015 and 2016) and Srebarna Lake (3-15 pairs). Since 2016 the species started breeding also in Hadzhidimitrovo fishponds – 1-2 pairs. A pair with 5 juveniles was recorded on 21.06.2010 in a small dam lake between Bogdantsi and Zafirovo, Silistra District.

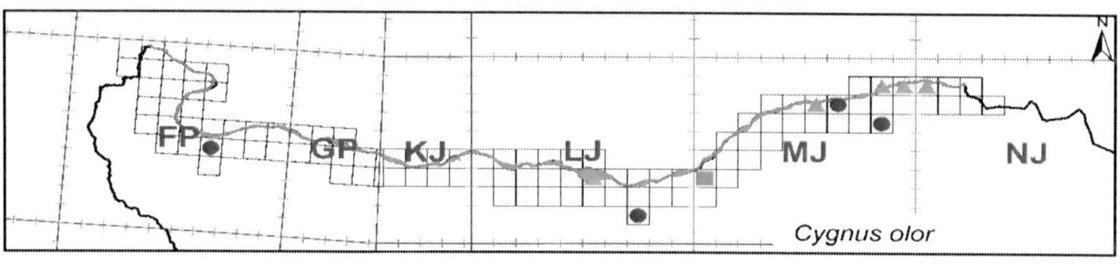

Figure 10. Breeding distribution of Mute Swan (*Cygnus olor*) along Bulgarian section of the Danube River (circles – new localities, triangles – confirmed published localities, squares – unconfirmed published localities).

The total numbers varied by years between 5 and 19 pairs with a clear positive trend. The most important breeding locality is Srebarna Lake where the numbers also increased during the period of the study.

Mute Swan was found as a breeder at the Mechka fishponds from 1993-1995 (Vassilev 2003), but it was not registered there as a nesting bird during the present study. However, the increase of its numbers is obvious for most of the others studied wetlands. In Srebarna

Lake the numbers of the species are increasing – from 4-5 pairs in 2001-2003 (Kambourova 2005) to 15 pairs in 2012. Certain growth of the entire population of the species in Bulgaria was reported recently – it was estimated at 12 - 36 pairs (Dereliev & Ivanov 2015) and even 20-50 pairs (Birdlife International 2015). Growth of the nesting Mute Swan populations was reported also for other Balkan countries as Serbia, Croatia and Romania (Birdlife International 2004, 2015).

Figure 11. Mute Swan (*Cygnus olor*) with three juveniles, Srebarna Lake, June 2010. Photo by Peter Shurulinkov.

Greylag Goose (*Anser anser*)

It is a rare species breeding regularly in low numbers in two of the studied wetlands – Persina marshes (2 - 9 pairs) and Srebarna Lake (6-8 pairs). Temporal breeding localities are also Garvan marsh (2006 – 2 pairs, 2014 -1 pair, 2015 - 2 pairs, 2016 - 2 pairs) and Kalimok fishponds (2012 and 2013 – 2 pairs). The total numbers of the breeding pairs of the species varied by years between 4 and 14 pairs and showed stability for the study period.

In the past, the Greylag Goose was much more abundant in the study area. Twelve pairs were found to nest in Srebarna reserve in the 1960s (Paspaleva & Michev 1968). The numbers of the species there decreased and its population was estimated recently at one to three pairs (Petkov et al. 2007). The results from the present study showed a stable population of 6-8 pairs at Srebarna Lake. Additionally to the present known breeding sites until 1994, Greylag Goose has been registered to nest also in Mechka fishponds but afterwards the species ceased to breed there (Vassilev 2003). For the period 1998 -2001

10-15 pairs of Greylag Geese were recorded in Persina marshes (Shurulinkov et al. 2005). Afterwards their numbers on the island were reported to be 5-11 pairs (Petkov et al. 2007a), which is higher than the present numbers (2-9 pairs). Thus for a long-term period the numbers of the Greylag Goose pairs in that major locality decreased. Thus despite the stability of the population of the species during the terms of the present study overall long-term trend for shows decrease. Decrease of the total national population of the species was also reported for the period 2001-2011 (Ivanov & Dereliev 2015). The most recent estimates of its numbers showed figures of 20-30 pairs (Nankinov et al. 2004) or 15-25 pairs (Birdlife International 2015).

Figure 12. A pair of Greylag Geese (*Anser anser*) with juveniles, Srebarna Lake, June 2011. Photo by P. Shurulinkov.

Ruddy Shelduck (*Tadorna ferruginea*)

The species is distributed in some wetlands situated only in the lowest part of the studied Danube section, in Silistra and the eastern parts of Rousse Districts. In 2006 a pair with juveniles was observed in the western parts of the Kalimok fishponds. A pair was observed in the same area in May 2005. In all the years of the study 1-2 pairs of that species were registered to breed successfully in a dam lake situated close to the Danube River, between the villages Zafirovo and Kolarovo, Silistra District. In 2012 solitary pairs of Ruddy Shelducks also nested in other dam lakes in the vicinity of the Danube River, at the villages of Antimovo, Chernogor and Nova Popina, all in Silistra District.

Until 1994 the Ruddy Shelduck bred in the Mechka fishponds but afterwards it disappeared from that locality (Vassilev 2003). A rapid decline of the breeding population of the species in the districts close to the Danube River was reported after 2000 (Ilieva et al. 2007). In the past, the species was also a breeding bird at Srebarna Lake (Michev 1968); recently, breeding at the lake was again confirmed for the species with a population of 4 pairs (Petkov et al. 2007). During our study and earlier, the species was not observed at Srebarna Lake, which is in accordance with the investigations of Kambourova (2005, 2012). During the last ten years, the species has small but stable population in the dam lakes in Silistra District.

Mallard (*Anas platyrhynchos*)

Common and numerous species recorded in all wetlands included in the present study (Figure 13). The total numbers varied between 84 and 148 breeding pairs, showing a positive trend. The most important wetlands for the species are the marshes on Persina Island (16% of all observed pairs, n = 588), Srebarna Lake (15%), Hadzidimitrovo fishponds (14%) and Kalimok fishponds (15%).

The overall trend of Mallard population in Bulgaria was evaluated as stable (Iankov ed. 2007) but it is not based on any national or even regional counts. It is not clear whether the short-term increasing trend for the species in the study area is a local event or it spreads to the other parts of Bulgaria and neighboring countries. The recent data from Serbia showed a decreasing trend in the breeding Mallards (Sciban et al. 2015). Increasing trends of the breeding numbers of the Mallards were reported for the populations in Northern Europe (Dalby et al. 2013).The long term trend (during last 30 years) in the study area is not clear due to the lack of precise counts.

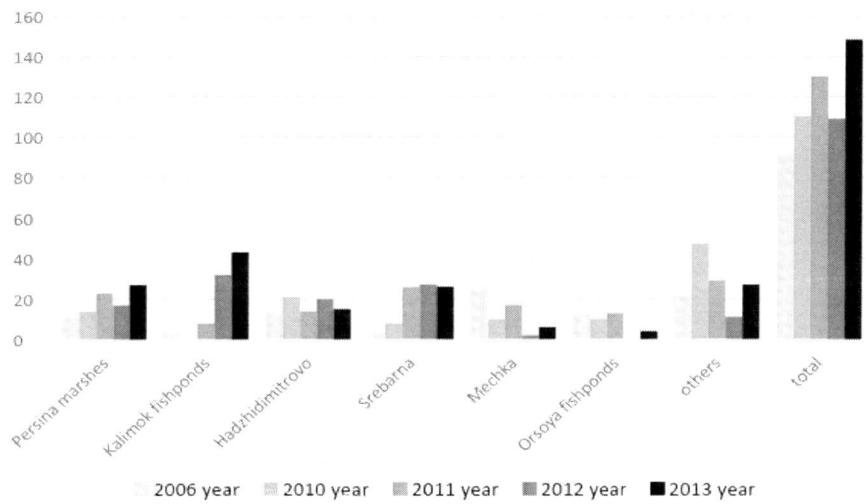

Figure 13. Distribution and numbers of Mallard (*Anas platyrhynchos*) along the Bulgarian section of the Danube River by years. Numbers on ordinate are in breeding pairs.

Gadwall (*Anas strepera*)

It is a widely distributed species but has a tiny population. It was found in 12 out of 14 studied wetlands (Figure 15). The total numbers varied by years between 30 and 51 pairs, showing a negative overall trend. Gadwall was more numerous in the central part of the study area. Three of the wetlands – marshes of Persina Island, Mechka fishponds and Hadzidimitrovo fishponds – contained 68% of all pairs of the species (n = 208). One of the most important places for the breeding and moulting of Gadwall along the Danube, the Mechka fishponds, was left without water during the breeding period in 2012 and a great part of its territory was ploughed. This had a very serious negative impact on the numbers of Gadwall and contributed strongly for the detected negative trend.

It must be taken in account that on the islands and along the temporal river branches of the islands there are additional breeding pairs of Gadwalls. Their numbers remain unclear.

During the 1980s, the national population of the Gadwall was estimated at 30-50 pairs; for the period 1997- 2002 the estimate was 50-80 pairs (Petkov 2011a). Later estimations increased to 100-140 pairs (Nankinov et al. 2004) and 70-120 pairs (Kostadinova & Gramatikov 2007). Despite the short term decrease found in the Danube wetlands the long-term trend for the area and for Bulgaria as a whole for the period 1983-2013 should be considered as positive. Increase of the Gadwall population was also reported in other countries in Central and Southeastern Europe, including Serbia, Romania, Hungary, Slovenia, Slovakia and the Czech Republic (Birdlife International 2004, 2015)

Figure 14. Gadwalls (*Anas strepera*). Photo by Ivailo Zafirov.

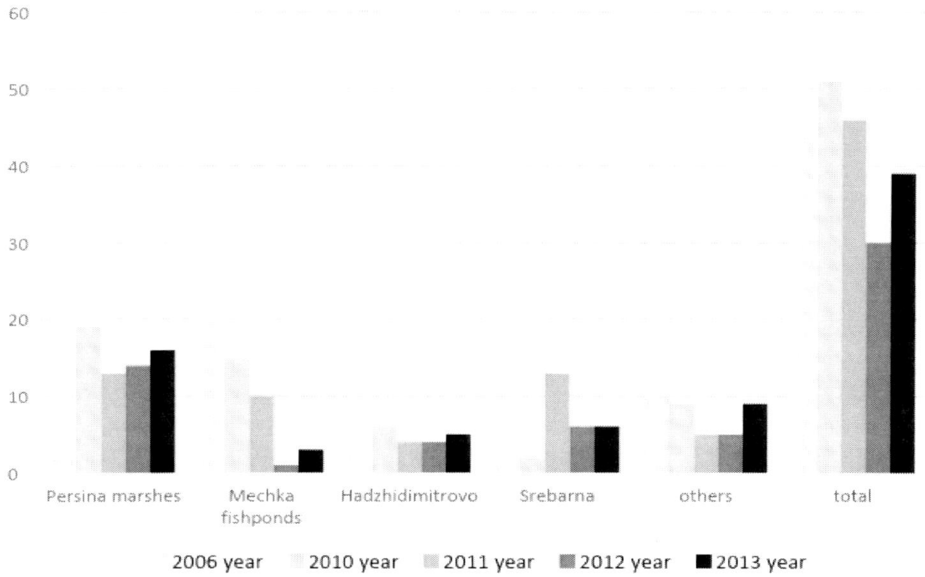

Figure 15. Distribution and numbers of Gadwall (*Anas strepera*) along the Bulgarian section of the Danube River by years. Numbers on ordinate are in breeding pairs.

The numbers of breeding Gadwalls recorded in the present work show that in the Danube wetlands are concentrated not less than 40% of the national population of the species.

Shoveler (*Anas clypeata*)

It is a very rare and an irregularly nesting species. The only permanent breeding site of Shovelers are the marshes of Persina Island: in 2006 - 4 pairs, in 2010 - 3 pairs, in 2011 - 3 pairs, in 2012 - one pair, and in 2013 – 11 pairs. In all others wetlands, eight areas of possible or probable breeding of Shovelers were registered: Dobri Dol marsh (in 2010 -3 pairs, in 2013 - 5 pairs); Srebarna Lake (in 2012 - 3 pairs, in 2013 - 2 pairs); Mechka fishponds (in 2006, one pair); Kalimok fishponds (in 2013 - 2 pairs); Hadzidmitrovo fishponds (in 2010 - 1 pair, 2017 - 3 pairs, 2018 - 1 pair); and Orsoya fishponds (in 2010, one pair). The total numbers of the species along the Bulgarian section of the Danube River can be estimated at 3-20 breeding pairs. We cannot consider all observed pairs as breeding because some of them can be non-breeding summer visitors.

The population of the Shoveler in Bulgaria was estimated at 12-25 pairs (Kostadinova & Gramatikov 2007). The species has been recorded to nest in the former Svishtov marsh in the last decade of 19th century (Reiser 1894). The species has never been a regular or numerous breeder in Bulgaria and the Lower Danube region (Nankinov et al. 1997).

Teal (*Anas crecca*)

Probably, separate pairs nest in the wetlands along the study area. The following observations of Teals give evidence to support this: Persina marshes – May-June 2006 – one individual, May-June 2007 – 9 adult individuals, May-June 2008 – 15 individuals, but without proof of breeding. On 22.06.2005, five males and one female were seen in a flooded meadows at Slivo Pole, Rousse District.

Teal is not confirmed as a breeding species in the study area. It has been observed during the breeding period but without sure proof of breeding. The following references support that statement: in July 1958, one female was observed in Srebarna Lake (Paspaleva-Antonova 1961); during the period 1980-1990, one female was again observed in the same lake in summer period (Iankov ed. 2007); on 22.07.2008, five adult individuals were observed in Garvan marsh (Shurulinkov & Tsonev 2009).

Garganey (*Anas querquedula*)

Widely distributed breeding species which was registered in all studied main wetlands. The total numbers were not high, varying between 38 and 81 pairs by years (Figure 16). A trend of decrease of the numbers of the Garganeys was recorded in some wetlands of the western part of the study area and for the Mechka fishponds. The most preferred wetlands for the Garganey are similar to those of the Gadwall – Persina marshes (22% of all observed, n = 291), Mechka fishponds (14%) and Hadzhidimitrovo fishponds (14%). 50% of all observed pairs of the species were concentrated in these three wetlands.

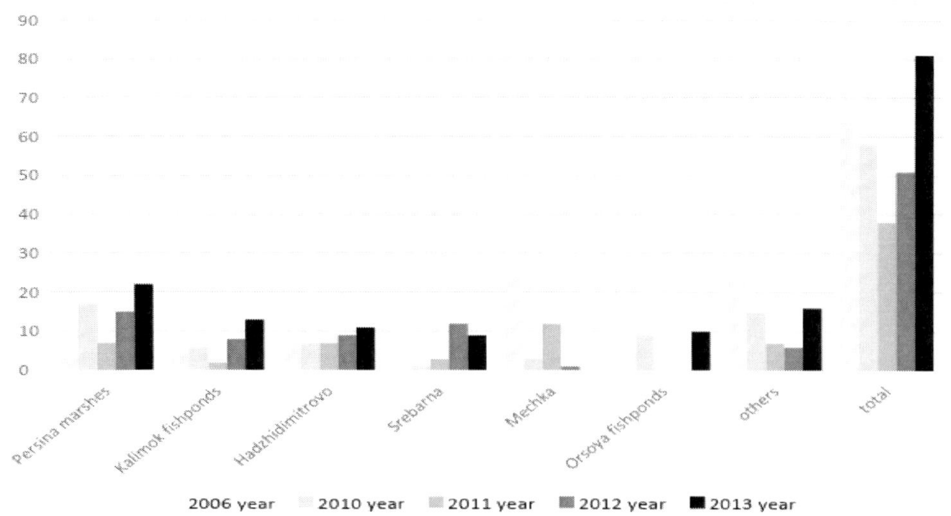

Figure 16. Distribution and numbers of Garganey (*Anas querquedula*) along the Bulgarian section of the Danube River by years. Numbers on ordinate are in breeding pairs.

The difference between the estimated numbers of Garganey in Srebarna Lake by Kambourova (2005, 2012) (10-23 pairs) and the present study (1-12 pairs) suggests a decrease in the population size in this key wetland. A long-term trend showing a decrease in the numbers of Garganeys is reported for many European countries – including Danube countries as Hungary, Austria, Slovakia and Ukraine – and for Europe as a whole (Poysa & Vaananen 2014, Birdlife International 2015).

Common Pochard (*Aythya ferina*)

Common species which was found to nest in 11 of the main studied wetlands. It was not present only in wetlands covered from 80 to 90% or more by reed and other water vegetation (Kaykusha and Ostrov marshes). The total numbers varied between 72 and 141 pairs (Figure 17) showing a slight decrease. Common Pochard didn't breed at Orsoya fishponds in 2011 and 2012 because of low water levels and even lack of water during the nesting period. The species is most numerous at Persina marshes (28% of all pairs, n = 592), Srebarna Lake (22%) and Hadzidimitrovo fishponds (15%).

Although the trend of the numbers of the Common Pochard registered in the present work is negative, if we compare the data with older published sources an increase can be detected. For example, according to a complete census of Common Pochards for the whole of Bulgaria made in 2002, the number of the recorded pairs was 125 (Petkov 2011b), which is less than the numbers counted at the Danube alone in 2006 and 2010. A long-term positive trend is reported for the Common Pochard population in Bulgaria and four other European states: Slovakia, Estonia, Belgium and United Kingdom. The species is generally decreasing in the rest of Europe (Fox et al. 2016).

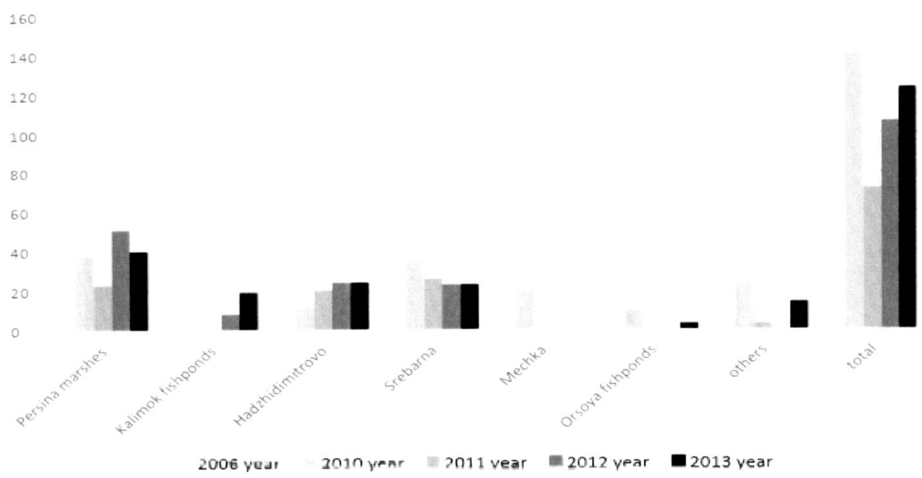

Figure 17. Distribution and numbers of Common Pochard (*Aythya ferina*) along the Bulgarian section of the Danube River by years. Numbers on ordinate are in breeding pairs.

Red-Crested Pochard (*Netta rufina*)

The species was considered extinct in Bulgaria as a breeding bird during the last 30 years. In May and June, 2013 one male Red-crested Pochard was observed at Persina marshes and another male – at Kalimok fishponds. We hope that these new observations can be a sign for future recolonization of that species in its former breeding range.

In the past the Red-crested Pochard has been a rare but regular breeding bird in the marshes of Bulgarian section of the Danube – at Srebarna Lake and in the marshes on Persina Island (Paspaleva - Antonova 1961, Paspaleva & Michev 1968, Ivanov 1985c, Petkov 2015c). Positive trends in the breeding numbers of the species recorded in Central Europe (Switzerland, Austria, Czech Republic, Slovakia, Croatia and Germany) and in Russia (European Communities 2007) give us a hope that soon the species will reoccupy its former breeding range along the Lower Danube.

Figure 21. Red-crested Pochard (*Netta rufina*), male. Photo by Svetoslav Spasov.

Moorhen (*Gallinula chloropus*)

Common and abundant species in all studied wetlands. The total numbers varied by years between 65 and 119 pairs. That value probably is lower than the real one because of the difficult detection and counting of the birds living in dense reed and bulrush massifs. The numbers of Moorhens showed an increasing trend. The highest numbers were recorded at Persina Island, Hadzhidimitrovo fishponds, Mechka fishponds and Srebarna Lake.

Coot (*Fulica atra*)

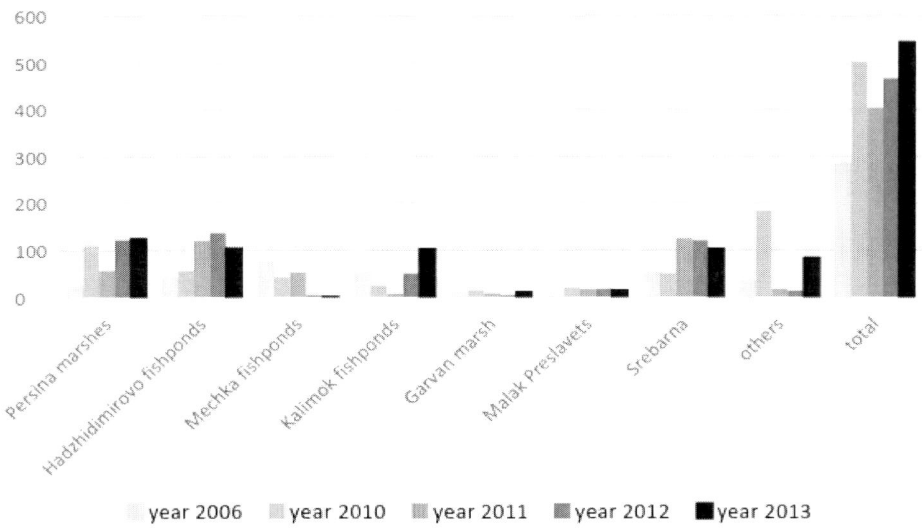

Figure 22. Distribution and numbers of Coot (*Fulica atra*) along the Bulgarian section of the Danube River. Numbers on ordinate are in breeding pairs.

That species is the second most numerous water bird species nesting along the Bulgarian section of the Danube River after the Whiskered Tern. Breeding of the Coot was recorded at all studied wetlands. The total numbers varied between 290 and 552 pairs. The numbers showed increasing trend (Figure 22). The number of detected pairs was highest at Srebarna Lake (20% of all pairs, n = 2203), Hadzhidimitrovo fishponds (21%) and Persina island marshes (20%).

Black-Winged Stilt (*Himantopus himantopus*)

It is an irregularly breeding species in low numbers. Breeding was recorded in seven wetlands - Dobri Dol marsh- 2010 - 3 pairs, 2018 – 2 pairs; marshes on Persina Island – 2010 - 4 pairs, 2011 – one pair, 2012 – 2 pairs, 2013 - 5 pairs; Hadzhidimitrovo fishponds - 2012 - 3 pairs, 2013 - 2 pairs, 2017 - 2 pairs, 2018 - 1 pair; Mechka fishponds- 2012 - 3 pairs, 2013 - 16 pairs; Kalimok fishponds - 2013 – 13 pairs; Baley floodplain – 2010 - 9 pairs and Slivo Pole flooded meadows – 22.06.2006 – 2 pairs with 2 juveniles. The total numbers varied by years between one and 36 pairs. The highest numbers were counted in 2013. Furthermore, breeding of the species was recently recorded in flooded agricultural lands close to Belene town -2005 - 2 pairs (Shurulinkov & Tsonev 2009), in Kalimok fishponds – 2009 - 34 pairs, in the vicinity of Brashlen pig farm – few pairs in May, 2005. In the past the species has bred in Persina Island marshes, former Karaboaz and Svishtov marshes (Michev 1985, Nankinov et al.1997).

Pied Avocet (*Recurvirostra avosetta*)

Very rare species breeding sporadically and irregularly in few wetlands along the Bulgarian section of the Danube River (Figure 23). Cases of probable or possible breeding of Pied Avocets were recorded three times in the following areas: "Ivata" floodplain between Zagrazhden and Dabovan villages, Pleven District from May - June 2010 - 2 pairs; in Hadzhidimitrovo fishponds 2012 - 2 pairs (incubating) and 2017 - 2 pairs and at Slivo Pole, Rousse District - June 2006 -1 individual. Additionally, one pair was recorded in May 2005 in flooded areas to the west of Belene town (Shurulinkov & Tsonev 2009). Until 1997 Pied Avocet was not recorded as a breeding bird in the wetlands of the Bulgarian section of the Danube River (Nankinov et al. 1997). Later on, some isolated and sporadically breeding pairs of Pied Avocets were found at Kalimok fishponds and Srebarna Lake (Dimitrov et al. 2007). In the last few years breeding of small colonies of Pied Avocets was proved for the temporary flooded wetlands at Dolno Tserovene and Yakimovo villages, Montana District (Shurulinkov et al. 2013) and on some sand islets of the main Danube course (Shurulinkov et al. 2016).

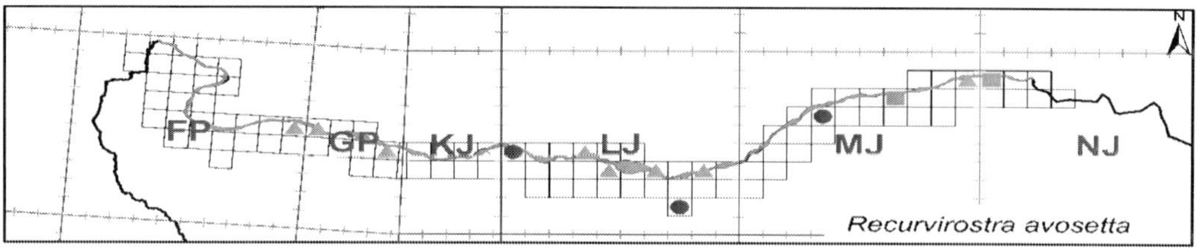

Figure 23. Breeding distribution of Pied Avocet (*Recurvirostra avosetta*) along Bulgarian section of the Danube River (circles – new localities, triangles – confirmed published localities, squares – unconfirmed published localities).

Northern Lapwing (*Vanellus vanellus*)

Common and widely distributed nesting species. Except in 10 of the studied main wetlands, the Lapwing was found in agricultural lands close to the Danube River, flooded or dry. The total numbers varied between 12 and 61 pairs, much higher in years with a wet spring and summer such as 2010 and 2013. The numbers showed high annual fluctuations but no any clear trend.

Green Sandpiper (*Tringa ochropus*)

Rare species along the Danube River during the breeding period. Its breeding was not proved in the studied wetlands but it is considered as possible. Two to three pairs of Green

Sandpipers were registered in the marshes of Persina Island in the period 2010-2013. Two pairs were seen at Hadzhidimitrovo fishponds in 2010 and 2013 and one pair – in 2012 and 2017. Solitary pairs were also observed at Kaykusha marsh and Mechka fishponds in 2010.

Little Ringed Plover (*Charadrius dubius*)

It is a rare and scarce breeding bird in the marshes along the Danube River although in the main course of the river it is very common. Breeding pairs were seen in Hadzhidimitrovo fishponds (2010 - 3 pairs, 2012 – 7 pairs, 2013 -1 pair, 2017 - 3 pairs); in the marshes of Persina island (2011 – 3 pairs, 2012 – 1 pair, 2013 - 2 pairs); in Mechka fishponds – in 2012 - 8 pairs; and in Kalimok fishponds- in 2013 - 2 pairs. Habitats suitable for that species are not available every year in the wetlands along the Danube River.

Black-Headed Gull (*Chroicocephalus ridibundus*)

The species nests regularly only in Srebarna Lake but with highly fluctuating numbers. It usually nests in mixed colonies with terns. The numbers in Srebarna Lake varied by years as follows: 2006 – 90 pairs, 2010 - 20 pairs, 2011 - 5 pairs, 2012 - 16 pairs, 2013 - 29 pairs. Numerous colonies also existed at Kalimok fishponds in 2013 - 47 pairs. In June, 2009 a colony of 80-100 breeding pairs of the species was found there. Cases of breeding of solitary pairs or small colonies of Black-headed Gulls were registered in Kalimok fishpond in 2012 -1 pair, in Malak Preslavets Lake - 2012 -1 pair, in the marshes of Persina island – 2013 - 5 pairs, 2014 – 7 pairs, 2018 - 8 pairs. These cases indicate irregular breeding at some places. In 2011 and 2012 Black-headed Gulls were breeding in a small dam lake at Nova Popina village with numbers of 15-20 pairs. In 2013 the total number of the Black-headed Gulls in all studied wetlands was 81 breeding pairs.

Only few pairs of Black-headed Gulls have nested at Srebarna Lake for the period 1959-1961 (Paspaleva-Antonova 1961). For the same locality the numbers in the period 2001- 2006 varied between 5 and 40 pairs (Kambourova 2005, Kambourova 2012). Breeding of the species has also been reported for Orsoya fishponds and Kalimok fishponds (Kambourova & Antonov 2007).

Whiskered Tern (*Chlidonias hybrida*)

Widely distributed and numerous breeding species (Figure 24). It forms significant colonies in the marshes, lakes and fishponds, especially in years with high water levels during the spring and summer periods (as 2010 and 2013). Breeding of the species was registered in all studied main wetlands and in Vidin marsh; in 2010, 3 pairs, and in 2013, 4 pairs.

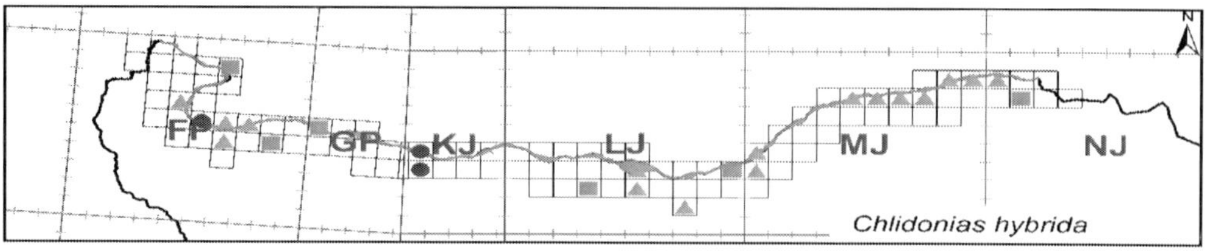

Figure 24. Breeding distribution of Whiskered Tern (*Chlidonias hybrida*) along Bulgarian section of the Danube River (circles – new localities, triangles – confirmed published localities, squares – unconfirmed published localities).

The total numbers varied by years. The range was between 339 and 1637 breeding pairs. There is no clear trend in the numbers because of the significant fluctuations (Figure 25). The highest numbers were recorded in Kalimok fishponds (24% of all counted pairs, n = 5022), the marshes in Persina Island (24%) and in Malak Preslavets Lake (12%). The colonies of the Whiskered Terns are attractive place for breeding of other water birds as Black-headed Grebes, Black-headed Gulls, Black Terns and White-winged Terns.

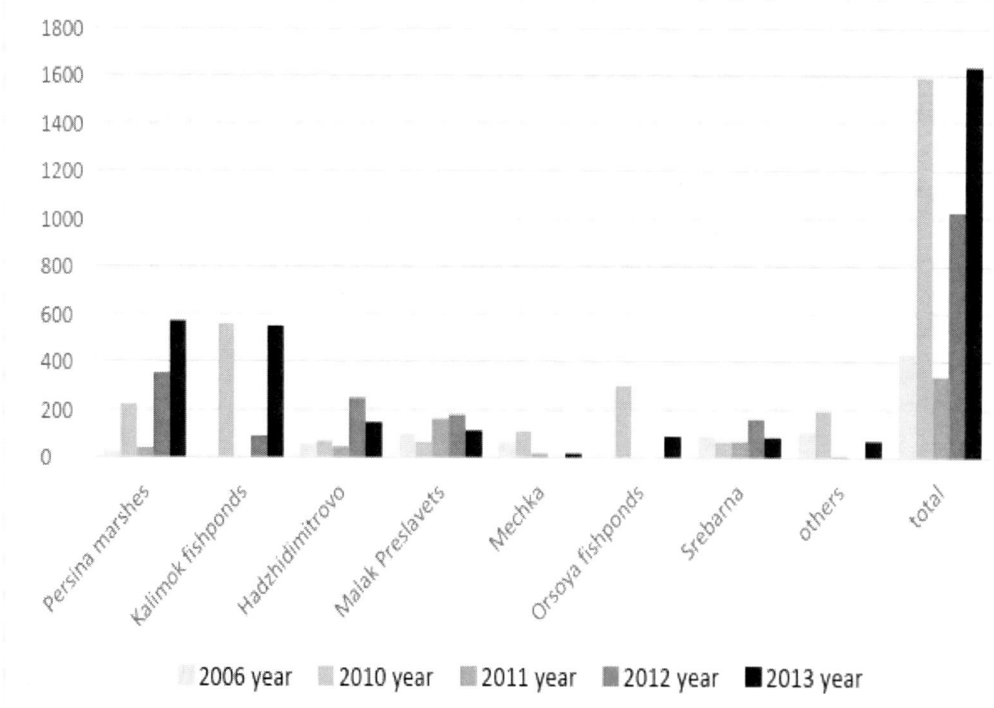

Figure 25. Distribution and numbers of Whiskered Tern (*Chlidonias hybrida*) along the Bulgarian section of the Danube River. Numbers on ordinate are in breeding pairs.

Figure 26. Colony of Whiskered Terns (*Chlidonias hybrida*) in Peschina marsh, Persina Island, June 2010. Photo by Peter Shurulinkov.

Figure 27. Nest of Whiskered Terns (*Chlidonias hybrida*) with eggs, Peschina marsh, Persina Island. Photo by Peter Shurulinkov.

The population of the Whiskered Terns in Bulgaria was estimated at 400 breeding pairs in 1997 and the biggest colonies reached up to 160 pairs at Persina Island (Nankinov et al.1997). In 2007 the numbers of the species at the national level were estimated between 220 and 800 breeding pairs (Kutsarov & Stefanov 2007). The results from the present work clearly showed that the numbers of the breeding Whiskered Terns have increased during the recent years. The numbers along the Bulgarian section of the Danube River alone should be accepted at levels of 300-1650 breeding pairs with separate colonies exceeding 500 pairs in particular years (in Persina marshes and Kalimok fishponds). This evaluation is close to the national population numbers for Bulgaria. The nesting of the species outside the Danube stream is a local and irregular event, typical for the Danube plain and the colonies there are not larger than 20-30 pairs (Shurulinkov et al. 2005, Shurulinkov et al. 2013).

Black Tern (*Chlidonias niger*)

Irregularly separate pairs or small groups of pairs nest inside the colonies of the Whiskered Terns. Such cases were recorded as follows: Hadzidimitrovo fishponds in 2006 – 1-5 pairs, in 2010 - 1-3 pairs, in 2013 - 4 pairs; and Mechka fishponds – in 2010 -6 pairs, at Srebarna Lake – in 2012 – one pair. Much larger colonies of Black Terns were registered only twice: at Kalimok fishponds in 2009 and 2013 with 70 and 48 breeding pairs, respectively.

In the past the Black Tern has been a common and numerous breeder along the Bulgarian section of the Danube River, much more numerous than the Whiskered Tern (Pateff 1950, Paspaleva-Antonova 1961). After 1970 its numbers started to decrease; only 1-3 pairs remained in Srebarna Lake, from the formerly numerous population of the species (Ivanov 1985d). During the last 20 years, the Black Tern became an even scarcer, irregularly nesting species in several sites, including the Malak Preslavets and Srebarna Lakes (Nankinov et al. 1997). During the period 1993-1995, 3-5 pairs have nested in the Mechka fishponds (Vassilev 2003) and in 2004 one pair was observed among the Whiskered Terns of a colony in the Hadzidimitrovo fishponds (Shurulinkov et al. 2005). Recently the numbers of the Black Terns in Srebarna Lake were evaluated at 6-9 pairs (Petkov et al. 2007).

White-Winged Tern (*Chlidonias leucopterus*)

A numerous colony of that species was found in 2013 at Kalimok fishponds – 37 breeding pairs. Cases of breeding of small numbers of pairs of White-winged Terns in the colonies of Whiskered Terns were recorded at Hadzidimitrovo fishponds in 2006 – 2-3 pairs and in Persina island marshes in 2009 - 4-8 pairs.

There are few concrete data of breeding of the species along the Bulgarian section of the Danube in the past. It includes the following: Srebarna Lake in 1956 (Cramp 1985) and in 2000 -2 pairs, as well in the Kalimok fishponds in 2005 - 4 pairs (Kutsarov 2007). The first and the only big monospecific colony of the White-winged Terns in Bulgaria (60 pairs) was found in 2010 at Chairya floodplain in Southern Dobroudzha (Shurulinkov et al. 2010). In the present study we report a second similar case for the country.

Common Tern (*Sterna hirundo*)

Breeding of that species was registered only at two wetlands – Srebarna Lake and Persina marshes. In 2006 in Srebarna Lake 275 pairs were counted. Small colonies numbering 5-6 pairs also existed there in 2012 and 2013. In Persina marshes the species started to breed annually since 2014 with a population of 23-36 pairs. The species is a much more common breeder on the sandpits and small islands in the main course of the Danube River (Shurulinkov et al. 2016). But in years with very high water levels the sandpits and islands are flooded and Common Terns nest only in Srebarna Lake and Persina marshes.

The species was also registered as nesting in Srebarna Lake during 1956-1961 and 2001-2003 (Paspaleva-Antonova 1961, Donchev 1967, Kambourova 2005). In 2007 the numbers of the species in Srebarna were estimated at 40 pairs (Petkov et al. 2007). Obviously, the colony is formed there not annually and it shows highly fluctuating numbers.

CONCLUSION

The most common and numerous breeding water birds along the Bulgarian Danube study area were the Whiskered Tern (*Chlidonias hybrida*), Coot (*Fulica atra*), Ferruginous Duck (*Aythya nyroca*), Mallard (*Anas platyrhynchos*), Common Pochard (*Aythya ferina*), Garganey (*Anas querquedula*) and Great Crested Grebe (*Podiceps cristatus*).

Short term (2006-2013) positive trends in numbers were recorded for ten species: Red-necked Grebe, Little Grebe, Bittern, Little Bittern, Mute Swan, Mallard, Ferruginous Duck, Moorhen, Coot and Whiskered Tern. Negative trends for the period 2006-2013 were recorded for the Gadwall and Common Pochard. A long-term (last 30 years) trend indicating an increase in numbers can be confirmed for the Mute Swan, Great Crested Grebe, Ferruginous Duck, Gadwall, Common Pochard and Whiskered Tern; conversely, a decrease in numbers can be confirmed for the Greylag Goose, Garganey and Black Tern. For the rest of the studied species the numbers were stable, highly fluctuating or too low for analysis of the trend. For some species as Mallard, Moorhen and Coot the long-term trend remains unclear.

The most important wetlands for the studied birds were the marshes on Persina Island (22% of all recorded pairs), Srebarna Lake (19%), the Kalimok fishponds (16%), and the Hadzhidimitrovo fishponds (14%) (Figure 28).

It should be taken in account that these figures don't include the colonially breeding herons and cormorants in some of these wetlands (Srebarna, Kalimok).

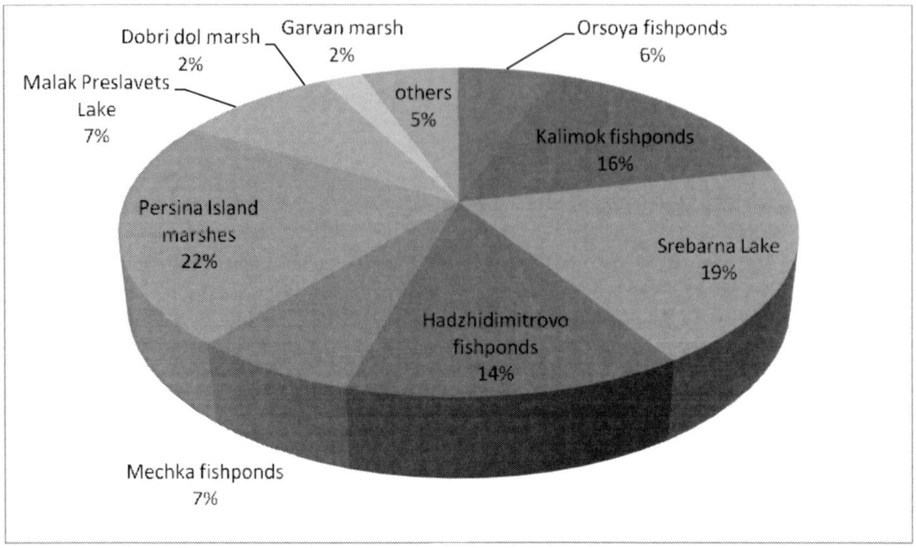

Figure 28. Importance of the studied wetlands for the breeding water birds – percentages of total numbers of breeding pairs counted.

ACKNOWLEDGMENTS

The authors are grateful to the World Wildlife Fund (WWF) – Danube - Carpathian program and Directorates of "Persina" and "Roussenski Lom" Nature Parks who kindly supplied funding and equipment for the terrain expeditions along the Danube River. We also personally thank Georgi Georgiev ("Roussenski Lom" Nature Park) and Stela Apostolova (WWF) for their assistance in part of the field expeditions.

REFERENCES

Antonov A. 1997. Orsoya fishponds. In: Kostadinova I., (Ed.): *Important bird areas in Bulgaria.* BSPB, Conservation Series Book 1, pp. 53-54.

Birdlife International 2004. *Birds in Europe: population estimates, trends and conservation status.* Cambridge, UK, Birdlife International (Birdlife Conservation Series 12).

Birdlife International 2015. *European Red List of Birds.* Luxembourg: Office for Official Publications of the European Communities.

Boev N. 1985. Bittern (*Botaurus stellaris*). In: *Red Data Book of PR Bulgaria.* BAS, Sofia.

Boev Z. & Petkov N. 2015. Bittern (*Botaurus stellaris*). In: Golemansky V. et al. (Eds.): *Red Data Book of Republic of Bulgaria.* vol. 2 Animals, BAS and MOEW, Sofia.

Cheshmedzhiev S. & Petkov N. 2014. Action Plan for the protection of the Ferruginous Duck in Bulgaria (2014-2013). Sofia, Bulgarian Society for the Protection of Birds.

Cramp S. 1985. *The Birds of the Western Palearctic.*vol.3. Waders to Gulls. Oxford University Press.

Dalby L., Söderquist P., Christensen T., Clausen P., Einarsson A., Elmberg J., Fox A., Holmqvist N., Langendoen T., Lehikoinen A., Lindström A., Lorentsen S., Nilsson L., Pöysä H., Rintala J., Sigfússon A., Svenning J. & Dalby L. 2013. The status of the Nordic populations of the Mallard (*Anas platyrhynchos*) in a changing world. *Ornis Fennica* 90: 2-15.

Dereliev S. & Ivanov B. 2015. Mute Swan (*Cygnus olor*). In: Golemansky V. et al. (Eds.). *Red Data Book of Republic of Bulgaria.* vol. 2 Animals, BAS and MOEW, Sofia, p.251.

Donchev S. 1967. Contribution to the knowledge on the water birds in Bulgaria. *Izvestia na Zoologicheskia Institut s muzej* [*Reports of the Institute of Zoology with Museum*] 23: 79-95.

European Communities 2007. Management Plan for Red-crested Pochard (*Netta rufina*) 2007-2009.*Technical Report* - 005-2007.

Fox A.D., Caizergues A., Banik M., Devos K., Dvorak M., Ellermaa M., Folliot B., Green A., Grunberg C., Guilleman M., Halland A., Hornman M., Keller V., Koshelev A., Kostiushyn V., Kozulin A., Lawicki L., Luigujoe L., Muller C., Musil P., Musilova P., Nillson L., Mischenko A., Poysa H., Sciban M., Sjenicic J., Stipniece A., Svazas S.& Wahl J. 2016. Recent changes in the abundance of Common Pochard *Aythya ferina* breeding in Europe. *Wildfowl* 66: 22-40.

Geissler K. 1962. Ornithologische Beobachtungen im Naturschutzgebiet von Srebarna (Bulgarien) [Ornithological observations in nature protected area of Srebarna (Bulgaria)]. *Der Falke* [*Falcon*] 9 (10): 327-331.

Golemansky V., Beron P., Zivkov M., Popov A., Popov V., Beschkov V., Deltchev C., Michev T., Spassov N., Stoev P. & Dobrev D. 2015. *Red Data Book of Bulgaria.* vol. 2 Animals. BAS, MOEW.

Hibaum G., Michev T., Vassilev V. & Uzunov Y. 2001. *Management Plan for the "Srebarna" biosphere reserve.* CLGE, BAS, Sofia.

Iankov P. (ed.) 2007. *Atlas of the breeding birds in Bulgaria*. Bulgarian Society for the Protection of Birds, Conservations Series Book 10**.** Sofia.

Ilieva M., Zehtindzhiev P., Kutsarov Y. & Georgiev D. 2007. Ruddy Shelduck (*Tadorna ferruginea*). In: Iankov P. (Ed.) *Atlas of the breeding birds in Bulgaria.* BSPB, Conservations Series Book 10, pp. 94-95.

Ivanov B. 1985. Colonially nesting birds at Belene Island. In: International symposium under project 8-MAB (UNESCO) *Protection of nature territories and their gene pool.* Paper of reports.vol.1.Blagoevgrad, 23-28.09.1985, Sofia, BAS. pp. 296-305.

Ivanov B. 1985a. Black-necked Grebe (*Podiceps nigricollis*). In: *Red data book of PR Bulgaria.* BAS, Sofia, p.43.

Ivanov B. 1985b. Red-necked Grebe (*Podiceps griseigena*). In: *Red data book of PR Bulgaria.* BAS, Sofia, p.44.

Ivanov B. 1985c. Red-crested Pochard (*Netta rufina*). In: *Red data book of PR Bulgaria.* BAS, Sofia, pp.63-64.

Ivanov B. 1985d. Black Tern (*Chlidonias niger).* In: *Red data book of PR Bulgaria.* BAS, Sofia, pp.117-118.

Ivanov B. 2015. Black-necked Grebe (*Podiceps nigricollis*). In: Golemansky V. et al. (Eds.) *Red Data Book of Republic of Bulgaria.* vol.2, Animals, BAS, MOEW, Sofia, p.66.

Ivanov B. & Dereliev S. 2015. Greylag Goose (*Anser anser*). In: Golemansky V. *et al.* (Eds.) *Red Data Book of Republic of Bulgaria.* vol.2, Animals, BAS, MOEW, Sofia, p.164.

Ivanov B., Petkov N. & Vassilev V. 2015. Red-necked Grebe (*Podiceps griseigena*). In: Golemansky V. et al. (Eds.) *Red Data Book of Republic of Bulgaria.* vol. 2 Animals. BAS, MOEW. Sofia, p.156.

Kostadinova I. (Ed.) 1997. *Important bird areas in Bulgaria.* BSPB, Conservation Series, Book 1.

Kostadinova I. & Gramatikov M. (Eds.) 2007. *Important bird areas and Natura 2000 in Bulgaria.* BSPB. Conservation books, Book 11, Sofia.

Kambourova N. 2005. The recent status of breeding bird communities of Srebarna Biosphere reserve, North-east Bulgaria. *Acrocephalus* 26 (125): 81-97.

Kambourova N. 2012. Ornithofauna of Biosphere Reserve "Srebarna" Lake. In: Uzunov Y., Georgiev B., Varadinoiva E., Kambourova N., Pehlivanov L., Vasilev V. (Eds.) *Ecosystems of Biosphere Reserve Srebarna Lake*. pp. 129-147.

Kambourova N. & Antonov A. 2007. Black-headed Gull (*Larus ridibundus*). In: Iankov P. (Ed.) *Atlas of the breeding birds in Bulgaria.* BSPB, Conservations Series, Book 10, pp. 278-279.

Kutsarov Y. 2007. White-winged Tern (*Chlidonias leucoptera*). In: Iankov P. (Ed.) *Atlas of the breeding birds in Bulgaria. BSPB*, Conservations Series, Book 10, pp. 296-29.

Kutsarov Y. & Stefanov T. 2007. Whiskered Tern (*Chlidonias hybrida*). In: Iankov P. (Ed.): *Atlas of the breeding birds in Bulgaria.* BSPB, Conservations Series, Book 10, pp. 292-293.

Michev T. 1968. New data on the avifauna of the Srebarna nature reserve, Southern Dobrudzha. *Izvestia na Zoologicheskia Institut s muzej* [*Reports of the Institute of Zoology with Museum*] 27: 13–21.

Michev T. 1981. Dynamics of the numbers of the Dalmatian Pelican (*Pelecanus crispus* Bruch) in Srebarna reserve, Southern Dobrudzha. In: *Regional symposium under project 8 MAB,* UNESCO, 20-24.10.1980, Blagoevgrad. pp. 516-527.

Michev T. 1985. Black-winged Stilt (*Himantopus himantopus*). In: *Red Data Book of PR Bulgaria*. BAS, Sofia. 112 p.

Nankinov D., Simeonov S., Michev T. & Ivanov B. 1997. *Fauna of Bulgaria.* vol.26. Aves part 2. BAS, Sofia.

Nankinov D., Dutsov A., Nikolov B., Borisov B., Stoyanov G., Gradev G., Georgiev D., Popov D., Domuschiev D., Kirov D., Tilova E., Nikolov I., Ivanov I., Dichev K., Popov K., Karaivanov N., Todorov N., Shurulinkov P., Stanchev R., Aleksov R., Tzonev R., Ivanov S.,Marin S., Staikov S., Nikolov S., Dalakchieva S.& Nikolov H. 2004. *Breeding totals of the ornithofauna in Bulgaria.* Green Balkans, Plovdiv.

Paspaleva - Antonova M. 1961. Study on the avifauna of Bulgarian Danube stream. PhD thesis. Institute of Zoology with Museum, Bulgarian Academy of Sciences.

Paspaleva M. & Mitschev T. 1968. Ornithological investigations along Bulgarian Danube section. pp. 445 - 454. In: *Limnological Reports of X-th Aniversary Meeting of Danube Research.* Bulgaria, Sofia, 18-20.X.1966.

Petkov N. 1997. Current status of the Ferruginous Duck, *Aythya nyroca* in Bulgaria. MSc. thesis. Faculty of Biology, Sofia University "St. Kl. Ohridski", Sofia.

Petkov N. 2000. Population trends of breeding Ferruginous Ducks in Bulgaria. *TWSG News* 12: 44-49.

Petkov N. 2011. Habitat characteristics assessment of the wetlands with breeding Ferruginous Duck, *Aythya nyroca* and Pochard *A. ferina* in Bulgaria. *Acrocephalus* 32 (150-151): 127-134.

Petkov N. 2015. Ferruginous Duck (*Aythya nyroca*). In: Golemansky V. et al. (Eds.) *Red Data Book of Republic of Bulgaria.* vol. 2 Animals, BAS, MOEW, Sofia. p.256.

Petkov N. 2015a. Gadwall (*Anas strepera*). In: Golemansky V. et al. (Eds.) *Red Data Book of Republic of Bulgaria. vol.* 2 Animals, BAS, MOEW, Sofia. p.74.

Petkov N. 2015b. Pochard (*Aythya ferina*). In: Golemansky V. et al. (Eds.) *Red Data Book of Republic of Bulgaria.* vol. 2 Animals, BAS, MOEW, Sofia, p.255.

Petkov N. 2015c. Red-crested Pochard (*Netta rufina*). In: Golemansky V. et al. (Eds.) *Red Data Book of Republic of Bulgaria.* vol. 2 Animals, BAS, MOEW, Sofia. p.37.

Petkov N., Iankov P. & Kurtev M. 2007. Srebarna. In: Kostadinova I., & Gramatikov M. (Eds.). *Important bird areas and Natura 2000 in Bulgaria.* BSPB. Conservation books Book 11, Sofia, pp. 226-228.

Petkov N., Iankov P. & Todorov E. 2007a. Belene Island complex. In: Kostadinova I. & Gramatikov M. (Eds.). *Important bird areas and Natura 2000 in Bulgaria.* BSPB. Conservation books Book 11, Sofia, pp. 173-176.

Poysa H. & Vaananen V. 2014. Drivers of breeding numbers in a long-distance migrant, the Garganey (*Anas querquedula*): effects of climate and hunting pressure. *Journal of Ornithology* 155 (3): 679-687.

Reiser O. 1894. *Materialen zu einer Ornis balcanica. II. Bulgarien.* [*Materials on the Balkan ornithofauna. II. Bulgaria.*] Wien, In Commission Carl Gerolds Sohn.

Sciban M., Rajkovic D., Radisic D., Vasic V. & Pantovic U. 2015. *Birds of Serbia – critical list of species.* Institute for Nature Conservation of Vojvodina Province and Bird Protection and Study Society of Serbia. Novi Sad.

Simeonov S., Michev T. & Nankinov D. 1990. *Fauna of Bulgaria,* vol. 20. Aves. BAS, Sofia.

Stanev S. 1956. Kaykusha marsh. *Priroda i znanie* [*Nature and knowledge*] 6 (56): 22.

Striburni V. 1930. In the kingdom of our water birds (Kara-boaz). *Lovec* [*Hunter*] 30 (10): 203-206.

Shurulinkov P., Nikolov B., Statkov P., Hristova I., Ralev A., Dinkov H., Hristov K. & Hristov I. 2006. Erster sicherer Brutnachweis des Scwarzhalstauchers nach uber einem Jahrungsdert in der Sofia-Ebene, Bulgarien [First confirmed breeding record of the Black-necked Grebe during the last century in Sofia district, Bulgaria]. *Ornithologische Mitteilungen* [*Ornithological observations*] 58 (11): 376-379.

Shurulinkov P., Tsonev R., Nikolov B., Stoyanov G. & Assenov L. 2005. *The Birds of Central Danube Plain.* Green Balkans Federation. Sofia, 120 pp. (in Bulgarian).

Shurulinkov P. & Tsonev R. 2009. Die Wasservogel der unteren Donau-Niederungen wahrend der Sommerfluten 2005 und 2006 [Water birds of the Lower Danube flow during the summer flooding in 2005 and 2006]. *Ornithologische Mitteilungen* [*Ornithological observations*] 61 (10): 317-324.

Shurulinkov P., Daskalova G., Ralev A. & Elenkova V. 2010. New data on the breeding of White-winged Tern (*Chlidonias leucoptera*) in Bulgaria. *Acta zoologica bulgarica* 62 (3): 301-306.

Shurulinkov P., Daskalova G. & Tzonev R. 2013. Breeding Waterbirds in Temporally Flooded Wetlands in Northern Bulgaria. *Acta Zoologica Bulgarica* 65 (2): 207-215.

Shurulinkov P., Daskalova G., Michov S. & Koev V. 2016. Distribution, numbers and breeding of terns and waders on the sand islands along the Bulgarian-Romanian section of the Danube. *North-Western Journal of Zoology* 12 (1): 65-77.

Vassilev V. 1999. Study on the avifauna of Mechka fishponds. Report. Biology and geography. In: *Jubilee scientific conference 25 yea*rs Shumen University "Ep. K. Preslavski", Shumen.

Vassilev V. 2003. Comparison of the nest ornithofauna in the four water basins of North-Eastern Bulgaria. *Annuaire de'l Univ. Sofia "St. Kliment Ohridski"* 95 (4): 241-246.

Yeatman L. 1976. *Atlas des oiseaux nicheurs de France de 1970 a 1975* [*Atlas of the breeding birds of France for 1970-1975*]. Paris, Societe Francaise d' Ornithologie [French Ornithological Society].

In: Biodiversity of the Bulgarian-Romanian Section
Editor: Shurulinkov, Hubenov et al.
ISBN: 978-1-53615-663-8

Chapter 11

STATUS AND DISTRIBUTION OF DIURNAL BIRDS OF PREY AND THE BLACK STORK ALONG THE BULGARIAN SECTION OF THE DANUBE RIVER

Svilen Cheshmedzhiev[1,*], Peter Shurulinkov[2] and Girgina Daskalova[2]
[1]Bulgarian Society for the Protection of Birds, Sofia, Bulgaria
[2]National Museum of Natural History, Sofia, Bulgarian Academy of Sciences, Sofia, Bulgaria

ABSTRACT

A total of 28 diurnal birds of prey species and Black Storks were recorded during intensive research along the Bulgarian section of the Danube River between 2008 and 2018. Out of these, 20 species are breeding. The Danube River and its neighbouring hills, forests, steppes, and the lowest sections of its tributaries hold significant populations of species of high conservation importance; including the Lesser Spotted Eagle (*Clanga pomarina*),40-50 breeding pairs (br.p.); Booted Eagle (*Hieraaetus pennatus*),10-15 br.p.; Short –toed Eagle (*Circaetus gallicus*),50-70 br.p.; European Honey Buzzard (*Pernis apivorus*),17-25 br.p.; Levant Sparrowhawk (*Accipiter brevipes*), 35-60 br.p.; Long-legged Buzzard (*Buteo rufinus*),75-130 br.p.; White-tailed Sea Eagle (*Haliaeetus albicilla*),12 br.p.; Marsh Harrier *(Circus aeruginosus*), 25-40 br.p.; and Black Stork (*Ciconia nigra*),80-150 br.p.. During migration, the most numerous species are the Marsh Harrier, Black Stork, European Honey Buzzard, Common Buzzard (*Buteo buteo*), Eurasian Sparrowhawk (*Accipiter nisus*) and Lesser Spotted Eagle. Migration is stronger during autumn and in central and eastern parts of the study area.

* Corresponding Author's Email: svilen.cheshmedjiev@bspb.org.

Keywords: birds of prey, breeding populations, migration numbers, Danube

INTRODUCTION

Diurnal birds of prey and the Black Stork have not been comprehensively studied in the Bulgarian section of Danube River. Concrete data for a significant number of species of raptors recorded in study area has been published in several ornithofaunistic studies in the past (Paspaleva-Antonova 1961, Shurulinkov et al. 2005). Valuable data on this group of birds can also be found in studies on particular species such as the Marsh Harrier (Petrov & Michev 1985), White-tailed Eagle (Todorov et al. 2015), and Black Stork (Nankinov 1997). Some particularly interesting observations on birds of prey along the Danube are also mentioned in the Red Data Book of PR Bulgaria (Botev & Peshev eds. 1985), the Red Data Book of Bulgaria (Golemansky et al. eds. 2015), Fauna of Bulgaria (SIMEONOV et al. 1990), and in the Atlas of the Breeding Birds of Bulgaria (IANKOV ed. 2007). Despite the existence of these valuable contributions to research, a comprehensive overview of the status, distribution and numbers of all of the birds of prey along the Bulgarian Danube has yet to be published. With this study, we seek to contribute to the existing body of research on birds of prey, and the Black Stork, in the Bulgarian Danube, and present data collected on the distribution and numbers of all diurnal birds of prey and Black Storks along Danube.

MATERIAL AND METHODS

Territorial Scope

The entire territory of the Bulgarian section of Danube River (470 km) falls into 109 standard 10 X 10 km Military Grid Reference System (MGRS) squares that are spatially identical to UTM squares. This field survey covers virtually the entire territory, inclusive of 104 full or partial 10 km UTM squares.

Scope of Species

In the present study, all breeding, migrating and wintering diurnal raptors species were included. The Black Stork was also included as data for the species can be obtained with methods similar to those used for diurnal raptors, and on account of its high conservation status. Every locality was entered in the database and reflected on the map on the basis of a specific observation or observations.

Completeness of the Coverage of Species per MGRS (UTM) Square

Despite attempts to identify all breeding, migrating and wintering diurnal raptors in the relevant square, not all squares were equally well-surveyed due to various reasons. In some squares, the entire species composition was recorded, while in others there were gaps in information during the field surveys. The systematic surveying and monitoring of a considerable part of the territory of the Danube from 2005 to 2018 and the detailed knowledge of the bird fauna in regions such as the Orsoya fishponds, SPA Zlatiata, Ostrovsko marsh, Nikopolsko plateau, "Persina" Nature Park, SPA Belene Islands Complex, Vardim Island, Hadzhidimitrovo fishponds, Mechka fishponds, "Roussenski Lom" Nature Park, Kalimok – Brashlen Protected area, and Srebarna lake, among others, enabled a very critical attitude towards the determination of the percentage of the coverage of the species composition.

Period of Study

Most of the data about the contemporary breeding, migrating and wintering distribution of diurnal raptors species concern the period from 2008 to 2018, when the field surveys were carried out. In order to complete the information, some data between 2000 and 2007 were also included in the findings. The trends in distribution and numbers of breeding species are for the period 2014 to 2018.

Methodology – Breeding Diurnal Raptors and Black Stork

The purpose of the methodology is collecting actual data about diurnal raptors and Black Stork breeding in the area. The type of methodology is point count, i.e., a point-based monitoring. The visual observations were carried out using methods recommended by the Ministry of Environment and Waters of the Republic of Bulgaria (http://eea.government.bg/bg/bio/opos/activities-results/Metodika_monitoring_ptitsi.pdf). Monitoring of raptors and Black Stork was carried out by sightings from observation points. The observer had to actively seek out flying or staggering birds for 3 hours. The statistical unit is the UTM square (10x10 km); in the squares a variable number of points with good visibility is selected to cover the main habitats. The minimum distance between points is 2 km. During the fieldwork, the observers sought the raptors and Black Storks with the help of binoculars and optical telescopes. The observation were carried out during the day (April – July) between 10:00 17:30 local time in sunny and clear weather conditions.

The main survey technique used for the investigation of the White-tailed Eagle if certain territory was occupied was observation from a stationary point (Bibby et al. 1998), situated in the vicinity of the potential nesting area and utilizing vantage points with good visibility. The best period of implementing this method was after incubation, when the adult eagles bring food in the nest very often in order to feed the eaglets (April to May). In order to confirm the breeding occupancy of each potential site, field work was carried out at least twice in the breeding period between March and July 2012 to 2018. A breeding territory is considered occupied if a pair of adult birds is observed in the area during the breeding season or an active nest is found. We also collected field data about the habitat and tree species where the nests were located during the field work.

Methodology – Soaring Passage Migrants

The purpose of the methodology is collecting actual data about soaring birds (raptors, storks, pelicans) migrating over the studied area. Monitoring of migrating soaring birds is carried out by sightings from observation points during 60 days in spring (15 March – 15 May) and during 90 days in autumn (August - October). The visual observations were carried out using methods, recommended by the Bulgarian Ministry of Environment and Waters (http://eea.government.bg/bg/bio/nsmbr/prakticheskо-rakovodstvo-metodiki-za-monitoring-i-otsenka/Metodika_monitoring_migrirashi_reeshti_vidove_ptitsi.pdf). The observation points were manned daily by one or more observers, between 08.00 and 18.00 h EET. Each observer used binoculars and a telescope, and covered air space with a diameter up to 7-8 km. Species not identified with certainty were logged as a species pair where possible (e.g., Pallid Harriers *Circus macrourus* and Montagu's Harriers *Circus pygargus*) or simply to genus. It was also difficult to separate migrants from some resident lingering birds which hunted in the area of the observation points, such as Marsh Harriers *Circus aeruginosus*, Eurasian Sparrowhawks *Accipiter nisus,* Common Buzzards *Buteo buteo* and Common Kestrels *Falco tinnunculus*. Collected data included the number of individuals, time, distance and direction from the observation point, direction, height, and character of flight. Direction of flight, as well as an estimate of distance from the observation point, was determined using particular landmarks at known distance/direction. The height of flight was estimated visually. When possible, migratory flocks were counted in tens or hundreds, while estimates were made of the larger flocks of storks and pelicans. Whenever possible, repeat counts of the same flock were made and the mean was noted. The observation points for the field survey of migrating soaring birds was organized by the Bulgarian Society for the Protection of Birds (BSPB) at the following stationary points:

- Slivata, Montana district – N 43.758299° E 23.047973°
- Zlatia, Montana district – N 43.771768° E 23.508714°
- Vabel, Pleven district – N 43.660494° E 24.888431°
- Peschina marsh, Pleven district – N 43.667485° E 25.207793°
- Kaykusha marsh, Veliko Turnovo district – N 43.595966° E 25.204941°
- Vardim, Veliko Turnovo district – N 43.609356° E 25.466406°
- Hadzhidimitrovo fishponds, Veliko Turnovo district – N 43.508235° E 25.496562°
- Ivanovo, Rousse district – N 43.691206° E 25.966925°

RESULTS AND DISCUSSION

Black Stork *(Ciconia nigra)*

It is a breeding and migrating species for the Danube region. The breeding territories of solitary pairs were localized at the following Danube islands: Ibisha, Kicha, Kozloduy, Lakat, Persin, Vardim, Batin, Brashlen and Golyam/Malak Kosuj Islands. Breeding of the species was also proved at Kutovo Island and it is very probable on Bogdan (Sechan) Island. Further breeding pairs were found in the lowest parts of the valleys of the rivers Ogosta, Iskar, Vit, Osam and Yantra. In Vidin district, Black Storks territories were found at Vartop (Peter Shurulinkov, Girgina Daskalova, Rossen Tzonev – pers. communication), Gramada and Dimovo. The species also nests in the Nikopolsko plateu, "Roussenski Lom" Nature Park, and, in the Silistra district at Garvan, Zafirovo, Srebarna and Sitovo. Single birds were also seen during the breeding period at Ostrov and Slivo pole (Figure 1).

The Black Stork prefers old forests, inland cliffs and exposed rocks near to rivers or other wetlands. In the study area the breeding population can be estimated at 80-150 pairs.

The spring migration of the Black Stork starts in mid-March (the earliest observed on 11.03.2014 in Vardim Island (1 ind. migrating to the northwest) and continues until mid-May (SHURULINKOV et al. 2005). The latest flock was observed on 30.05.2017 in the Hadzhidimitrovo fishponds – 5 ind. migrating to the north (CHESHMEDZHIEV 2017). Autumn migration starts from the second half of August and continues until mid-October. In this period, small flocks of 3 to 24 ind. were observed. In the 1950s a flock of 300 ind. was observed at Belene Island during autumn (NANKINOV 1997).

During the field studies of the spring and autumn migration of birds, organized by BSPB, Black Storks were observed as follows: Peschina marsh - September - October 2008 - 344 ind., September - October 2009 - 402 ind.; Ivanovo - August - October 2009 - 25 ind.; Kaykusha marsh - August - October 2009 - 169 ind.; Vabel - August - October 2009 - 98 ind., April - May 2010 - 6 ind.; Zlatia - August - October 2009 - 505 ind.; Slivata -

March - April 2011 - 62 ind.,, August - October 2011 - 125 ind.; Hadzhidimitrovo fishponds - April - May 2017 - 32 ind., August - October 2017 - 32 ind.; and Vardim - April - May 2017 - 7 ind., August - October 2017 - 26 ind.

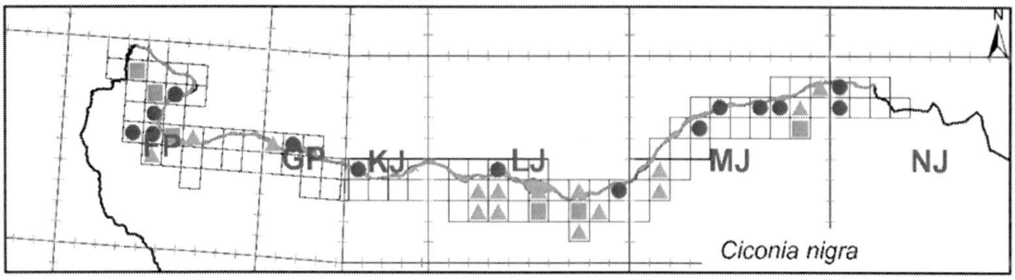

Figure 1. Breeding distribution of the Black Stork (*Ciconia nigra*) along the Bulgarian section of the Danube River (circles –new localities, triangles – confirmed published localities, squares – unconfirmed published localities).

Osprey *(Pandion haliaetus)*

It is a rare migrating species. Osprey have been often observed during spring and autumn migration.

Around the middle of the 19th century, it was relatively common along the Danube and bred near Vidin (Finsch 1859). On 12.05.1945, a nest with three eggs was found near Svishtov (Arabadzhiev 1962). During that time, the species bred at Kerkeneza Island near Vidin (Stoyanov 1948). Patev (1950) also mentioned that it was a rare breeding species along the Danube. In 1965, a nest of an Osprey was found on Persin Island. The species was also found near Archar, Oryahovo, Ostrov, Svishtov, Mechka, Tutrakan, and Srebarna Lake. In the period 1990-2005 the Osprey did not breed along the Danube (Iankov 2007). An adult bird was observed on 02.07.2002 near Kaykusha marsh in a direct flight to the Danube, but breeding has not been proven.

During the autumn and spring migration (2005-2018), the Osprey was regularly observed at wetlands along the Danube, including Kutovo Island, Orsoya fishponds, Ibisha Island, Kozloduy Island, Ostrovsko marsh, Persin Island, Hadzhidimitrovo fishponds, Vardim Island, Mechka fishponds, the mouth of Yantra River near Krivina, Kalimok-Brashlen, Pozharevsko marsh, Malak Preslavets marsh, Garvansko marsh, and Srebarna Lake, among others.

European Honey Buzzard *(Pernis apivorus)*

Breeding and migrating species along the Danube. It is rare during breeding season and numerous during spring and autumn migration.

In the study area the estimate of the breeding population is 17-25 pairs. The breeding areas were localized at Dobri dol, Oryahovo, Slavovitsa, Lakat Island, Dragash Voyvoda, Lyubenovo, Byala voda, Lozitsa, Persin Island, Basarbovo, Krasen, Ivanovo, Nova Cherna, near Tutrakan, Malak Preslavets, Nova Popina, and Srebarna Lake (Figure 2). One pair was observed on 1.07.2005 at the mouth of the Ogosta River (Rossen Tzonev –pers. communication). In Vidin district, three individuals were seen on 9.06.2010 along the Skomlya River, at Izvor village, and a pair was recorded at Vartop village.

During migration the European Honey Buzzard is common. It can be observed mostly from the end of August to the end of September during the autumn, and in April until mid-May in the spring. During the field studies of the spring and autumn migration of birds, organized by BSPB, Honey Buzzards were observed as follows: Peschina marsh - September - October 2008 - 81 ind., September - October 2009 - 130 ind.; Ivanovo - August - October 2009 - 15 ind.; Kaykusha marsh - August - October 2009 - 100 ind.; Vabel - August - October 2009 - 145 ind., April - May 2010 - 5 ind.; Zlatia - August - October 2009 - 32 ind.; Slivata - March - April 2011 - 14 ind., August - October 2011 - 24 ind.; Hadzhidimitrovo fishponds - April - May 2017 - 79 ind., August - October 2017 - 66 ind.; and Vardim - April - May 2017 - 19 ind., August - October 2017 - 19 ind.

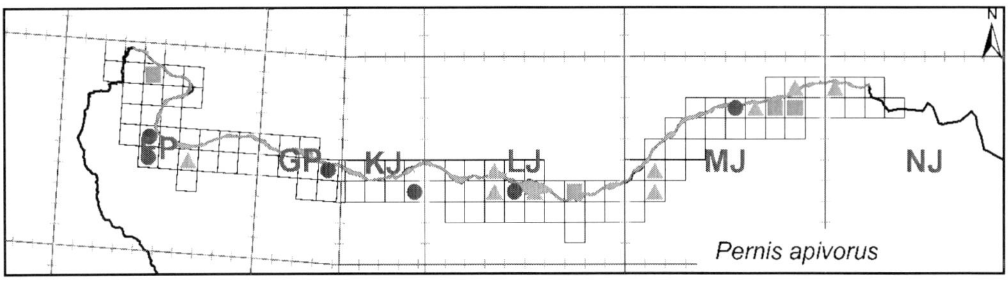

Figure 2. Breeding distribution of the Honey Buzzard (*Pernis apivorus*) along the Bulgarian section of the Danube River (circles –new localities, triangles – confirmed published localities, squares – unconfirmed published localities).

Black Kite *(Milvus migrans)*

Very rare breeding and migrating species along the Danube area. It inhabits forests and groups of trees among pastures and agricultural lands near wetlands.

In the 19th and the beginning of 20th century, the species was common and even numerous during the breeding season, mainly along the Danube River (near Lom, Svishtov, etc.). Around 1950, it was still quite common along the Danube (Patev 1950). After 1950, the national population showed decrease until 1985. After 1995, stabilisation of the population was reported (IANKOV 2007).

In the study area the estimation of the breeding population is 5-9 pairs. Breeding territories of Black Kites were localized around Debovo, Evlogievo, Brestovitsa, Bozhichen, Garvan and Bulgarka villages.

There were regular observations of the species during migration but in low numbers. During the spring the Black Kite was observed on 24.04.2000 in the region of Dabovan; during the autumn the species was observed on 19.09.2000 between Cherkovitsa and Somovit (Shurulinkov et al. 2005). Field studies of the spring and autumn migration of birds, organized by BSPB, established the species in the following locations: Peschina marsh October 2008 - 1 ind.; September - October 2009 - 1 ind. Kaykusha, August - October 2009 - 1 ind.; Vabel, August - October 2009 - 3 ind., April - May 2010 - 3 ind.; Slivata, March - April 2011 - 2 ind. August - October 2011 - 1 ind.; Hadzhidimitrovo fishponds, April - May 2017 - 3 ind.; August - October 2017 - 6 ind.; and Vardim, April - May 2017 - 1 ind. August - October 2017 - 3 ind. Three Black Kites were observed hunting at Kalimok fishponds (west) on 21.05.2005; they were probably late spring migrants.

Red Kite *(Milvus milvus)*

It is an extremely rare and not a breeding species. In the study area, the Red Kite was observed twice: on 20.09.2009 with 1 ad. at Kaykusha marsh (Cheshmedzhiev & Todorov 2009) and on 26.04.2010 with 1 ad. close to Vabel village, Nikopol Municipality (Cheshmedzhiev & Todorov 2011).

White–Tailed Eagle *(Haliaeetus albicilla)*

It is a breeding and wintering species. The White-tailed Eagle breeds on the Danube islands mainly on Black Poplars (*Populus nigra*). In the study area the estimation of the breeding population is 12 pairs, all along the Danube islands in the Bulgarian part of the river.

By the end of the 19th and the beginning of 20th century, the species often bred along the Danube from Lom to Srebarna Lake. After the middle of the 20th century, the numbers of breeding pairs dropped sharply; by 1980 only one nest was known to exist in Bulgaria: at Persin Island (Botev & Peshev eds. 1985). The gradual recovery of the population started around 1985 (Iankov 2007). Between 2006 and 2012, the numbers of breeding pairs in the country increased to 23, 11 of them recorded in the Danube islands (Todorov et al. 2015).

During a field survey organized by BSPB in the period 2014 - 2016, 33-35 breeding pairs were found in the country (Cheshmedzhiev 2016).

Figure 3. White-tailed Eagle (*Haliaeetus albicilla*). Photo: Ivailo Zafirov.

During the autumn and winter, the White-tailed Eagle is common along the Danube. It prefers wetlands with a big concentrations of waterfowl such as geese, cormorants and ducks. The most important wintering territories are the Danube islands, Peschina marsh, Martvo marsh, the Kalimok-Brashlen Protected area, and Srebarna Lake. Groups of up to 14 young eagles were observed during the winter in the region of Peschina marsh.

In 2010, a team comprised of BSPB and Persina NP carried out the first colour ringing of young eagles along the Danube River. Between 2010 and 2018, a total of 13 young birds were ringed and 5 of them were fitted with satellite transmitters.

Egyptian Vulture *(Neophron percnopterus)*

It is an extremely rare breeding and migrating species. In the study area, the Egyptian Vulture breeds in the region of Roussenski Lom Nature Park (1-2 pairs).

In the 19th and the beginning of 20th century, it was a numerous breeding species on rocks all over the country. The species bred along the Danube (on 19.06.1889 near Nikopol) and Yantra (2 pairs between Novgrad and Belyanovo) (Iankov 2007).

During migration the Egyptian Vulture was observed in the region of Nikopolsko Plateau between Lyubenovo and Vabel.

Golden Eagle *(Aquila chrysaetos)*

Golden Eagle is not a breeding species along the Danube River. There were several observations of young birds in the study area during the spring and autumn migration as well as in the winter at the following locations: Ivanovo - 14.09.2009 - 1 juv.; Peschina marsh - 25.09.2009 - 1 juv.; Vabel - 26.04.2010 - 1 juv.; Oresh - 10.08.2010 - 1 juv.; Batin Island - 12.03.2011- 1 ind. of second winter plumage; Tsarevets - 14.01.2014 - 1 juv.; and between Dragomirovo and Morava- 10.11.2018 - 1 juv.

Imperial Eagle *(Aquila heliaca)*

The Imperial Eagle is a globally threatened species not breeding along the Danube River. In the past the species had been found in the region of Obnova during the breeding period (Michev & Petrov 1979, Shurulinkov et al. 2005). It was also observed on 12.06.1965 near the Danube at kilometre 664 (Michev & Petrov 1979, Shurulinkov et al. 2005).

There was one observation of the Imperial Eagle in the studied region during migration; on 21.08.2009, 1 young bird was observed close to Ivanovo village in "Roussenski Lom" Nature Park (Cheshmedzhiev & Todorov 2011).

Lesser Spotted Eagle *(Clanga pomarina)*

It is a breeding and migrating species along the Danube region. The Lesser Spotted Eagle can be observed in natural forests of beech, oak or mixed, and in old shelterbelts and other woodland areas in the vicinity of spacious grass communities and agricultural land, where the birds hunt.

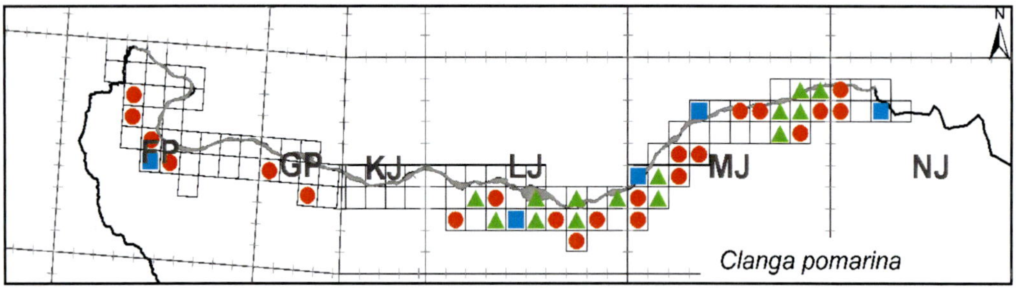

Figure 4. Breeding distribution of the Lesser Spotted Eagle (*Clanga pomarina*) along the Bulgarian section of the Danube River (circles –new localities, triangles – confirmed published localities, squares – unconfirmed published localities).

The Lesser Spotted Eagle is a breeding, migratory and transitory species in Bulgaria. In the past it was a common bird in the country but from 1950 to 1980 its numbers show an uninterrupted decrease (Simeonov et al. 1990). At present nearly 50% of its national population is concentrated in the Eastern Rhodopes Mountain, Sakar Mountain, Derventski

Vazvishenia Hills, and Strandzha Mountain. There are scattered records from the Dobroudzha Plain, Danubian Plain, Thracian Plain, the Western Rhodopes Mts., the Sredna Gora Mts., Mt. Vitosha, Mt. Plana, among others. During autumn migration, it concentrates in high numbers in the area of the Bay of Bourgas. The current breeding population for the whole of Bulgaria was estimated at 550-600 pairs (Cheshmedzhiev et al. 2018). In the study area the estimation of the breeding population was 40-50 pairs (Cheshmedzhiev et al. 2018), with a higher density in the Central and Eastern parts of the region. Breeding territories of the species were localized at the following locations: Kalenik, Voynitsa, Gaytantsi, Septemvriytsi, Zlatia, along the Ogosta River near Butan, along the Iskar River, along the Vit River near Kreta, along Osam river near Muselievo, Nikopolsko plateau near Vabel, Dragash Voyvoda and Novachene, Persin Island, Tatari, Tsarevets, Vardim, Sovata, Hadzhidimitrovo fishponds, Piperkovo, Batin, Trastenik, Obretenik, "Roussenski Lom" NP around Krasen, Koshov, Nisovo, Svalenik and Pisanets, at Nikolovo, Shtruklevo, Prossena, Kalimok-Brashlen, Staro selo, Pozharevo, Preslavtsi, Malak Preslavets marsh, Zafirovo, Glavinitsa, Garvan marsh, Nova Popina, Srebarna lake, and Sitovo (Cheshmedzhiev et al. 2018) (Figure 4).

During autumn, the Lesser Spotted Eagle migrates mostly from mid-September to early October; in spring, it migrates in April. During field studies of the spring and autumn migration of birds, organized by BSPB, the species was established in the following locations: Peschina marsh - September - October 2008 - 65 ind., September - October 2009 - 41 ind.; Ivanovo - August - October 2009 - 22 ind.; Kaykusha marsh - August - October 2009 - 37 ind.; Vabel - August - October 2009 - 51 ind., April - May 2010 - 5 ind.; Zlatia - August - October 2009 - 8 ind.; Slivata - March - April 2011 - 3 ind., August - October 2011 - 12 ind.; Hadzhidimitrovo fishponds - April - May 2017 - 12 ind.; August - October 2017 - 37 ind.; and Vardim - April - May 2017 - 11 ind., August - October 2017 - 27 ind.

Greater Spotted Eagle *(Clanga clanga)*

It is an extremely rare species. It was observed several times in the study area, mostly during the winter. During the period 2010-2017 the Greater Spotted Eagle regularly spent the winter in the region of the Persina marshes. One adult individual was observed on 13.11.2018 in flight over the Peschina marsh.

There were also several sighting during the spring and autumn migration: 29. 09. 2009 - Peschina marsh - 1 adult; 26.04.2010 - Vabel - 1 adult; and 09.04.2012 – Srebarna Lake - 1 subad.

During winter, the Greater Spotted Eagle was also observed in the region of Srebarna Lake (BSPB Database).

Steppe Eagle *(Aquila nipalensis)*

It is an extremely rare and not a breeding species for the country. In the study area the Steppe Eagle was observed once on 15.04.2017 (1 subad.) at Hadzhidimitrovo fishponds (Cheshmedzhiev 2017).

Booted Eagle *(Hieraaetus pennatus)*

It is a breeding and migrating species in the Danube area. The Booted Eagle can be observed in old forests close to open areas.

At the end of the 19th century and during the first decades of the 20th century, it was widely distributed in the woodland areas of the plains and mountains (Botev & Peshev eds. 1985). Around the middle of the 20th century, studies do not mention any change in the status of the species in the comparison with the previous period (Patev 1950). By 1980, the breeding population was estimated to be 50-100 pairs (Botev & Peshev eds. 1985) and 140-200 pairs in 2007 (Iankov 2007).

In the study area the Booted Eagle is a rare breeder. The estimation of its breeding population is 10-15 pairs. The breeding areas were localized mostly in the central and eastern parts of the studied Danube section (Figure 5). They include the following locations: Iskar River between Stavertsi and Bregare; near Zhernov; Muselievo; Novachene; Dragash Voyvoda; the Hadzhidimitrovo fishponds; between Piperkovo and Tsenovo; near Dolna Studena; Bozhichen; between Ivanovo and Koshov; near Staro Selo; and Malak Preslavets Lake and between Yordanovo and Slatina villages (last finding made by Michail Iliev - pers. communication). An adult specimen of the dark phase was observed on 18.06.2005 at "Boblata" forest close to Tutrakan (Rossen Tzonev - pers. communication). Another Booted Eagle (light phase) has been recorded on 12.06.2013 at Vardim village. In the western parts of the studied area a pair was observed between Kozloduy and Gorni Tsibar village, on 23.05.2014 (Hristo Dinkov - pers. communication).

Booted Eagle is rare for the Danube region even during migrations. During field studies of the spring and autumn migration, organized by BSPB, the species was observed in the following locations: Peschina marsh - September - October 2008 - 1 ind., September - October 2009 - 1 ind.; Ivanovo - August - October 2009 - 7 ind.; Kaykusha marsh - August - October 2009 - 2 ind., Hadzhidimitrovo fishponds - April - May 2017 - 3 ind., August - October 2017 - 5 ind.; and Vardim - April - May 2017 - 1 ind., August - October 2017 - 2 ind.

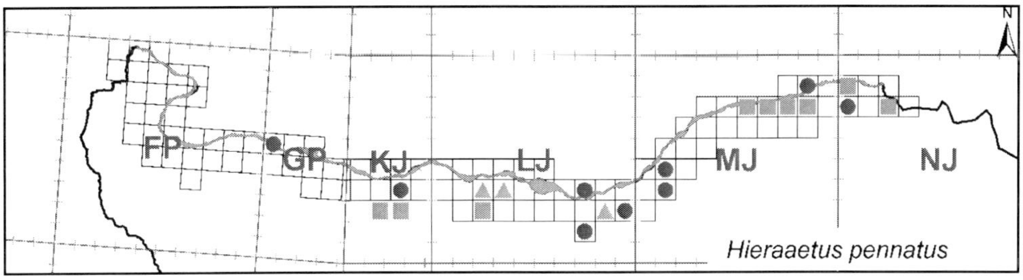

Figure 5. Breeding distribution of the Booted Eagle (*Hieraaetus pennatus*) along the Bulgarian section of the Danube River (circles –new localities, triangles – confirmed published localities, squares – unconfirmed published localities).

Short-Toed Eagle *(Circaetus gallicus)*

It is a breeding and migrating species in the Danube region. It inhabits open, dry, and rocky areas, and forests and river valleys. In the studied region the most important breeding areas for the species were Nikopolsko Plateau and "Roussenski Lom" Nature Park.

In the 19th and the first half of the 20th century, the species was reported as breeding on islands along the Danube (Svishtov) and near the Yantra River (Iankov 2007). In 1950, it was still quite common (Patev 1950). In 1990, estimations do not show more than 100 pairs for the whole country (Nankinov et al. 1990). After 1990, numbers and the occupied territories increased, a possible reason for that being the improvement of the food supply owing to the increase of abandoned lands all over the country (Iankov 2007).

In the study area, the estimation of the breeding population is 50-70 pairs. Breeding localities were recorded as follows: Voynitsa, Orsoya, along the Ogosta River between Hayredin and Sofronievo, Glozhene, Iskar River near Orehovitsa, Kreta, Debovo, Muselievo, Zhernov, Novachene, Dragash Voyvoda, Byala voda, Persin Island, between Vardim and Novgrad, Hadzhidimitrovo fishponds, Krasen, Bozhichen, Ivanovo, Pozharevo, Sitovo, and Srebarna Lake (Figure 6).

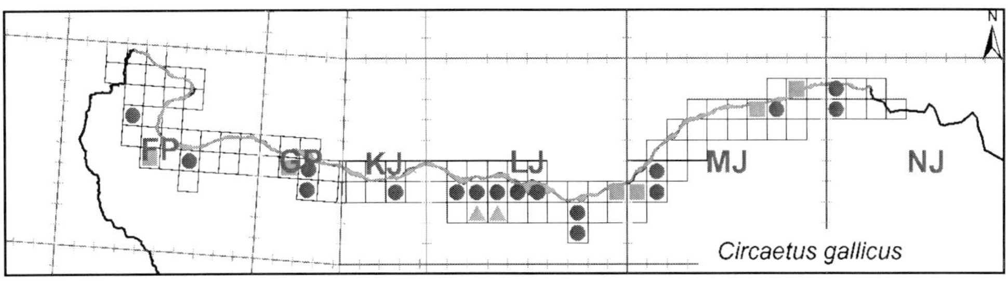

Figure 6. Breeding distribution of the Short-toed Eagle (*Circaetus gallicus*) along the Bulgarian section of the Danube River (circles –new localities, triangles – confirmed published localities, squares – unconfirmed published localities).

During the spring migration the Short-toed Eagle can be observed from March to April; in autumn, it can be observed from the second half of August to the end of September (Shurulinkov et al. 2005). Field studies of the spring and autumn migration of birds, organized by BSPB, established the species in the following locations: Peschina marsh - September - October 2008 - 7 ind. September - October 2009 - 6 ind.; Ivanovo - August - October 2009 - 3 ind.; Kaykusha marsh - August - October 2009 - 6 ind.; Vabel - August - October 2009 - 4 ind., April - May 2010 - 3 ind.; Zlatia - August - October 2009 - 5 ind.; Slivata - March - April 2011 - 5 ind.; August - October 2011 - 4 ind.; Hadzhidimitrovo fishponds - April - May 2017 - 2 ind., August - October 2017 - 4 ind.; and Vardim - April - May 2017 - 5 ind., August - October 2017 - 3 ind.

Marsh Harrier *(Circus aeruginosus)*

It is a breeding, migrating and wintering species along the Danube. It inhabits various lakes, marshes, dams, fishponds, and rivers with extensive reed beds. Most often the species can be observed hunting while flying on the edge of the reed beds, and also in open meadows and agricultural lands near the wetlands.

Marsh Harrier has been a very common breeding species at the end of the 19th and the first half of the 20th century along the Danube (Iankov 2007). Around the middle of the 20th century, it was also mentioned as a very common bird (Patev 1950). During the period 1950-1984, it was recorded at 38 breeding localities along the Black Sea Coast and along the Danube River, Thracian plain and Sofia Plain. Population numbers for the whole country were estimated at 40-60 breeding pairs (Petrov & Michev 1985). There has been an obvious decline that probably began before 1950. Possible reasons for this could be the large-scale drainage of the Danube marshes. After 1985, the national population showed gradual restoration (Bavoux et al. 1997).

In the study area the estimation of the breeding population is 25-40 pairs, mainly in the wetlands along the Danube. The breeding territories were localized at the following locations: Vidin marsh, Orsoya fishponds, Dobri Dol-Slivata marshes, "Kovachitsa" dam lake, "Shishmanov val" dam lake, Sofronievo fishponds, Ostrovsko marsh, lower Iskar River, Dabovan and Zagrazhden, Kulina voda, Peschina marsh, Martvo marsh, Kaykusha marsh, Hadzhidimitrovo fishponds, Karamanovo, Batin, Mechka fishponds, Kalimok-Brashlen Protected area, Pozharevo marsh, Malak Preslavets Lake, Garvan marsh, and Srebarna Lake.

During migration the Marsh Harrier is much more common. It is even numerous during autumn migration over central parts of the studied Danube section. The spring migration is from March to mid-May; the autumn migration is from August to the end of November, most often in September (Shurulinkov et al. 2005). During field studies of the spring and autumn migration of birds, organized by BSPB, the species was observed at the following locations: Peschina marsh - September - October 2008 - 208 ind., September - October 2009 - 189 ind .; Ivanovo - August - October 2009 - 8 ind.; Kaykusha marsh - August - October 2009 - 268 ind.; Vabel - August - October 2009 - 229 ind., April - May 2010 - 21 ind.; Zlatia - August - October 2009 - 8 ind.; Slivata - March - April 2011 - 9 ind., August - October 2011 - 31 ind.; Hadzhidimitrovo fishponds - April - May 2017 – 34 ind,, August - October 2017 - 212 ind.; and Vardim - April - May 2017 - 18 ind., August - October 2017 - 213 ind.

During winter, the Marsh Harrier was regularly observed along the Danube River and nearby wetlands, including Persin Island, Hadzhidimitrovo fishponds, Kalimok-Brashlen Protected area, and Srebarna Lake, and sand dunes along the Danube and other unfrozen small dams and fishponds.

Hen Harrier *(Circus cyaneus)*

It is a migrating and wintering species in the Lower Danube region. The Hen Harrier inhabits wet meadows, agricultural areas and pastures. It can be often observed during migration and winter. The autumn migration is from late October to December. The Hen Harrier is quite common during the winter. The spring migration is from March to early April (Shurulinkov et al. 2005). On 10.12.2010, 2 male and 7 female individuals were observed in the area of the Peschina marsh.

During the field studies of the spring and autumn migration, organized by BSPB, Hen Harriers were observed in the following locations: Peschina marsh - September - October 2008 - 1 ind., September - October 2009 - 4 ind.; Ivanovo - August - October 2009 - 22 ind.; Kaykusha marsh - August - October 2009 - 10 ind.; Vabel - August - October 2009 - 1 ind., April - May 2010 - 13 ind.; Zlatia - August - October 2009 - 1 ind.; Slivata - March - April 2011 - 1 ind.; Hadzhidimitrovo fishponds - August - October 2017 - 10 ind; and Vardim - August - October 2017 -10 ind.

Pallid Harrier *(Circus macrourus)*

It is an extremely rare species along the Danube. It was recorded at rare occasions during spring and autumn migration.

During field studies of the spring and autumn migration of birds, organized by BSPB, the species was observed in the following locations: Peschina marsh - 1 (1-st year) - on 08.09.2008 and 1 female on 03.10.2008; Zlatia - 1 male on 23.08.2009; Slivata- 1 female on 26.04.2011; Hadzhidimitrovo fishponds - 1 male, 1 female on 29.04.2017; 1 male on 04.10.2017 and 1 female on 05.10.2017; and Vardim - 1 male on 29.09.2017. One male Pallid Harrier was observed on 01.10.2018 near Nova Cherna.

Montagu's Harrier *(Circus pygargus)*

It is an extremely rare breeding species in the Lower Danube area. It can be often observed during spring and autumn migration. It inhabits wet meadows, agricultural areas with cereals and pastures.

The breeding areas were localized in SPA "Zlatiata"- 2 pairs and between Sovata and Hadzhidimitrovo - 1 pair. Also, one hunting male was observed on 2.08.2015 at Krushovitsa village, Miziya municipality, at 6 km. to the east of "Zlatiata" SPA (Peter Shurulinkov, Borislav Borisov –pers. communication).

Spring migration is mainly in April. A very early migrant was found on 2 March 1960 at Nikopol; the bird was ringed on 21.04.1956 by Kap-Bon in Tunisia (Paspaleva-Antonova 1961a). Autumn migration is from the mid-August to the end of September (Cheshmedzhiev 2017).

During the field studies of the spring and autumn migration of birds, organized by BSPB, the species was observed at the following locations: Peschina marsh - September - October 2008 - 15 ind., September - October 2009 - 14 ind.; Ivanovo - August - October 2009 - 6 ind.; Kaykusha marsh - August - October 2009 - 28 ind.; Vabel - August - October 2009 - 12 ind., April - May 2010 - 10 ind.; Zlatia - August - October 2009 - 8 ind., Slivata - March - April 2011 - 1 ind., August - October 2011 - 6 ind.; Hadzhidimitrovo fishponds - April - May 2017 - 8 ind., August - October 2017 - 4 ind.; and Vardim - April - May 2017 - 12 ind., August - October 2017 - 10 ind.

Northern Goshawk *(Accipiter gentilis)*

The Northern Goshawk is a breeding species along the Danube. It inhabits extensive deciduous and mixed forests on slopes with meadows and pastures among them. In some places, it nests on the Danube islands, including in poplar plantations. During winter the species can be found in open agricultural areas, around wetlands, and in urban territories.

In the study area the estimation of the breeding population is 40-70 pairs. The breeding territories were found at the following locations: Kalenik, Voynitsa, Medeshevci, Vartop, Lagoshevtsi, Kozloduy, Sarajevo, Mizia, Ostrov, Boril Island, Slavovitsa, Kreta, Muselievo, Zhernov, Asenovo, Lakat Island, Dragash Voyvoda, Lyubenovo, Novachene, Sanadinovo, Byala voda, Kulina voda, Persin Island, Oresh, Belyanovo, Beltsov, Batin Island, Brestovitsa, Volovo, Krasen, Bozhichen, Ivanovo, Staro selo, Pozharevo, Garvan, Bogorovo, and Glavan.

Eurasian Sparrowhawk *(Accipiter nisus)*

Breeding species along the Danube. The Eurasian Sparrowhawk is more common during migrations, as well as in the winter. During the breeding season the species can be found mainly in forested areas. In the autumn and winter it occurs in various habitats, including in the settlements.

In the study area, the breeding population can be estimated at 180-330 pairs. The species is breeding in almost all UTM 10x10 squares.

During migrations, the Eurasian Sparrowhawk can be observed mainly from September to mid- November and from mid-March to the end of April. During field studies of the spring and autumn migration of birds, organized by BSPB, it was observed in the following locations: Peschina marsh - September - October 2008 - 54 ind., September - October 2009 - 35 ind.; Ivanovo - August - October 2009 - 5 ind., Kaykusha marsh - August - October 2009 - 43 ind.; Vabel - August - October 2009 - 42 ind., April - May 2010 - 10 ind.; Zlatia - August - October 2009 - 3 ind.; Slivata - March - April 2011 - 19 ind., August - October 2011 - 17 ind.; Hadzhidimitrovo fishponds - April - May 2017 - 9 ind., August - October 2017 - 39 ind.; and Vardim - April - May 2017 - 10 ind., August - October 2017 - 25 ind.

Levant Sparrowhawk *(Accipiter brevipes)*

It is a rare breeding and migrating species for the Lower Danube region. It inhabits riparian forests and tree groups among pastures and agriculture lands. During migration single birds or small groups can be observed everywhere in the studied region.

In the study area the estimation of the breeding population is 35-60 pairs, with a higher density along the Danube River and the lower sections of the Ogosta, Iskar, Vit and Yantra Rivers.

The breeding territories were recorded at the following locations: the mouth of the Timok River, Slanotran, Archar, Orsoya fishponds, Gorni and Dolni Tsibar, Kozloduy, Kriva Bara, Sofronievo, Mizia, Ostrov, between Bregare and Orehovitsa, Krushovene, Iskar, Gulyantsi, between Muselievo and Evlogievo, Dragash Voyvoda, Belene, Persin Island, Svishtov, near the mouth of the Yantra River, Krivina, Beltsov, in the region of Mechka fishponds, Kalimok-Brashlen Protected area, Tutrakan, Pozharevo, Malak Preslavets marsh, Garvan marsh, Vetren, Srebarna Lake, and Sratsimir (Figure 7). Solitary Levant Sparrowhawks were observed at Debovo on 13.05.2005 and Nikopol on 19.05.2005 (Peter Shurulinkov, Rossen Tzonev - pers. communication).

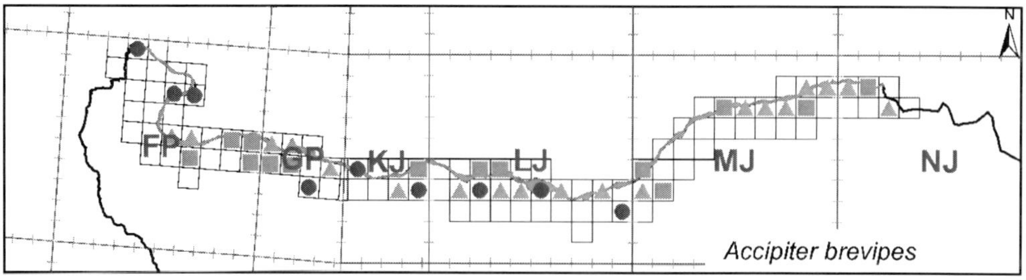

Figure 7. Breeding distribution of the Levant Sparrowhawk (*Acci[iter brevipes*) along the Bulgarian section of the Danube River (circles –new localities, triangles – confirmed published localities, squares – unconfirmed published localities).

During migration the Levant Sparrowhawk can be observed mostly during August and September and in April- May in the spring. During field studies of the spring and autumn migration of birds, organized by BSPB, the species was observed at the following locations: Peschina marsh - September - October 2008 - 7 ind., September - October 2009 - 7 ind.; Ivanovo - 1 ind. on 10.09.2009; Kaykusha marsh - August - October 2009 - 5 ind.; Vabel - August - October 2009 - 4 ind.; April - May 2010 - 3 ind.; Zlatia - August - October 2009 - 5 ind.; Slivata - August - October 2011 - 10 ind.; Hadzhidimitrovo fishponds - April - May 2017 - 10 ind., August - October 2017 - 26 ind.; and Vardim island - April - May 2017 - 3 ind., August - October 2017 - 4 ind.

Common Buzzard *(Buteo buteo)*

It is a common breeding, migrating and wintering species. It breeds practically in all forests in the Danube plain (Shurulinkov et al. 2005). The Common Buzzard inhabits forests near open areas, mosaic meadows, pastures, and agriculture lands with trees and riparian forests. One of the most numerous and widespread birds of prey in Bulgaria.

In the study area the Common Buzzard is breeding in each UTM 10x10 square. The breeding population of the species can be estimated at 320-560 pairs.

The autumn migration starts from the second half of September and lasts until December with a peak in October. The spring migration is well presented during March and April (Shurulinkov et al. 2005). During the field studies of the spring and autumn migration of birds, organized by BSPB, the following Common Buzzard numbers were counted: Peschina marsh - September - October 2008 - 51 ind. September - October 2009 - 128 ind.; Ivanovo - August - October 2009 - 37 ind.; Kaykusha marsh - August - October 2009 - 136 ind.; Vabel - August - October 2009 - 86 ind. April - May 2010 - 34 ind.; Zlatia - August - October 2009 - 46 ind.; Slivata - March - April 2011 - 74 ind. August - October 2011 - 37 ind.; Hadzhidimitrovo fishponds - April - May 2017 - 9 ind., August - October 2017 - 275 ind.; and Vardim - April - May 2017 - 14 ind., August - October 2017 - 316 ind.

During winter the Common Buzzard is very common. It can often be observed perching on trees near motorways and roads.

Rough-Legged Buzzard *(Buteo lagopus)*

It is a wintering species along the Danube River and nearby territories, but with a relatively low numbers. It can be observed from the end of September to March. The earliest record of that species in autumn was made on 21.09.2017 (1 ind. at Peschina marsh), and the latest was on 14.03.2002 at Obnova (2 ind.) (Shurulinkov et al. 2005). Mostly single birds were observed.

Long-Legged Buzzard *(Buteo rufinus)*

It is a common breeding species. The Long-legged Buzzard inhabits pastures, meadows and agriculture areas near rock massifs or forests. Its nests can be found on rocks, trees and electric poles. It can be often observed hunting at European ground squirrel (*Spermophilus citellus*) colonies. The species can also be observed during the winter.

In the study area the estimation of the breeding population is 75-130 pairs, with a higher density in the central and eastern parts (Nikopolsko Plateau and "Roussenski Lom" Nature Park). Breeding territories in the western part of the study area were at the following locations: Kalenik, Voynitsa, Gaytantsi, Shipot, Septemvriytsi, Brusartsi, Zlatia, Gorni Tsibar, Kozloduy, Butan, Galovo, Gorni and Dolni Vadin, Bregare, Iskar, Brest, Milkovitsa, Lenkovo, Zhernov, Muselievo, Evlogievo and Debovo. Breeding territories in the central part of the studied region were localized at: Vabel, Dragash Voyvoda, Lyubenovo, Lozitsa, Byala voda, Novachene, Dekov, Kulina voda, Tatari, Tsarevets, Bulgarsko Slivovo, Sovata, Hadzhidimitrovo, Alekovo, Belyanovo, Beltsov, Piperkovo, Batin Island, Tsenovo, and Brestovitsa. In the eastern part of the study area it was found at the following locations: Trastenik, Volovo, in "Roussenski Lom" Nature Park at Basarbovo, Krasen, Bozhichen, Ivanovo and Koshov, around Shtruklevo, Chereshevo, Zvanartsi, Pozharevo, Tarnovtsi, Chernogor, Zafirovo, Popina, Sitovo, Bulgarka, and Glavan.

Outside the breeding season the species occurs mainly in the places where it breeds. The Rough-legged Buzzard is common between August and November. From the beginning of December until March, its number decreases significantly (Shurulinkov et al. 2005).

Peregrine Falcon *(Falco peregrinus)*

The Peregrine Falcon is an extremely rare breeding species along the Danube River. Several pairs were found during the breeding period in the region of "Roussenski Lom" Nature Park.

Outside the breeding season solitary Peregrine Falcons were observed at the following locations: Zlatia – 13.08.2009; Ivanovo – 23.10.2009; close to Svishtov – 2.11.2011; Vardim Island – 5.03.2012; close to Vidin – 14.01.2017; Hadzhidimitrovo fishponds – 10.10.2017; and to the west of Kozloduy – 26.09.2018. During this time, Peregrine Falcons can be observed close to the areas with large concentrations of waterfowl or pigeons.

Saker Falcon *(Falco cherrug)*

It is an extremely rare species for the country and for the study region. The Saker Falcon is a globally threatened species. It is not a breeding species in the Bulgarian part of the Lower Danube River but can be observed irregularly during the period of migration and wintering. In the past the Saker Falcon bred very often along the Lower Danube (Patev 1950). On 25.07.1988, one bird was observed at Devetashko plateau (Shurulinkov et al. 2005).

There was a single observation of Saker Falcon in the studied region during autumn migration: one adult was observed in "Roussenski Lom" Nature Park on 17.08.2009, 24.08.2009 and 11.09.2009. In 2014, two Hungarian birds fitted with satellite transmitters spent 3 weeks in the study area.

Hobby *(Falco subbuteo)*

The Hobby is a common breeding and migrating species along the Danube River. It inhabits riparian forests, groups of trees, often poplars, and pastures and agriculture areas, usually close to wetlands. During the migration, single birds or small groups can be observed everywhere in the studied region.

In the study area the estimation of the breeding population is 70-110 pairs with the highest density along the Danube, Ogosta, Iskar, Vit, Osam and Yantra rivers. The breeding territories were found in the following locations: Timok River mouth, Archar, Orsoya fishponds, Dobri dol, Slivata, Smirnenski, Dolno Linevo, Dolni Tsibar, Ibisha island, Kozloduy, Glozhene, Butan, Oryahovo, Ostrov marsh, Bregare, Baykal, Gigen, Iskar, Krushovene, Dabovan, Brest, Dolni Vit, Gulyantsi, Somovit, Muselievo, Lakat, Vabel, Dragash Voyvoda, "Hisarlaka" area close to Belene, Dekov, Persin Island, Tatari, Tzarevetz, Vardim Island, Krivina, Dzhulyunitsa, Batin, Mechka, Basarbovo, Ivanovo,

Nikolovo, Brashlen, Nova Cherna, Tutrakan, Pozharevo, Malak Preslavets marsh, Garvan marsh, Vetren, and Srebarna Lake.

During migration the Hobby can be observed from the end of August to the end of September and in April. During field studies of the spring and autumn migration of birds, organized by BSPB, the following numbers of Hobbies were counted: Peschina marsh - September – October 2008 – 42 ind., September – October 2009 – 13 ind.; Ivanovo - August – October 2009 – 5 ind., Kaykusha marsh - August - October 2009 – 39 ind., Vabel - August – October 2009 – 8 ind., April – May 2010 – 2 ind.; Zlatia - August - October 2009 – 7 ind.; Slivata - March – April 2011 – 17 ind., August – October 2011 – 22 ind.; Hadzhidimitrovo fishponds - April – May 2017 – 6 ind., August – October 2017 – 22 ind.; and Vardim island - April – May 2017 – 5 ind., August - October 2017 – 19 ind.

Common Kestrel *(Falco tinnunculus)*

It is the most common and numerous Falcon in Bulgaria. The Common Kestrel is a numerous breeding species along the Danube. It inhabits open areas, pastures, meadows, agriculture lands close to rocky areas, rocky canyons, groups of trees, forest edges, and urban territories. The species often occupies and nests in abandoned corvid nests.

In the study area the Common Kestrel nests in every 10x10 sq.km. UTM square. The estimation of the breeding population is 420-750 pairs with the highest density in the region of Nikopolsko plateau and "Roussenski Lom" Nature Park.

During the winter the Common Kestrel can be found in open areas in the vicinity of its breeding territories.

Red-Footed Falcon *(Falco vespertinus)*

It is an extremely rare breeding species in the country. Several pairs breed successfully in the region of South-Eastern Bulgaria during the recent decades. In the study area the Red-footed Falcon probably breeds in the region of Selanovtsi; several hunting male and female birds were observed in the mid-June and July 2018. Recently the Red-footed Falcons has been observed during the breeding period in several locations in the Lower Danube plain (Svishtov – Belene lowland, Odarne, Nova Cherna, Oryahovo), but breeding pairs have not been found.

During migration the species can be observed from the beginning of September to mid-October and from mid-April to the end of May. During the field studies of the spring and autumn migration of birds, organized by BSPB, the species was observed as follows: Peschina marsh - September – October 2009 – 5 ind.; Ivanovo - August – October 2009 – 22 ind.; Kaykusha marsh - August - October 2009 – 2 ind.; Vabel - August – October 2009

– 1 ind., April – May 2010 – 5 ind.; Zlatia - August - October 2009 – 5 ind.; Slivata - August – October 2011 – 1 ind.; Hadzhidimitrovo fishponds - April – May 2017 – 25 ind., August – October 2017 – 12 ind.; and Vardim Island - April – May 2017 – 43 ind., August - October 2017 – 9 ind.

Merlin *(Falco columbarius)*

Merlin is a rare wintering species along the Danube. In autumn it was observed the earliest on 04.11.2018 in the region of the Svishtov-Belene lowland and latest in spring on 24.02.2017, close to Peschina marsh.

In winter, single birds were observed in the region of SPA "Zlatiyata", Selanovtsi, Nikopol, Baykal, Gulyantsi, Belene, Svishtov, Novgrad, Slivo pole, Nova Cherna, Tutrakan, and Srebarna Lake.

Acknowledgments

We would like to thank to our friends and colleagues who joined us during field studies or shared their observations from the region: Zornitsa Petkova, Vensislav Panev, Vladimir Mladenov, Ralitsa Georgieva, Angel Angelov, Georgi Bardarov, Rossen Tzonev, Emil Todorov, Irina Mateeva, Todor Natzkov, Dimitar Nedelchev, Ventseslav Petkov, Michail Iliev, Krasimir Kirov, Hristo Dinkov, Ivan Hristov and Vladimir Petrov.

References

Arabadzhiev I. 1962. *Birds of Prey in Bulgaria*. Sofia. Nauka i izkustvo [Science and art], 175 pp. (in Bulgarian).

Bavoux C., Burneleau G. & Nicolau-Guillaumet P. 1997. Marsh Harrier (*Circus aeruginosus*). In: Hagemeijer E. & Blair M. eds. *The EBCC Atlas of European breeding birds: their distribution and abundance*. T& AD Poyser, London.

Bibby C., Burgess N., Hill D. & Mustoe S. 1998. *Bird census techniques*. Second Edition, 215 p.

Botev B. & Peshev T. (Eds.) 1985. *Red Data Book of PR Bulgaria*. vol. 2 *Animals*. BAS. Sofia. (in Bulgarian).

Cheshmedzhiev Sv. 2012. *Field survey of migrating soaring birds in the region of Slivata*. Report. Svishtov, 2012.

Cheshmedzhiev Sv. 2016. *Current breeding distribution and conservation of White-tailed Eagle Haliaeetus albicilla in Bulgaria*. Report. Svishtov.

Cheshmedzhiev Sv. 2017. *Field survey of breeding birds in the region of SPA Hadzhidimitrovo fishponds and SPA Vardim Island.* Report. Sofia.

Cheshmedzhiev Sv. 2017. *Field survey of spring and autumn migration of soaring birds in the region of SPA Hadzhidimitrovo and SPA Vardim Island.* Report. Sofia.

Cheshmedzhiev Sv., Ratarova V., Ferdinandova V., Arkumarev V., Batalov D., Demerdzhiev D., Vasilev N. & Petkov N. 2017. *National Action Plan for conservation of Lesser Spotted Eagle (Clanga pomarina) in Bulgaria 2018 – 2027.* Sofia, BSPB, EFA, MOEW, p. 10-14.

Cheshmedzhiev Sv. & Todorov E. 2008. *Field survey of breeding and migrating soaring birds at Persin Island.* Report. Svishtov.

Cheshmedzhiev Sv. & Todorov E. 2009. *Field survey of breeding and migrating soaring birds at Persin Island.* Report. Svishtov.

Cheshmedzhiev Sv. & Todorov E. 2011a. *Field survey of migrating soaring birds of the territory of the Persina Nature Park.* Report. Svishtov.

Cheshmedzhiev Sv. & Todorov E. 2011b. *Field survey of migrating soaring birds in the region of SPA Zlatiata,* Nikopolsko plateau and Roussenski Lom Nature Park. Report. Svishtov, 2011.

Finsch O. 1859. Beitrage zur Ornithologischen Fauna von Bulgarian besonderes Berucksichtungs des Balkans [Contribution to the ornithofauna of Bulgaria as special attention to the Balkans]. *Journal fur Ornithologie* [*Journal of Ornithology*] 7: 378-387. (in German).

Golemansky V., Beron P., Zivkov M., Popov A., Popov V., Beschkov V., Deltchev C., Michev T., Spassov N., Stoev P. & Dobrev D. (Eds.) 2015. *Red Data Book of Bulgaria.* vol. 2 *Animals.* BAS, MOEW.

Iankov P. (Ed.) 2007. *Atlas of Breeding Birds in Bulgaria. Bulgarian Society for the Protection of Birds*, Conservation Series, Book 10. Sofia, BSPB, 122, 124, 128, 132, 138, 140, 148, 156, 164, 168, 172, 178, 186, 188 pp.

Michev T. & Petrov T. 1979. On the distribution of the Imperial Eagle (*Aquila heliaca* Savigny) in Bulgaria. *Proceedings of the Museums of Southern Bulgaria* 5: 65-77. (in Bulgarian).

Nankinov D. 1997. Past and present status of the Black Stork, *Ciconia nigra* (L.) (Aves: Ciconiiformes), in Bulgaria. *Crystal (Zool.)* 4: 3-25.

Paspaleva-Antonova M. 1961. Studies on the avifauna of the Bulgarian Danube River. PhD-Thesis, Institute of Zoology, Bulgarian Academy of Science, 145 pp. (in Bulgarian).

Patev P. 1950. *The birds of Bulgaria.* Sofia, Institute of Zoology with Museum, BAS. 106 pp. (in Bulgarian).

Petrov T. & Michev T. 1985. Breeding distribution, numbers and conservation of the Marsh Harrier, *Circus aeruginosus* (L., 1758) in Bulgaria. In: *International Symposium under*

*Project 8 –MAB (UNESCO). Conservation of the nature territories and their genetic fund.*vol.1. Blagoevgrad, 23-28.09.1985, Sofia, BAS: 306-313. (in Bulgarian).

Shurulinkov P., Tzonev R., Nikolov B., Stoyanov G. & Assenov L. 2005. *Birds of the Middle Danube Plain*. Green Balkans, Sofia, 120 pp. (in Bulgarian).

Simeonov S., Michev T. & Nankinov D. 1990. *Fauna of Bulgaria*. vol. 20. Aves. BAS, Sofia. 350 pp. (In Bulgarian).

Stoyanov N. 1948. *The vegetation on our Danube islands and its economy use*. Sofia, BAS, 188 pp. (in Bulgarian).

Todorov E., Daskalova G. & Shurulinkov P. 2015. Current breeding distribution and conservation of White-tailed Eagle, *Haliaeetus albicilla* (L.) in Bulgaria. *Acta zoologica bulgarica* 67 (1): 3 - 10.

In: Biodiversity of the Bulgarian-Romanian Section
Editor: Shurulinkov, Hubenov et al.
ISBN: 978-1-53615-663-8

Chapter 12

MAMMALS IN DANUBIAN REGION OF BULGARIA: SPECIES COMPOSITION, DISTRIBUTION, HABITATS, AND CONSERVATION

Vasil Popov**, Sirma Zidarova and Ivan Pandourski
Institute of Biodiversity and Ecosystem Research,
Bulgarian Academy of Sciences, Sofia, Bulgaria

ABSTRACT

A review of the biodiversity of mammals in the Danubian-Dobrudzha region (DDR) of the Lower Danubian Province (LDR) of Bulgaria was made. Seventy-four mammalian species occur in the area. Recent data on the distribution of 57 species were used to compile models of habitat suitability throughout the country. The modeling results for LDR are presented in detail. Analyses have been made to assess the territorial differentiation of habitat suitability and species richness within particular ecological groups (orders or group of orders). The suitability of habitats and species richness are unevenly distributed in the area, and spots with high suitability and species richness largely overlap in the ecological groups. They are mainly associated with forests, wetlands, and karst areas. The spots with high species diversity are connected with ecological corridors and "stepping-stones." They follow the river valleys and are formed mainly by the few preserved riparian forests. The role of the Natura 2000 network in the area has been assessed. It covers most of the identified patches of high habitat suitability and species richness as well as the eco-corridors. Habitat suitability for most mammalian species is higher in the protected areas.

---

* Corresponding Author's Email: vasilvpopov@gmail.com.

INTRODUCTION

The Danubian region of Bulgaria is a lowland river stretch with a large floodplain area, which includes a wide range of riverine habitats. In the course of the second half of the last century, wide spaces of floodplain have been dyked and converted into agricultural fields, fishponds, and territory for hybrid poplar plantations. These zones are roughly 73% of the former floodplain which was formerly 5-10 km broad along the Romanian – Bulgarian span and trimmed to merely a few hundred meters. As a result from disconnection from the river changes a fall or substantial contraction of essential ecological functions took place, comprising hydrological functions such as water retention, self-purification, groundwater supply, biogeochemical serves such as nutrient retention and recycling and ecological functions between them the critical function as an environment for plants and animals (spawning, feeding, nesting etc.), as a source of biodiversity, storage for genetic resources, as eco-corridor and area with high bio-productivity (Schneider 2002, Schneider & Günther-Diringer 2004, Dister 2007). Additionally, the human infrastructure, the expansions of urban areas, the establishment of new transport routes and power network as well as the grown intensification of the land-use set increasing press on the natural environments in the area. Consequently, the isolation of the fragments of the natural habitats is one of the major perils to biodiversity in the Lower Danube, especially having in mind that the region is part of a significant ecological corridor that connects the Black Sea, Steppic, Pannonian, Continental, and Alpine regions. Thus, the increase of fragmentation reduces the ecological connectivity of high-value ecosystems in Europe (Frank 2017). The insurance of adequate and functioning environmental corridors joining several biogeographical regions will be more important than ever in future in order to mitigate the negative effects of long-term habitat displacement and climate change.

Many of the remaining natural habitats along the Danube fall within protected areas according to the Flora-Fauna-Habitat Directive (NATURA 2000 92/43/EWG) and form the backbone of this mega ecological corridor. But, because the amount of territory allocated to these areas is limited to support self-sustaining wildlife populations, our primary demand – now and in future – will be to set up and support habitat connectivity and contiguous ecosystems in order to protect biodiversity in the ages to come.

An essential element of this effort is the assessment of the state of biodiversity, particularly in its spatial aspects. This is a laborious and costly activity that takes a long time. In this context, given the urgent need for such information, it is reasonable to focus on groups of organisms that are relatively well known and can serve as indicators of the state of biodiversity as a whole. Mammals are a group of animals that meet these requirements. Although important aspects of their taxonomy, ecology, and distribution on the territory of Bulgaria are comparatively well researched, the specific information for particular areas of the country is scarce (Peshev et al. 2004). This is largely the case for the Danubian area.

The aim of the present study is to: review the available data and assess the biodiversity of mammals; provide more detailed information on mammals of international conservation interest; identify areas of importance for the conservation of mammalian biodiversity; evaluate ecological connectivity of these territories; evaluate the Natura 2000 network from the point of view of protecting the hot spots of mammalian biodiversity and linking them with ecological corridors and "stepping stones."

MATERIAL AND METHODS

Identification of areas of importance for conservation demands knowledge on the spatial distribution of biodiversity. As mentioned above the information is often grossly insufficient, especially for the study area. Analyses, therefore, rely on species for which we do have valid distributional information as 'proxies' for biodiversity as a whole. Even for these proxy species, distributional data may be constrained to scattered records separated by large areas of unsurveyed territory. Statistical modeling of such sample data in relation to remotely mapped environmental variables – e.g., topography, climate, substrate or landcover variables – produces a robust means of filling geographic gaps in the coverage of data. When connected with a geographic information system (GIS) technology, such modeling enables biological distributions to be hypothesized across broad territories, thereby giving geographically complete information for conservation treatment and other environmental studies. For the purposes of this study, models of 57 species were used. They were based on maxent model algorithm (Phillips et al. 2006). The set of variables included was different for the particular taxonomic groups - small terrestrial mammals, bats, and medium-sized and large mammals - in view of the various factors affecting their distribution. Common to all groups were Digital Elevation Model (DEM), topographic wetness index (TWI) and land cover based on satellite imagery from the Enhanced Thematic Mapper Plus (ETM+) sensor onboard Landsat 7 satellite. In order to eliminate the effect of the inevitable sampling bias and spatial autocorrelation of the field data, an approach proposed by Phillips et al. (2009) was used. More modeling details can be found in Popov (2015, 2018).

Each species model generated a mapped layer indicating the predicted probability of that species occurring within each and every grid cell in the region. Most often this probability can be considered also as habitat suitability (HS). All models had a high level of predictive accuracy, and the species-environment relationships registered by the models corresponded very well to the known ecology of the studied species. It was shown that raw Landsat ETM+ data integrated with topographical variables have the potential to produce models that have a high degree of ecological realism and reflect both the general type of distribution of each species in the country and peculiarities associated with fine-scale local conditions (Popov 2015, 2018). The modeling results were used to obtain species richness, threats, and connectivity maps.

By averaging the habitat suitability within several ecological groups (terrestrial small mammals - insectivores and rodents, bats, carnivores and ungulates), information was obtained about the territories that support high species diversity for each group as well as the existence of eco-corridors or "stepping-stones" between them.

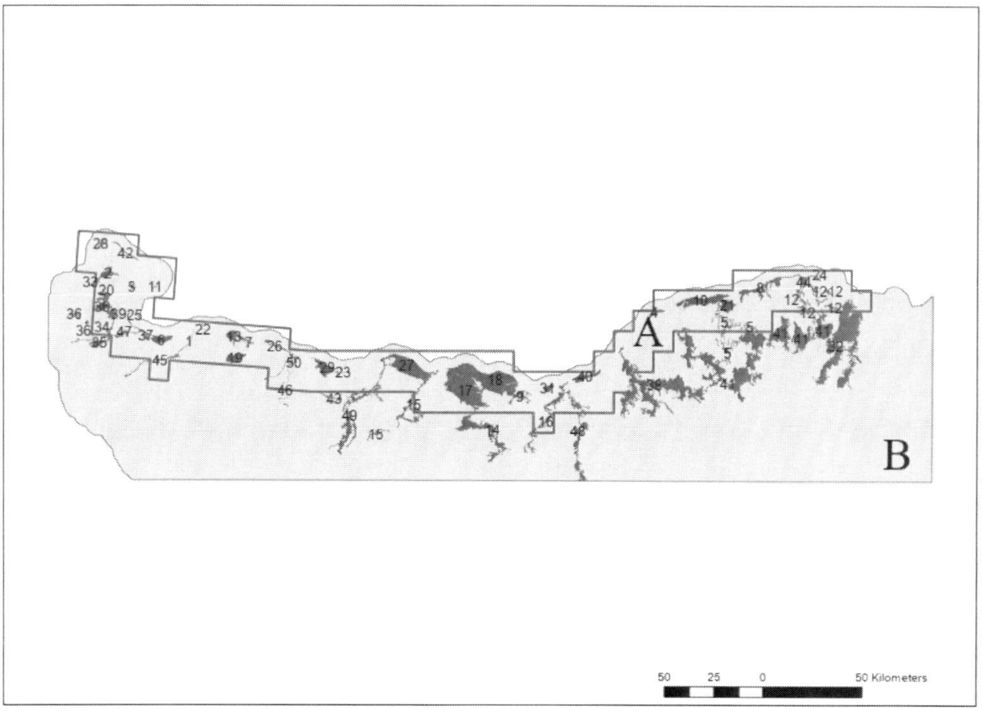

Figure 1. Spatial Scope of the Study - the Danube-Dobrudzha Region (A) and the South-Danubian Ludogorie Region (B) of the Lower Danubian Landscape Province of Bulgaria. Natura 2000 zones in red: 1. BG0000519, Mominbrodsko blato; 2. BG0000507, Deleyna; 3. BG0000524, Orizishteto; 4. BG0000529, Marten - Ryahovo; 5. BG0000171, Ludogorie - Boblata; 6. BG0000182, Orsoya; 7. BG0000527, Kozloduy; 8. BG0000530, Pozharevo - Garvan; 9. BG0000576, Svishtovska gora; 10. BG0000377, Kalimok - Brashlen; 11. BG0000552, Ostrov Kutovo; 12. BG0000169, Ludogorie - Srebarna; 13. BG0000199, Tsibar; 14. BG0000239, Obnova - Karaman dol; 15. BG0000181, Reka Vit; 16. BG0000233, Studena reka; 17. BG0000247, Nikopolsko plato; 18. BG0000396, Persina; 19. BG0000336, Zlatia; 20. BG0000340, Tsar Petrovo; 21. BG0000180, Boblata; 22. BG0000526, Dolno Linevo; 23. BG0000528, Ostrovska step - Vadin; 24. BG0000534, Ostrov Chayka; 25. BG0000532, Ostrov Bliznatsi; 26. BG0000533, Ostrovi Kozloduy; 27. BG0000335, Karaboaz; 28. BG0000525, Timok; 29. BG0000334, Ostrov; 30. BG0000587, Varkan; 31. BG0002018, Ostrov Vardim; 32. BG0000106, Harsovska reka; 33. BG0000339, Rabrovo; 34. BG0000498, Vidbol; 35. BG0000521, Makresh; 36. BG0000500, Voynitsa; 37. BG0000497, Archar; 38. BG0000608, Lomovete; 39. BG0000522, Vidinski park; 40. BG0000232, Batin; 41. BG0000168, Ludogorie; 42. BG0000631, Novo selo; 43. BG0000627, Konunski dol; 44. BG0000241, Srebarna; 45. BG0000503, Reka Lom; 46. BG0000508, Reka Skat; 47. BG0000518, Vartopski dol; 48. BG0000610, Reka Yantra; 49. BG0000613, Reka Iskar; 50. BG0000614, Reka Ogosta.

Species richness was estimated based on the classification of modeled probability surfaces at a threshold level of 0.5 (for more details see Popov 2018). The maps obtained on this basis reflect the presence-absence of each modeling species. Summing up these maps makes it possible to obtain maps of the species richness.

The map of vulnerability was obtained by summing the presence/absence maps of the species with conservation status NT and VU according to IUCN (2018) with the first category given a weight of 1 and for the second one of 2.

In order to assess the potential of the Natura 2000 network, the habitat suitability rasters were averaged and compared on three levels - the whole country, DDR and the network of Natura 2000 in DDR.

Study Area

According to the regional differentiation of the landscapes in Bulgaria (Velchev et al. 2003), the study area (Figure 1 A) comprises the Danubian-Dobrudzha region (hereafter DDR) of the Lower Danubian Province. When presenting and analyzing the data, it is dealt with in a wider context, including part of the Southern-Danubian and Ludogorie Region (SDLR) (Figure 1 B) with a view to assessing the area in the context of the rest of Lower Danubian Province.

The Lower Danubian Province is characterized by a moderate continental climate. It has the greatest seasonal temperature amplitude (hot summer, but cold winter) and the greatest drought compared to the other parts of the country. The openness of the Danubian Plain to the Northeast and the East allows for more frequent access of colder air masses, and there, especially in Dobrudzha, the climate is similar to the steppe one in more eastern parts of Europe. The relief is flat or hilly. In many places, the rivers have been cut into hard limestone and canyon valleys have been formed that tear apart the plateau ridges. The weathering processes create in many places typical karst areas with numerous cliffs, caves, and rock niches. The rugged terrain in these areas is a prerequisite for preserving patches of (semi)natural vegetation. These peculiarities, combined with the presence of rivers, karst springs, lakes, and marshes, are particularly typical of the eastern and middle parts of the DDR and are true oases of wildlife in the monolithic agricultural landscape. Characteristic of the area is the well-developed loess cover on the flat interfluve areas (placors). A characteristic feature of the loess is its water permeability, which further enhances the effect of the drier climate on the biota. In accordance with these climatic and geomorphological features, the zonal primary vegetation in the DDR is represented by xerothermic broad-leaf woods - the most widely distributed type of zonal forests. They are composed of pubescent oak (*Quercus pubescens*), Italian oak (*Q. frainetto*), Turkey oak (*Q. cerris*), and sometimes Oriental hornbeam (*Carpinus orientalis*). Zonal grassland communities occupy areas where the natural development of forest vegetation is hampered, for example, well-drained placors on a loess substrate. These are the most southwestern patches of the eastern steppe, which along with tree vegetation in the reliefs depressions form a forest-steppe landscape. Mesophilous deciduous forests of the Middle European type have a more limited distribution and are associated with damp places. These are forests consisting of common hornbeam (*Carpinus betulus*), sessile oak (*Quercus petrea*), narrow-leafed ash (*Fraxinus angustifolia*), field elm (*Ulmus minor*), field maple (*Acer campestre*), Tatarian maple (*Acer tataricum*), lime trees (*Tilia* spp.), and others. Along the Danube river, there are patches of marsh vegetation. The natural zonal vegetation is destroyed on most of the territory. Secondary vegetation is developed in its place. The secondary tree vegetation is represented by artificial plantations of poplars and conifers. Secondary herbaceous vegetation on damp soil is represented by meadows. Eroded terrains are overgrown with *Chrysopogon gryllus, Dichanthium ischaemum, Limnanthes alba*, and *Festuca ovina*. In general, the secondary herbaceous vegetation has a characteristic tendency for xerophytisation, which leads to the penetration of steppe species into these communities. Thus, elements of a zonal type (steppe) penetrate deep into the territory of another zonal type - secondary grasslands within deciduous forests. The agricultural crops form the largest part of secondary vegetation.

Because of the strong anthropogenic changes in this sub-region, there are not many opportunities for developing a network of protected areas. Still, there are several natural parks, reserves, protected areas and natural landmarks that are important for preserving the species diversity of mammals and especially their habitats. The importance for preserving the species diversity of bats has the caves declared natural monuments or protected sites, among which the most significant in this respect are Magura (Rabisha, Vidin), Nanin stone (Muselievo, Pleven), Orlova Chuka (Pepelina, Russe). A total of 50 zones of the Natura 2000 ecological network (Figure 1) fall entirely or partly on the territory of DDR.

RESULTS AND DISCUSSION

Biodiversity of Mammals

The area has not been the subject of special mammalian studies. The scattered data available in the literature until 2000 was systematized in Peshev et al. (2004). New information has been accumulated as a result of regional studies mainly related to projects for the management of protected areas. Most of the data for bats, medium-sized and large mammals, and some rodents were collected within the scope of the project 'Mapping and Determining the Nature Conservation Status of Natural Habitats and Species - Phase I', activity 4, project DIR - 59,318-1-2, run between 2011 and 2013 by the Ministry of Environment and Waters (MOEW). For the species included in Annex 2 of the Habitat Directive, the results were published online (http://natura2000.moew.government.

bg/Home/Reports?reportType=Mammals [Accessed: 2018-12-11]). Accompanying information concerning other species is archived in the compiled databases available in MOEW. As a result of these studies, the species composition has been specified and supplemented (Table 1).

Table 1. Species composition of mammals in Danubian-Dobrudzha region (DDR), their conservation status, and habitat suitability. 1. Species composition based on literature analysis up to 2000 for Bulgaria (Peshev et al. 2004) and 2005 for Romania (Murariu 2005); 2. Species established on the basis of literary sources or unpublished data after 2000; 3. Species that are likely to occur in the area but there are no reliable documented data; ? Species whose presence is not conclusively proven

Species	Bulgaria	Romania	Conservation Status (IUCN 2018)	Mean Habitat Suitability
Erinaceus concolor	1	1	LC	0.41
Talpa europaea	1	1	LC	-
Sorex araneus	1	1	LC	0.22
Sorex minutus	2	1	LC	0.35
Neomys anomalus	1		LC	0.34
Crocidura leucodon	1		LC	0.48
Crocidura suaveolens	1		LC	0.52
Rhinolophus ferrumequinum	1		NT (Eu)	0.43
Rhinolophus hipposideros	1	1	NT (Eu)	0.3
Rhinolophus euryale	1		VU (Eu)	0.25
Rhinolophus blasii	1		VU (Eu)	0.09
Rhinolophus mehelyi	1	1	VU	0.4
Myotis myotis	1	1	LC	0.32
Myotis blythii	1		LC	0.46
Myotis bechsteinii	2		Vu (Eu)	0.35
Myotis nattereri	2		LC	0.27
Myotis emarginatus	2	1	LC	0.41
Myotis mystacinus	2		LC	0.06
Myotis aurascens	2		LC	0.39
Myotis alcathoe	2		DD	0.14
Myotis brandtii	3		LC	0.077
Myotis daubentonii	1	1	LC	0.43
Myotis capaccinii	1		Vu	0.22
Myotis dasycneme	?		NT (Eu)	-
Plecotus auritus		1	LC	0.09
Plecotus austriacus	1		LC	0.52

Table 1. (Continued)

Species	Bulgaria	Romania	Conservation Status (IUCN 2018)	Mean Habitat Suitability
Barbastella barbastellus	2		NT	0.17
Nyctalus noctula	1	1	LC	0.42
Nyctalus lasiopterus	1	1	DD (Eu)	-
Nyctalus leisleri	1		LC	0.22
Pipistrellus pipistrellus	1	1	LC	0.41
Pipistrellus nathusii	2	1	LC	0.41
Pipistrellus pygmaeus	2	1	LC	0.34
Pipistrellus kuhlii	1	1	LC	0.48
Hypsugo savii	1		LC	0.31
Eptesicus serotinus	1		LC	0.37
Vespertilio murinus	1	1	LC	0.103
Miniopterus schreibersii	1	1	NT	0.3
Sciurus vulgaris	1	1	LC (Eu)	0.17
Spermophilus citellus	1	1	VU (Eu)	-
Nannospalax leucodon	1	1	LC (Eu)	-
Dryomys nitedula	1		LC	0.38
Glis glis	1		LC	0.38
Muscardinus avellanarius	1	1	LC	0.08
Apodemus flavicollis	1	1	LC	0.39
Micromys minutus	1	1	LC	0.41
Apodemus agrarius	1	1	LC	0.45
Apodemus sylvaticus	1	1	LC	0.56
Apodemus uralensis	2	1	LC	-
Rattus rattus	1		LC (Eu)	-
Rattus norvegicus	1	1	LC	-
Mus spicilegus	1		LC	0.33
Mus musculus	1	1	LC (Eu)	-
Mesocricetus newtoni	1		NT	-
Cricetulus migratorius	1		LC	-
Cricetus cricetus	1		LC	-
Arvicola amphibius	1	1	LC	-
Microtus arvalis-levis (rossiameridionalis)	1	1	LC	0.52
Microtus subterraneus	1	1	LC	0.28
Ondatra zibethicus	1	1	LC	-
Lepus europaeus	1	1	LC	0.2
Meles meles	1	1	LC	0.33
Lutra lutra	1	1	NT	-
Mustela nivalis	1	1	LC	-
Mustela putorius	1	1	LC	0.16
Mustela eversmanii	1		LC	-
Mustela erminea		1	LC (Eu)	-

Species	Bulgaria	Romania	Conservation Status (IUCN 2018)	Mean Habitat Suitability
Martes foina	1		LC	0.31
Vormela peregusna	1		VU (Eu)	-
Canis lupus	2		LC	0.19
Canis aureus	1	1	LC	-
Vulpes vulpes	1	1	LC (Eu)	0.3
Nyctereutes procionoides	1		DD	-
Felis silvestris	1	1	LC	0.29
Cervus elaphus	2		LC	0.05
Capreolus capreolus	1	1	LC	0.13
Sus scrofa	1	1	LC (Eu)	0.21

Seventy-seven mammal species have been found along lower Danube in Bulgaria and Romania. There are 74 species registered on the Bulgarian part of the Lower Danube area. One species of bat, *Myotis dasycneme*, was reported from the vicinity of Rousse (30.09.1999, Krivina, 25.96/43.83, over Danube river), based on a recording of the echolocation call (Limpens 2000). Although its presence in the area is likely (Popov & Lakovski 2019) in regard to possible confusion with other species this record is considered doubtful until supported by another line of evidence (Görföl et al. 2017).

Forty-four species have been registered on the Romanian bank of the river (Table 1). Perhaps the list is incomplete because it lacks many common and widespread mammals. It should be noted that on the Romanian coast there are two species that are not registered on the southern bank of the Danube or in Bulgaria - *Plecotus auritus* and *Mustela erminea*. These are species with more northern distribution in Europe. In Bulgaria, *Plecotus auritus* is established only in the mountains (Popov 2018).

Some species rare for the area, as well as species that are of particular interest to conservation, deserve special attention.

Nine bat species represent international conservation importance (Table 1). Two of them are typical forest dwellers. In general, the area does not provide the optimal conditions for them - the average habitat suitability is low (Figure 2, Table 1).

Current data about their occurrence are from the western and eastern part of the area or near it, where forests are relatively well preserved (Figure 2). This is most relevant to *Myotis bechsteinii*. The remaining bat species are mainly associated with cave shelters. As the karst areas occupy a relatively small part of DDR, the HS for most of the cave dwelling species is lower compared to the adjacent SDLR. As is to be expected, they all have a high degree of modeled habitat suitability in proximity to several of the big caves in the area - Magura, Nanin kamak, Gubarnika, and Orlova Chuka. For *Miniopterus shreibersii* and *Myotis capaccinii* the model habitat suitability (Figure 2, Table 1) shows a spatial differentiation similar to that of the forest species of bats, and near the Danube River and the tributaries there are small spots with relatively high habitat suitability. There are connecting corridors between the significant cave shelters for these and other cave species of bats in Northern Bulgaria. Many of these sites fall within the Natura 2000 network (Figure 2). Some species of the genus *Rhinolophus* have generally low HS in the area - *Rh. hipposideros, Rh. euryale,* and *Rh. blasii*. Conversely, *Rh. ferrumequinum* shows a comparatively higher average HS in the Lower Danube region. This species, apart from caves, often find shelter in other roost types, including those of anthropogenic origin. According to the model, one of the rarest horseshoe bat in Bulgaria, the Mehely's horseshoe bat (*Rh. mehelyi*) has a relatively great HS in the area. However, the exact registrations are scarce (Figure 2). This is largely due to the fact that it is difficult to distinguish it from *Rh. euryale* and *Rh. blasii*, that is why in the field reports it often falls into the group of "undetermined middle-sized horseshoe bats."

According to the models for the four species of terrestrial mammals of particular conservation importance, the HS in the area is low. These species, besides the otter, are representatives of the steppe faunistic complex (Popov 2007). In the past, during the Late Pleistocene, under conditions of a drier and continental climate, they were much more widespread on the territory of the country and in DDR. There are concrete palaeontological data for European ground squirrel and Romanian hamster (Popov 2018). The reduction of the number of the last species is well traced on the basis of stratified Holocene subfossil materials near the area (Mitev 2006). The reasons for their population reduction are both natural and anthropogenic. The natural causes are related to the Holocene climate change, which has led to a reduction in the steppe habitats. With the development of agriculture, the few preserved steppe habitats have been turned into arable land not suitable for these species (Popov 2018). According to the recent records of Romanian and common hamsters (*Mesocricetus newtoni* and *Cricetus cricetus*), both species still occur in the periphery of the central and the western parts of the DDR (Simeonovska-Nikolova & Dekov 2013, Nedyalkov et al. 2015).

The European ground squirrel (*Spermophilus citellus*) is closely associated with pastures since wild (Pleistocene) and domestic (Holocene) ungulates maintain an optimal for the species grass cover structure. In recent decades, pastoral farming in this country is in regress, which also contributes to the reduction of the populations of species. Today, the European ground squirrel is a vulnerable species. It is relatively widespread in DDR (Figure 3). Some of its colonies are under the protection of Natura 2000 sites: Pozharevo - Garvan (BG0000530), Yantra River (BG0000610), Persina (BG0000396), Karaboaz (BG0000335), Ostrovska step - Vadin (BG0000528), Ostrov (BG0000334), and Tsibar (BG0000199). The species is also recorded in the surroundings of Nova Cherna and Bosilkovtsi villages (authors unpubl. data).

The marbled polecat (*Vormela peregusna*), listed as vulnerable in the IUCN Red List, is another rare and cryptic mustelid species whose distribution in Bulgaria is known mostly by accidental registrations. There are some former records in the eastern and central parts of DDR (Spassov & Spiridonov 1985). A recent road kill of a marbled polecat near Srebarna Managed Nature Reserve and an observed individual in the field station

"Kalimok" near Nova Cherna village (unknown year but after 2000) confirm the species' current presence in the area.

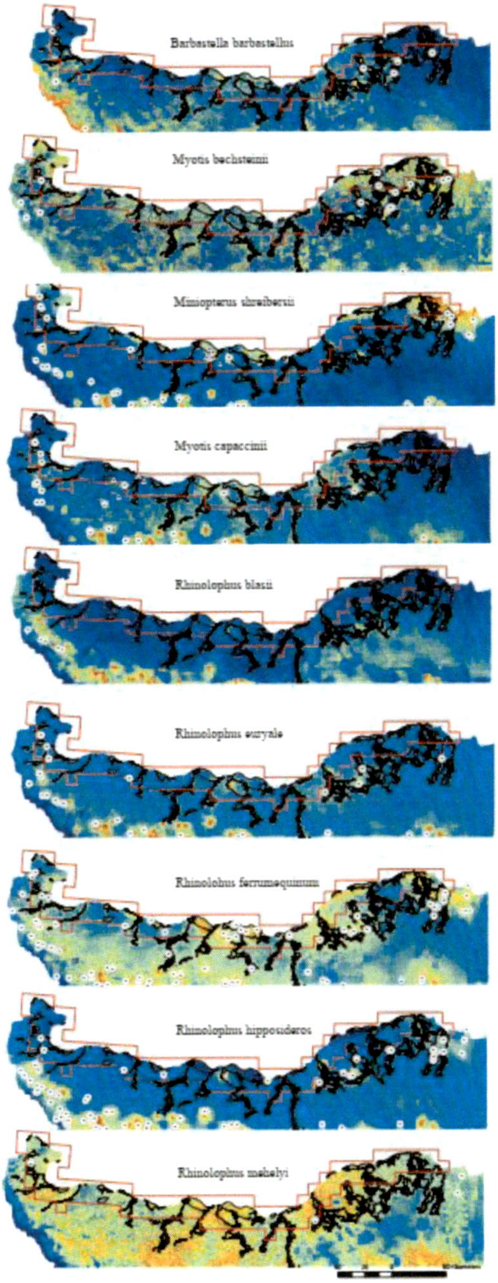

Figure 2. Habitat suitability (low – blue, red – high) and up-to-date recordings of bat species of conservation significance in Lower Danubian Landscape Province of Bulgaria. Circles – data from the project 'Mapping and Determining the Nature Conservation Status of Natural Habitats and Species - Phase 1', activity 4, project DIR - 59,318-1-2, run between 2011 and 2013 by the Ministry of Environment and Waters (MOEW); triangles – authors' data; red polygon - Danube-Dobrudzha Region; Natura 2000 zones outlined with black lines

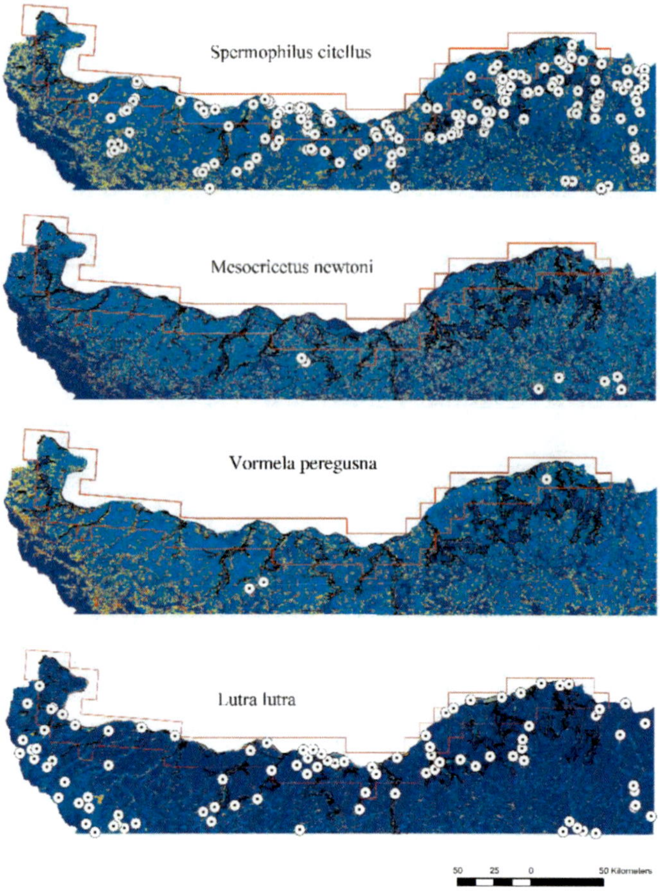

Figure 3. Habitat suitability and up-to-date recordings of terrestrial mammals of conservation significance in Lower Danubian Landscape Province of Bulgaria (for map symbols see Figure 2).

Otter (*Lutra lutra*) has specific habitat requirements - it is spread along the banks of running and stagnant waters. In the recent past, the main factors contributing to decreasing its numbers are human persecution and water pollution. In recent decades the population has recovered.

Particular attention deserves certain species that, although not endangered globally, are rare for the region or for the country. The steppe polecat (*Mustela eversmanni*), which in Bulgaria is extremely rare belongs to this group. It is known to occur in the area of DDR (Spassov & Spiridonov 1985). The secret way of life, the embarrassed identification and the low density of the steppe polecat have led to a lack of distribution data all over the territory of Bulgaria. However, the two recent records of the species within European ground squirrel colonies near village of Obnova (Tzonev R., Shurulinkov P., Chakarov N., 08.06.2004, personal communication, N43° 27` 26.8," E24° 58' 31.4") and near city of Knezha (Koshev & Genov 2008, N43°29.92' E 24°05.89`) confirm its presence in the region, as well as the importance of the European ground squirrel as a main food source for this predator.

The grey wolf (*Canis lupus*), which is rare in the lowlands of the country, is known to appear in low numbers along Danube river. Two recent records of hunted wolves prove its current presence in the western as well as in the central parts of the DDR: a killed female in 2013 - 4 km southeast of Dragomirovo village (Svishtov Municipality) and a killed male in 2012 - 5 km west of Zheglitsa village (Vidin Municipality) (Tsingarska E., personal communication). According to the official results of spring game counts 24 individuals (definitely overestimated) inhabited the area of Rousse in 2014 (Markov 2014). Undoubtedly the status of this species in DDR needs clarifying. What can be said for the moment is that although rare and in low number, the species is still hunted in the region.

The DDR provides some suitable lowland habitats for the red deer (*Cervus elaphus*). The species occurs in low numbers in reed beds and forest patches in the protected area "Kalimok-Brushlen," "Persina" Natural park, State hunting reserve "Dunav," "Ludogorie" Natura 2000 site (BG0000168) (authors unpublished data). The red deer population has declined critically in the region mainly due to poaching.

Analysis of pellets of a long-eared owl (*Asio otus*) in the area of the town of Silistra revealed the presence of pygmy field mouse (*Apodemus uralensis*) (Dipchikova 2016) – a species previously known only from two localities in western Bulgaria (Peshev et al. 2004). The specific conditions of the wetlands in the area of Silistra favor some typical mountainous species such as the pygmy shrew (*Sorex minutus*), the hazel dormouse (*Muscardinus avellanarius*) and the barbastelle (*Barbastella barbastellus*) (Dipchikova 2016).

Areas of Importance for the Conservation of Mammalian Biodiversity

Habitat Suitability

The modeled habitat suitability makes it possible to assess the ecological significance of DDR compared to the rest of the country and also to assess the role of the Natura 2000 network in this respect. For this purpose a comparison of the habitat suitability for each species for which a model is available on two levels is made - the survey area (DDR) and the Natura 2000 network in this area. The basis of comparison is the average HS of each species for the territory of the whole country. In Figure 4, the results of these comparisons are presented as differences in regional (DDR) or local (Natura 2000 network) average HS for each species with respect to the whole country. The species are ranked on the basis of DDR differences (Figure 4). These that show negative biases have a greater HS in the area under investigation. All of them are terrestrial small mammals and bats. Most of these species are characteristic for the lowlands of the country and are tolerant to the anthropogenic landscape transformations - well adapted to arable lands, settlements, etc. Species for which the average HS in the studied area is lower than in the rest of the country have positive deviations. These are some species of bats and small terrestrial mammals that are characteristic of the mountains. Most of the medium-sized and large mammals also belong to this group. They find better living conditions in areas with more rugged relief in mountainous and semi-mountainous regions of the country, where forests are better preserved and their habitats are not so highly fragmented. In the zones of the Natura 2000 network, HS for most species is higher than in DDR (Figure 4). As a whole, the Natura 2000 zones offer better conditions for the species that prefer hilly areas with preserved forest habitats. These results indicate that the network of protected areas covers the most valuable habitats for most mammal species in DDR.

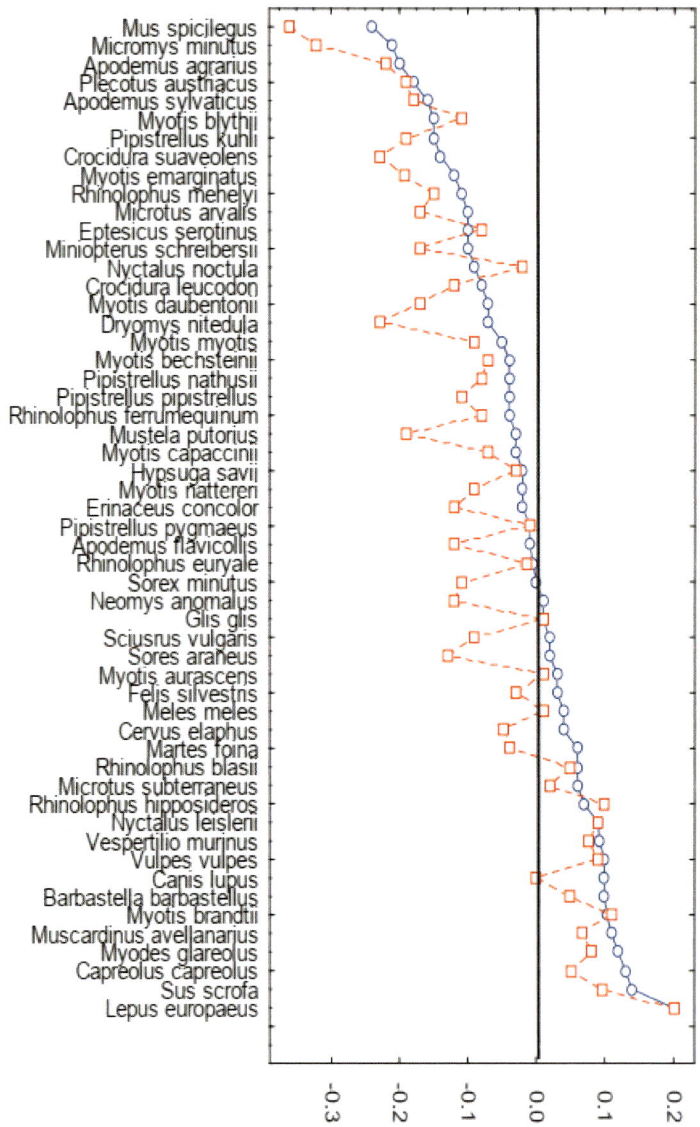

Figure 4. Deviations from the average for the country HS of mammalian species - with circles - for the study area (DDR), squares - for the Natura 2000 zones in this landscape subregion.

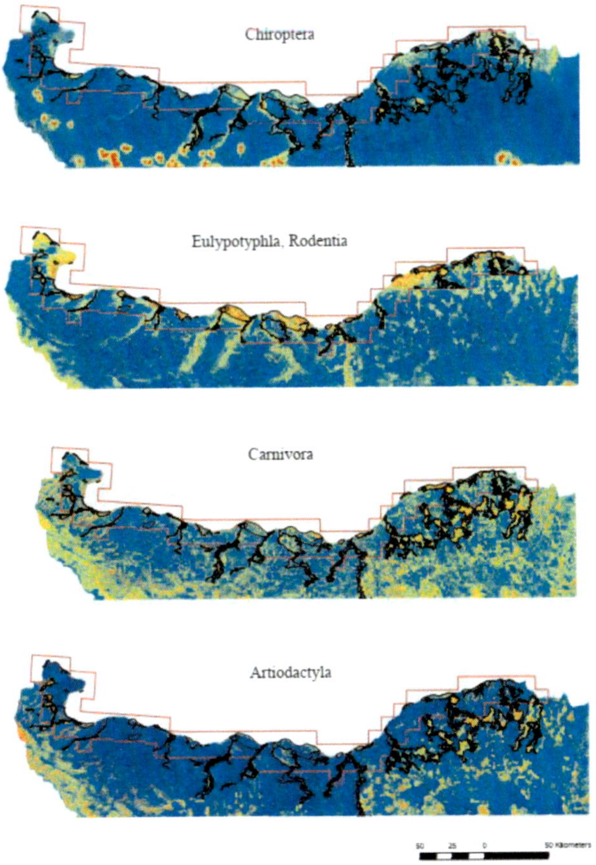

Figure 5. Mean habitat suitability by order (low – blue, red – high) of mammals in Lower Danubian Landscape Province of Bulgaria (for map symbols see Figure 2).

Average habitat suitability by order allows estimating the spatial differentiation in this respect in the context of the ecological requirements of each group (Figure 5).

For bats HS in the area is not high. As it can be expected, the highest HS exists around the several large caves in the area. However, small patches of suitable habitats along the Danube coast can also be noticed. These spots can be seen as "stepping stones." Along the valleys of some Danube tributaries, the habitat suitability is also high. They can be regarded as ecological corridors that provide the link between the relatively isolated patches of high habitat suitability along the Danube and those located in the south where, due to the rugged relief, the forest cover is relatively well preserved and where there are karst areas with numerous caves.

Small terrestrial mammals show the highest habitat suitability in DDR (Figure 5). The location of the spots, corridors, and "stepping stones" is similar to that of bats and coincides with riverbeds and riparian and other types of forests.

In carnivores and especially in ungulates, the average HS in the area is low (Figure 5). Most high suitability patches occur in the westernmost and eastern parts of the area.

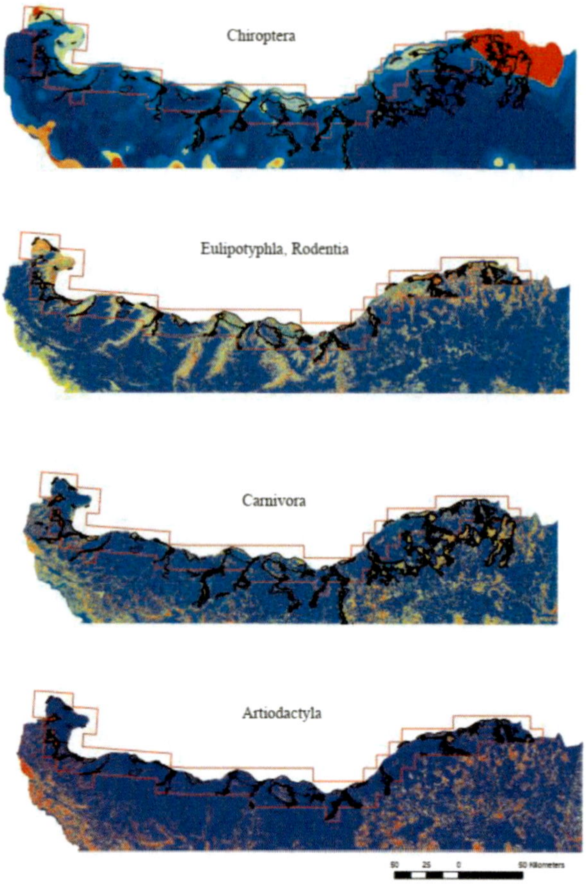

Figure 6. Species richness by order (low – blue, red – high) of mammals in Lower Danubian Landscape Province of Bulgaria (for map symbols see Figure 2).

Species Richness

The species richness of bats is unevenly distributed (Figure 6). The largest is in the eastern part of the area. This is related to the wider spread of deciduous forests there combined with numerous river canyons and their accompanying rock cliffs and a number of karst cavities. In the rest of the region, there is a relatively great species richness in the areas with preserved riparian forests. Most of them, except the most western ones, are within the Natura 2000 network (Figure 6).

The areas of high species richness in terrestrial small mammals are relatively evenly distributed along the Danube coast (Figure 6). It is the lowest in the areas where natural habitats are replaced by arable land.

Like in bats, the species richness of carnivores is highest in the western and eastern parts of the area (Figure 6). This is obviously related to the rugged terrain and the better preservation of forest habitats. In the middle part of the area, there are spots of high species diversity on the territories of some protected areas with relatively well-preserved forests.

The situation with the ungulates is similar, and it is very low in the anthropogenically affected areas (Figure 6).

Vulnerability

The vulnerability map is presented in Figure 7. Most of the endangered species are bats inhabiting caves. For this reason, the areas in the vicinity of several large caves in the area are of the greatest importance for their preservation. In addition, the importance of rivers and wetlands where the combination of increased humidity, preserved patches of forests and karst phenomena provide good conditions for the existence of all endangered mammal species. Most of these sites fall within the existing Natura 2000 sites (Figure 7).

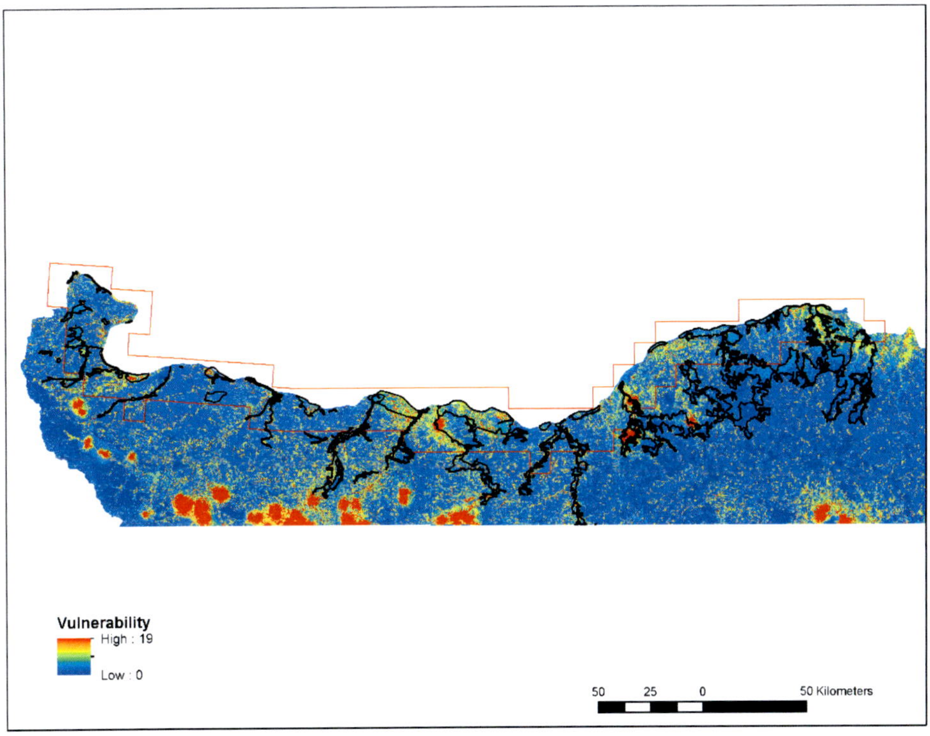

Figure 7. Map of vulnerability of mammals in Lower Danubian Landscape Province of Bulgaria (for map symbols see Figure 2).

Conclusion

The Lower Danube region is one of the strongest anthropogenically affected in Bulgaria. In large areas, natural habitats are destroyed or heavily modified. The wetland drainage and the loss of associated mammalian habitats have a well pronounced negative impact on ecosystems and mammalian biodiversity. Large areas are occupied by farmland, not very suitable for must mammalian species. Nevertheless, the preserved natural and semi-natural habitats make it one of the regions in Bulgaria with the great diversity of mammals. The relative share of species with IUCN conservation status and species with limited distribution in Bulgaria is high. The karst landscapes provide shelter for many species of cave-dwelling bats. Particularly significant in this regard are the caves Magura, Nanin kamak, Orlova chuka, and Gubarnika. The few preserved wetlands are also of great importance for maintaining the high species diversity of mammals. Here, in the conditions of higher humidity, in the context of the cooler climate of northern Bulgaria, there are moisture-loving and mesophilic species, which are more typical for the mountains and rare in the low parts of the country. Areas with preserved habitats and high mammalian diversity are small and highly fragmented. However, analyzes show that most of them are connected with eco-corridors and "stepping-stones." The existing network of protected areas of Natura 2000 is important for preserving the species diversity of mammals and especially their habitats. According to standard forms, 24 of the Natura 2000 zones are of particular importance for the conservation of mammalian species with priority conservation status according to the IUCN red list, including more than 5 species. These areas are Ludogorie (BG0000168), Lomovete (BG0000608), Reka Yantra (BG0000610), Reka Vit (BG0000181), Nikopolsko plato (BG0000247), Vidbol (BG0000498), Vartopski dol (BG0000518), Reka Iskar (BG0000613), Harsovska reka (BG0000106), Makresh (BG0000521), Voynitsa (BG0000500), Tsar Petrovo (BG0000340), Kalimok - Brashlen (BG0000377), Obnova - Karaman dol (BG0000239), Rabrovo (BG0000339), Archar (BG0000497), Persina (BG0000396), Reka Lom (BG0000503), Ludogorie - Boblata (BG0000171), Batin (BG0000232), Karaboaz (BG0000335), Deleyna (BG0000507), and Ludogorie - Srebarna (BG0000169). Some zones in the eastern part of DDR comprise the typical for the region deciduous forests, combined with rock cliffs, caves and other karst phenomena, and riparian forests - Ludogorie (BG0000168), Lomovete (BG0000608), Reka Yantra (BG0000610), Reka Vit (BG0000181), Nikopolsko plato (BG0000247), Vidbol (BG0000498), Vartopski dol (BG0000518), Reka Iskar (BG0000613), Harsovska reka (BG0000106), Makresh (BG0000521), Voynitsa (BG0000500), Tsar Petrovo (BG0000340), Obnova - Karaman dol (BG0000239), Rabrovo (BG0000339), Reka Lom (BG0000503), Ludogorie - Boblata (BG0000171), and Ludogorie - Srebarna (BG0000169). Other protected areas are of great importance for conservation of wet habitats characterized by high species diversity of mammals - Kalimok - Brashlen (BG0000377), Archar (BG0000497), Persina (BG0000396), Ludogorie - Srebarna (BG0000169), etc. The most significant eco-corridors along the Danube River and in the valleys of its tributaries are also covered by the Natura 2000 network of protected areas - Timok (BG0000525), Novo selo (BG0000631), Ostrov Bliznatsi (BG0000532), Reka Lom (BG0000503), Ostrovi Kozloduy (BG0000533), Reka Ogosta (BG0000614), Karaboaz (BG0000335), Reka Vit (BG0000181), Studena reka (BG0000233), Reka Yantra (BG0000610), Ludogorie - Boblata (BG0000171), and Ludogorie - Srebarna (BG0000169).

Acknowledgments

This work has been carried out in the framework of the National Science Program "Environmental Protection and Reduction of Risks of Adverse Events and Natural Disasters," approved by the Resolution of the Council of Ministers № 577/17.08.2018 and supported by the Ministry of Education and Science (MES) of Bulgaria (Agreement № ДО-230/06-12-2018).

References

Abramov A., Kranz A. & Maran T. 2016. *Vormela peregusna. The IUCN Red List of Threatened Species 2016*: e.T29680A45203971. http://dx.doi.org/10.2305/IUCN.UK.2016-1.RLTS.T29680A45203971.en. Downloaded on 20 November 2018.

Coroiu C., Kryštufek B., Vohralík V. & Zagorodnyuk I. 2008. *Spermophilus citellus. The IUCN Red List of Threatened Species 2008:* e.T20472A9204055. http://dx.doi.org/10.2305/IUCN.UK.2008. RLTS.T20472A9204055.en. Downloaded on 20 November 2018.

Dipchikova S. 2016. Food analysis of the long-eared owl (Asio otus Linnaeus, 1758) in the area of the town of Silistra. Diploma work/thesis, University of Forestry, Sofia, 75 p.

Dister E. 2007. Le Danube et ses plaines alluviales - une vue d'ensemble [The Danube and its alluvial plains - an overview.] In: Trémolières M. & Schnitzler A. (eds.) *Protéger, restaurer et gérer les zones aluviles. Pourqui et comment?* [*Protect, restore and manage the alluvial zones. Why and how?*]: 13 - 25. Lavoisier, Paris.

Frank G. 2017. The Danube River as European ecological corridor and its further development as Trans European Green Infrastructure by DANUBE parks CONNECTED. *Danube News* 35 (19): 11 - 14.

Görföl T., Dombi I., Barti L., Bücs S., Jére C., Pocora V., Pocora I., Szodoray-Parádi F., Paunović M., Karapandža B. & Csősz I. 2017. A review of the occurrence data of the pond bat (*Myotis dasycneme*) in its southern distribution range. *North-Western Journal of Zoology* (2017): e174702

Koshev Y. & Genov P. 2008. New record of steppe polecat (*Mustela eversmanni* Lesson, 1827) in Northwestern Bulgaria. *Historia naturalis bulgarica* 19: 183-184.

Limpens H. 2000. *Report on the program of bat detector training workshop in Bulgaria and Croatia in 1999 and in Ukraine, Georgia, Slovenia, Romania Moldova in 2000*. Report to the German Federal Agency for Nature Conservation Consult & Project Management, Wageningen.

Maran T., Skumatov D., Abramov A. & Kranz A. 2016. *Mustela eversmanni. The IUCN Red List of Threatened Species 2016*: e.T29679A45203762. http://dx.doi.org/

10.2305/IUCN.UK.2016-1.RLTS.T29679A45203762.en. Downloaded on 20 November 2018.

Markov G. 2014. Distribution and numbers of wolves (*Canis lupus*) in Bulgaria: what is going on? *Journal of Academy of Sciences of Moldova, Life Sciences* 1(332): 59-65.

Mitev I. 2006. Subfossil finds of birds and mammals in accumulations of the food of Eagle Owl (*Bubo bubo* (L., 1758) (Aves: Strigiformes) from the valley of Roussenski Lom river. *Historia naturalis bulgarica* 17: 137-151. [In Bulgarian with English summary]

Murariu D. 2005. The state of the mammals (Mammalia) along the Danube, between Gârla Mare and Călăraşi (Romania). *Travaux du Muséum National d'Histoire Naturelle "Grigore Antipa"* [*Works of the National Museum of Natural History "Grigore Antipa"*] 48: 427 - 445.

Nedyalkov N., Raykov I., Koshev Y., Atanasova I., Dekov O., Kolev I., Raykova V. & Staneva A. 2015. *Current status, distribution and conservation of Mesocricetus newtoni and Cricetus cricetus (Mammalia: Cricetinae) in Bulgaria.* Annual Zoological Congress of "Grigore Antipa" Museum 18-21 November 2012 Bucharest, Romania, Book of Abstracts p.222 (poster 105) pdf.

Peshev T., Peshev D. & Popov V. 2004. *Fauna Bulgarica.* 27. Sofia, "Marin Drinov," Publishing House of the Bulgarian Academy of Sciences, 633 p. (In Bulgarian).

Phillips S., Anderson P. & Schapire R. 2006. Maximum entropy modeling of species geographic distributions. *Ecological Modelling* 190: 231-259.

Phillips S., Dudík M., Elith J., Graham C. H., Lehmann A., Leathwick J. & Ferrier S. 2009. Sample selection bias and presence-only distribution models: Implications for background and pseudo-absence data. *Ecological Applications* 19: 181-197.

Popov V. 2007. Biogeographical and ecological spatial patterns of terrestrial mammals in Bulgaria. In: Fet V. & Popov A. (eds.). *Ecology and Biogeography of Bulgaria. Monographiae Biologicae*, Vol. 82: 9 – 38. Springer. ISBN: 978-1-4020-4417-5.

Popov V. 2015. Presence-only Habitat Suitability Modelling UsingUnclassified Landsat ETM+ Imagery: Fine- resolution Maps for Common Small Mammal Species in Bulgaria. *Acta zoologica bulgarica* 67(1): 51 - 66.

Popov V. 2018. Bats in Bulgaria: Patterns of Species Distribution, Richness, Rarity, and Vulnerability Derived from Distribution Models. In: *Bats*, pp. 39 - 62. Ed. H. Mikkola. IntechOpen. DOI: 10.5772/intechopen.70927 ISBN: 978-1-78923-399-5, Print ISBN: 978-1-78923-398-8.

Popov V. 2018. Pliocene-Quaternary Small Mammals (Eulipotyphla, Chiroptera, Lagomorpha, Rodentia) In Bulgaria: Biostratigraphy, Paleoecology, and Evolution. In: Huard G. & Gareau J. (eds) *The Pleistocene.* pp. 109 - 235. Editors: Nova Science Publishers, Inc. ISBN: 978-1-53613-728-6.

Popov V. & Lakovski K. 2019. The southernmost postglacial record of Pond bat (*Myotis dasycneme*, Chiroptera, Mammalia) – Varteshka cave, NW Bulgaria. *Acta zoologica bulgarica* 71 (1): 57 - 62.

Schneider E. 2002. The ecological functions of the Danubian floodplains and their restoration with special regard to the Lower Danube. *Large Rivers* 13 (1-2), *Arch. Hydrobiol.Suppl.* 141 (1-2): 129 - 149.

Schneider E. & Günther-Diringer D. 2004. Ecological and restoration potential of the Lower Danube floodplains. Problems and perspectives. 3rd European Conference on River Restoration 2004. Zagreb, Croatia, 17-21 May 2004.

Simeonovska-Nikolova D. & Dekov O. 2013. Some aspects of the behavior and defensive vocalization of the Romanian Hamster, *Mesocricetus newtoni. Acta zoologica bulgarica* 65 (4): 461-468.

Spassov N. & Spiridonov Z. 1985. Steppe polecat. *Mustela eversmanni* Lesson, 1827. In: *Red Data Book of the Republic of Bulgaria*, Volume 2 – Animals. Sofia. BAS.

Velchev A., Todorov N. & Penin R. 2003. Regional differentiation of landscapes in Bulgaria. In: *Collection of Scientific Papers. Natural Sciences - Geography*. Shumen, University Publishing House "Bishop Konstantin Preslavski," 35-48. (in Bulgarian).

About the Editors

Peter Shurulinkov, PhD
Associate Professor, Natural Museum of Natural History,
Bulgarian Academy of Sciences
Sofia, Bulgaria
Email: p.shurulinkov@gmail.com

Associate Professor Dr. Peter Shurulinkov was born in Sofia, on 25.12.1975. Studied in National School for Mathematics and Life Sciences in Sofia 1989 -1993. Afterwards (1993-1998) studied in Sofia University, specialty “Zoology and Antropology”, specialization on verterbrate zoology. Became PhD student in the Institute of Zoology, Bulgarian Academy of Sciences in 2000. In 2003 obtained PhD degree with thesis on the blood parasites of wild birds in Bulgaria. In 2009 declared as Associate Professor in zoology in the Institute of Zoology. Since 2010 working in the National History Museum, Bulgarian Academy of Sciences as a curator of the bird collection. Interests in ornithology, parasitology, ecology and nature conservation. Member of “Balkani” Wildlife Society since 1993. Author of 107 scientific articles and two books in the fields of ornithology, ecology and parasitology. Active bird ringer since 1990. Married, with two children.

Zdravko Hubenov, PhD
Professor, Natural Museum of Natural History,
Bulgarian Academy of Sciences
Sofia, Bulgaria
Email: zkhubenov@abv.bg

Prof. Dr. Zdravko Hubenov graduated at the Sofia University, Faculty of Biology (Zoology speciality) in 1976. He is PhD (Entomology) since 1985. Workplace: Institute of Zoology of the Bulgarian Academy of Sciences since 1978. Head of the Laboratory of Entomology and the Taxonomy, Faunistics and Zoogeography Department of the Institute of Zoology. Current workplace: National Museum of Natural History since 2010. Areas of research: zoology, entomology, zoogeography, malacology, faunistics, protection of the faunistic diversity of the invertebrates, exploration of the fauna of the protected areas, protected species and invasive species, assessment of areas with conservation importance and conservation problems of the fauna. He has greatly contributed to the study of the Bulgarian Diptera fauna and malacofauna. There are over 170 scientific publications, among which extensive articles on the faunistic diversity of Bulgaria, a review of the Black Sea fauna, the Bulgarian malacofauna and a characteristic of the Diptera from many mountains.

Stoyan Beshkov, PhD
Associate Professor, Natural Museum of Natural History,
Bulgarian Academy of Sciences
Sofia, Bulgaria
Email: beshkov@nmnhs.com

Associate Professor Dr. Stoyan Beshkov graduated Sofia University "Sveti Kliment Ohridski", Faculty of Biology, Department of Ecology and Protection of Nature. With parents biologists from his childhood he is trained to love and live in harmony with nature. He has dedicated his life to science, nature and nature protection and study Lepidoptera fauna of Bulgaria, Turkey, Tunisia, Italy and Balkan Peninsula. He is the author of five books and more than 100 scientific publications on the fauna of Bulgaria, Macedonia, Albania, Greece, Serbia, Montenegro, Turkey and Italy. Several taxa are bearing his name; some more are described by him. He proposed several Natura 2000 sites for Bulgaria and worked on biodiversity in majority of Natura 2000 sites and Parks in Bulgaria. Free thinking and intransigent to destroying of Nature he always expressed and uncompromisingly defend his position as environmentalist. Idealistic, stranger to secular stir, he is living in poverty to be able to travel in the most remote places of Bulgaria and

Balkan Peninsula where he spends more than 70 nights in his car during his field work. At present he is Head of the "Invertebrates" Department of National Museum of Natural History, Sofia.

Georgi Popgeorgiev, PhD

Associate Professor, Natural Museum of Natural History,
Bulgarian Academy of Sciences
Sofia, Bulgaria
Email: georgi.popgeorgiev@gmail.com

Associate Professor Dr. Georgi Popgeorgiev completed his MS in Biology at the Plovdiv University "Paisii Hilendarski," Bulgaria in 2001. In 2009 he completed his PhD, titled "The Herpetofauna in Burned Areas of Eastern Rhodopes and Sakar Mountains: Rates of Development." Georgi's professional career is focused on the study, analysis and conservation of amphibians and reptiles in Bulgaria. He has participated in multiple scientific and conservation projects as a team leader and expert. So far he has over 49 scientific publications and has co-authored over 15 books and chapters. He has been a member of the National council for biodiversity to the Ministry of the Environment. He is an associate professor at the National Museum of Natural History-Sofia and works for the Bulgarian Society for the Protection of Birds / Birdlife Bulgaria. Georgi is also a member of major professional organizations such as the Societas Europaea Herpetologica.

INDEX

#

A

B

C

D

E

F

G

H

I

K

L

M

N

O

P